T0237688

# Roots of Caribbean Identity

# Roots of
# Caribbean Identity

**Peter A. Roberts**

CAMBRIDGE
UNIVERSITY PRESS

PETER A. ROBERTS is Professor of Creole Linguistics at the Cave Hill campus of the University of the West Indies in Barbados. He has held Visiting Professorships at the University of Tennessee (Knoxville) and at the University of Puerto Rico (Rio Piedras), and he was a Senior Fulbright Fellow at the John Carter Brown Library, Brown University. He is a former President of the Society for Caribbean Linguistics, the author of the books *West Indians and Their Language* (1988, 2007), *CXC English* (1994) and *From Oral to Literate Culture: Colonial Experience in the English West Indies* (2000), and he is the co-author of *Writing in English* (1997). Roberts has been in the fields of Creole language studies and applied linguistics for over three decades and he has written several articles about language in the Caribbean region.

# CAMBRIDGE
## UNIVERSITY PRESS

University Printing House, Cambridge CB2 8BS, United Kingdom

Cambridge University Press is part of the University of Cambridge.

It furthers the University's mission by disseminating knowledge in the pursuit of education, learning and research at the highest international levels of excellence.

www.cambridge.org
Information on this title: www.cambridge.org/9780521727457

© Cambridge University Press 2008

First published 2008

*A catalogue record for this publication is available from the British Library*

ISBN 978-0-521-72745-7 Paperback

Cover image: *Roots and Rhythm* by Bernard S. Hoyes (© 1983)
Typeset by Edgerton Publishing Services

Every effort has been made to reach copyright holders of material in this book. The publisher would be pleased to hear from anyone whose rights they have unwittingly infringed.

Cambridge University Press has no responsibility for the persistence or accuracy of URLs for external or third-party internet websites referred to in this publication, and does not guarantee that any content on such websites is, or will remain, accurate or appropriate.

# Contents

# Preface

The Spanish fascination with gold in the New World is best known through the notion of Eldorado, a fictitious kingdom of gold said to be located on the Amazon River. However, even before the fantasy of Eldorado, the Spanish fascination with gold had revealed itself right from the first encounters with the indigenous inhabitants of the islands (Antilles) and the vision of them as beings in a 'golden worlde'.[1] This vision had two aspects – one material and the other philosophical.

In relating the first European encounters with the 'Indians', Peter Martyr spoke profusely of the Spaniards' vision of gold and specifically mentioned in Hispaniola 'the Region of **Cipanga**, otherwise called *Cibana*', of which he said 'This Region is full of mountaynes and rockes and in the middle backe of the whole Ilande is great plentie of Golde' (1628: 19). In his profusion about gold, Martyr mentioned 'stones of golde as bigge as the head of a child' (1628: 19). The constant repetition of the words 'gold' and 'golden' leaves no doubt in the reader's mind that for the Spaniards a 'golden worlde' meant, in part, a world literally teeming with gold.

At the same time, Martyr (1628:15) described the indigenous inhabitants of Hispaniola using 'golden' in a metaphorical sense:

> ... so that if we shall not bee ashamed to confesse the trueth, they seeme to
> live in that golden worlde of the which olde writers speake so much,
> wherein menne lived simply and innocently without enforcement of lawes,
> without quarrelling, judges, and libelles, content onely to satisfie nature,
> without further vexation for knowledge of things to come.

It was the appearance of innocence in the morals and behaviour of the indigenous inhabitants of Hispaniola that gave rise to this philosophical aspect of the vision; it was essentially a contemplation of the human condition involving a contrast of cultures.[2]

For the Spanish the possession of gold and a state of bliss seemed to be inseparable. For the indigenous inhabitants gold did not have any special functional or decorative significance, which was what in itself led the Spanish to conceive of them as idyllic. In a clash of cultures and

---

1. In fairness to the Spaniards, the writers who most created this view of the New World (e.g. Peter Martyr and even Columbus himself) were not Spaniards.
2. This may have been sincere as well as being, consciously or unconsciously, an attempt to justify Spanish removal of gold from the islands.

values in which one side, the outsider/newcomer, desired wealth, much of it to send back home, and the resident, probably as a result of the bounty of Nature, was less acquisitive and more contented with life, the newcomer was more driven and quickly overpowered the resident, who dramatically declined under the onslaught. However, the victory of the outsider did not destroy the vision of the 'golden worlde', because such a vision is part of human consciousness, exemplified most forcefully in the concept of heaven. The image of the Antilles as the 'golden worlde' lived on even when the gold was gone in most places and when the indigenous inhabitants were virtually forgotten. Nature at its most propitious exercised a powerful influence on the spirit and lives of those who came from elsewhere to the Antilles. As colonisation evolved, the contrast between cultures with varying philosophies and degrees of acquisitiveness persisted and made the Caribbean islands, because of their beauty and bounty, the central theatre of the human struggle between acquisitiveness and enjoyment, between those who saw them for what they could get out of them and for those who saw them as home.

From 'discovery' to 'independence' the people of the islands of the Caribbean went through a continuous process of configuration and reconfiguration that was substantially outside their control because they were buffeted by more powerful forces and because they did not dominate the media that broadcast their identity. They were forced to see themselves through the eyes of others. Characteristics were ascribed to them that they often came to accept even though initially they themselves had no conscious awareness of them. What they were acutely conscious of was their own internal dissimilarities, as a result of the way in which they were composed, structured and governed. Identity across the whole society, assuming common ethnicities, linguistic traditions, supernatural beliefs and myths, was not a part of these islands. Thus, one may reasonably ask 'What importance could such nations have and what insights into the human condition could they provide?' One answer is that they are the blossoms of hybrid, cross-fertilised plants which survived continuous transplantation from elsewhere and which are overcoming the smell of the dung heap of slavery that tainted them. As societies today, they represent the triumph of the human spirit over the human physical condition against a backdrop of natural beauty.

It is a story of intellectual struggle, contradictory visions and reverse views that this book deals with as it relates how conflicting ethnicities coalesced or were forged into varying identities. It uses race, place and language as central elements in the examination of identity development. The chapters are set out in chronological sequence stretching from the time of the European encounters with the indigenous inhabitants in the fifteenth century to the nineteenth century, when the consciousness of national identities was evident among the Caribbean peoples. The work encompasses the islands that were held by the three main colonial powers – the Spanish, the French and the English. It uses a wide range of Spanish, French and English primary source materials and includes extensive citation from these sources to let the voices of the past speak for themselves. One of the intentions in the use of citations across these languages is to help to reduce the effect that language barriers have had and thus to provide greater transparency across the historical record.

A work of this scope must be selective and cannot reflect the entire developmental history of all or any one of the Caribbean islands. It chooses to focus exclusively on islands in the Caribbean not only because islands in themselves, being geographically distinct, challenge identity formation but also paradoxically because the name *Antillia*, which has remained (as *Antilles/Antillas*) with the islands up to today, exotically brought the islands together as one in European mythology from as early as the fifteenth century. The general picture of the developing Caribbean is revealed through discussion of selected features, which in several cases challenges received views. The book also contains reproductions of images from historical sources as illustrations or to facilitate the understanding of early constructions of identity.

One of the greatest difficulties in writing this book has been to use appropriate terminology that reflects each time period but at the same time does not perpetuate myths and damaging concepts. In the attempt not to be anachronistic, it may seem justifiable to use the contemporary term 'Indian', for example, to refer to the people of the new lands that Columbus and other Europeans encountered after crossing the Atlantic, but to do this is to create more problems than one is resolving. Among the possible alternatives are 'Amerindians' and the more recent 'Neo-Indians'. An illustration of the difficulty appears in the *Introduction* of Hulme and Whitehead (1992), where within the space of a few pages the terms used are 'native Americans', 'Caribbean islanders', 'native islanders' and 'indigenous inhabitants', as well as explanatory devices such as 'so called' and terms in inverted commas.

The decision taken here is to replace the contemporary, general name 'Indian' with the descriptive term 'indigenous inhabitant'. The notion 'Carib' is addressed as a topic within the book and so does not remain problematic. The term 'Arawak' is not used because it was not a contemporary term during the period of the encounter and in any case it is incorrect when used to refer to inhabitants of the islands. In the case of the term 'negro', which was used over a long time to refer to Africans and their descendants, that is maintained because it is not usually ambiguous or misleading. Note that it is still in popular use in some Caribbean countries to contrast with Indian (from India) without necessarily being regarded negatively.

# Acknowledgements

The author and publishers wish to thank the following for permission to use illustrations:

Bibliothèque Nationale de France and M. Moleiro Editor (www.moleiro.com), Figure 2.2

British Library, Figures 2.1, 2.5, 4.2, 4.3, 5.1, 7.1 and 7.3

CENDA (Centro de Comunicación y Desarrollo Andino), Figure 7.5

Library of Congress, Figure 2.3

National Library of Jamaica, Figures 4.1 and 7.2

New York Public Library, Figures 2.4 and 3.1

The Hakluyt Society, Figure 3.3. The Hakluyt Society was established in 1846 for the purpose of printing rare or unpublished Voyages and Travels. For further information please see their website at: www.hakluyt.com

# Timeline

| Date | Events of colonial and political importance | Milestones in printed literature |
|---|---|---|
| 1578 | | Montaigne's *Des Cannibales* features the **bon sauvage** |
| 1592 | Spain establishes a lasting settlement in Trinidad | |
| 1601– | | Herrera y Tordesillas' *Historia general de los hechos . . .* |
| 1603 | Devastation of the 'banda del norte' (Hispaniola) | |
| 1605 | French adventurers and others begin to populate Tortuga as well as the *banda del norte* | |
| 1609 | | Garcilaso de la Vega's version of **criollo** in *Los Commentarios Reales* |
| 1610 | | Shakespeare's Caliban in *The Tempest* |
| 1625 | St Kitts colony founded (by both England and France) | |
| 1627 | Barbados colony founded by the English | |
| 1635 | The French colonise Guadeloupe and Martinique | |
| 1654 | | Du Tertre's *Histoire generale des isles . . .* |
| 1655 | England takes Jamaica from Spain | |
| 1657 | | Ligon's *History of . . . Barbados* |
| 1658 | | De Rochefort's *Histoire naturelle et morale* |
| 1665 | | Breton's *Dictionaire Caraibe Francois* |
| 1678 | | Exquemelin's *De Americaensche Zee-rovers* |
| 1685 | | French *Code Noir* |
| 1695 | Spain recognises French claim to Saint Domingue | |
| 1707 | | Sloane's *A voyage to the islands* |
| 1722 | | Labat's *Nouveau voyage* |
| 1771 | | Cumberland's play *The West Indian* |
| 1774 | | Long's *History of Jamaica* |

| Date | Events of colonial and political importance | Milestones in printed literature |
|------|---------------------------------------------|----------------------------------|
| 1776 | US declaration of independence | |
| 1788 | | Abbad y Lasierra's *Historia . . . de Puerto Rico* |
| 1789 | The French Revolution begins | |
| 1793 | | Edwards' *The history, civil and commercial . . .* |
| 1791– 1804 | Haitian Revolution | |
| 1796 | | Moreau de Saint-Méry's *Description . . . de l'isle Saint-Domingue* |

# Introduction

## In search of colonial identity in the Caribbean

The identity of any human can be specified by using, singly or in combination, many different universal criteria, e.g. sex, age, religion. However, in the modern, political world human beings are automatically identified with a country, and national identity is generally established by place of birth or place of residence. Very strongly tied to the feature *place* has been the feature *race*, so much so that race was once, without reflection, subsumed under place, as for example 'white' under European and 'black' under African. In fact, the traditional construction of ethnic identity (European, Chinese, Indian, African) is inalienably tied to a historicity of race that evolved in the Old World. Another major feature of identity is language. In a great many cases across the world, and especially in Europe, the name of the language (e.g. *le français*), the national designation (e.g. *les français*) and the name of the country (*la France*) are virtually the same, which suggests that place, people and language are closely allied in the formulation of national identity.

Political identity and the notion of 'home' overlap, but 'home' embodies a psychological factor of attachment, which probably issues from the basic animal instinct of territoriality, but is more an emotional bond created through experience of a place. Because the human being does not necessarily remain in one place throughout a lifetime, 'home' is variable and may be place of birth, place of residence or may be defined by the popular notion 'where the heart is'. The factor of place, as it relates to 'home', may also be defined by using the term 'habitat', which speaks to a compatibility of human being and place as well as a formative influence of place on human being and the reverse. Race, for example, is perceived by many to be a factor of habitat, though there is no absolute proof that this is so. Even more complicating is the fact that race itself is a construct made up of several components, which also vary depending on genetic combination.

Language is in part a universal human factor and in part a factor of place: human language manifests itself primarily in speech as distinct languages, each of which is geographically determined. As a factor of place, language can sharply distinguish between insider and outsider through difference in accent, idiom, structure and word. Language therefore establishes bonds between all communities of human beings but at the same time sets up barriers between communities. Human beings, however, are not restricted to use of a single language and can cross barriers.

Cultural identity, then, as opposed to political identity and involving the features, place, race and language, results from a coalescence over time of highly variable factors. Moreover, the matter of coalescence over time and continuity is not easily resolved.

In the Caribbean throughout the colonial period, the three features – place, race and language – were quite separable elements subject to considerable variation. Neither 'Caribbean', 'Antillean', 'American' nor 'West Indian' was used as the name of a language in the way that French, English and other European national designations are. Place names in the New World could not be transferred to people or language because the 'outsiders' had only recently arrived, and for them there was no historical link between person and place. Moreover, whether Caribbean nations have identities that are typologically different from those of European nations is not just a matter of how they emerged, but also a matter of who authored the historical record in each case. For the Caribbean islands, there is no Homeric type of record in verse purporting to capture the people's accumulated oral account of them-selves. Neither is there any record equivalent to that of Garcilaso de la Vega, the Inca, giving an epic account from a personal point of view of the history of his own people. The available record of the evolution of colonial identity in the Caribbean is a written one, which is principally a third-party, European record. Even then, it is not a full and sufficient record, for societies in the Caribbean started off as economic ventures, which meant that social and cultural information in the records was incidental to economic information. Furthermore, when Europeans began to write about the New World, they already had an exotic vision of people, places, things and events elsewhere that had been formed by Greek mythology and beliefs about 'the Indies'. It was a vision that was modified but never really disappeared from the imagination of Euro-peans, especially in the case of the islands. The economic view and the exotic view of the islands complemented each other throughout the colonial period, thereby maintaining an external viewpoint as the well-spring of their identity.

Colonial societies in the Caribbean were artificial, in the sense that they were imposed on the land and controlled from outside. In addi-tion, the islands themselves seemed to have encouraged movement, nomadic and migratory, even before the advent of Columbus. There-after, though rigid systems remained in place, the rate of population replacement was consistently high, especially where sugar and slavery thrived. In spite of the fact that names were given to every island, there was no early consciousness of them being new and separate 'nations' or of their inhabitants having new and separate identities. In terms of population genetics, it must have taken some time for the stage of genetic equilibrium to be reached when each island could be said to have a homogeneous population. Moreover, even if the people had had some consciousness of a new identity, there was no easily available way for them to record this consciousness.

· In time, however, societies in the Caribbean evolved into familiar entities, each with its own sense of identity, each with its own peculiar population, but all using the word *creole* in some way to identify themselves or elements within their societies. Terms such as 'mixed', 'hybrid', 'mutated', 'syncretic', 'blends', 'remodelling', 'corruptions', 'borrowings', 'imitations', 'fragmented' and 'plural' have been applied to different Caribbean societies or features in them. They all in some way relate to theories on identity and theories of emergence of new identities, which provide a framework within which the development of colonial identity in the Caribbean can be examined.

# Identification of identity

## Identity and name

The notion of 'identity' in human society is based on two fundamental factors – the perception of sameness/difference and the instinctiveness of man to be a social being. The perception of sameness logically implies the perception of difference, which in turn implies that those who are perceived as different are treated differently. In fulfilling the need to associate with others, humans, probably through the inevitable reality of birth and upbringing as well as through practical experience, come to associate with those who have a high measure of sameness and come to be separated from others who are different. Consequently, social organisation leads to conflict, which in itself causes sameness and difference to become even more important.

Sameness in a population has been regarded as a fact in population genetics and expressed in the concept of 'genetic equilibrium'. Savage explains thus:

> Each population is essentially a unit with a common body of genetic material ... the hereditary conservation of DNA and genes is a populational characteristic, and ... if all other factors remain constant the

> frequency of particular genes and genotypes will be constant in a
> population generation after generation. (1969: 39)

Wells, in his investigation of the history and development of human beings, seems to lend some support to this when he remarks:

> We have used the Y-chromosome for most of our studies of human
> migration. This is because the Y shows greater differences in frequency
> between populations than most other genetic markers. (2002: 175)

However, Wells, in an interesting (Y-chromosome) genetic analysis (for
a television audience) he did on four people living in London, England,
showed that three (Irish/Scottish, Japanese, Pakistani) of the four men
fitted the pattern of their fellow countrymen, but the fourth man, an
'Afro-Caribbean', 'turned out to have an M173 Y-chromosome, the ca-
nonical European lineage' (2002: 184), even though 'The other, non-Y
markers we tested revealed him to be otherwise genetically African'
(2002: 184–5). Wells goes on to explain:

> The reason our Afro-Caribbean man had a European Y-chromosome was
> that ... one of his male ancestors must have had a European father ... it is
> likely that this occurred ... during the era of slavery. (2002: 185)

What is remarkable about Wells's analysis and conclusion is that even
though 'science' showed the man to be genetically complex (which is
not earth shattering in itself), the man is classified as *Afro-Caribbean*
and *otherwise genetically African,* no doubt because of his colour and
physiological features. This shows the importance of physical appearance as a measure of sameness in classification, even among scientists.

The constant association with and experience of sameness leads to a
recognition of one's own features as normal and those of others as
abnormal/strange/foreign. The best example of the recognition and
identification of difference is with language. All human beings recognise
speakers from outside their community by their speech, and all those
who are different are said, in the case of English speakers, to have an
'accent', with the implication that the speaker himself/herself does not
have one. Of course, all speakers have an 'accent' when judged from the
standpoint of persons outside their community. Besides language, most
other characteristics that are 'foreign/strange' usually have a negative
valuation given to them as a result of normal, human bias. While intellectually the bias in valuation can be understood by human beings, it
does not prevent people from automatically regarding their own as
superior. To some extent this reaction may be a creature of the innate
emotions of human beings. At the same time, when one group is seen to
be or made to look clearly superior in some respect, the instinct for self-

preservation causes the 'inferior' group to concede superiority. This is typical in situations where one group dominates another.

Sameness among human beings is commonly judged under certain basic categories – how people look, how they sound, where they were born and bred and how they behave. The main features that dominate in the way people look are colour and race. How a person sounds is in essence what language or variety of language they speak; this immediately links them to a community and separates them from others. Most people in the world today have a national label assigned to them according to where they are born, e.g. Japanese, Italian, Brazilian, Kenyan, and every human being in the modern world must be a citizen of some place. Place of birth and residence of course govern other areas such as movement, dress, work routines and food. Behaviour is the biggest category in judgements of identity, one that covers a wide array, including supernatural practices, entertainment, sports and games, and educational practices. While behaviour may in some objective way be the best criterion for judging sameness, it is the senses of sight (colour/race) and sound (language) that provide the initial and usually most deep-seated conclusions about sameness and difference in identity.

The giving of a specific name to a group is preceded by the perception or assumption of sameness across a number of individuals and the perception or assumption of difference between them and others. Naming of a group, however, can be done by the group itself or by others, which means that a name may be a reflection of a shared experience of sameness across individuals, or, on the other hand, it may be a projection of beliefs, values and desires on to people without them having any prior consciousness of identity. Whether the one name prevails over the other is a matter of who controls the dissemination of information. Differences in names may also be resolved when 'foreigners' become more familiar with 'natives'. The naming of identities may therefore be an evolving process when it is foreigners who are doing the naming, since initial crude generalisations and mistakes disappear as foreigners move beyond initially striking 'primary' features, and come to identify and distinguish 'secondary' or cultural features. Such knowledge is gained through direct contact with natives through some kind of language adjustment for purposes of communication.

## Naming nations and civilisations

Nation in its etymological sense highlights place of birth as the most important factor in 'national' identity, and the concept 'native' links a person to a place. The concept of 'nation' flourished among sedentary peoples who only left home to migrate or to go to war, when they were adult. It therefore also embraces the notion of 'home', which apparently does not have the same degree of importance in the case of most nomadic people. Yet, 'sedentary' and 'nomadic' are relative and not absolute in their contrast since migration is an integral part of human development and history.

Nationality, or the notion of belonging to a specific political entity or state, is a more modern concept, which can be regarded as the least absolute factor of 'nation' and thus may be deemed 'tertiary' in a hierarchy of criteria. In fact, the nation state may be viewed as an economic necessity and is often made up of 'nations' in a more fundamental sense. The 'primary' criteria used in naming peoples and civilisations have been visible (racial) differences between human beings – similarities in colour of skin, hair and eyes, in height and physique. Cultural features, including language, can be termed 'secondary', that is not as absolute as physical features that are beyond the control of the individual and cannot in themselves be tempered by will, personality or psychology.

That race is a primary criterion for establishing identity is seen in the fact that it moves beyond 'nation' to establish 'civilisations'. For instance, Elliott justifies the concept 'Europe/ European' thus:

> First, we mean that Europe is a distinctive historical entity, in the sense that, although it is a composite of peoples of very different origin, it has developed a civilization which can be clearly differentiated from the civilizations, say, of China or Islam. Secondly, its component parts, although differing widely in character, have enough shared experiences and features in common for the elements of unity to outweigh the elements of diversity. (1998: 20)

What he does not point out, however, is that the starting point for the classification is race, and in fact he could easily have said that Europeans are white, as opposed to Chinese, who are not, with Islam being essentially associated in the minds of people with those who look like the Prophet Mohammed.

The primacy of race in establishing national identity can also be expressed in a negative way, as seen in the comment of Davy:

> This admixture of races, this state of society, moreover, has no wise been favorable to the formation of a representative form of government, or of any kind of self government or independent local rule; – the absence of which in turn has tended to cramp the faculties and feelings and to check patriotism and public spirit. ([1854] 1971: 306–7)

The obvious implication here is that mixture or co-existence of races is inimical to the notion of identity in a nation.

A variation of the same idea of genetic relatedness is the notion of 'family' as the basis for the 'nation'. This approach is adopted by the missionary pastor, M. B. Bird: 'a nation is whatever its families are; the domestic circle well formed, so also will be the nation, hence it will naturally follow, that untrained families will form an untrained nation' (1869: 317). In this argument, grouping by genetic sameness/blood relationships at the first level or base (i.e. the nuclear or extended family) is seen as indispensable for meaningful and successful groupings at higher levels. The 'nation' is seen as a collection of families, and it is presumably the interweaving of families by 'marriage' that would lead to cooperation and not conflict. This structuralist approach therefore preserves genetic relatedness, even if not racial identity, as the basis for the 'nation'.

Another characteristic of this view of the family-based nation is that it accords the mother a central position:

> The mothers of a nation give it form
> And shape. (Bird 1869: 349)

This is a traditional Christian position, which sees the role of the mother as domestic and home-based: 'a large family, fully and minutely cared for, by an entire and assiduous attention to all the endless wants of domestic life, would quite absorb every moment of a mother's care' (Bird 1869: 349). In this formulation all females are regarded as potential mothers. It fits into the Christian view of marriage as the base for the family and fidelity as the basis for a good marriage. It seems at first to be a position in which happiness is at the core, but this is a view that sees social responsibility (what Bird called *training*) as more important than individual choice. Marriage was often used strategically for political purposes to strengthen nations. The Christian nation was essentially European in its conception and was one that in practice, because choice was controlled by strict social and racial factors, created strength but also allowed for distinctions between families, classes and races.

Wells in his discussion of 'nation' and 'nationality' is extremely cynical about what he interprets to be deliberate nineteenth-century attempts to create 'nations' out of disparate groups. He argues that what determines the coming into being of the modern nation state is a political intent to dominate others:

> We may suggest that a nation is in effect any assembly, mixture, or
> confusion of people which is either afflicted by or wishes to be afflicted by
> a foreign office of its own, in order that it should behave collectively as if

its needs, desires, and vanities were beyond comparison more important than the general welfare of humanity. ([1920] 1951: 982)

Implicit in Wells' hostility to the formation of modern nations is the idea that they are violating some natural principle of grouping according to sameness in race and other clearly observable human features. He makes this point even more clearly when he says:

> Oriental peoples, who had never heard of nationality before, took to it as they took to the cigarettes and bowler hats of the West. India, a galaxy of contrasted races, religions and cultures, Dravidian, Mongolian, and Aryan, became a 'nation'. (1951: 983)

Wells' point is that such political creations are not spontaneous as they are not based on racial and cultural identities among the people who are so brought together. Wells sees a difference between 'natural' nations and 'artificial' nations, and his obvious hostility to 'artificial' nations means that he believes that groupings should be natural.

Stalin, writing around the same time as Wells, rejected a common race as an essential feature of a nation and was very clear in what he thought to be the essential features:

> A nation is a historically evolved, stable community of language, territory, economic life, and psychological make-up manifested in a community of culture. ([1935] 2003: 8)

Probably in reaction to Stalin's use of *language* as one of the primary features of a nation, Hobsbawm argues:

> National languages are therefore almost always semi-artificial constructs and occasionally, like modern Hebrew, virtually invented. They are the opposite of what nationalist mythology supposes them to be, namely the primordial foundations of national culture and the matrices of the national mind. (1990: 54)

Recognising the difficulty of defining a nation, Hobsbawm assumes the position that it is more profitable to begin with 'nationalism' than with 'nation' because 'Nations do not make states and nationalisms but the other way round' (1990: 10).

Smith, in the following definition, sets out his view of the fundamental features of national identity:

> A nation can therefore be defined as **a named human population sharing an historic territory, common myths and historical memories, a mass, public culture, a common economy and common legal rights and duties for all members.** (1991: 14)

Smith, while identifying the formative core of nations as *ethnie*, goes on to differentiate between a nation and an *ethnie* or ethnic community by saying that the latter has the following six main attributes:

1 A collective proper name

2 A myth of common ancestry

3 Shared historical memories

4 One or more differentiating elements of common culture

5 An association with a specific 'homeland'

6 A sense of solidarity for significant sectors of the population. (1991: 21)

In arguing for the critical importance of *ethnie*, Smith says:

> so **ethnie**, once formed, tend to be exceptionally durable under 'normal' vicissitudes and to persist over many generations, even centuries, forming 'moulds' within which all kinds of social and cultural processes can unfold. (1986: 16)

In his earlier and underlying thesis, Smith argued that:

> the 'core' of ethnicity, as it has been transmitted in the historical record and as it shapes individual experience resides in this quartet of 'myths, memories, values and symbols' and in the characteristic forms or styles and genres of certain historical configurations of populations. (1986: 15)

The fact is, however, that a virtually inescapable element in the formation of *ethnie* is race, even though *ethnie* is seen as extending beyond race.

In the structural hierarchy – nation > ethnie > myths – which is implicit in Smith's thesis, the time factor is critical not only because of the durability of myths but also in the formation of these common myths. Smith speaks of the durability of *ethnie* under *'normal' vicissitudes*, but in the case of development of identity in the Caribbean, it would be difficult to sustain an argument that normalcy prevailed fundamentally. So, unless one is saying that identity in the Caribbean is only marginal, myths cannot be the starting point for a discussion of the development of identity there.

In a later work, Smith identified the key institutional and cultural dimensions of nations and nationalism, which should form the basis for discussion of identity, as 'the state; territory; language; religion; history; rites and ceremonies' (1998: 226–7). The present work, recognising the pitfalls in and disagreements about the starting points for discussion of the highly variable concept of 'identity', nevertheless uses the features 'place/ecology' (territory), 'language' and ethnicity more as 'race' than as religion, rites and ceremonies. While one may disagree about the relative importance of these features historically or synchronically in a matrix of features, there is hardly any work that has not found it necessary to discuss them.

# Theories about the emergence of new identities

Human societies are not static – not only are identities constantly being modified but new ones also come into being. The two fundamental processes through which new identities come into being are genetic change and ecological change. This is so because one of the most common human realities is that social and economic pressures in a population often result in migration, which in turn leads to contact across ethnic groups. In such cases there is both change of environment and infusion of new genetic material into a population. Since genetic and ecological factors can be modified and combined in several ways and are weighed against each other, the result is that a variety of theories have been put forward to explain the coming into being of new identities. Such theories also concern themselves with the psychological effects of contact, in terms of tendencies towards hostility or refusal to mix as opposed to tendencies towards accommodation.

## Transmission theory

Transmission theory, which operates within the framework of creation or procreation, is substantially genetic in nature. Transmission theory is essentially a matter of creating another identity in one's own image either by transfer of characteristics, by edict or by model. This is a theory that has a long history, seeing that it is part of a theological explanation of man in which man is created in the image of God. In this explanation what is created is not identical to the parent and, in religious terms, is seen as flawed to some extent or at least is not as perfect as the original. One of the entrenched views of New World, for example, is that Europeans created 'America':

> It was the Europeans, too, as Edmundo O'Gorman took pleasure in reminding us many years ago, who 'invented' America. They invented it as a name, they invented it as a concept, and finally they invented it as a historical entity. (Elliott 1998: 20–1)

Elliott expands on this idea by saying:

> These European settlers were creating America, an America which can be regarded historically as an extension of Europe, in a way that Asia and

Africa could never be. This was a continent imagined, invaded, occupied, and developed – or exploited – by Europe; and for all the other elements – indigenous, African, and more recently Asian – which have gone into its making, that original European imprint has been so strong and all-pervasive as to mark it for all time. (1998: 22)

The name *America* was a case of a name being imposed from outside without reference to what or who it was naming: it was merely commemorating the name and deed of an outsider who in no way characterised the place or people so identified. It was a matter of branding as one's own a place that one had discovered. It was a case of instinctively projecting oneself on to one's territory, trying to leave some lasting mark of identity on it.

Even so, because of rival 'creations' and territorial markings, it took some time for the name *America* to be accepted over others, as Elliott points out:

Although the name [America] appeared, by courtesy of Martin Waldseemüller, as early as 1507, it took the best part of three centuries for it to gain universal recognition. For the Spaniards until well into the eighteenth century, it remained **Las Indias**. But already by the later seventeenth century, 'America,' long established as the name of a continent, was beginning to give **las Indias** a run for its money. (1998: 21)

Ironically, the natives of the New World came to accept, or had no choice but to accept, the name of their place of birth thus conferred on them because it was the Europeans who controlled the dissemination of information. This theory of creation and transmission, then, when it is set within a general context of European empire and colonialism, renders the non-European groups, including the indigenous inhabitants and their culture, ineffective and marginal. As Elliott says: 'It was the Europeans who endowed these peoples with their first, specious, unity, by labelling them indiscriminately with the brand-name of Indian' (1998: 20).

The general identity *Indian* was thus accepted by Europeans as a label for all the indigenous peoples of the New World – this was different from the Old World, in which distinctions were made between Europeans, Africans and people in Asia. *Indian* was a non-discriminating identity that did not connote a specific civilisation, for this would have had to be based on identity in race and behaviour. It was merely the projection of a belief (a mistaken one) onto others who had little power to contradict it.

Likewise, the European colonies themselves are seen to fit into the different moulds of their colonial masters:

the same set of manners will follow a nation, and adhere to them over the whole globe, as well as the same laws and language. The Spanish, English, French and Dutch colonies are all distinguishable even between the tropics. (Hume 1963: 210, cited in Elliott 1998)

The differences between them are seen to be the result of the fact that the manners of the European colonists who acted as the models in the colonies were not all the same – what they did and how they behaved in the colonies varied. The umbilical link therefore preserved similarities between mother country and colony that were quite clear evidence in support of the transmission theory.

Thus, certain behaviours in the New World were associated with one set of European colonists rather than another and were critical in the formation of the respective societies. For example, Bailyn, in considering British colonial societies, asks the question: 'Were the British simply xenophobic? Were they innately racist – more racist than other European people?' (1992: 25). He poses these questions after pointing out that

> Miscegenation, race mixing, was common in Spanish and Portuguese America from the beginning; by the end of the colonial period forty percent of the Latin American population was **mestizo** (mixed Spanish and Indian blood). The French in Canada lived in similar intimacy with the natives, and an important population of **métis** resulted. But not the British. In the British colonies miscegenation is known to have happened – but only occasionally, here and there, marginally. (Bailyn 1992: 25)

There even seemed to have been some measure of acceptance, among the British, of this view of avoidance of race mixing and a tacit justification for it. Coleridge, for example, claimed that: 'The French and Spanish blood seems to unite more kindly and perfectly with the negro than does our British stuff' ([1826] 1970: 141).

Another example of association of a certain behaviour with one set of Europeans, and consequently their colonies, was the association of *idleness culture* with Iberians. This Elliott rejects, saying 'In early Modern England and the Iberian Peninsula alike, the work ethic and the idleness culture existed side by side, and both together made the Atlantic crossing' (1998: 27). So, following Elliott's argument, the assignment of *idleness culture* almost exclusively to Iberian New World colonies was a matter of distortion caused by bias and colonial conflict. Yet, it was a general belief, which, like a name, came to be accepted through constant repetition as reality. Transmission theory was not therefore just a post facto, academic explanation, but a part of the consciousness of the colonists and used with both negative and positive intent.

## Contrastive theory

While transmission theory can be explained as creation in one's own image, contrastive theory is a matter of creating 'otherness' either for others or for oneself. There are two different versions of contrastive theory, the one focusing on creating for others and the other on creating for oneself. The first may therefore be interpreted as a variation of transmission theory. This version is identified as a characteristic of Spanish colonialism by Adorno:

> *Como proceso cultural, la creación de la alteridad parece ser una exigencia y una inevitabilidad del sujeto, sea éste colonizador o colonizado. Los discursos creados sobre – y por – el sujeto colonial no nacieron sólo con el deseo de conocer al otro sino por la necesidad de diferenciar jerárquicamente el sujeto del otro: el colonizador de las gentes que había tratado de someter y, al contrario, el colonizado de los invasores que lo querían sojuzgar. Vista así, la alteridad es una creación que permite establecer y fijar las fronteras de la identidad.* (1988: 66–7)

[As a cultural process, the creation of otherness seems to be a requirement and an inevitability of everybody involved, be that person the coloniser or the colonised. The discourses created about, as well as on behalf of, the colonial person were not born solely out of the desire to know the other but also through the need to make a hierarchical distinction between oneself and the other: between the coloniser and the peoples he tried to subjugate and, on the other hand, between the colonised person and the invaders who wanted to subjugate him. Seen in this way, otherness is a creation which permits the establishment and fixing of the frontiers of identity.]

Adorno had previously said in relation to the Spanish culture of otherness:

> *la búsqueda de semejanzas y la elaboración de comparaciones, por un lado, entre el amerindio y el hebreo, y por otro, el amerindio y el moro o el morisco, revelan los procesos de fijar la alteridad apoyándose en la semejanza.* (1988: 63)

[the search for similarities and the working out of comparisons, on the one hand, between Amerindian and Hebrew and, on the other, between Amerindian and Moor or Moorish, reveals the processes of fixing otherness through likeness.]

Though the creation of *alteridad* is said by Adorno to be a necessity and an inevitability for both the coloniser and the colonised, she highlights the coloniser, arguing that, for the Spanish, it was first a matter of propagating the idea of cultural difference between the Spanish on the one hand and on the other those who aspired or presumed to be

Spanish or came under Spanish control, and second the likening of all of the latter.

The other version of contrastive theory called *identity defined by alterity* by Gikandi (1996: 44), highlights the colonised. In explanation of this version, Gikandi cites Colley (1992: 6), who says about the development of British identity:

> The sense of a common identity here did not come into being, then, because of an integration and homogenisation of disparate cultures. Instead Britishness was superimposed over an array of internal differences in response to contact with the Other, and in response to the conflict with the Other. (Gikandi 1996: xviii)

The processes identified in the two versions of contrastive theory may be seen as occurring sequentially – the coloniser creating otherness and then the colonised reacting against the coloniser – in which case they may be seen as belonging to the same, overall process of growth and development, with the latter sometimes identified as 'post-colonial'.

## Environmental theory

Environmental theory sees group and national identity as fashioned by local circumstances. Stalin argued, for example, that:

> If England, America and Ireland, which speak one language, nevertheless constitute three distinct nations, it is in no small measure due to the peculiar psychological make-up which they developed from generation to generation as a result of dissimilar conditions of existence. ([1935] 2003: 7–8)

Here environment is seen as a major influence on psychological make-up, which would suggest that psychological make-up should be sub-sumed under 'territory' in Stalin's formulation of a nation.

However, in environmental theory the factor that is most often identified as having the greatest influence is climate and this specific aspect of it is therefore referred to as *ecological determinism*. Environmental theory is explicitly opposed to transmission theory, as is clear in the words of Elliott: 'It could be argued therefore that, as far as settlement of the land is concerned, national characteristics are less significant than the nature of the local environment' (1998: 30). Elliott sees the environment as a dominant force that, in the case of America, was able over a period of time to re-shape disparate peoples into a single people:

> The American environment, with its abundance of land, created over time a new, and distinctively American, people, whose shared characteristics came to blur, and ultimately cancel out, the diversity of their origins. (1998: 24)

Elliott's version of adaptation is that the environment caused common characteristics to come to the fore and to dominate differences. As such, it is different from that version of adaptation which states that the environment creates characteristics.

An instance of this latter version is given by Diop ([1981] 1991: 216–18), who points out that the Greek physician, Galen, who lived in the second century AD, identified hilarity or the strong propensity for laughter as one of the fundamental traits of black people and regarded it as a 'permanent trait due solely to the sun'. In this case, the environment was believed to have created the characteristic, which presumably was then passed on genetically to the offspring, a belief consistent with what later the scientist Lamarck (1744–1829) claimed about environmental adaptations. Similarly, 'rhythm' is today widely believed to be a characteristic of Africans and their descendants, not the diurnal (light and darkness) rhythm that affects many living things including human beings, but a kind of innate spontaneous rhythm, presumably, which causes the black person to be able to coordinate movement perfectly with musical beats. For such a belief to make sense, such 'rhythm' must be thought to have developed initially because drumming, which was a part of African culture, caused Africans to develop an ability, which then was passed on genetically to their offspring.

Diop himself dismisses the idea of permanence in psychological traits in relation to Africans. He does, however, contrast Africans and Europeans, but explains the difference between European and African psychological traits as the result of a difference in social structures:

> this communicative gaiety, which goes back to Galen's epoch, instead of being a permanent trait due solely to the sun, is a result of the reassuring communally securing social structures that bog down our people in the present and in a lack of concern for tomorrow, in optimism, etc., whereas the individualistic social structures of the Indo-Europeans engendered anxiety, pessimism, uncertainty about tomorrow, moral solitude, tension regarding the future, and all its beneficial effects on the material life, etc. (Diop [1981] 1991: 218)

This view seems to represent the modern consensus, that is that the environment, in its broadest sense, brings about uniformity in people who dwell together, but that it does not result in permanent or genetically transferable traits in people. Where offspring have characteristics similar to those of the parents, they are most likely acquired in the same way that they were acquired by the parents. A corollary of this is that, as the elements of environments change, identities change.

Within the environment framework one specific interpretation of the New World has been to see it, especially in the case of colonial North

America, as a frontier engendering a frontier mentality. However, Bailyn (1984, 1992) has turned this notion on its head and has regarded the North American colonies not as a frontier, but as a periphery. He concludes his analysis by saying:

> In every aspect of life, in the cities and in the countryside, colonial North America was, in the sensibilities of the creoles who lived there, a peripheral world, a diminished world, which might one day recover the metropolitan standards left behind. (Bailyn 1992: 40)

Bailyn's definition of *periphery* is quite explicit:

> For those on the outer perimeter the natural orientation is inward, back toward the familiar, realistic central core, not outward toward some ideal future. One is nostalgic for the fully developed reality left behind; one feels inferior, not superior to the past; one's natural point of reference is an earlier 'home,' whether one ever actually experienced it or not. (1992: 37)

Bailyn's analysis, while it challenges the *frontier* notion, is not new in itself – it is effectively a more graphic presentation of the notion of the umbilical link between mother country and colony.

## Assertion of identity – intent and mode

Assertion of identity is not a spontaneous proclamation of sameness for its own sake. The need for a conscious assertion of identity arises out of a perceived erosion of identity or a threat of loss of identity through loss of life, property, privilege or home. In history it has often been accompanied by a call to arms. Accordingly, it is seldom a bottom-up explosion of sentiment, but more often a top-down infusion of spirit and motivation. It is part of the portfolio of leaders rather than the natural expression of all followers. Note, for example, that the unification of almost the whole of the Apennine peninsula into one country (Italy) is in no small measure due to the vision of Garibaldi.

Assertion of identity is political in intent and may either follow or precede actual experience of identity. Tonkin (1992: 130) makes this point clear when she says:

> Delineating and supporting social identities or ethnicities requires innovative and active work by many people which then often has to be repeated over and over; it is not an automatic consequence of living or working together. Indeed in 'nation building', as in all historically oriented legitimations, massive political effort which includes 'intentional rhetorics' and symbolic evocation is undertaken to convince people of a social identity which they may not otherwise experience as such.

However, what Tonkin presents as a general practice without time limits and almost as an accepted necessity today, Wells vehemently denounces as dangerous political agitation in the nineteenth century. Wells ([1920] 1951: 982–5) further claims that acute political hostility towards others is not an inevitable consequence of group formation but is the result of this kind of political agitation. The difference between the viewpoints of Tonkin and Wells highlights what one may simplistically regard as the good and bad intents in asserting nationalism. One may see the former as the promotion of identity for the ultimate good of the people themselves and the latter as the pursuit of exclusivity in order to gain advantage or establish dominance over others.

In the matter of mode of asserting identity, choice is related to level of technological development as well as to intensity of desire for control and power. While Tonkin, whose focus was on a predominantly pre-literate West African society, speaks of *intentional rhetorics* as a mode of asserting identity, Wells specifies that in the nineteenth-century European political arena nationalism 'was taught in schools, emphasized by newspapers, preached and mocked and sung into men'. The difference between the two is not just a reflection of technological development in the West African society in question but rather the importance of empire to Europeans in the nineteenth century. However, although various means can be said to be used to foster nationalism, basically they can be separated into oral and written modes. The use of the latter is of course more modern and its success depends in large measure on the extent of literacy in the society. In addition, although the oral mode is natural and used by all, the written mode among literate nations is regarded as more sophisticated and often takes precedence in judgements over the oral.

# The factor of time – continuity and discontinuity in identities

Conceivably, the life of an identity may be seen as terminal, continuous or discontinuous, but the most common formulation of the life of an identity is in keeping with a view of Nature, that is that it is terminal but part of a bigger cyclical process. In the evolution of the life of the identity there is the idea of a coming together first followed by a fragmentation later, which in several accounts of histories of peoples represents what is the 'natural' process of rise and fall or the continuously repeated parent–child relationship. In this cycle the umbilical cord, which at first is the lifeline, later increasingly becomes a restraint that has to be severed to allow the children to go off independently to start their own families. Of course, either 'parent' or 'child' may object to or try to prevent separation. Note, for instance, that in the middle of the eighteenth century there was a notion that a colony did not have a raison d'être of its own: *'Une colonie n'existe point pour elle; elle existe pour la Nation qui l'a fondée'* ['A colony does not exist for itself, it exists for the nation that founded it'] (Saintard 1756: 232). This was a notion that in itself suggested unease on the part of both 'parent' (European country) and 'child' (American colony) and foreshadowed the start of the quest for independence in the Americas. Bear in mind, however, that the quest for political independence did not necessarily always result in separation.

Looking backwards in time, Elliott makes the following general statement about the process of development in the Americas:

> Yet if, as I believe, we can see the eighteenth century as a century of convergence in the history of the Americas, I also believe that the succeeding century witnessed a new, and sharpening, divergence. (1998: 41)

What Elliott is alluding to is the fact that at the end of the eighteenth century, after two centuries of European consolidation through colonisation, the Americas started to break up into several independent nations. Yet, while the severing of the umbilical cord or the idea of the apple dropping from the tree naturally allows for the coming into being of a new individual, unless there are changes in environment, in the general sense, the emergence of new group identities is not proclaimed.

It is when there are breaks in transmission as well as major changes in environmental factors and relations between individuals that statements are made about the coming into being of a new identity as opposed to the continuation of the old one. When, for example, a number of people move away from their home and form a community elsewhere in a different environment, in time they proclaim a new identity among themselves, especially if the group comes into contact with other people. If a subject group overpowers the dominant group and assumes control of the society, then a new identity is seen to come into being. If a group is invaded and over-run by others, there is also a tendency to assert that, because of new relations between people, those invaded cease to be what they were previously and in time become absorbed into the dominant group or are changed by the dominant group.

It is not surprising, therefore, that Cheikh Anta Diop, in his definition of cultural identity, identifies *historical continuity* as one of the factors that contribute to the formation of the cultural identity of a people. He defines what he calls the *historical factor* as 'the cultural cement that unifies the disparate elements of a people to make them into a whole, by the particular slant of the feeling of historical continuity lived by the totality of the collective' ([1981] 1991: 212). In the case of the language, which he sees as another of the major elements of cultural identity, Diop argues, using Europe as the example, that one has to look not at the current language situation in Europe, where there are a great number of different languages, but at linguistic history to see that European linguistic unity (and therefore cultural unity) comes out of a uniform source – the proto-language called 'Indo-European', which is the parent of and link between the family of European languages. Diop thus suggests that knowledge of *historical continuity* as evidenced in historical links and relations in language bolsters feelings of cultural identity.

Clifford, however, in relation to a case of a native American group in North America, argues for maintenance of community identity, despite lack of historical continuity and in the face of major changes in the elements of identity:

> Groups negotiating their identity in contexts of domination and exchange persist, patch themselves together in ways different from a living organism. A community, unlike a body, can lose a central 'organ' and not die. All the critical elements of identity are in specific conditions replaceable: language, land, blood, leadership, religion. Recognised, viable tribes exist in which any one or even most of these elements are missing, replaced, or largely transformed. (1988: 338)

This, however, was not the legal opinion in this case, which was decided on the notion of continuity in the major elements. While such difference of opinion indicates that there are no absolute criteria for determining when a new identity comes into being or when an old one ceases to exist, it indicates that there is the possibility of arguing for an identity while allowing for discontinuities.

In at least one current theory the idea of central organs, rootedness, a persistent driving force or an essential quality is denied in the construction of the identity of individuals, according to Harris:

> The existence of composite selves, as distinct from an anarchistic, diffractured, or ephemeral self, is denied by postmodernists as a feature definitive of personhood because such notions of identity are often grounded on conceptions of the subject as having some hidden essence. ([1993]1995: 373)

Harris also argues:

> Rather, the world of decentered, diffractured, and doxical persons with
> transparent and constantly changing identities both (a) defines personal
> identity in postmodern culture and (b) represents the traits of what
> persons are as agents. ([1993] 1995: 368)

This type of identity is properly meant to apply to individuals in western, urban society today, but it can also be applied to a continuity of uprooted people who are transported or move from place to place, from situation to situation, not entirely dissimilar in their experience to the native American group referred to by Clifford (1988).

This kaleidoscopic identity is also tied, in its formulation, to the contemplation of Nature by Sartre in his essay 'The mobiles of Calder', in which he says:

> In his mobiles chance probably plays a greater part than in any other
> creation of man. The forces at work are too numerous and too
> complicated for any human mind, even that of their creator, to foresee all
> possible combinations ... the time, the sun, heat and wind will determine
> each particular dance. Thus the object is always midway between the
> servility of statues and the independence of natural events. Each of his
> evolutions is an inspiration of the moment; it reveals his general theme but
> permits a thousand personal variations. It is a little hot-jazz tune, unique
> and ephemeral, like the sky, like the morning ...
>
> Valéry said that the sea is a perpetual renewal. One of Calder's objects
> is like the sea – and equally spellbinding: ever changing, always new. A
> passing glance is not enough; one must live with it and be bewitched by it.
> Then the imagination can revel in pure, ever-changing forms – forms that
> are at once free and fixed:
>
> ... his mobiles are at once lyrical inventions, technical, almost
> mathematical combinations and the perceptible symbol of Nature: great
> elusive Nature, squandering pollen and abruptly causing a thousand
> butterflies to take wing and never revealing whether she is the blind
> concatenation of causes and effects or the gradual unfolding, forever
> retarded, disconcerted and thwarted, of an Idea. (Sartre [1948] 1966:
> 124–7)

The mobile is not simply a creation of man, but is a symbol of Nature and as such a symbol of man.

The mobile recalls the Greek idea of man as a marionette in the hands of the gods, but for Sartre the mobile is not under the control of the creator and is *midway between servility and independence.*

Such a formulation of identity can be applied to the Caribbean colonial situation. What it signals is that in an analysis of cultural identity the factor of time is most disturbing to those who seek 'cultural equilibrium' (parallelling 'genetic equilibrium' in the population), because change is constant. Proof of cultural identity, however, does not require demonstration that any single or number of features be present in all the members of a community, and it does not require demonstration of the permanence of any feature. What it requires is that on a feature matrix constructed using the features of all members of a community there must be a significant overlap and, though the feature matrix changes over time, overlap remains a reality. Identity and continuity are thereby established, together with the kaleidoscope and the absence of essence.

It must be borne in mind, however, that this approach cannot in itself remove contention and bias simply because the historical record is not unequivocal and the starting point for 'indigenous' is often arbitrary. This is starkly illustrated in González ([1980] 1993) and Forte (2005). These are two works that paradoxically use different starting bases in the reconstructing of national identity – González featuring Puerto Rico as the country in question, and Forte featuring Trinidad. González, using a Spanish-speaking island, assumes 'the extermination of the Indian population' and posits that 'the first Puerto Ricans were in fact black Puerto Ricans'. Forte, using an English-speaking island, discounts the notion of the extermination of the indigenous population and stays within the Spanish Caribbean tradition by 'forging a sense of national indigeneity, in part and sometimes indirectly, via the figure of the heroic Amerindian, the first root of the nation, the territorial predecessor even if not the biological ancestor of the true national' (2005: 28).

# Written and oral modes of representation

Knowledge of a historical record is available through oral and written modes of expression, which are based on what Bourdieu calls *symbolic power* and what he analyses as follows:

> a power of constituting the given through utterances, of making people see and believe, of confirming or transforming the vision of the world and, thereby, action on the world and thus the world itself, an almost magical power which enables one to obtain the equivalent of what is obtained through force ... What creates the power of words and slogans, a power capable of maintaining or subverting the social order, is the belief in the legitimacy of words and of those who utter them. (1991: 170)

According to Bourdieu, it is not the symbol or word itself that embodies power, but 'the belief in the legitimacy of words and of those who utter them'. Therefore, whether it be by the politician, the historian or the writer, words uttered can transform those who hear or read them because the hearer believes in the legitimacy of those 'leaders' or shapers of thought who produce them.

Diop underlines the idea of a people's consciousness of their own history, their own historical continuity, and argues that consciousness of one's history becomes a reality 'from the moment that it [a people] becomes conscious of the importance of the historical event to the point where it invents a technique – oral or written – for its memorization and accumulation' (p. 214). In other words, Diop associates consciousness of history with the recording (by the people themselves) of history, and recording of history is thus seen as the most important means of establishing the historical continuity of a people. Diop's analysis validates Bourdieu's view of symbolic power but without distinguishing between leadership and followers in the process. So, if Diop's view of the importance of the historical record is taken together with Bourdieu's view of symbolic representation, then the political significance of the recording of one's own history and asserting one's own identity become even clearer.

However, when Diop presents *oral* or *written* almost as two equivalent alternative modes of representation, it masks the sequential relationship in evolution between the two and thus the greater significance of the written record in modern society. The accepted view of the superiority of the written was expressed four centuries ago by Gaspar de Villagrá in the Prologue to his *Historia de la Nueua Mexico* (1610):

> No greater misfortune could possibly befall a people than to lack a historian properly to set down their annals; one who with faithful zeal will guard, treasure, and perpetuate all those human events which if left to the frail memory of man and to the mercy of the passing years, will be sacrificed upon the altars of time. (Translation from Delgado-Gomez 1992: 3)

The difference between the oral and the written in establishing cultural identity is explained by Clifford (1988) in his account of the Mashpee trial, a court case in the United States in which a group was trying to establish their claims as a native American tribe. In what he labels as 'the hierarchical distinction between oral and literate', Clifford says:

> The Mashpee trial was a contest between oral and literate forms of knowledge. In the end the written archive had more value than the evidence of oral tradition, the memories of witnesses, and the intersubjective practice of fieldwork. In the courtroom how could one give value to an undocumented 'tribal' life largely invisible (or unheard) in the surviving record? (1988: 339)

The court imposed a literalist epistemology. Both sides searched the historical records for the presence or absence of the word and institution **tribe.** (1988: 340)

As a result, Clifford concluded that:

History feeds on what finds its way into a limited textual record ... Anthropology, although it is also formed and empowered by writing, remains closer to orality ... For even though the origin of evidence in an archive may be just as circumstantial and subjective as that in a field journal, it enjoys a different value: archival data has been found, not produced, by a scholar using it 'after the fact.'

The distinction between historical and ethnographic practices depends on that between literate and oral modes of knowledge. History is thought to rest on past – documentary, archival – selections of texts. Ethnography is based on present – oral, experiential, observational – evidence. Although many historians and ethnographers are currently working to attenuate, even erase this opposition, it runs deeper than a mere disciplinary division of labor, for it resonates with the established (some say metaphysical) dichotomy of oral and literate worlds as well as with the pervasive habit in the West of sharply distinguishing synchronic from diachronic, structure from change. (1988: 340–1)

In western societies, then, the written record, probably because it is externalised from the individual and is visible, creates the illusion of being objective reality, as opposed to the oral record, which remains in the memory of the individual and thus appears to be constantly subject to subjective variation. Of course, the written record is not in essence any more truthful than the oral since both proceed from the same source and both can be used, abused, manipulated and rejected according to the intentions, abilities and power of individuals or groups. Hayden White speaks of 'Fictions of factual representation' (1978: 121–34) and Foucault says the following about truth:

truth isn't the reward of free spirits, the child of protracted solitude, nor the privilege of those who have succeeded in liberating themselves. Truth is a thing of this world ... Each society has its regime of truth, its 'general politics' of truth – that is, the types of discourse it accepts and makes function as true; the mechanisms and instances that enable one to distinguish true and false statements; the means by which each is sanctioned; the techniques and procedures accorded value in the acquisition of truth; the status of those who are charged with saying what counts as true. (2000: 131)

There is no necessary link between writing and truth; what happens is that when some idea, thought or opinion is externalised from the individual and converted into a written document, it seems to acquire truth and authority, if it did not have it before, and especially if there are no competing documents to refute it.

Goody (1977: 37, 44, 49–50) argues, however, that writing allows for reflection, accumulation of knowledge and criticism, all of which bring about a movement forward. Goody's argument may be logical in itself and cogent within a consideration of philosophical development, but it does not address reality, in the sense that written information is not necessarily as powerful in day-to-day activities as the spoken word, or, as he himself puts it, 'the oral form is intrinsically more persuasive because it is less open to criticism (though not, of course, immune from it' (p. 50). The demagogue is often more powerful than the philosopher. Goody's argument also does not address the problem of ignorance or access to writing on the part of one or more groups affected, and therefore absence of a true dialectic.

When writing is available to only one side in an argument, even if there are dissenting opinions on that one side, writing can often consolidate misinformation based on vast ignorance. It can also privilege what is initially selected (out of the many possible elements that could have been selected), thus distorting the picture, effecting a certain path of reasoning and conferring importance on this element, which may be further consolidated by spirited and prolonged dialectical discourse on the very element. In this way, the ecological determinants of literacy can distort the kind of power of literacy argued for by Goody. As Street argues:

> Literacy ... is a socially constructed form whose 'influence' depends on how it was shaped in the first place. This shaping depends on political and ideological formations and it is these which are responsible for its 'consequences' too. (1984: 65)

So, since control of literacy and control of access to publication delimit the power of literacy, the distinction between philosophy and propaganda has never been clear. This is especially evident in the European creation of the New World and especially of the Caribbean islands. In the present work, it is hoped that as a result of looking more intently at a range of primary sources in a non-canonical way a more representative view and understanding will emerge.

The problem of propaganda in image creation is especially relevant, as it relates to language, in the use of 'actual speech' to characterise an individual or group. One of the central problems in the use of primary sources from centuries ago to reconstruct language development (as part of identity development) is to assess the validity and representativeness of citations of speech given especially by writers who were non-native speakers of the language in question.

The problem arises in the first place without prejudice simply because what is cited is not an objective (e.g. electronic) recording of what was said. Since it is also unlikely that what is cited would have been written down at the time it was said, it is clear that the elements of distance and memory affect the accuracy of the citation. Furthermore, the conversion of oral to written is problematic even for the best-trained experts in the field. At best, then, writers merely create an impression of what was actually said and as, in the case of the historical Caribbean, this was riddled with prejudice and bias, the impression created was removed to varying degrees from reality.

Lalla recommends that 'A framework for evaluating language representation in a text must ultimately integrate the evaluation of text, writer and audience' (2005: 16)' In explanation of this recommendation, she goes on to say that:

> Realism is essentially notional, an illusion constructed of such elements as credibility and objectivity (which in the case of discourse can only be assessed in connection with some profile of the producer of the text), of other elements such as authenticity and verisimilitude (which must rest on analysis of the text itself), and of vividness (which has much to do with the profile of the audience, as elements of the discourse must be directed to effect strong perceptual impressions). (2005: 17)

Lalla's framework is not only sound for cases where one has to make 'representative' selections to illustrate arguments but is also helpful in cases, such as the present work, in which one has to discuss misconceptions about language, such as 'corruption', which were an integral part of the historical record.

# Ladinos, Caribs, Tainos, Criollos

## 'Otherness' – overarching yet divisive

Early sixteenth-century literature on the islands of the New World was exclusively 'current' literature of discovery, that is reports on voyages, visits and stays in one location or another, giving information about the flora, fauna and human inhabitants of the islands. Even works called 'general histories', such as Gonzalo Fernández de Oviedo y Valdés's *La historia general de las Indias*, the first 19 volumes of which were published in 1535, were a combination of personal experience, information given to the author by explorers and information from written sources. In the second half of the sixteenth century, when the attention of writers was directed more towards accounts of exploration of the mainland, comments on the islands were introductory and incidental. It was not until the seventeenth century that writers (e.g. Herrera 1601) were sufficiently removed from the early encounters on the islands that they could begin to assess the forces in the formation of island society following Columbus's arrival and to construct for themselves an interpretation of people and events in island society. In other words, some time had to pass before writers could be said to be writing 'histories' of the early Spanish settlements in the New World. It is not that there was a radical difference in vision between the early writers and the later ones, but at least the geography of the New World had become separated from Asia after Magellan's expedition and the early fantasies were beginning to recede, thereby allowing a clearer picture of what continued to be called 'the West Indies'.

Even though people began leaving Spain for the settlements in Hispaniola and the neighbouring islands within a decade after Columbus's first journey, the islands were not a major feature of interest in the lives of Spaniards generally, and even less so in other parts of Europe. However, as the wealth of the New World became more apparent, greater attention was shifted in that direction especially by other European nations that were overcoming their own internal problems and wanted a better picture of the new lands. Thus, French and English translations of early works became increasingly important and writing about the New World became more purpose oriented, that is nationalistic, economic and moralistic, as competitiveness among European nations increased. Writers of the seventeenth century and after, in constructing their own visions of the New World, used the accounts of the sixteenth century, but what they selected out of them and how they re-presented them were obviously determined by contemporary forces. For example, French and English condemnation of Spanish behaviour and actions in the islands was nothing more than self-serving. On the other hand, they accepted many of the details of Spaniards' accounts of the culture of the indigenous inhabitants, because they had no first-hand knowledge to dispute them.

In order to understand the vision fashioned by later writers of differ-ent nationalities and the way in which this vision in turn shaped identi-ties and concepts of ethnicities in the islands, it is necessary to understand the first hundred years of settlement by the Spanish in the islands, the relationship between these settlements and Spain, and the role of these settlements in the wider history of the Americas. From the the beginning of the first hundred years, there has been a continuity in ideation, ideology and culture in European expansion in the New World because language was at its core. The continuity and its meshing with language are embodied in four designations – **Ladinos**, a word which spans a considerable period of history and a wide range of people; **Caribs** and **Tainos**, concocted names for the indigenous inhab-itants of the islands, the former of which has lived on in the name of the region; **Criollos**, a word which, though unclear in its etymology, is New World in its meaning and durable and positive in its spread.

While these names did not at the time identify fixed political group-ings either in the minds of the people so identified or in the minds of Europeans, yet, from a cultural point of view, one may argue that all four names and identities were the product of the colonial culture of 'otherness' developed by Spain. It was first this propagation of cultural difference between on the one hand the Spanish and on the other those who aspired or presumed to be Spanish or came under Spanish control, and second the likening of all of the latter that was the basis for the term *ladino*. In the case of *Carib*, it was both a matter of zeroing in on a sub-grouping within the general term 'Indian' to identify those who were more 'other' than the rest and targeting a group that showed itself to be more distinctive than the rest. In the case of *criollo*, the earliest Spanish literature saw the person so identified as a new and different Spanish person over there (*allá*).

# 'Ladino' and the naming of 'otherness' in the Old World

In 1492, after crossing the Atlantic, Columbus and his diverse crew encountered people who apparently welcomed them to the islands. One of the first steps the Europeans took in the subjugation process was linguistic; it was related in the following way by Peter Martyr:

> Thus making a league of friendship with the King, and leaving thirty eight men to search the island, he departed to Spain, taking with him ten of the inhabitants to learn the Spanish tongue, to the intent to use them afterward for interpreters. (Anghiera 1555: 4)

Having been forced to provide interpreters for the Europeans, these friendly people unwittingly played a role in facilitating the adoption of the Spanish language and in creating what was to become Latin America. For them, the year 1492 was an infamous one – it was the dawn of their own annihilation. Their societies gave way in the face of those established by the invaders from across the ocean and, while they themselves in general did not last long enough to become *ladinos* through contact with the Spaniards, their relatives on the mainland to the west did.

*Latino* and *ladino* are Spanish words, but they are also words that went far beyond Spain in a long history, with usages that changed over time, principally because they were used to represent peoples who had undergone, or who resulted from, a certain process of mixing or change through culture contact. In its etymology, the word *latino* is merely a variant of *Latinum*, which was used to refer to people (and their language) from an area of Italy that included Rome. With the rise of the Roman Empire, the adjective 'Latin' was most closely associated with the language of the Roman people, but it also came later to be associated with the Christian Church that had its base in Rome to distinguish this Church from the Greek Church. Originally, people in Europe who were identified as 'Latin' were the product of the 'romanisation' of Europe. The word 'Latin' in its earliest extended reference was associated with western Europe and it gradually came to refer to the people of the Iberian peninsula, speaking the two close languages Spanish and Portuguese, even more so than its original geographical source language in Italy. Apparently, the reason for this was that the Iberian provinces were thought to be the best in the Roman Empire, providing it with great emperors, and in the post-colonial period they led the way in looking back to the mother country for cultural models.

The word *ladino* is generally regarded as having derived from *latino* and came to embody the notions of hispanisation of people, conversion to Catholicism and the mixture of Spanish with other languages. The cultural matrix of the word *ladino* was the contact between Arabs/ Moors and Iberians in the Iberian peninsula, which lasted for seven centuries. It was a situation in which politics and religion caused many to have to change from one ethnic group to another and it also brought about extensive mixture within ethnic groups. The mixture that characterised religion, sex and politics in this situation was reflected in language and specifically in the meanings of words emerging in that context.

In its earliest cited uses from the Middle Ages (given by Kany 1960: 22), *ladino* refers more to the result of a process than to a characteristic. In one case, that is *moro tan ladino que semejava christiano* [Moor so ladino that he seemed Christian], the idea of process is clear, and in the other, *un moro latinado* [a Latinised Moor], *ladino* is presumed to be a variant of *latinado*.

Elcock (1960: 276–7) presents a fuller explanation of the Moorish role in the evolution of *ladino*. He claims that the Romance speech (*latini*) used among the Moors (especially the women) lost its literary backing (i.e. Classical Latin), became associated with the lower classes and as a result the word *ladino* (a derivative of *latini*) developed a pejorative meaning – 'sly', 'roguish'.

Somewhat in contrast to Elcock, Kany demonstrates a shift or extension in the meaning of the word from language to people when he says 'ladino … applied in medieval times to a Spanish-speaking Moor' (1960: 22). This could have been quite a normal development, seeing that name of language and name of people are in many cases identical.

In his discussion of the Arabic word *muwallad* and its putative derivative *muladí*, Forbes (1993: 140, 141) cites three sources that cast some light on the use and meaning of these words:

> He [a rich Portuguese landowner] was descended from a northern family 'which had passed over to Islam and had become **muwallad** (converted) … Later on … many Christians were converted to Islam and became muwalladun …' (Oliveira Marques 1972 1: 32, 69, 71)

> the Hispano-Arabic word **muladí**, used in Spain for a 'Christian Spaniard who made himself a renegade by becoming a Muslim'. (Corominas 1954)

> **Muladi** was: 'a term which the Arab historians used to designate the indigenes of the Iberic peninsula, children of indigenes converted to Christianity'. (*Grande Enciclopédia Portuguesa e Brasileira*)

The word *ladino* is here seen to be not just an unmotivated variant of *latino* but one that came under the influence of muwallad (sing.)~ *muwalladun* (masc. pl.) and *muladí*, embodying the notion of ethnic and religious conversion. It should be noted that, whereas in the beginning the conversion from Christian to Muslim was the required move, at the end of the fifteenth century the reverse was the case, which meant that *ladino* came to be more closely associated with being Latinised.

In the context of the Americas, a variation in the basic meaning of *ladino* (that is, a shift from a name to an ability) as well as extension in the meaning of the word are to be found in the account of Pedro Simón (1627), who said in a glossary at the end of his work: '*Ladino, se llama el que sabe bien la lengua estraña de la suya, y con metaphora se llama ladino el que es resabido en qualquier trato*' ['ladino is the name given to anyone who knows a language other than his own very well, and by extension ladino is a person skilful in any area']. This same kind of explanation, associating positive connotations with the word, is preserved and repeated three centuries later by Esteban Deive:

> El vocablo **ladino** es una transformación de **latino** y se aplicó originalmente en España a los antiguos habitantes que aprendían a hablar el latín con elegancia, haciéndose luego extensivo a quienes mostraban habilidad y destreza en cualquier asunto y oficio. (1978: 153)
>
> [The word ladino is a transformation of latino and in Spain originally referred to the former inhabitants who learned to speak Latin elegantly. The meaning was extended to refer to anyone showing skill and dexterity in any area.]

The admiration for competence in a foreign language, especially the language of colonial masters or of the elite, transferred easily from the European context to the American – in the same way that inhabitants of the province of Spain were keen to emulate their role models in Rome, people in the Americas were also keen to emulate the Castilians.

Notwithstanding the element of admiration, *ladino* was a term applied by Spaniards to 'others' when they adopted it from its Moorish context and extended its use to refer to foreigners living among them for some time who spoke a foreigner variety of Spanish. The main element of meaning was the acquisition and use of Spanish by those who had their own native language, as is exemplified later in the colonial context in the following: '*un capitán inglés, llamado don Juan, bien ladino en nuestra lengua*' ['an English captain named Don Juan, well versed in our language'] (Simón [1627] 1987: 639). In addition, as part of the Latinisation process, the word also involved the idea of conversion to Roman Catholicism.

# Latinisation and the coming of age of the Spanish language

The word *ladino* and its predecessor as well as variant *latino* represent the encapsulation of a process and the identification of a type in the history and world of western Europeans. The process and type clearly started with the movement of the Romans westward and continued across the Atlantic, through the islands and on to the mainland of the Americas. The idea of a superior culture and of the exposure of barbarous people to this culture for their own salvation was a fundamental philosophy in this movement. This crusading philosophy was supported on one side by the Roman Catholic religion and on the other by the printing press.

At the heart of this process was language, originally Latin, and it was through language that acculturation was slowly and very effectively achieved. A fairly early and perceptive view of the role of the Latin language in the original Latinisation process in Western Europe is given by Brerewood:

> For first, concerning Ambassages, suites, appeals, or whatsoever businesse of the Provincials, or forraines, nothing was allowed to be handled or spoken in the **Senate** at **Rome,** but in the Latine tongue. Secondly, the Lawes whereby the provinces were governed, were all written in that language, as being in all of them, excepting onely municipall Cities, the ordinarie Roman law. Thirdly, the Prætors of the Provinces, were not allowed to deliver their Judgements save in that language … Fourthly the generall schooles, erected in sundry Cities of the Provinces, whereof we finde mention in **Tacitus, Hierome,** and others (in which the **Roman** tongue was the ordinary and allowed speech, as is usuall in universities till this day) was no small furtherance to that language. (1614: 16)

This practice of conducting all official business, including formal education, in a single official language was the model taken by Europeans across the Atlantic, because language was believed to be the single most powerful unifying force in empire building. As Brerewood said, Latin continued to be used as the language of instruction in universities in the seventeenth century and its fundamental role as a conduit of European thought and values was quite evident.

In European expansion across the Atlantic, language policy in the educational system became a little more complicated because, although Latin itself had assumed an unassailable position of prestige and was therefore regarded as the foundation of culture, Spanish, for the Spaniards, was the language of empire. A double hierarchy began to establish itself in the minds of educated people in Spain – Latin was seen as the supra-national language of literature and logical thought, a kind of window to supreme knowledge; Spanish was seen as the national language and the language of current business. With the growth of printing, the written language was seen as higher in value, more permanent and influential than the spoken. Consistent with this second part of the hierarchy, classical Latin, the literate language of Rome, completely overshadowed vulgar Latin, the oral language that had actually produced Spanish and the other Romance languages. In like manner, literate Spanish came to overshadow the actual spoken Spanish dialects of the various parts of Spain, the language spoken by those who crossed the Atlantic. The classical to modern and literate to oral oppositions were seen not as different media but as scales of culture and intelligence.

During the sixteenth century the Romance languages were still emerging in status and structure. Because Spanish and French were known to have evolved from Latin, grammarians looked to Latin for rules of grammar and roots of words to give these derived languages a stamp of authority as well as to tie them more closely to their prestigious and presumed perfect parent. Words in Spanish and French were therefore given spellings closer to their classical roots in spite of their actual pronunciation, and the grammatical concepts of classical Latin were used as guides for written Spanish and French. This Latinisation of these two languages was a conscious practice and a major focus of writers in the rise and generalisation of printing in western Europe.

The Latinisation process generally, as part of the culture of Spain, was continued in the New World since the island colonies were nothing but extensions of Spain and Spanish interests. However, the attempts to reach the indigenous populations, whether for evangelical or secular reasons, brought about a modification in the Latinisation process and a recognition that effective, literate communication could certainly not be achieved by Latin or even by the exclusive use of the colonial or official language. In addition, the early transmission of printing into Mexico, while it did not immediately affect the islands, split the source of Spanish literate dominance and brought Spanish much closer to its American dependents.

In 1539 the first work printed in the Americas rolled off the press in Mexico, and in the 1550s universities started in Mexico and Peru. Although there were no comparable developments in the island colonies of Spain until the eighteenth century, it is quite likely that these latter were influenced by the intellectual activity on the mainland. Most of the material printed in the Americas in the sixteenth century was religious, was intended for the conversion of the indigenous inhabitants and therefore had translation of language as one of the major concerns, both as a part of the structure of the work and as a theme of religious education. Even though, according to Johnson (1988), 'Spain's first real **Index** of forbidden books [1558] ... warned against circulation of Christian doctrine in languages other than Latin', most of the religious works published in Mexico and Peru in the sixteenth century were in Spanish with translated versions intended for the local converts.

It was as a result of Spain's conquests in the New World that Spanish grew as a 'world' language, a fact that encouraged writers to use it in preference to Latin. Fernandez de Oviedo made it clear that he was interested in the reach of his message in explaining why he wrote in Spanish:

> *Algunos, que dicen ser mis amigos, han querido reprehenderme u honestamente desalabar o tachar lo que a mi honor dicen ellos más conveniente e de mayor auctoridad fuera, si como estas historias que en lengua mera castellana he escripto, fueran latinas ... y en fin ésta es regla universal: que todos los escriptores caldeos, hebreos, griegos e latinos, en aquella lengua escribieron en que más pensaron ser entendidos, y en que más aprovecharon a sus propios naturales. E pues la lengua castellana está tan ampliada e comunicada por tantos imperios e reinos, como lo está, no se han de tener en menos estima los que en ella escriben que los que escribieron en las otras. ([1555] 1959, tome 5, pp. 414–15)*

> [Some, who claim to be my friends, want to reprimand me or truly withhold praise from or censure what for my honour they say would be more convenient and more authoritative if these histories which I have written in mere Castilian were in Latin ... and in short it is the universal rule that all Chaldeans, Hebrews, Greeks and Latins wrote in that language in which they believed they would be more widely understood, and in which they brought greatest benefits to their own people. And so, the Castilian language is so widespread and used by so many empires and kingdoms, as it is, those who write in it do not have to be regarded as less esteemed than those who wrote in other languages.]

The writing of a massive, informative and influential work by Oviedo in Spanish dramatically increased the prestige of Spanish as a language of scholarship.

It is likely that in the colonies writers were under greater pressure to demonstrate their scholarship and education by writing in Latin, in effect becoming more conservative than writers in Europe. For instance, Acosta, while he was in Peru, wrote the first two volumes of his history in Latin, but when he returned to Spain he not only translated them *en vulgar* (into the common tongue) but continued the next five volumes in Spanish, a fact which he thought, like Oviedo, he should account for. The shift from Latin to Spanish was essentially a shift from an elitist, academic world to a more practical and wider market. With the improvement in the status of Spanish there was a corresponding improvement in the benefits of knowing Spanish and being *ladino*.

# Old World ladinos and their influence in the New World – the case of Africans

About the time of Columbus's first trip, Spain was beginning to become a unified nation by the union of Aragon and Castile, through the marriage of Ferdinand and Isabella, as well as by the reconquering of Granada from the Moors. The Moors, who had invaded the Iberian peninsula through southern Spain from North Africa (*Berbería*), had spent over seven hundred years there before they were in turn pushed back into Africa. Quite obviously, over such a long period their cultural and linguistic influence in Spain was considerable and, more significantly, many of the early migrants from Spain to the New World islands were from southern Spain. While Kany (1960: 22) said that *ladino* applied to a Spanish-speaking Moor in medieval times, later the word was applied more specifically to African slaves in Spain and became even more precise in its meaning with the transfer of African slaves to the New World, as Deive explains: '*Con la trata de esclavos, la voz* [ladino] *pasó a calificar al negro que había residido dos o más años en la península y que había sido cristiano*' ['With the slave trade, the word came to refer to a Negro who had lived two or more years in the [Iberian] peninsula and who had become Christian'] (1978: 153).

The movement of black Africans into Spain was a part of the Moorish/Arab invasion and lengthy occupation of Spain. It was also the deliberate taking of people directly from the West African coast to serve as slaves, mostly in Seville. The trade in West African ('Guinea') slaves and their use in Spain were noted at the time of the early Spanish migrations to the islands by Fernández de Enciso (1519):

> toda esta tierra desde cabo verde al cabo de las tres puntas se dize la costa de Guinea. en esta tierra se prenden los hermanos unos a otros y se venden. tambien venden los padres a los hijos a los que se los compran; y dan los alos de los navios a troque de paños de colores: y de manillas de cobre y de otras cosas. desta costa se traen a españa los esclavos negros y de toda la costa de africa que hazia al austro.

> [All this territory from Cape Verde to the Cape of Three Points is called the Guinea Coast. Across this territory brothers capture and sell each other. Parents also sell children to whoever will buy them. They give to those in the ships in exchange for coloured cloth, and copper manacles and other

things. From this coast and all along the coast to the south Negro slaves are taken to Spain.]

This supply of West Africans to Spain had been going on for some time and in fact had come to provide an indispensable part of the social structure of Spanish life. An indication of the presence of West Africans in Spain and the kind of climate that the long presence of the Moors in Granada created can be ascertained from the story of one well-known and distinguished African called Juan Latino, who rose from being a slave to become a Latin and Greek teacher and a poet in Granada in the sixteenth century. The Spanish name Latino that was given to the African could either have been in keeping with what Deive (1978: 153) calls the original meaning of the word *ladino* ('*los antiguos habitantes que aprendían a hablar el latín con elegancia*' ['the old inhabitants who learnt to speak Latin with elegance']) or was a stereotypical name denoting the process of hispanisation that he had undergone or was a complimentary name for this African teacher of Latin. However, though Juan Latino himself had become a scholar, the ladino Africans who went to the New World were not and spoke a variety of Spanish heavily influenced by their own native languages.

When the social and linguistic relationship between African slave and European overlord that had developed in Spain in the fifteenth century began to be extended into the islands after the journeys of Columbus, at first, around 1503, there was some opposition. The introduction of African-born slaves into Hispaniola drew protests from the first governor, as is pointed out by Herrera: '*Procurò que no se embiassen esclavos negros a la Española, porque se huian entre los Indios, y los enseñavan malas costumbres*' ['He tried to make sure that negro slaves were not sent to Hispaniola, because they fled among the Indians and taught them bad habits'] (1601: Decada 1: 180). This opposition was based not only on events in the islands themselves, but also on previous behaviour of ladino Africans in Spain. Deive (1978:153) argues, in relation to Hispaniola, that the ladino slaves rebelled because the transfer from Spain to Hispaniola resulted in a lowering of their status and an increase in laborious and difficult work. However, according to Sued Badillo (Sued Badillo and López Cantos 1986: 173), the problems with ladinos that flared up in Hispaniola and then in Puerto Rico were problems that had started in Spain itself.

The problem of ladino Africans mixing with indigenous inhabitants no doubt was a real source of fear among the Spanish, seeing that the two groups together constituted the majority in the early colonies. It seems as if this mixing led to extension of the term *ladino* from the

Africans, including the sub-Saharan ones, to the indigenous inhabitants, who, as part of the workforce, learnt to speak some kind of Spanish. What is significant about this extension is that, as was the case in centuries past with the Latinised people of Europe, the process identified as 'Latinisation' was transferred from what initially was undergone in the mother country to what later took place in the colonies, that is from Rome to Spain (in the case of Iberians) and then from Spain to Spanish America (in the case of the indigenous inhabitants). The Latinisation of Africans, however, did not fit into this pattern of extension or colonisation: they were migrants, as they continued to move or be removed from their homeland. Consequently, ladino Africans became accustomed to the reality that they had to be always armed with a language they could use to communicate with Europeans and others, the kind of language that was said to have become normal in *Berbería*.

While ladino Africans were the first blacks to be brought into the New World, in 1518 Africans were brought directly from West Africa for the first time to the Spanish possessions. The Spanish used the word *bozal* to describe the new African brought directly from Africa. The word *bozal* initially was not used by the Spanish with any specialised meaning; it was used to refer to young and inexperienced boys. The metaphorical meaning of 'inexperienced' was extended in the case of Africans to refer to a person *'que se expresa con dificultad en castellano'* ['who expresses himself with difficulty in Castilian'] (*Enciclopedia Universidad Ilustrada Europeo-Americana*). Deive quotes a 1528 ordinance in Hispaniola as giving the following definition for *bozal*: *'aquel que hubiere menos de un año que vino a esta isla, de Cabo Verde o Guinea, salvo si tal esclavo fuere ladino cuando allí viniere'* ['*bozal*: a slave from Cape Verde or Guinea who had been less than a year in this island, unless he was a ladino when he came there'] (1978: 154). What this shows is that the word *bozal* acquired a more precise meaning when it developed as a contrast to *ladino*.

Boyd-Bowman (1964: xii) makes the case that during the period 1493–1519 of every three colonists at least one was Andalucian, of every five one was a native of the province of Seville, and of every six one was an inhabitant or native of the city of Seville. From this conclusion, the speech of Andalucians, specifically those of Seville, emerges as the element that was most likely dominant in the evolution of early Antillean language. According to Pike (1967: 345), Seville in 1565 had one slave for every 14 inhabitants and also according to Pike's assessment: 'The ethnic variety that characterised sixteenth-century Seville set it apart from other Spanish centers and increased its similarity to the cities of the New World' (1967: 359). The long period of interaction between Iberians and Africans of various types no doubt had an effect on spoken Spanish in Seville and it is likely that the African element in fifteenth- and sixteenth-century Spanish society could have been partly responsible for some of the distinctive features of Sevillian dialect, bearing in mind that it had become a common practice for enslaved African to act in Spanish households as nurses and nannies and to be constant companions to their masters and mistresses. Ladino African women serving as nannies and domestic help in early Antillean society also must have had some influence on the language of children growing up in those households. So, whether indirectly through Sevillian dialect or directly, as domestic help, ladino African women were part of the development of early Antillean Spanish.

Boyd-Bowman (1976: 733, note) identifies the following as features of sixteenth-century 'Andaluso-Caribbean phonology':

neutralisation of /l/ and /r/ at the end of syllables, e.g. *la muhe* °(< *la mujer*); *e* °*papé* (< *el papel*);

the aspiration or loss of /s/ at the end of syllables, e.g. *la* °*muhére* (< *las mujeres*); *lo* °*papéle* (< *los papeles*);

/n/ at the end of syllables being articulated further back in the mouth (i.e. 'n' > 'ng'), e.g. *eη* E°*paña* (< *en España*);

merger into a weak /h/ of the Old Spanish phonemes /s/, /z/ and /h/.

The similarity between features of ladino Africans in Spain at that time (occurring in the plays of Lope de Rueda) and later features of Cubans of African descent is pointed out by Bachiller y Morales (1883a: 99). However, some of the features discussed by Bachiller y Morales as '*las desfiguraciones á que está expuesto, con la gente de color, el más hermoso idioma del mundo*' ['the disfigurations to which the most beautiful language in the world had been subjected by people of colour'] are the same as those given by Boyd-Bowman, who identified them as Andalucian. Bachiller y Morales's association of non-prestigious features of speech with Africans, which was intended to be disparaging, ironically serves to show African continuities.

Bachiller y Morales, characteristically for his time, was driven to identify variations from contemporary Castilian as distortions by Africans and their descendants, but his citation of historical and literary sources in Spain and Cuba shows that the variants were typical, to varying degrees, in the speech of ladino Africans in Spain, bozal Africans in Cuba and black Creoles in Cuba. This long association of Africans with these features of popular Spanish suggests that ladino

Africans were directly involved in the transmission of the features from the Old World to the New, and that for the Africans the features were related to their native languages, for there is a strong tendency in languages stretching along the coast from Senegal to Nigeria to have syllables ending in vowels rather than consonants.

# Old World ladinos and their influence in the New World – the case of Jews

As was the case with the Tainos, the year 1492 was also an infamous one for another ethnic group on the other side of the Atlantic, in Spain itself. In that year Jews were expelled from that country and, while some went east further into Europe, others went across the Atlantic. These Spanish (Sephardic) Jews and the language they spoke carried the name *Ladino*, which embodied the same notions that were applied to the Moors. The concept of *ladino* therefore established a link or similarity between Jews and Moors in the minds of Europeans before the American experience. Unlike the Tainos, however, the Ladino Jews over time turned their diaspora to their advantage by establishing an economic network and a network of communication across Europe and the Atlantic. This was partly because Jews, as a result of discrimination, tended to stick to their own and so preserved their cultural (including language) and religious characteristics more strongly. In fact, it is not accidental that the word *ladino*, which once embodied a concept with wide application, is today in popular usage predominantly associated only with the Sephardic Jews.

It was really in the seventeenth century and after that the Jews became more influential in the islands of the Caribbean, since they had networks of business contacts in the Americas and Europe, and also because they were in the forefront of changes in agricultural technology (Arbell 1981). So, by the end of the seventeenth century there were Jewish communities in Surinam, Curaçao, Barbados, Jamaica and St Thomas. The expertise that the Jews developed in sugar technology made them valuable to the European colonists and reduced the prejudice that these latter directed towards them.

While *ladino* had been a prominent term among the Jews themselves, especially in Europe, in the Americas it did not seem to be a term generally used by others to refer to them or their language. It is interesting to note, however, that, even beside this term, another connection was established between Jews and Africans in the Caribbean. The Iberian Jews who were forced to convert to Christianity but secretly maintained their Jewish way of life were disparagingly called *marranos* [pigs] by Christians. The Africans who escaped from European domination into the bush were called *maroons*. While the etymology of this latter word may in reality have nothing to do with pigs, this was exactly the sense given to the word by Dallas, who said: 'The rest of the fugitive negroes, now designated by the appellation of Maroons, or hog-hunters' (1803: 26). It was an etymology that Dallas adopted from Edwards (1796: iii, note), who got it from Long ([1774] 1970 2: 338 [note]). That this derogatory 'pig' etymology was associated with two ethnic groups, both of whom were otherwise called *ladinos*' does not seem to be totally accidental.

In the cultural construction of otherness Europeans established links not only between Jews and Africans but also between Jews and the indigenous inhabitants. The imagining of links between Jews and the indigenous inhabitants is explained by Elkin:

> Diego Durán, sixteenth-century Dominican missionary to the Aztecs, believed the Aztecs to be descended from the Ten Lost Tribes of Israel because of their cowardice. Furthermore, he 'could not help but be persuaded' by the similarity of their customs: 'the sacrifice of children, the eating of human flesh, the killing of prisoners of war, all of these being Jewish ceremonies' ...
> Confusion between Jew and Indian is apparent in documents of the period. In manuscript, and even in print, **iudío** easily becomes **indio** and vice versa ... These orthographic slips mirrored actual confusion in European minds between Indians and Jews. (1992: 28–9)

It is not surprising therefore that the term *ladino* also came to link Jews and the indigenous inhabitants.

Most likely the Ladino Spanish of the Jews who came directly from Spain or through Portugal to the Caribbean islands in the sixteenth century was not a major influence because their numbers were small. Their influence was felt more in the lower social groups in the early societies since the kind of discrimination that they had undergone in Europe and continued to undergo when they crossed the Atlantic made them second-class citizens. In addition, those Jews who left Spain (in contrast to the *conversos*) were of the working classes, and, as Renard explains: *'il s'agissait surtout de commerçants ou d'artisans citadins. Leurs moyens d'expression étaient forcément plus limités que ceux de l'élite intellectuelle'* ['it was especially merchants or urban artisans. Their speech was necessarily more limited than that of the elite'] (1966: 131). Enslaved Africans who belonged to Jews or associated with them were most likely affected by their Ladino Spanish. In any case, the 'mixed Spanish and Hebrew dialect' of the Jews fitted into the ladino pattern of the labouring classes of the society in which there were Spanish–Taino and Spanish–West African mixtures.

# New World ladinos – the conversion of the indigenous inhabitants

To provide some clarity about the popular but problematic terms 'Arawak' and 'Carib', used to refer to the various people living in the islands when the Europeans arrived, Rouse (1992) says the following:

> Columbus encountered Tainos throughout most of the West Indies. They inhabited the Bahamas and all of the Greater Antilles except western Cuba ... A second peripheral group, the Island-Caribs, lived on the islands from Guadeloupe southward ... (page 5)

> Ethnohistorians have grouped together the residents of various localities who shared a single language and had the same culture. The group is called Taino ... because several of its members spoke that word to Columbus to indicate they were not Island-Caribs ... Daniel G. Brinton (1871) preferred to call the group Island Arawak because it shared many linguistic and cultural traits with the Arawak Indians (also known as Lokonos), whose descendants still live in northeastern South America. His followers shortened the phrase to Arawak. That was a mistake. The Indians who called themselves Arawaks lived only in the Guianas and the adjacent part of Trinidad ... Columbus never met them. Moreover, they differed in both language and culture from all of the natives he did meet. (page 5)

> Carib = Ethnic group in the Guianas and around the Orinoco Delta during the Historic age [= post-Columbus] (page 175)

> There is reason to believe that the term **Carib** was being loosely applied to all warlike Indians in the small eastern islands, regardless of their cultural affiliation. (page 155)

In contrast to the personal, obviously *ad hoc*, early exchanges between indigenous inhabitants and Europeans, and the linguistic consequences thereof, the political response of the Spanish to the language barriers that they encountered was two-pronged. First, Columbus 'persuaded' some of the indigenous inhabitants to return with him to Spain. Initially, his primary reason for this was to convince his patrons of his 'discovery', but it also meant that the indigenous inhabitants became more familiar with the Spanish language and could therefore serve the purposes of the Spanish in subsequent expeditions. The value of these indigenous inhabitants (i.e. those who experienced life in the land of the white men) to the Europeans on the second voyage of Columbus can be discerned from the very mention of them and their role as interpreters in the 1494 account by the doctor on the expedition, Diego Alvarez Chanca (Major 1847: 56–7). The value of interpreters led to a practice of taking indigenous inhabitants back to Europe, as a strategy to overcome linguistic and cultural difficulties on future expeditions. It became common, so that in time the acculturation of the New World to European ways and language was in no small way facilitated by the Europeanised indigenous inhabitants themselves, because such natives, on their return home, fostered a desire among their own people for competence in the language of the foreigners, even if only for the easier acquisition of goods.

Fascination with the world and goods of the Spanish led the Tainos (i.e. indigenous inhabitants in the bigger Caribbean islands) to become increasingly involved with the Spanish and to learn elements of their language. From the early years onward, trade in trinkets, cloth and tools meant that the Spanish words for these items were adopted into the languages of the indigenous inhabitants and, as trade grew, more and more Spanish words and phrases were acquired by the various ethnic groups throughout the island chain. So, initially, Spanish spread among the indigenous inhabitants, through trade and interpreting, which in essence was the beginning of the Latinisation process among them. The emergence of Spanish-speaking indigenous inhabitants, called 'ladinos', can therefore be linked to the Spaniards' earliest commercial and political decisions in their encounters with the indigenous inhabitants.

The second response to the language barrier in the initial encounters was for the Spaniards to learn to converse with the indigenous inhabitants in their language. So, in his account of events in Hispaniola in 1494, Herrera said:

> *Una de las cosas provechosas que el Almirante* [Columbus] *hizo en aquellos principios para la conversion de la gente, fue procurar con mucho cuydado que assi sacerdotes como legos aprendiessen la lengua de los Indios.* (1601: Decada 1, 88)

> [One of the beneficial things that the Admiral did in the early days for the conversion of the people was to try diligently to get not only priests but also lay people to learn the languages of the Indians.]

This second response from the Spanish also became common later among other Europeans, especially as a tool of evangelisation. The Europeans sought to conquer the souls and minds of the Tainos by introducing Christianity to them in their own language. After the initial introduction of concepts in the catechism through pictorial means and literal linguistic substitutes (along the lines of the sixteenth-century catechisms of Jacobo de Testera in Mexico), converts of the Roman Catholics were shifted into Latin, the holy language of the Church, because decisions at the Council of Trent forbade the translation of the Bible itself into indigenous languages. The evangelising of the indigenous inhabitants in their own languages was therefore only the first step in Latinising them, that is doing to them what centuries before had been done to Western Europeans to make them Latin. This policy of the Spanish, no doubt, made them more sensitive to the cultures of the indigenous inhabitants in some ways, especially on the mainland, but the virtual annihilation of them in the islands negated any possible positive results, such as the preservation of, or greater role for, their languages in Antillean culture.

The two-pronged cultural and religious policy that the Spaniards used to lead the indigenous inhabitants away from their traditional ways was counterbalanced by the Spaniards' own need to deal with the practicalities of everyday life – to survive, to have daily food, to avoid poisons and other dangers, to find gold, to produce sugar – which required the knowledge and help of the indigenous inhabitants. Indigenous languages, or at least their words, were as a result virtually indispensable in the earliest years of the European–indigenous inhabitant encounter, as is argued by Tuttle (1976: 601). While words from indigenous languages may have survived, the languages themselves had little chance of survival when their speakers dwindled in number. Thus, Alvarez Nazario (1992: 75) argues that after a short period of Arawak–Spanish bilingualism in the Spanish Antilles, the more powerful language of the colonisers began to assert itself, thereby causing the virtual disappearance of the Taino language. So, beyond the words needed for practical purposes, which were borrowed into the Spanish language, survivals from the indigenous languages (i.e. their phonology and syntax) are seen to be minimal.

Unfortunately, the full truth of the language situation in the bigger islands in the sixteenth century and the influence of the languages of the majority, native population are not recoverable from European accounts. However, the admiration accorded to those who 'succeeded' with the language of the Spaniards was quite evident as was the fact that this 'success' was assumed to be accompanied by other positive characteristics. Two references from Juan de Castellanos illustrate this:

| | |
|---|---|
| Y *una Caribe India Catalina* | [And Catalina a Carib Indian |
| *De Peraluarez moço diligente,* | a diligent youth of Peraluarez |

| | |
|---|---|
| *Muger de gran razon e ya ladina* | a woman of great intelligence and |
| (1589: 208) | already 'ladina'] |

| | |
|---|---|
| *Fue Enriquillo, pues, indio ladino* | [Enriquillo, then, a 'ladino' Indian |
| *que supo bien la lengua castellana,* | who knew the Castilian language |
| *era gentil lector, gran escribano,* | well, was a keen reader, great notary, |
| *y en estas islas tuvo grande mano.* | and had a major role in these |
| (1589: 49) | islands] |

The latter reference is to that native of Hispaniola who was to become the most romanticised figure in the literature of Santo Domingo and the other Spanish islands. This was a case of a child, according to the story of Las Casas, who had seen his parents destroyed by the Spanish invaders in the first decade of the sixteenth century, had been raised by the Spanish, yet had retained pride in his heritage and had eventually risen up in rebellion against them. It is not only the romanticised figure of Enriquillo that reverberated in the cultural history of Santo Domingo as well as Cuba and Puerto Rico, but also the admiration for the conversion from *indio* to *ladino* through the *lengua castellana*.

# Ladinos and the linguistic beginnings of the Latinisation process in America

The first stage in the evolution of Spanish America started with the coming together of Old World Latinised people, ladinos of various races, West Africans and indigenous inhabitants in the mines and elsewhere in the islands of the Caribbean at the beginning of the sixteenth century. It can be regarded as a double process – a process of Americanisation of the ladinos and others who came from the other side of the Atlantic, simultaneously with the Latinisation of native Americans and subsequently Creoles. In this process generally, natives of the Americas who came into contact with Europeans were increasingly made to feel that Europe held the key to civilisation because it was a place already made. On the other hand, most of those coming from the other side of the Atlantic believed that the Americas were a new world without history and culture, a place to be feared or fashioned. In this process, the acts of the Europeans were viewed against a background of culture, while those of the native inhabitants were seen as natural, that is determined by the elements. This was the European way of seeing things and it became the reality through literature.

Older European literature (of the 'discovery' period and thereafter) promoted the notion of the adoption of Spanish and Portuguese words by the indigenous inhabitants and their spread throughout the islands as well as in South America. Acosta noted, for example:

> *los Indios no tienen en su lengua vocablos propios para estos animales* [brought from Spain], *sino que se aprovechan de los mismos vocablos Españoles, aunque corruptos ... las* [cosas] *que de nuevo recibieron, dieronles tambien nombres de nuevo, los quales de ordinario son los mismos nombres Españoles, aunque pronunciados a su modo.* ([1590] 1591: Lib. 4, Cap. 34)

> [the Indians had no words of their own in their language for these animals, unless they made use of Spanish words themselves, although corrupted ... all the new things they received were also given names again, which normally are the same Spanish names, although pronounced in their own way.]

In fact, Acosta came up with a simple method of identifying what things the indigenous inhabitants had before the arrival of the Spaniards and what things they did not: he surmised that old or known things had a native name and new things were given a Spanish name ([1590] 1591: Lib. 4, Cap. 34, p. 181).

It seems reasonable to suppose that Acosta's thinking on the subject must have been aroused by a noticeable number or prominence of Spanish words in the speech of ethnic groups in Peru (where he was). So, the hispanisation of the vocabulary of indigenous languages is presented as starting with the introduction of words for cats, dogs, horses and other four-legged animals that were brought from Spain; it continued with agricultural products like sugar and the industry that surrounded their production; it included minerals like silver, and mining operations; and it extended to all the goods that the Spaniards imported, which became part of American life.

Language modification among the indigenous inhabitants was not a matter of Latinisation exclusively. Pelleprat, having had more time than Acosta to reflect on the process of vocabulary expansion, identified onomatopoeia in addition to borrowing as a method used:

> *ils appellent **Vacca**, les **Taureaux**, & les **Genisses**; **Cabáïo ou Caválle**, les chevaux: **Sombrero**, les chapeaux. **Camicha**, les chemises, & même tous les habits dont nous nous servons: & **Carta**, du papier, ou un livre. Pareillement ils nomment, **Tintin**, un marteau, à cause du bruit qu'il fait: **Ikirilicátopo**, une poulie; & **Corótoco**, un coq, ou une poule, pour la même raison.* (1655: part 3, 12)

> [they call bulls and heifers 'vacca'; horses 'cabáïo' or 'caválle'; hats 'sombrero'; shirts and all the clothes we wear 'camicha'; paper or a book 'carta'; Similarly, they call a hammer 'tintin' because of the noise it makes; a pulley 'ikirilicátopo'; and a cock or a hen 'corótoco' for the same reason.]

In the case of borrowed Spanish words, Pelleprat claimed that they were changed only very slightly. Acosta's comment on the subject, *pronunciados a su modo*, although derogatory in tone, does not suggest drastic restructuring. However, these claims differ from what the evidence shows in the case of the Island-Caribs (in the smaller islands), which in turn suggests that substantially restructured Spanish words in the mouths of indigenous inhabitants went unrecognised by some Europeans.

The role of the various ethnic groups among the indigenous inhabitants, while not as significant as that of the native inhabitants of Gaul, Iberia and other regions of the Roman Empire in determining the distinctiveness of modern Romance languages, was significant enough to lead Kany to state:

> These various substratum languages coloured the Spanish spoken in each region and were deciding factors in the division of Spanish America into five linguistic zones: the Caribbean zone with Arawak and Carib; the Mexican zone (including Central America) with Nahuatl and Maya-Quiché; the Andean zone with Quechua and Aymara; the River Plate zone with Tupi-Guaraní; the Chilean zone with Mapuche. (1960: 3)

However, in addition to the influence of substratum languages locally, the path of early Spanish colonisation suggests that in the borrowing and assimilation there were also transmissions from the geographical bases outward. In other words, some words from the earliest Spanish colonies were generalised across later colonies, while words from each locality affected the local dialect of Spanish.

Some of the words that were generalised across Spanish America and came to be adopted in the European languages themselves were from the language of the Tainos. Words such as *battata, cazabi, canoa, ma(h)iz* and *jucca*, which were highlighted by Martyr in 1516, came to have general currency, partly through natural face-to-face transmission and partly because of the immense influence of this printed early account of Columbus's encounters with the indigenous inhabitants in the islands. When Tuttle (1976: 603) makes the claim that 'Spanish became the vehicle for diffusing many arawak designations on the mainland', he is essentially arguing that the Spanish in use in the islands of Hispaniola, Puerto Rico and Cuba had incorporated useful native words and that this variety of Spanish was used in the colonisation of the mainland.

However, it was not just a matter that the islands were used as bases for the conquest of the mainland but also that the people of the islands were directly involved in the expeditions. Sued Badillo, for instance, speaks of '*el drenaje continuo de jóvenes, muchos de ellos criollos ya, reclutados para expediciones de conquista y guerra a otras tierras*' ['the continuous drain of young men, many of them by that time Creoles, recruited for expeditions of conquest and war in other lands'] (Sued Badillo and López Cantos 1986: 128). In other words, the expeditions were made up not only of Spaniards who were using borrowed words, but also of locally born whites for whom these indigenous words were normal.

In addition to white Creoles, there were, according to Sued Badillo (Sued Badillo and López Cantos 1986: 131, 133, 134), creole blacks and bozal Africans who were taken from the bigger Caribbean islands to the mainland. Both groups, the creoles and the bozales, would have used words from the languages of the various ethnic groups in the islands and presumably would have spread them to the areas to which they went. However, it is most likely that the influence of local languages gradually overshadowed features transmitted from languages of the early island Creoles.

The influence of local indigenous languages, which led Kany (1960) to identify five linguistic zones of Latin-American Spanish, was heightened by the constant increase in the number of Creoles and the variety of them. The early encounters in the islands were principally between adults – indigenous inhabitants, Europeans, ladino Africans and bozal Africans – but it did not take long for the results of miscegenation to become numerous, highly visible and an increasingly significant element in the population in the islands and later on the mainland. It does not appear as if children of mixed race were distinguished as a group from their non-white parents unless there was a difference in dominant language use. In other words, they were included in the term *ladino* when they spoke Spanish predominantly, even if they were physically indistinguishable from some of those who did not. So, the term *ladino* in later centuries came to include both concepts – 'mestizo' and 'Spanish-speaking'.

European literature noted the role of the Spaniards in the spread of words, from the languages of the people of the islands whom they first met, to mainland America as well as back to Europe. This was usually interpreted as validation of the functional role of the indigenous languages. What the indigenous inhabitants did in their own languages and in their own versions of Spanish has often been treated as bastardisation or corruption of the Spanish language. Linguistically, in both cases, it was a matter of borrowing and assimilation, with inevitable modification in pronunciation and structure. Thus, each side played a role in converting Spanish into a more complete instrument, not only for themselves but generally for Spanish conquest.

The two currents – Americanisation and Latinisation – which seemed to be flowing in different directions, constituted the essence and foundation of modern Latino culture. It can be said that Americanisation was the undercurrent and Latinisation the overlay in the Spanish island colonies – the one non-prestigious, oral, everyday and practical, the other prestigious, official and literate. The two began to converge from early on because the printing press was introduced into Spanish America in the first half of the sixteenth century, where it began to develop its own American traditions and in a sense compete with and reduce direct European influence. It is probably because of the early foundation of its own literate culture that Spanish America seemed to be less bound by the umbilical cord to Spain than was the case in other later European colonies in the Americas.

# Naming the New World islands

## The myth of Antillia

The names for places in the New World that emerged in the works of the early writers prevailed because they represented specific concepts for Europeans and were linked to their traditions. They demonstrated by their survival the power of the Latinisation-turned-hispanisation process in the literary sphere. Of the three names, 'West Indies', 'Caribbean' and 'Antilles', the one that seems most innocent but was actually the most romantic and written about is 'Antilles'. The name 'Caribbean', deriving from the people named 'Caribs' by the Europeans, has a high level of notoriety and hysteria in its associations (as will be seen), and the name 'West Indies' is widely known to be the perpetuation of an error. The reasons for the emergence and survival of these names, so full of myth and error, are to be found in the works of the earliest writers on the New World, and of all these early works it was the 1516 *Decades* of Martyr that was the most influential and played the major role in linking the islands to Europe as part of the (mythological) history of Spain and the Moors. Without doubt, European naming of places and people of the region was for them both an act of domestication and one of civilisation.

'Antilles' was the earliest of the three names, or more properly its predecessor 'Antilia'; it has an intriguing history with its own significant dates. It predates Columbus and is said to go back to visions of the Middle Ages. The most comprehensive single work on the origin and history of the word 'Antilles' is Cortesao (1954), which uses 'The nautical chart of 1424' to present a comprehensive survey of theories and works on the subject. The following are some conclusions by Cortesao (1954):

> 4. The tradition of the existence of lands beyond the straits of Hercules, more or less far away in the Atlantic, is very old ... Therefore, when, after

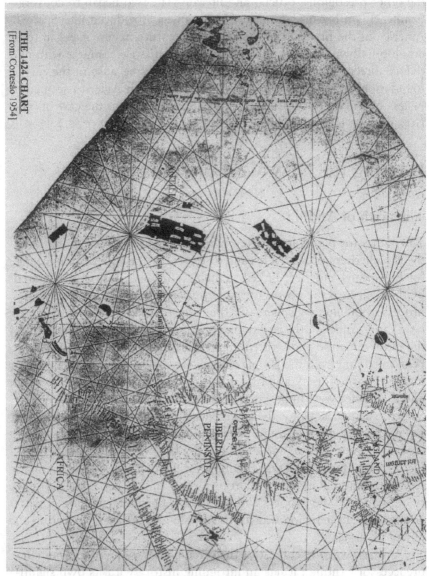

**Figure 2.1** The 1424 Chart. Armando Cortesâo, 1954. In *The nautical chart of 1424 and the early discovery and cartographical representation of America: A Study of the History of Early Navigation and Cartography.* Coimbra: University of Coimbra. British Library.

more than a thousand years of inactivity, the Atlantic voyages south and westward were recommenced with the expeditions to the Canaries, the traditional knowledge of the existence of lands far away beyond the Straits of Hercules became more vivid than ever before.

8. The 1424 Chart is the first known document in which the word *Antilia* appears ...

9. The 1424 Chart is the first known document in which the island *Antilia* is depicted ...

11. The Antilia Island was from the beginning associated with the eighth-century legend of the 'Island of the Seven Cities', as shown by the fact that the first time it was represented, in the 1424 Chart, it had seven mysterious names written on it which correspond obviously to the seven cities of the legend.

13. After the 1424 Chart the group of four islands is represented, in whole or in part, and sometimes with Antilia called Island of the Seven Cities, in at least nineteen maps and two globes of the fifteenth century ... (pp. 105–107)

As Cortesao (1954: 69) states, the legend of Antilia was known to Columbus and it was also known to the writers at the beginning of the sixteenth century. This is clear from the way in which Martyr (Anghiera 1530) referred to the name; he said in an almost familiar manner: *'sed cosmographorum tractu diligenter considerato, Antiliae insulae sunt illae & adjacentes aliae ...'* ['but the description of the cosmographers well considered, these [Hispaniola and Cuba] and the adjacent ones are the islands of Antilia']. Although Martyr cannot be said to have independently revived the word Antilia to name the now factual lands as opposed to the previously mythical ones, his reference at the beginning of the very influential Decades did in great measure serve to promote the name as one with a plural or regional reference, as opposed to being the name of a single island, as represented on the 1424 Chart.

It is not that Martyr *post facto* changed the reference from singular to plural. The 1455 map of Presbyter Bartholomeus de Pareto bears the title *Antillia*, which is clearly intended as the name of the more than 20 islands indicated, stretching from in front of Portugal in a vertical line (north/south) down to Africa. However, up to the end of the fifteenth century there was no unanimity of view that the word Antilia was either a single island or a group of islands. This is evident in Table III in Cortesao (1954: 68), which is called 'The Antilia Group of Islands in Fifteenth-Century Cartography' and which shows variation between the singular and plural references.

The word 'Antilia' itself was considered to be European. Pizzigano, the map maker who was the first to record the word in 1424, was Venetian. There is no argument anywhere that the word could have come into the written literature, as did others, relayed through the mouths of returning explorers. The basic etymology of the word 'Antilia' appears to have been agreed on by European writers over the years, although there is some argument about which specific European language should take credit for the original word. The etymology given for the word was, according to a Greek/Latin morphological evolution, *ant = before/opposite* + *ilia = island*. What is significant in all 21 maps referred to in Cortesao's Table III is that, in spite of the variation in spelling, none of the cartographers gave the name a plural form. In fact, it was not until there was an attempt to explain the etymology of the name that a plural form appeared. This occurred on the c.1519 Lopo Homem map of the North Atlantic (referred to by Cortesao 1954: 69), which had a suggestive variant of the name – ANTE‹YLLAS‹. It seems more than coincidence that this was just a few years after Martyr's plural reference.

In spite of the suggestivity of the Lopo Homem map, Antillia had not become a generally accepted designation for the islands in 1519. The reality of the many islands and their geographical disposition had been emerging from reports about the settlements the Spanish had established. The Spanish had set themselves up first in Hispaniola and then ventured forth from there to establish themselves in the neighbouring islands of Puerto Rico in 1508, Jamaica in 1509 and Cuba in 1512. It was from these island settlements that explorations were made further afield to the north, west and south. In 1519, Fernández de Enciso gave a comprehensive picture of the islands, including the names of the smaller islands to the south and a detailed positioning of them. From it the early sixteenth-century reader could visualise the great number of islands, some much bigger than others, in a relatively small area, as well as their dispersal in the shape of a bow. The view given in Fernández de Enciso's picture was that the (sea) area referred to as 'Paria' was dotted with islands to the east running vertically, whose inhabitants were culturally different and hostile to the Spanish, while to the north running horizontally were the bigger islands settled by the Spanish Christians, and even further to the north the tiny islands of the Bahamas, whose native inhabitants were docile. There was no attempt by Fernández de Enciso to use Antillia as a comprehensive term to identify the islands.

**Figure 2.2** *ANTE ᐓYLLAS ᐸ 'Ante Illas', carta maritima portuguesa del Atlas Miller, atribuida a Pedro Reinel* 'Portuguese maritime map from the Miller Atlas, attributed to Pedro Reinel' c. 1519. Bibliothèque Nationale de France.

Even as late as the beginning of the seventeenth century, Herrera (1601), for example, associated 'Antilla' with imagination and mythology when he said:

*Un vezino de la isla de la madera, el año de 1484. pidio al Rey de Portugal licencia para yr a descubrir cierta tierra que juraba que via cada año, y siempre de una manera, concordando con los de las islas de los Azores: y de aqui sucedio, que en las cartas de marear antiguas, se pintaban algunas islas por aquellos mares, especialmente la isla que dezian Antilla, y la ponian poco mas de docientos leguas al Poniente de las islas de Canaria, y de los Azores, la qual estimaban los Portugueses, que era la isla de las siete ciudades, cuya fama y apetito ha hecho a muchos por codicia, desviar y gastar muchos dineros sin provecho.* (Decade 1, p. 5)

[A citizen of the island of Madeira in the year 1484 asked the King of Portugal permission to go and discover certain land which he swore he saw every year and always in the same way, in concordance with the islands of the Azores. As a result, in the maps of the old sea routes some islands were drawn in those seas, especially the island called Antilla, and it was located little more than two hundred leagues to the west of the Canary Islands and the Azores. The Portuguese considered it to be the island of the seven cities, whose fame and lure caused many a person through greed to go astray and waste great sums of money fruitlessly.]

In fact, the islands collectively, for Herrera, were associated with the 'Indians' who inhabited them. He therefore did not follow Martyr's lead and associate the old myth with the new islands.

The mythical name was revived and restored by the French in the seventeenth century – first by Bouton (1640) and then in the (pseudo-) etymological form 'Antisles de l'Amerique' by Du Tertre (1654). In 1658 De Rochefort not only preserved the plural form and restored the older spelling but gave it an etymology that had some literary history. It was at the beginning of his description of the area that de Rochefort used the name 'Antilles' and gave it the following explanation:

*Entre le continent de l'Amerique meridionale, & la partie orientale de l'ile de Saint Jean Porto-Rico, il y a plusieurs Iles, qui ont la figure d'un arc, & qui sont disposées en telle sorte, qu'elles font une ligne oblique au travers de l'Ocean.*

*Elles sont communément appellées, les Antilles de l'Amerique. Que si l'on demande la raison de ce nom là, il est à croire, qu'elles ont été ainsi nommées, parce qu'elles font comme une barriere au devant des grandes Iles, qui sont appellées, les Iles de l'Amerique: Et ainsi il faudroit écrire, & prononcer proprement Antiles, ce mot étant composé de celuy d'Ile, & de la particule Grecque ἀντ, qui signifie à l'opposite. Neantmoins l'usage a obtenu, que l'on écrive & que l'on prononce Antilles.* (1658: 1–2)

[Between the continent of South America and the eastern part of the island of St John Puerto Rico there are several islands in the figure of a bow and they are situated in such a way that they form an oblique line across the [Atlantic] Ocean.

They are commonly called 'the Antilles of America'. And if one were to
ask the reason for this name, it is to be believed that they were so named
because they form like a barrier in front of the big islands, which are called
'the Islands of America'. And so one should write it and properly
pronounce it as 'Antiles', this word being made up of the word 'ile' and the
Greek particle ἀντ, which means 'opposite'. Nevertheless, in normal usage
one writes it and pronounces it 'Antilles'.]

Except for the name *Antilles*, de Rochefort's version comes directly
from the 1640 French translation of Jean de Laet's *Nieuwe wereldt ofte
beschrijvinghe van West-Indien*:

> Depuis le costé Oriental de l'isle de S. Jean Porte rique, jusques au
> Continent de l'Amerique Meridionale, il y a plusieurs petites Isles, qui
> disposees en arc font comme une barre à travers la mer. ([1633] 1640:
> 25)

> [From the east coast of the island of St John Puerto Rico to the continent
> of South America there are several small islands, which, in the figure of a
> bow, form like a barrier across the sea.]

At the head of the same chapter, de Laet gives the name of the small islands as *Isles Canibales* and *Isles des Canibales*, and the title of Book 1,
which describes both the big and the small islands, is *Des Isles de
l'Ocean*. What de Rochefort did, therefore, was to combine the words
of his predecessors; he replaced the local, post-Columbus name for the
small islands with the mythical Latin name, not necessarily to restore a
European bias but because the explanation given by de Laet apparently
seemed to him to be a perfect complement to Du Tertre's *Antisles de
l'Amerique*. Contrary to what he said, the only evidence to support the
idea that the islands were at the time commonly called *les Antilles de
l'Amerique* was the use of the name in the works of the two French
writers who preceded him, and that in itself was a clear case of the one
(Du Tertre) copying the other (Bouton).

As the name 'Antilles' and its new etymology took hold, there was
disagreement whether the presumed prefix (ante) meant 'before' or
'opposite', and whether the islands were before (= in front of) or opposite the bigger islands (= the Greater Antilles) or India or Europe. This
meant, in fact, that there was disagreement about what islands were
comprehended under the term 'Antilles'. The disagreement is pointed
out by de Rochefort:

> Quelques uns, comme Linscot en son Histoire de l'Amerique, prenant le
> nom d'Antilles en une signification plus generale, le donnent aus quatre
> grandes iles, l'Espagnole, ou Saint Domingue, Cube, Jamaique, & Porto
> Rico, aussi bien qu'a ces Vint-huit. (1658: 2)

[Some, like Linscot in his History of America, using the name 'Antilles' with a more general meaning, apply it to the four big islands, Hispaniola or Santo Domingo, Cuba, Jamaica and Puerto Rico, as well as these twenty eight.]

De Rochefort's words were translated into English a few years later by Ogilby:

> The islands called de barlovento, by which are understood Hispaniola, Cuba, Jamaica, and Boriquen, as also the Lucaies, with the Caribes, and lastly the isles called de sotavento, viz. Margareta, Cabagua, and Tabago, are by some comprehended all under the general name of the Isles Antilles; though others reckon the Antilles to be the same with the Caribes onely. (1671: 314)

Martyr's comment on Antilia, which in 1596 Linschoten (mentioned by de Rochefort above) obviously followed, did not allow for the interpretation of 'Antilles' as smaller islands in front of bigger ones, because Martyr specifically said that Cuba and Hispaniola as well as the adjoining ones were the islands of Antilia. Part of the development of this idea of islands in front of another land mass can be attributed to the very Linschoten, who referred to the islands as *'les Antilles qui gisent devant la terre ferme, & lui servent comme de couverture & defense'* ['the Antilles which lie in front of Terra Firma and serve as a cover and protection for it'] (Linschoten 1619: 1).

The origin of the name was still a subject of discussion in the eighteenth century, again in the account of a French writer. Charlevoix, after commenting on the views of his own countrymen, came out in favour of the Spaniard Herrera's interpretation:

> *Le nom d'Antille, que j'ai dit être general à toutes ces Isles, a exercé plus d'un écrivain, & donné lieu à bien des fables. Le Ministre Rochefort le fait venir de la particule Grec ἀντ, le P. du Tertre de la Latine* **ante***. Comme qui diroit, selon le premier, Isles opposées au Continent, selon le second, Isles, qu'on rencontre avant que d'arriver à la terre ferme. Antoine Herrera, un des plus exacts & des plus judicieux écrivains, qui ayant parlé du nouveau monde, croit avec plus de fondement que ce nom a été donné aux premiers Isles, qu'on a découvertes dans l'Amérique, à cause d'une Isle imaginaire, qui se trouvoit marquée sur d'anciennes cartes sous ce même nom, & qui le devoit peut-être à la fameuse Thulé des Poëtes. (1730, tome 1, pp. 2–3)*

[The name 'Antille', which I have said is general to all the islands, has exercised the mind of more than one writer and has spawned many stories. Minister Rochefort derives it from the Greek particle ἀντ and Father DuTertre from the Latin 'ante'. Some would say, following the former, 'islands opposite the continent' and others, following the latter, 'islands that one comes to before reaching the mainland'. Antoine Herrera, one of the most exact and judicious writers, who, having spoken of the New

World, believes with more reason that this name was given to the first islands discovered in America because of an imaginary island which was marked on old maps with the same name and which was due perhaps to the famous Thulé of the poets.]

Charlevoix did not give the impression that his interpretation had become the accepted one, but by that time it must have become difficult to use etymology (i.e. before/opposite + island) to explain a name that had existed before the islands of the New World came to be known to Europeans. The etymology 'before islands' was not based on reality, but was a simplistic attempt to explain the (French) plural form of the name. It was, though, a logical attempt to explain the structure *ante* + *ilia*, in which *ilia* and the place being referred to (Antilia) could not be the same. In other words, Antilia, by normal derivation, would have had to mean 'a place in front of/opposite an island' and since it was clear from the 1424 map that Antilia was itself an island, the derivation had to mean 'an island in front of/opposite another island'. Yet, most derivations try to make *ante* adjectival or descriptive (i.e. 'the before islands', 'the opposite islands') in order to make both *ante* and *ilia* describe the same place. However, neither the Romance languages nor English has the prefix (< preposition) *ante* as adjectival, which rules out most of the traditional attempts at explaining the origin of Antilia. It is quite likely that 'Antilia' (the pre-Columbus word) was no more than a single, indivisible word with the normal geographical suffix *-a*, that is a single word not made up of root parts.

The identification of the New World archipelago of islands with 'Antilia' brought them within the historical knowledge and traditions of western Europe. For Iberians and the very early writers (e.g. Martyr) it was a projection that had some measure of fantasy in it. For later French and English writers, who were unaware of the Spanish historical fantasy about the 'Island of seven cities', it was a case of a logical name being given to describe the geography of the islands.

## Making a case for 'India' and the rise of 'America'

In the same way that he played a role in presenting 'Antilia' as the name for the islands, Peter Martyr was also one who initiated the spread of 'West India' as the name for the newly discovered lands on the other side of the ocean. Peter Martyr knew Columbus and others who went on the voyages to the New World and stoutly defended Columbus in his belief about India and the naming of the new lands 'West India'. As Sued Badillo (1995: 75) concludes, from evidence provided by Gil (1989), Hulme (1979), Todorov (1982) and Pastor (1983), 'Columbus found innumerable clues to suggest that the Caribbean was the Indian Ocean and its islands those described by Marco Polo or attributed to him'. Martyr, in turn, sought to corroborate what Columbus believed.

One of the features of Martyr's writing in this and a few other cases was that he kept repeating his point. One can only surmise that he was making excuses for ignorance, he was trying to convince himself or he was succumbing to prevailing opinions. The fact that Martyr regarded Columbus's view as reasonable is illustrated in the following passages:

> albeit the opinion of Christophorus Colonus (who affirmeth these islands to be part of India) doth not in all points agree with the judgement of ancient writers as touching the bigness of the Sphere and compass of the Globe, as concerning the navigeable portion of the same being under us, yet the popinayes and many other things brought from thence, do declare that these islands favour somewhat of India, either being near unto it, or else of the same nature. (p. 3)
>
> Preparing therefore three ships, he made haste toward the island of Johanna or Cuba, whither he came in short space: and named the point therof, where he first arrived, Alpha and O, that is, the first and the last: for he supposed that there had been the end of our East, because the sun falleth there, and of the West, because it riseth there. For it is apparent, that Westward, it is the beginning of India beyond the river of Ganges, and Eastward, the furthest end of the same: which thing is not contrary to reason, for as much as the cosmographers have left the limits of India beyond Ganges undetermined, whereas also some were of the opinion, that India was not far from the coasts of Spain, as we have said before. (Anghiera 1555: 13)

So, Peter Martyr, in this early and influential work, gave credibility to the name 'West India' by using what seemed to be rational and navigational arguments.

It was not only Martyr who supported Columbus in his belief, for Lopes de Gomara, writing in 1552, introduced another possibility, which was used to exonerate Columbus and perpetuate his error. In a section titled 'The first discovering of the West Indies', Lopes de Gomara told the story, supposedly well known, of 'the pilot of the caravell that was first driven by forcible wind to an unknown land in the west Ocean', a pilot who had suffered misfortune on the voyage but managed to make it back and tell Columbus of his discoveries. Lopes went on: 'the pilot called the same India because the Portugales so called such lands as they had lately discovered eastward. Christopher Colon also after the said pilot, called the west lands by the same name' (Anghiera 1555: 311). It is of interest to note that in a similar story referred to by Cortesao (1954: 72) relating to the year 1447, the intention was to account for the name 'Antiles' and not 'West India'.

Lopes had two other sections dealing with what was essentially an explanation of Columbus's error – 'Of the colour of the Indians' and 'Why they were called Indians'. In the first he commented:

> and tawny like unto the west Indians which are all together in general either purple, or tawny like unto sodden quinces, or of the colour of chestnuts or olives, which colour is to them natural ... (Anghiera 1555: 310)

In the second is the following:

> Some think that the people of the new world were called Indians because they are of the colour of the East Indians. And although (as it seemeth to me) they differ much in colour and fashions, yet it is true that of India they were called Indians. India is properly called that great province of Asia in which great Alexander kept his wars: and was so named of the river Indus: and is divided into many kingdoms confining with the same, from this great India (called the East India) came great companies of men as writeth Herodotus: and inhabited that part of Ethiopia that lieth between the sea Bermeia (otherwise called the red sea or the gulf of Arabia) and the river of Nilus: all which regions that great Chieftain prince Prester John doth now possess. The said Indians prevailed so much, that they utterly changed the customs and name of that land, and called it India:by reason whereof, Ethiopia also hath of long time been called India ...
>
> After this also, of latter days our West India was so called of the said India of Prester John where the Portugales had their trade. (Anghiera 1555: 311)

Two features of identity used in the above – Prester John and olive colour – also appear in the following account of Africa by Richard Eden himself:

> In the East syde of Afrike beneth the redde sea, dwelleth the greate and myghtye Emperour and Christian Kynge Prester Iohan, well knowen to the Portugales in theyr vyages to Calicut.... . This myghty prynce is cauled David The[e] emperour of Ethiopia ... The chiefe citie of Ethiope where this great Emperour is resydent, is cauled Amacaiz beinge a fayre citie, whose inhabitants are of the colour of an olyve. (Eden 1555: 344)

The above citations from Martyr, Lopes and Eden can be said to have represented an influential body of written opinion intending to validate Columbus's view that the new lands were India. The reasons given vary – similar fauna to that of India, navigational justification (Martyr); the account of a predecessor, an explanation of the population and position of Prester John's India (Lopes); the colour of the inhabitants of Amacaiz in Prester John's empire (Eden). However, the very fact that there was a need to justify the name indicates that there was some doubt about its validity. Such doubt and reasoning appearing in the literature of discovery 60 years after Columbus's voyages indicate that a clearer picture of the globe had not yet been developed by that time and that this state of ignorance allowed the name 'West India' to have credibility and to grow in use.

A later explanation given for the use of the name 'India', by Acosta at the end of the sixteenth century, appealed not to arguments about precision but to metaphorical extension:

> *porque el uso y lenguaje nuestro nōbrando Indias, es significar unas tierras muy apartadas y muy ricas, y muy estrañas de las nuestras. Y asi los Españoles igualmente llamamos Indias al Piru, y a Mexico, y a la China, y a Malaca, y al Brasil: y de qualquier partes de estas q vengan cartas, dezimos que son cartas de las Indias, siendo las dichas tierras y Reynos de immensa distancia y diversidad entre si: Aunque tampoco se puede negar, que el nombre de Indias se tome de la India Oriental: y porque cerca de los antiguos, essa India se celebrara por tierra remotissima: de ay vino, q estotra tierra tan remota quando se descubrio, la llamaron tambien India, por ser tan apartada como tenida por el cabo del mundo, y assi llaman Indios a los que moran en el cabo del mudo.* (1591: 34)

[the use of the name 'Indies' in our language signifies places that are very far away, rich and very different from ours. And so the Spaniards equally apply the name 'Indies' to Peru, Mexico, China, Malaca and Brasil. And when letters come from anywhere in these places, we say that they are letters from the Indies, the said places and kingdoms being great distances away and very different from each other. Still it cannot be denied that the name 'Indies' came from East India. And because among the ancients that India was regarded as a very remote place, that is the reason that this other very remote land when it was discovered they also called it India, being so far as if at the end of the world, and so they called people who lived at the end of the world 'Indians'.]

The idea that any place that was remote, rich and exotic was called India was probably a meaning that developed after Columbus's voyages and had become current during Acosta's time. So, even though Acosta attributed this reasoning to the original naming, there is no indication in the early accounts that the use of the name 'India' was a metaphorical extension of the use of the word. However, such metaphorical extension was no doubt a powerful factor in the maintenance of the name.

The persistence of the name in the later period after the error became evident was also integrally tied up with political developments. From the arguments and passages above (in Anghiera 1555), it is clear that the name 'West India' referred to all of 'the new found lands in the

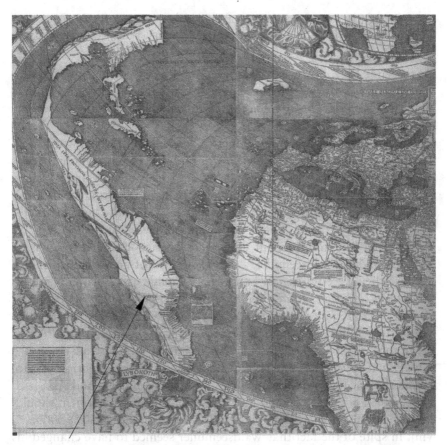

**Figure 2.3** The Waldseemüller 1507 Map; section showing the name *America* appearing on a map for the first time. B*rochure for Exploring the early Americas*. The Jay I. KislakCollection at the Library of Congress. Martin Waldseemüller. Detail of *Universalis Cosmographia Secundum Ptholomaei Traditionem et Americi Vespucii Aliorumque Lustrationes*, St. DiJ: 1507. Woodcut. Geography and Map Division, Library of Congress (142).

west Ocean'. More than a century later the area that the name referred to was still the same, for John Ogilby, 'His Majesty's Cosmographer' and 'Geographick printer', writing in 1671, said the following:

> The length of the West Indies is generally computed to be from the utmost south of Terra Magellanica [tip/bottom of South America], to the farthest north of Estotiland [an imaginary tract of land near the Arctic Circle in North America], about six thousand English miles, reaching from about sixty degrees of northern, to fifty three of southern latitude; the breadth from St. Michael or Piura westward [the westernmost part of South America in Peru], to Parabaya [Paraíba], a town on the coast of Brasile

eastward [the easternmost region of South America in Brazil], three thousand nine hundred miles, and the whole compass thirty thousand. (p. 126)

At the same time, however, the name 'America' was displacing 'West India/West Indies' as the common name, because the very title of Ogilby's book was *America: being the latest, and most accurate description of the New World.*

In fact, from the beginning 'America' was in competition with 'West India' as the name for the New World, both the result of ignorance. Whereas Columbus thought that he had reached India, Martin Waldseemüller, 'an obscure scholar living in St. Die in the Province of Lorraine in France' (*Antilia and America* 1955: 3), initially believed that Vespucci had discovered a new world, and argued that it should be named after him and thus put the name 'America' on what today is South America. Waldseemüller's famous words, of great consequence because they were written in Latin, the language of scholarship, were: '*& quarta orbis pars (qua quia americus invenit Amerigen: quasi Americi terram: sive America nucupare licet)* [and the fourth part of the world (which because Americus discovered it, it is proper to call Amerigem, that is, the land of Americus or America)] (1507, translation in *Antilia and America* 1955: 7). Part of Waldseemüller's intention was to counterbalance what had happened in the past – Europe and Asia had been named after women and he thought that the new continent should be named after a 'man of intelligence' (Waldseemüller [1507] 2003: viii). In spite of the fact that Waldseemüller seemed to have changed his mind subsequently (in that he omitted the name 'America' from his 1516 map), his suggestion that the new continent be named 'America' was duly followed by map-makers and most others, except the Spanish. Delgado-Gomez points out that although the word 'America' was used in a Spanish printed text for the first time by Fernández de Enciso in 1519, it was a name which 'in Spain remained rare until the nineteenth-century – the word "Indies" [las indias] being preferred' (1992: 12).

In the early years when the two names were in competition with each other, it seemed as if one was regarded as official and the other as popular, for De Laet ([1633] 1640) starts off his work by saying: '*Notre resolution est descrire en ce lieu toute l'Inde Occidentale, appellee vulgairement Amerique*' ['Our intention here is to describe the whole of West India, commonly called America']. Later on, the two names underwent a mutually exclusive contraction in their reference, with 'America' being increasingly associated with the mainland to the north and 'West Indies' being used to refer to the middle portion of the New

World, which included the 'Antilia' islands. An early eighteenth-century definition of the 'West Indies' is given by Atkins:

> For a general Idea of the West Indies, we may understand by that Term, all the Continent, Sea, and Islands, from Terra Firma to Florida, or from near the Equinoctial to 28° of N. Latitude; and if you include Bermudas, to 32°. The main Land in this Circuit divided into Spanish Provinces, is more peculiarly called the Spanish West Indies, they possessing all, unless to the Southward in Guiana and Paria, where there are a few English, Dutch, and French, interspersed on the Rivers and Coast of Oronoko, Surinam, and Amazons. (1735: 220)

By the second half of the eighteenth century, the term 'West Indies' no longer referred to the American mainland, but was being used to refer to the islands. This is clear from the following:

- In 1767 John Singleton published a book with the title *Description of the West Indies*, which referred specifically to the islands.

- In 1778 in the book *West India Merchant* there is the following: 'The West Indies, while the harmony of the empire subsisted, were supplied from America [= the British North American colonies]' (p. 122).

These last two references are from the period of US independence from Britain, by which time the New World had taken on a completely different look and the different regions had, as a result of European expansion, become more partitioned and viable as entities (i.e. countries), instead of being a vague, unknown mass thought to be the other side of India. As British North America became more prominent and distinct, the English increasingly used the term 'the West India islands' to refer to their non-mainland possessions, and this eventually became the 'West Indies'.

# Naming New World people

## Caribs as cannibals, fierce and man-eating

'Antillia' and 'West India' were names given by the Europeans out of their own knowledge to the islands and lands across the ocean. The names given to the places did not inevitably and exclusively result in the names for the people, because the Europeans to some extent realised that the people had names for themselves, even if they were not meaningful to Europeans. In any case, familiar names for places linked those places to European conceptions of the world, while exotic and strange names for people made them fit the European vision of other people as savages, heathens and barbarians. What resulted therefore was a combination and confusion of names – those projected from the place names given by the Europeans and those the Europeans believed to be designations the indigenous inhabitants used among themselves. As would be expected in a situation where communication was initially minimal, the designations that the Europeans perceived to be those of the indigenous inhabitants were in several cases really products of the European mind. Specifically in the case of the names for the earliest New World people Europeans encountered, the basic categories used to establish their identity were behaviour and geography. Indeed, for the people who came to be called Caribs, behaviour and geography were contending explanations of their name.

Since claims about consumption of human body parts as a feature of regular diet come out of contentious and highly suspect 'eye-witness' accounts, the creation of this kind of identity by European writers for indigenous inhabitants is now regarded as politically motivated. In other words, names were used to signal whether one was friend or foe. So, part of the mythology that came to be associated with people unfriendly to the European explorers was that they ate their enemies. It is possible that such a conclusion by early Europeans could have result-ed from observation of ceremonial practices and also from the sight of unburied bodies, the practice of keeping parts of dead relatives in the house (De Vries [1655] 1853: 163–4), cremation or killing of prisoners by burning. There is some support for this kind of conjecture in Vespucci's description of burial rites in which sick people were taken to places apart to die; food was left with them and they were not buried when they died (Young 1893: 124–5).

Part of the mythology about cannibalism must have resulted from the way Columbus went about getting his information. Note the fol-lowing from Columbus's first letter (15 February 1493):

> *yluego que lege alas īdias ēla primera isla q̃ halle tome pforza algunos dellos pa ra que deprēdiesen yme diese notia delo que auia enaquellas partes easi fue que luego ētendirō y nos aellos quando por lengua oseñas:* (Obregón 1991: 60)

> [When I arrived in the Indies, I took some of these people by force in the first island I found, so that they might learn our language and give me news of what existed in those parts. And it so happened, for later they understood us and we them, either by speech or by signs.] (Lucia Graves in Obregón 1991: 66)

In such a situation where, in addition to the fact that coercion would have led to poor communication between Columbus and his 'interpret-ers', there is known to have been diversity in language, animosity and hostility between ethnic groups, the meanings that Columbus interpreted no doubt were in many cases erroneous.

In any case, the people of the 'west Ocean' who did not welcome the Europeans became victims of abuse by European writers. From the start wild imaginings of European writers became 'literate' evidence. For instance, in the first known book written in English on the New World (printed by John of Doesborowe in 1511) the vision of unfamiliar people as wild beasts is put forward as an eye-witness account:

but that lande is not nowe knowen for there have no masters wryten therof nor it knowethe and it is named Armenica [America]/ there we sawe many wonders of beestes and fowles yat we have never seen before/ the people of this lande have no kynge nor lorde nor theyr god But all things is comune/ this people goeth all naked But the men and women have on theyr heed/ necke/Armes/knees/and fete all with feders bounden for there bewtyness and fayreness. Theese folke lyven lyke bestes without any resonablenes and the wymen be also as comon. And the men hath conversacyon with the wymen/ who that they ben or who they fyrst mete / is she his syster/ his mother/ his daughter/ or any other kyndred. And the wymen be very hoote and dyposed to lecherdnes. And they ete also on[e] a nother. The man etethe his wife his chylderne/ as we also have seen and they hange also the bodyes or persons fleeshe in the smoke/ as men do with us swynes fleshe. And that lande is ryght full of folke/ for they lyve commonly. iii.c. [300] yere and more as with sykenesse they dye nat/ they take much fysshe for they can goen under the water and fe[t]che so the fysshes out of the water. and they werre also on[e] upon a nother/ for the olde men brynge the yonge men therto/ that they gather a great company therto of towe partyes/ and come the on[e] ayene the other to the felde or bateyll/ and slee on[e] the other with great hepes ... they take the other prysoners And they brynge them to deth and ete them/ and as the deed is eten then sley they the rest And they been than eten also. (Arber 1885: xxvii)

[but that land is not yet known, for no masters have written about it or know it. The name of it is America. There we saw many wonders of beasts and fowls that we have never seen before. The people of this land have no king or lord or god. But all things are [held in] common. This people go about naked. But the men and women wear feathers on their heads, necks, arms, knees and feet to appear beautiful and attractive. These folks live like beasts without 'reasonableness' and the women are also as common. And the men have intercourse with the women whoever they are or whoever they first meet, be she his sister, his mother, his daughter or any other family. And the women are very hot and disposed to lechery. And they eat one another. The man eats his wife or his children, which we have also seen, and they also hang the body's or person's flesh to smoke, as men do with swine's flesh. And that land is right full of people, for they commonly live three hundred years or more because they do not die of sickness. They catch a lot of fish because they can go under the water and

catch the fish out of the water in that way. And they also war against each
other, for the old men bring the young men together there. They gather a
great company of two parties for that reason and they go against each
other to the field or battle and they slay each other in great heaps ... they
take the other prisoners and they put them to death and eat them and as
the dead are eaten then they slay the rest and they are then eaten also.]

This vision of people of the New World was fed by religious concepts
of heathens and, having been 'corroborated' by first-hand accounts, it
now became even more factual for Europeans.

The early and simple division of the peoples of the islands into two
groups according to whether or not they were friendly to the Europeans
started with the first letter of Columbus (1493). Ironically, in trying to
discredit the preconceived idea of monsters in those places, Columbus
himself contributed to it when he said:

> asique mostruos nohe hallado ninoti cia saluo de vnaysla que es aqui enla
> segunda ala eutrada delas yndias q̄ es poblada devna iente que tieuē en
> todas las yslas por muy ferozes los qualles comē carne vmana. (Obregón
> 1991: 61)

> [Thus, monsters I have not found or heard of, except in an island which is
> the second at the entrance to the Indies, which is populated by a people who
> are considered in all the islands as very ferocious, who eat human flesh.]

This was where the association of cannibalism with Caribs took root,
for even though the island which Columbus referred to was not named
in the printed 1493 Barcelona version of the letter, another version of
the same letter did have the word 'Carib' in it – (instead of 'vnaysla que
es aqui enla ...' in the version above) 'una isla que es Carib, la ...'
(Varela 1992: 224).[1]

Besides this reference in the text, an illustration *accompanying* a
later (Basel) version of the letter featured what seemed to be a division
of the peoples of the islands into generous on the one hand and fearful
on the other. In spite of the fact that Columbus was speaking of the
same set of people when he said that at first they were fearful and later
they were generous, the illustration created the impression that he was
contrasting two sets of people.[2]

Martyr also helped to propagate the picture of an ethnic contrast, in
the process converting one set into hostile, man-eating natives, whom
Columbus said he had only heard about. This notion of a broad divi-

---

1. The Obregón version of the letter follows the text of the sole extant copy printed in
   Barcelona, which is today housed in the New York Public Library; the Varela transcription cites the Archivo General de Simancas (Estado) in Spain as its primary source.
2. This explanation continues to be repeated up to today, as for example in Delgado-
   Gomez (1992: 9): 'The two groups of Indians shown react differently to the arrival of
   the Europeans: one flees in panic; the other engages in an exchange of gifts.'

sion into good and bad natives was later strengthened by Herrera's account, which, in reference to happenings in 1520, said:

*El Licenciado Rodrigo de Figueroa, despues de aver hecho diligente pesquisa sobre los Indios que comian carne humana, y en que tierras se hallavan, para que so color de cautivarlos, no se tomassen otros, declarò por auto judicial, que eran Caribes, todos los Indios de las Islas que no estavan pobladas de Christianos, salvo las de Trinidad, Lucayos, Barbudos, Gigantes, y la Margarita: todos los demas dixo que eran gentes barbaras, enemigos de Christianos, repugnantes a la conversion dellos, y tales que comian carne humana, que no querian admitir los predicadores de nuestra santa Fe catolica:* (1601, Decade 2, Book 10, p. 328)

[Licentiate Rodrigo de Figueroa, after having made a thorough investigation of the Indians who ate human flesh and the lands where they lived, in order not to take others, under the pretext of capturing them, declared by judicial edict that all the Indians in the islands where there were no Christians, except for Trinidad, Lucayos, Barbudos, Gigantes (= Aruba and Curacao) and Margarita, were Caribs. All the rest, he said, were barbarous people, enemies of Christians, repugnant to conversion and such as would eat human flesh, who refused to allow in the preachers of our holy Catholic faith.]

The Catholic Church, having had great success in Mexico, was not going to tolerate unfriendly indigenous inhabitants who refused to be christianised and so all such indigenous inhabitants were lumped together, regarded as uncivilised, branded as cannibals, and became fair game to be captured and enslaved. An official act of the Church therefore converted a prejudiced response into reality, thereby not only assuring the simple division into Caribs and Tainos, but also ensuring that the name 'Canibales' would come to mean 'eaters of human flesh'.

This division of the islanders into two was probably also supported by Fernández de Enciso's account (1519), which identified the people of the 'Lucayos' (Bahamas) as non-meat eaters:

*pero no tienen en ellas ninguna carne. su comer es pescado y rayzes y pan de rayzes: y cogollos de yerva. si los llevan a otras partes y les dan carne a comer mueren se si la carne no es muy poca.*

[However, there is no meat there. Their food is fish and roots and bread from roots and shoots of grass. If they are taken elsewhere and given meat to eat, they die unless the meat is very little.]

These Bahamians were the friendly natives who had welcomed Columbus and this account made it clear that they were not to be confused with others elsewhere who ate human flesh.

The European distinction between the two was essentially one between those in the bigger islands and the Bahamas on the one hand (in which the Spanish managed to set up their first colonies fairly easily) and those in the smaller islands that ran from north to south down to South America (in which the people were not as accommodating). In Martyr's 1516 division of the peoples of the islands, those who welcomed the Europeans were called 'Taini'. In fact, in the First Decade Martyr actually uses the term 'Taino' and gives its meaning as 'a good man' (Anghiera 1555: 4) and 'noble men' (Anghiera 1555: 9). In the latter case he makes a contrast between the Taini on the one hand and the Canibales on the other. Thus, this designation, which really was not a name initially but became one subsequently, was based on behaviour, not geography.

Ironically, the name of the first set, the Taini, who were the ones to encounter Columbus first, virtually disappeared in English accounts, while the second (Canibales) became very prominent in the literature generally. The second set were described as eaters of man's flesh; Martyr actually used for this second set two alternative names – Caribs and Canibales. In his references to them in the text itself when he introduced them or re-introduced them he said 'Caribs or Canibales', but in the graphic and gory detailed descriptions that followed each time he only used the name 'Canibales'. As a result of this, the word 'canibales' quickly became synonymous with 'men who eat men's flesh', and displaced the word 'anthropophagi' used by Martyr himself. The word 'Carib', on the other hand, did not acquire that meaning, although the people it referred to continued to be negatively stereotyped. Thus, the use of the word 'canibale' with its new meaning was one of the earliest examples in European vocabularies of Americanisation (i.e. an old word displaced by a new one), the reverse of Latinisation.

Martyr did not explain why these New World people had two names. In fact, it was not Martyr who gave these names to the world, although he may be credited with juxtaposing them as alternatives. 'Carib' first appeared in a version of Columbus's 1493 letter (see Varela 1992: 224) in which the inhabitants of the island called Carib were said to be fierce and to eat human flesh. The two words '*Camballi*' and '*Carabi*' occur in Vespucci's (1505–6) account and a modification of their spelling also occurs on the *Admiral's Map, From the Strassburg Ptolemy of 1513*, which Martyr clearly used. In recounting his first voyage, Vespucci said:

> sapemo costoro erano una gente che sidicono Camballi molto efferati ch mangiono carne humana. (Young 1893: 97)

> [we learned that those were a people who are called Camballi, very savage, who ate human flesh.]

**Figure 2.4** Depiction of Columbus's first encounter with indigenous inhabitants. Christopher Columbus De insulis inventis epistola. Basel 1493. New York Public Library.

*cichiamanano in lor lingua Carabi che uuol dire huomini di gran sauidoria.* (Young 1893: 92)

[they called us in their language *Carabi,* which means men of great wisdom.]

There are no other occurrences of these terms (*'Camballi'* and *'Carabi'*) in the text, which appeared about ten years before Martyr (1516) used both words as alternative names for the same people. The most likely explanation for the juxtaposition of the two 'names' was that one (*'Carib'*) was the name apparently used by Columbus and the other (*'Camballi'*) was the one used by Vespucci.

The change in spelling from *'Camballi'* to *'canibales'* can be accounted for by referring to the *Admiral's Map*[3] of 1513. On this map, in what may be construed as the eastern side of the Tocantins River (in what is today northeastern Pará) in Brazil, the word *'Canibales'* occurs as if it were an area where people with such a name lived. Also on the map in the area that could be construed to be the southern 'Caribby' islands, there is an island identified as *'y. de los canibabales'* [*y* = *ysla* = island]. The change in spelling was essentially a change from an Italian version to a Spanish one. Sued Badillo presents a similar argument to account for the change in the spelling:

> We are strongly inclined to believe that the term was first introduced by Columbus phonetically but that the paleographical transcription of it led to the new variant **canibalos – canibales.** The letters M, N, and I were easily confused in Spanish paleography. (1995: 76)

Though Martyr's spelling of Vespucci's *'Camballi'* can be accounted for in this way, his association of *'Camballi'* and *'Carabi'* was contradictory, seeing that Vespucci used the first to refer to 'bad' natives and the second to refer to 'good' Europeans and also that for Vespucci the word *'Carabi'* seemed to be a common noun, whereas in Columbus's account it was a name.

Martyr's change would have to follow from his own explanation that 'Carib, in the universall languages of those countries, signifieth, stronger than the rest, and from thence they are called Caribes' (Anghiera 1628: 296). The meaning 'stronger then the rest' that he associated with *'Carabi'* was equally applicable to *'camballi'*, if one follows the dictionary of the Carib language in Dominica, compiled in the 1640s by the priest Raymond Breton. Breton gives *'ka'* as a common prefix and the root *'balli'* (1665: 72) with a variant *'niballi'* (1666: 394) as

---

3. Although the Admiral's Map is supposed to be Columbus's map (i.e. a map made according to Columbus's information), Young (1893) disagrees and suggests that it is based on Vespucci.

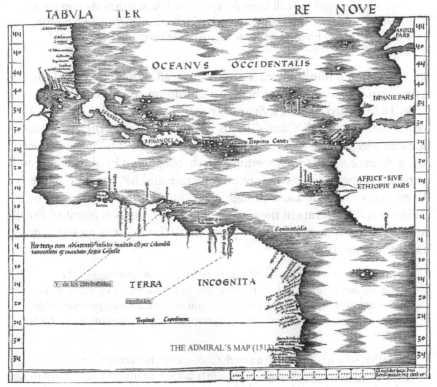

**Figure 2.5** The Admiral's Map (1513). British Library, Map Division

meaning 'valiant, hardy, strong, tough'. Vespucci's '*kaniballi*' or '*camballi*' could therefore have been a positive word for the people of the smaller islands (Lesser Antilles). There is evidence that what the literature did subsequently (e.g. Thevet 1575: 957) was to establish *post hoc* a semantic and cultural link between 'valiant', 'tough', 'cruel' on the one hand and 'cannibal' on the other to explain the initial association of the two words '*caribe*' and '*canibalei*'.[4]

In contrast to this explanation was that of Marmol, who only mentioned the exploits of the Spaniards in the New World in passing on his account of Africa – both his spelling of the name of the indigenous inhabitants of the islands and his explanation of it were somewhat different:

> *los naturales de la tierra las llamavan las Quiribas, por estar pobladas de hombres de guerra grandes flecheros, que tiran con una yerua, que al que hirieron muere raviando como perro dañado.* (1599: fol. xlii.4)

---

4. Taylor (1958) has also argued that '*caribe*' and '*caníbal*' have the same source, though different from the meaning 'stronger than the rest'. His argument is that 'our "cannibal" and "Carib" go back to Arawakan designations describing some tribe or clan as **manioc people**' (p. 157).

[the natives of the land call them the Quiribas, because they are populated by archer warriors who shoot arrows which have on them a herb poison which when they hit anyone causes them to die raving like a wounded dog.]

So, more than a hundred years after European writers first used the terms, variations and contradictions in meaning continued.

In 1627 Pedro Simón, a Spanish priest working in the area of Venezuela, explained 'Caribe' in his glossary as 'cosa aspera, brava, y de mala digestion, y assi llaman con este nombre a los Indios indomitos, y destas calidades' ['something harsh, tough and hard to digest and so they call the Indians who are indomitable and have these qualities']. A few years later, De Laet, though accepting 'Caribe' as a name for indigenous inhabitants in the Caribbean islands, in his account of Brazil continued to support Vespucci's point that 'Carabi' was an indigenous name for Europeans:

> Caraiba (que quelques-uns ont pris pour un Enchanteur) signifie la puissance par laquelle se font les miracles, voila pourquoi ils ont nommé les Portugais, & les nomment encore aujourd'hui Caraiba, pource qu'ils faisoyent beaucoup de choses qui surpassoyent leur entendement. ([1633] 1640: 476)

[Caraiba, which some see as an enchanter, means the power miracles are made by; that is why they called the Portuguese Caraiba and still do up to today, because they did many things which surpassed their understanding.]

However, unless De Rochefort a few years later was himself trying to resolve the problem in a *post hoc* fashion, the meanings that he gave for the word account for its reference to both Europeans and indigenous inhabitants: 'Ce mot de **Caraibes** signifie en leur langue, des **Gens ajoutez**, ou **survenus subitement** & à l'improviste, des **Etrangers**, ou **Hommes forts** & **vaillants**' ['This word Caraibes in their language means additional people or people coming suddenly and unexpectedly, strangers, or strong and valiant men'] (1658: 337). The last meaning is consistent with that generally given for the word 'Carib' and the preceding ones consistent with Europeans, from the point of view of the indigenous inhabitants of the islands. It seems likely that the variety of meanings of the word 'Carib' and the contradictions involved were in part caused by its widespread geographical occurrence and its use as a common descriptive word.

## Caribs as *Caribana* people, spreading across *Terra Firma*

Beyond the apparent contradictory references of the word 'Carib' to both Europeans and indigenous inhabitants, another contradiction exists in the proposed origins of the word. In De Laet's words (cited above) the '*quelques-uns*' to whom he was referring was specifically Thevet, who by explanation (1575: 913) and elaboration related '*Caraibe*' to '*prophetes*', but who (no doubt following Martyr's early account) also gave the following explanation: '*Ces Caraibes, ou Canibales, ont prins leur nom d'une riviere, qui se dit Caribane, laquelle est au goulphe d'Vrabe*' ['These Caribs or Cannibals get their name from a river called Caribane which is in the Gulf of Urabá'] (p. 957). Martyr himself had earlier said: 'Sailing forward from hence, he came to the East coasts of Vraba, which the inhabitants call Caribana, from whence the Caribes or Canibales of the islands are said to have their names and original' (Anghiera 1555: 53).[5]

This geographical source for the word 'Carib' was repeated by other writers, for example, Linschoten:

> *& de là on vient au Havre de Caribana d'ou est derivé le nom de Caribes ou Canibales mangeurs d'hommes.* ([1596] 1619: 23)

> [from there one comes to the harbour of Caribana whence derives the name of Caribs or Cannibals eaters of men.]

> *La commune opinion est que les Insulaires mangeurs de chair humaine sont issus des environs de Caribana pres d'Vraba & de Nombre de Dios, lesquels on appelle Caribes ...* ([1596] 1619: 25)

> [The common opinion is that the eaters of human flesh in the islands called Caribs came from around Caribana near Urabá and Nombre de Dios.]

The Spanish writer, Herrera, mentioned the place but did not associate it with 'Carib' when he said: '*segun se entendio despues, no era Babeque la Espanola, sino la Tierra firme, porque por otro nombre la llamavan Caribana ...*' ['according to what was understood later Babeque was not Hispaniola but Terra Firma, because they called Caribana by another name'] (1601: 32). The question that arises out of this, then, is whether 'Carib' derived from a common noun that referred to some human characteristic or from a place name. The same question

---

5. De Rochefort, in his contention that the inhabitants of the islands came from Florida, rejected this view, saying: '*Il y en a qui les font deriver du havre de Caribana, & qui pretendent qu'ils en sont issus. Mais cette opinion n'est fondée que sur la seule rencontre des mots Caribana & de Caribes, sans aucun autre fondement*' ['There are those who say that they came from the harbour of Caribana and claim that that is where they originated. However, that claim is founded solely on the similarity if the words "Caribana" and "Caribs" and nothing else'] (1658: 327).

arises in the case of '*Canibale*', which, following Breton's dictionary, referred to a human characteristic ('valiant, hardy, strong, tough'), but which, according to Sued Badillo, originated as a place name: 'The **Canibales** (Spanish plural) or **Camballi - Chamballi** (Italian plural) were according to Marco Polo, the inhabitants of **Canbalu** or ancient Peking, the city where the Great Khan supposedly lived' (1995: 76). As this explanation illustrates, there is a further complication of the matter in that, in contrast to the notion of a common etymology for the two words deriving from the usage of the indigenous inhabitants, there are arguments that the two words came out of different, pre-1492, European fantasies.

In this explanation of '*Canibale*', where the name of the people derived from the name of a place, while there is no literary 'evidence' from Marco Polo himself that the inhabitants of the city of the Great Khan were cannibals, such an association easily developed because Marco Polo had identified the inhabitants of Cipango as cannibals, and Cipango and Canbalu were not distinct in the minds of early writers, who might only have heard about, not read, Marco Polo's account. So, '*Camballi*' could have been both a projection from India on to the New World as well as an exotic distortion of the meaning and connotation of the word.

In the case of '*Caraibe/Caribe*', the pre-1492, classical allusion is Homer's description of the pair, Scylla and Charybdis, the last of which (a place name) can be construed to be the source of the New World word. The description in the *Odyssey* (Book 12) is:

> We then sailed on up the narrow strait with wailing. For on one side lay Scylla and on the other divine Charybdis terribly sucked down the salt water of the sea. Verily whenever she belched it forth, like a cauldron on a great fire she would seethe and bubble in utter turmoil, and high over head the spray would fall on the top of both the cliffs. But as often as she sucked down the salt water of the sea, within she could all be seen in utter turmoil, and round about the rock roared terribly, while beneath the earth appeared black with sand; and pale fear seized my men. (Homer 1966 1: 449)

There is a strong similarity to this in Martyr's description of the sea near the area identified as the origin of the people referred to as *Canibales*. Martyr's words were:

> The flowing of the sea, raged and roared there, with a horrible whirling, as we read of the dangerous place of Scylla in the sea of Sicily, by reason of the huge & ragged rocks reaching into the sea, from which the waves rebounding with violence, make a great noise and roughness on the water, which roughness or reflowing, the Spaniards call Resacca. (Anghiera 1555: 58)

It seems a little suspicious that Martyr mentions one half of a well-known pair, especially when not long before he was talking about '*Caribes*'. Martyr's earlier identification of the specific geographical source for '*Caribe*' as Caribana (on the east coast of the Gulf of Urabá in modern Colombia) is really what allowed for the association of the name in the classical allusion with the New World place. This kind of explanation added a bit of glamour and increased the myth by identifying Charybdis with a place in the New World. The association was also buttressed by orthography experts who, assuming that Charybdis and Carib were related, promoted the spelling *Charib* (rather than *Carib*) as the name of these indigenous inhabitants.

Negative characterisation of the so-called 'Caribs' was an obvious feature of colonisation, but an additional reason for it could have been the presence, in the neighbourhood of their reputed place of origin, of a group referred to in Anghiera 1628 as 'blacke moores' ('black Africans'). In his account of the New World, Martyr related that there were in the Darien mountains, not too far from the area of Caribana, Africans who had come across before Columbus and who were exceedingly fierce and cruel:

> There is a region not past two days journey distant from Quarequa, in which they found only blacke moores, and those exceeding fierce and cruell. They suppose in tyme past certaine blacke moores sayled thither out of Ethiopia to robbe, and that by shippewracke or some other chaunce, they were dryven to these mountaynes. The inhabitantes of Quarequa live in continuall warre and debate with these black men. (Anghiera 1628: 96, back)

The presence of Africans in the area was also later repeated by Ogilby, whose account was heavily based on that of Martyr:

> not one negro to be found, except a few near the River Martha, in the little territory Quarequa, which must by storm be drove thither from the Guinny Coast ... (1671: 40)
>
> Amongst the slain were found several negroes, which suffering shipwreck had been cast on the American shore, and maintain'd a continual war with the Quaraquanan ... (1671: 69)

The same black people were mentioned in the 1653 Drake account, although they were there interpreted to be maroons: 'the **Symerons** (a black People, which about eighty years past, fled from the **Spaniards** their masters, by reason of their cruelty, and since growne to an Nation under two Kings of their own' (Drake 1653: 7)). One wonders then whether this conflict between *fierce*, black Africans and indigenous inhabitants fed the idea that it was from the area where the black Africans lived that the reputed man-eating natives (i.e. the Caribs) originated. Additionally, the (mis)interpretation of the black Africans to be maroons rather than pre-Columbus arrivals raises the question about a similar interpretation of the origins (usually said to be around 1656) of the Black Caribs in St Vincent. The naming of these people 'Symerons' (from Spanish 'cimarrones') also raises the question of a connection between them and the Maroons in Jamaica, whose beginnings are usually associated with the English take-over of this Spanish colony in 1655. In short, it seems as if there could have been a continuity in ideation from the Symerons to the *fierce* Caribs to the Black Caribs and the Jamaican Maroons.

The notion of 'Caribana', as source or home of the Caribs, grew as their reputation as fierce cannibals took hold of the European mind. On the Admiral's Map of 1513 '*canibabales*' was associated with one of the small islands off what is now Venezuela and '*Canibales*' with an area of northeastern Brazil. In Martyr's 1516 text *Caribana* was identified as an area in what is now northwestern Colombia, while some seventy years later (in Hackluyt's 1587 version of Martyr) *Caribana* was identified as the area stretching from Panama to the bottom of the 'Antilles' archipelago. However, less than a hundred years later, on a 1656 French map, *Caribana* had become a wide expanse of the north coast of South America stretching from the bottom of the archipelago across to Brazil. At the same time, the small (now West Indian) islands were being generally referred to as the 'Caribbee/Caribby Islands' (see, for example, De Vries [1655] 1853: 119). In other words, within two hundred years of Columbus's first encounter with the indigenous inhabitants, the concept of 'Caribbean' was extensive, but in its orientation, extending from Panama to Brazil and including the southern part of the island chain, it still contrasted with the four original Spanish island colonies and, in that sense, can be seen as 'non-Spanish' or 'other'. Indeed, this mentality pervaded the 1553 'eyewitness' account of Cieza de Leon in which the distinction between conquering, Christian Spaniard and uncivilised native is everywhere. It is this account that would also have been substantially responsible for the extension of the area of *Caribana*, for in it almost all the ethnic groups in northwestern South America were described as cannibals.

## Tainos as *guatiaos* and people of the 'golden worlde'

In contrast to the natives of the region who conjured up the vision of uncivilised people who ate human flesh were those who were identified as *Taini* ('good and noble men') and *guatiaos* (those who had pledged a bond of perpetual friendship with the Spanish). It was first Martyr (Anghiera 1628: 11) who spoke of a bond of friendship and then more explicitly Las Casas, who, in describing the early encounters with the indigenous inhabitants in the first ten years of the sixteenth century, said:

> A éste, como a señor principal y señalado, el capitán general dio su nombre, trocándolo por el suyo, diciendo que se llamase adelante Juan de Esquivel, y que él se llamaría Cotubano, como él. Este trueque de nombres en la lengua común desta isla se llamaba ser yo y fulano, que trocamos los nombres, guatiaos, y así se llamaba el uno al otro guatiao. Teníase por gran parentesco y como liga de perpetua amistad y confederación. Y así el capitán general y aquel señor quedaron guatiaos, como perpetuos amigos y hermanos en armas. Y así los indios llamaban Cotubano al capitán, y al señor, Juan de Esquivel. (Las Casas [Bk. 2, Chap. 8] 1986 2: 37)

> [To him, as the principal and distinguished person, the captain general gave his name, exchanging it for his own, saying that henceforth he would be Juan de Esquivel and that he would be Cotubano, like him. This exchange of names in the common language of this island is called guatiaos when anyone and I exchange names, and so they called each other guatiao. It is regarded as a great relationship and a bond of perpetual friendship and a league. And so the captain general and this person remained guatiaos, as perpetual friends and brothers in arms. And so the Indians called the captain Cotubano and their leader Juan de Esquivel.]

The word 'guatiaos' was repeated and explained again by Las Casas in Book 2, chaps 16 and 46. These references constitute the earliest evidence in the New World of the institution and naming of strategic political alliances across ethnicities and, at the same time, evidence of deliberate division of the indigenous inhabitants (i.e. Caribs vs guatiaos) for political reasons.

Citing evidence from *Nueva Biblioteca de Autores*, vol. xxv, p. dliv, Henriquez Ureña says that from early in the century the term 'guatiao' had become a generally used one: '*En 1516 ya usan la palabra guatiao como genérica los padres jerónimos que gobernaban las Indias desde Santo Domingo*' ['In 1516 the Hieronymite priests who governed the island from Santo Domingo were already using the word guatiao as a generic'] (1938: 96, note 1). Henriquez Ureña also points to evidence (in the *Coleción de documentos … de Indias,* 1, 278, 328 and 379–385) of the contrast between those indigenous inhabitants called '*caribes*' and those called '*guatiaos*'.

Interestingly enough, some linguistic research suggests that the two words '*guatiaos*' and '*tainos*', the latter of which has come to be treated as a name, may actually have had the same root. According to Bachiller y Morales: '*Uno de los elementos más repetidos del haitiano, ya lo* *indicó Pedro Mártir de Anglería, que era el artículo **gua** que entra en la composicion de las palabras. Tiene significacion propia: este, esta, esto, como los demostrativos ...'* ['One of the most repeated features of Haitian [Taino language in Hispaniola], Peter Martyr of Angleria already identified it, was the article "gua" which forms part of words. It has its own meaning: "this", like the demonstratives ...'] (1883: 139). Taylor (1977: 19), for his part, relates the Taino prefix *'gua'* to Island Carib *'oua'* (= 'our'). Bachiller y Morales goes on (p. 144) to give **tai, tiao** as alternatives having the meanings *hermano, amigo, bueno,* and the combination **gua-tiaos** as having the meaning *Estos hermanos* (p. 139). Taylor (1977: 19) identifies the suffix *'-no'* in *Taino* as a plural-forming suffix. On this basis, then, *guatiaos* and *taino* can be taken to be substantially the same, which suggests that the separation of the two words was due to a lack of understanding of what the indigenous inhabitants were saying.

The word *guatiaos,* used to designate those who had established a bond of fellowship with the Spanish, was known throughout the region[6] and survived, with some modification, much longer than the people so identified by the Spanish. In the writings of the French, specifically Montaigne, it was reinterpreted as the French word *moitié.* In his essay 'Des Cannibales' (circa 1578), Montaigne, talking about the views of three indigenous inhabitants from Brazil whom he had spoken to when they were brought to France by French explorers in the middle of the sixteenth century, said:

> *Ils dirent ... (ils ont une façon de leur langage telle, qu'ils nomment les hommes **moitié** les uns des autres) qu'ils avoyent aperçeu qu'il y avoit parmy nous des hommes pleins et gorgez de toutes sortes de commoditez, et que leurs moitiez estoient mendians à leurs portes, décharnez de faim et de pauvreté.* (Tilley and Boase 1954: 43)

[They said ... [they have a fashion in their language such that the men name one another *moitié*] they had noticed that there were among us people who were full and stuffed with all sorts of commodities whereas their *moitiés* were beggars at their doors emaciated from hunger and poverty.]

Montaigne's phrase, *de leur langage,* specifies that the concept came from the indigenous inhabitants. The fact that they were from Brazil shows how widespread the word and cultural practice were.

---

6. This view is supported by evidence of lexical similarities across the Arawakan languages of the bigger and smaller islands given by Taylor (1977: 19), who argues that the Taino word *guatiao* 'appears to be cognate with IC [Island Carib] *ouatíaon* 'our friend' (a formalized relationship)'.

As the decimation of the indigenous inhabitants was revealed to other European nations, it contradicted the notion of a perpetual bond of friendship, and as these nations themselves began to consider a challenge to the Spanish monopoly, the indigenous inhabitants were promoted by some writers as sympathetic and innocent victims of Spanish barbarity. This in turn encouraged a conception of the indigenous inhabitant as the antithesis of the corrupted European and the model against which decadence could be contrasted. The contrast between indigenous inhabitant and European had really been first suggested in 1516 in Martyr's *Decades*:

> A fewe things content them, having no delight in such superfluities, for the which in other places menne take infinite paynes, and commit manie unlawfull actes, and yet are never satisfied, whereas manie have too much, and none enough. But among these simple soules, a fewe clothes serve the naked: weightes and measures are not needfull to such as cannot skill of craft and deceite, and have not the use of pestiferous money, the seed of innumerable mischieves: so that if we shall not be ashamed to confesse the trueth, they seem to live in that golden worlde of the which olde writers speake so much, wherein menne lived simply and innocently without enforcement of lawes, without quarrelling, judges and libelles, content onely to satisfie nature, without further vexation for knowledge of things to come. Yet these naked people also are tormented with ambition, for the desire they have to enlarge their dominions: by reason whereof, they keep war and destroy one another, from which plague I suppose the golden worlde was not free. (Anghiera 1628: 15)

Martyr adverted to the same idea a little later –'so that (as we have sayde before) they seem to live in the golden worlde' (Anghiera 1628: 25) – which obviously made it stand out, and, as happened in many other cases, Martyr's statement was repeated by later writers. It was Montaigne principally,[7] in his essay 'Des Cannibales' (c. 1578), who rekindled and developed in France the idea of the *'bon sauvage'*, or Nature at its most pure and unspoiled:

> *car il me semble que ce que nous voyons par experience en ces nations là, surpasse, non seulement toutes les peintures dequoy la poësie a embelly l'age doré, et toutes ses inventions à feindre une heureuse condition d'hommes, mais encore la conception et le desir mesme de la philosophie. C'est une nation, diroy je à Platon, en laquelle il n'y a aucune espece de trafique; nulle cognoissance de lettres; nulle science de nombres; nul nom de magistrat, ny de superiorité politique …Les paroles mesmes qui*

---

7. Schäffer (1988: 26) claims that de Léry's book (1578) was also 'highly influential on the developing concept of the Noble Savage in France, and it is often mentioned in the same breath with the famous essay "Des cannibales" by Michel de Montaigne'. Boucher (1989: 14) makes the same point: 'With Léry (and Montaigne) was born the French version of the exemplary savage'.

*signifient le mensonge, la trahison, la dissimulation, l'avarice, l'envie, la detraction, le pardon, inouies.* (In Tilley and Boase 1954: 33–4)

[for it seems to me that what we see by experience in these nations surpasses not only every painting with which poetry embellished the golden age as well as all the fictitious inventions of a happy human condition but also the conception and desire of philosophy itself. It is a nation, I would tell Plato, in which there is no kind of trading, no knowledge of literature, no science of numbers, no magistrates, no political superiority ... The very words which mean lies, treachery, double dealing, avarice, envy, detraction and pardon are unheard of.]

Even the eating of human flesh (which was accepted as factual) was explained away as excusable and as having almost noble precedents (as, for example, in Marco Polo's accounts). In addition to the three whom he met in 1562, Montaigne formed his view of the indigenous inhabitants from information given to him by a man who had spent ten or twelve years in Brazil. He heard the indigenous inhabitants respond to questions about the things that they found peculiar about France, and this no doubt caused him to try to assess the European world through their eyes.

On the other hand, even though Shakespeare read Montaigne's *Essays* and used some words from 'Des Cannibales' in *The Tempest*[8] (1610), Shakespeare's Caliban was not like Montaigne's *bon sauvage*. He was the opposite and revived the half-human, grotesque image that had been presented to Europeans by the earliest reports from the New World. Shakespeare's Caliban was a Creole and a slave, born of a North African woman from Algiers who had been taken to a New World island. Caliban was also a ladino in that he had been taught language by an Italian. Apparently, Shakespeare also intended him to be a Carib by giving him the name Caliban, even though there was no suggestion of him being a cannibal. This creole slave-cum-Carib character was intended to be physically, morally and intellectually the antithesis of the magnificent Italian or Latin, Prospero. So, while the Italian Anghiera (Martyr) suggested that indigenous inhabitants seemed to live in the 'golden worlde' and the Frenchman Montaigne raised them to a pinnacle through idealisation, the Englishman, Shakespeare, consigned New World people – the Creole, the slave and the Carib – to the lowest level of humanity.

It is possible that Shakespeare could have been influenced in the choice of the name Caliban by the mythology of his own country. In

---

8. Montaigne: '*C'est une nation, diroy je à Platon, en laquelle//il n'y a aucune espece de trafique*' (*et seq.*); Shakespeare (*Tempest* II, i): 'for no kind of traffic// Would I admit' (*et seq.*).

the St George and the Dragon story there is a character Kalyb, an enchantress, who is supposed to have stolen St George soon after his birth and kept him for 14 years (see Helm 1981: 4). This obviously is not distinct from the Greek myth given by Homer in the *Odyssey* in which Calypso, also an enchantress, is supposed to have captured Ulysses and imprisoned him for seven years. The phonetic similarity between Calypso, Kalyb and Carib is quite clear. In addition, the notion of enchantress or the idea of having unusual powers is trans- ferred not only from the Greek myth to the British, but also from these to the New World word *Carabi* by Vespucci (*'huomini di gran sauido- ria'* ['men of great wisdom']) and by De Laet (*'ils faisoyent beaucoup de choses qui surpassoyent leur entendement'* ['they did many things which surpassed their understanding']) in their explanation of the indigenous inhabitants' view of the Europeans. It is therefore not entirely surprising that Shakespeare's Caliban's mother was also an enchantress. So, even if Caliban was not of the 'golden worlde', he was of the exotic world.

What is quite evident from all this then is that European literature identified the word 'Carib' and cultivated it as an extension of their own beliefs and traditions. Likewise, Martyr's association of the Tainos with inhabitants of the 'golden worlde' and Montaigne's idealisation of them were projections of beliefs of the Old World into the New. Martyr's reference to 'that golden worlde of the which olde writers speake so much' went back to the ancient Greek poet Hesiod and his notion of the first or golden age of the world when people were innocent (Hesiod: *Works and Days* 106–20). The simple and general image of the natives of the islands that emerged therefore was one that embodied a contrast between two types – one hospitable and the other hostile. This vision of a hospitable people in a golden world (i.e. a world of leisure) was the one that survived and came to characterise the region.

To summarise the development of the names for the region and its people, then, one can say that the Italian Vespucci, through the agency of Waldseemüller, gave the New World his name '*Amerigo*', introduced '*camballi*' into the region and also helped to popularise '*Carabi*', a word probably coming from the indigenous inhabitants but converted to fit an Old World image. '*Carib*' apparently was first given, as the name of an island, by Columbus even though '*Carabi*', as a name for the natives of the small islands, had no prior validity among the native inhabitants themselves. Breton (1665: 229) explained that word and '*Galibi*', to which it is etymologically related, as names that, according to the natives of Dominica, the Europeans called them. In short, then, Columbus, Vespucci and other early writers projected cannibals on to the region, and Martyr, who was no explorer, confused 'cannibals' and 'Caribs' and also revived 'Antilia'. The abiding concoction 'West India' can be credited solely to Columbus. Neither 'West Indians' nor 'Americans', as new terms derived from new place names, was immediately used to refer to the indigenous inhabitants, although a few examples of such use occurred in early works – in Eden's (1555) translation of Martyr there is the phrase 'tawny like unto the west Indians' (p. 310); in De Lery (1578) South American natives were occasionally referred to as '*Ameriquains*' and later Ogilby used 'Americans' in 'The natives of this island, though wholly rude of acquir'd knowledge or literature, like the generality of the Americans, yet they took care to instruct their children in the original and antiquities of their country ... in songs call'd areitos' (1671: 320). Quite clearly, then, the identification of the places and people of the New World through the names **West India, America** and **Carib** was a re-fashioning of Old World concepts as well as images from Old World (Greek) mythology.

# The Americanisation process – criollos, creolisation and *mestizaje*

The name '*criollo*' did not derive from a place name and so had no geographical entity to sustain and bolster it. It had no exotic literary tradition to support it. There is no early bibliographical information that explains its etymology; it simply appeared on the scene to represent a reality that had come into being. Yet, it has survived to capture an identity of convergence and to retain a notion of spice and liveliness that none of the other alternative terms has done. The fact of continuity, repetition and borrowing tempered by local modification cannot be better symbolised than in '*criollo*', which emerged in the sixteenth century to identify a combined process of Latinisation and Americanisation.

From the earliest years, American experience, or lack of it, was signalled as critical in the lives of Europeans in the Americas by new words coming into general use that attested to the need to mark a hierarchy of expertise in things American, and especially to distinguish the newcomer from the more experienced person. Oviedo was the first to signal this when he said '*los que nuevamente vienen á ellas, á los quales en estas Indias llamamos chapetones, y en Italia les dicen visoños*' ['those who have recently come there whom we call "chapetones" in these Indies, and in Italy they call "visoños"'] (1555, 2: 175). Some years later, Friar Pedro Simón (1627), in repeating what

Oviedo said, developed the contrast further between '*chapeton*' and '*baquiano*':

> *Chapeton, es lo mismo que visoño en la guerra, son los recien entrados en las Indias, y que aun no se les entiende de sus tratos, y modos, llaman chapetones a los que han venido en una flota, hasta que llegan otros en otra, en quien parece se traspassa el nombre.* (Simon's Glossary)

> [*Chapeton* is the same as *visoño* in war: it is those persons who have recently arrived in the Indies who still do not understand their dealings and ways. Those who are the last batch to have arrived by boat are called *chapetones* until the next batch arrives, to whom apparently the name is transferred.]

> *Baquiano, es hombre versado en las cosas, y tratos de las Indias, como son los que han estado ya algun tiempo en ellas, y sabe todos los modos, como se vive en ellas al contrario chapeton.* (Simon's Glossary)

> [*Baquiano* is a person versed in the things and affairs of the Indies, like those who have there for some time, and knows all the customs, like a person who lives there, the opposite of *chapeton*.]

Such a contrast must have been even more evident in the early years when populations were being increased only by successive arrivals from the Old World. Though these two words did not find their way into other colonial languages as did '*criollo*', they underlined the need for distinction in experience. Distinction was also necessary for identification of children born in the Americas of Old World parents, for they were beings who could neither be classed as '*chapeton*' or '*baquiano*'.

The first appearance in print of the word '*criollo*' is in many works identified as that in José de Acosta's *Historia natural y moral de las Indias*, which was first published in Seville in 1590: '*Esta fruta, dezian algunos Criollos (como allá llaman a los nacidos de Españoles en Indias) que excedía a todas las frutas de España*' ['This fruit, said some *Criollos* (that is what children born of the Spaniards in the Indies are called) exceeds all the fruits of Spain'] (1591, Lib. 4, cap. 25, p. 167). However, Alvarez Nazario (1992: 19 [note 11]) points out an earlier reference from the previous year occurring in Juan de Castellanos (1589), in an elegy written in honour of Juan Ponce de Leon, a white Puerto Rican involved in an expedition to seize the island of Trinidad:

> *Criollo de San Joan que conocemos,*
> *De parte principal ilustre abuelo.* (Juan de Castellanos 1589: 203)

> [Creole of San Juan we know, with an illustrious grandfather.]

In spite of the continued accreditation of the first citation of '*criollo*' to Acosta, it is clear that Acosta himself got the word from an earlier work, that by Juan López de Velasco (1571–1574). The heading of the section in which the word occurs in López de Velasco is '*De Los Españoles Nacidos en Las Indias*'. This is clearly repeated in Acosta's '*como allá llaman a los nacidos de Españoles*', as well as López de Velasco's words '*los que nacen dellos, que llaman criollos*'. Acosta obviously came upon the word in his early years in the New World, since he went to Peru in 1570, spent 17 years there, and taught there and elsewhere in the New World for some time before he returned to Spain where he wrote his *Historia*. López de Velasco gave no information about the origin of the word. Even though he mentioned that there were many *mulatos* there, the way in which he presented the word '*criollo*' suggests that it referred to whites only.

On the other hand, Arrom (1971a: 13), using documentation from the period 1578–85 by María Teresa de Rojas (1947), concludes firstly that '*criollo*' was current in Cuba at the same time and was also used to refer to enslaved Africans. As evidence he identifies dates and citations from the documentation as follows:

3$^{rd}$ April, 1579: given as a guarantee for a mortgage *un esclavo negro, llamado Juan Mayor, criollo* [a black male slave, named Juan Mayor, creole];

19$^{th}$ June, 1579: sale of a slave described as *criollo desta ysla* [creole of this island].

It was Garcilaso de la Vega the Inca (1609) who, in an explanation of new names for various racial groups, was the first to say that '*criollo*' originated with the Negroes:

*A los hijos de español y de española nacidos allá dicen criollo o criolla, por decir que son nacidos en Indias. Es nombre que lo inventaron los negros, y así lo muestra la obra. Quiere decir entre ellos negro nacido en Indias; inventáronlo para diferenciar los de los que van de acá, nacidos en Guinea, de los que nacen allá porque se tiene por más honrados y de más calidad, por haber nacido en la patria, que no sus hijos porque nacieron en la ajena, y los padres se ofenden si les llaman criollos. Los espanoles, por la semejanza, han introducido este nombre en su lenguaje para nombrar los nacidos allá. De manera que al español y al guineo nacidos allá les llaman criollos y criollas.* (BAE 1963 133: 373)

[The children of a Spanish man and a Spanish woman born there they call creole, that is to say, born in the Indies. It is a name the Negroes invented, and that is what the record shows. It means among them a Negro born in the Indies. They invented it to differentiate between those born in Guinea who came there and those who were born there because it is considered more honourable and of greater quality to have been born in the fatherland, as opposed to their children because they were born in a foreign land and the parents are offended if they are called Creoles. The Spaniards, similarly, introduced this name into their language to name those born there. And so both the Spaniard and the Guinean born there are called Creoles.]

Garcilaso left Peru in 1560 at the age of 21 and never returned. If his explanation was based on his own knowledge, then it means that the word '*criollo*' had been in existence for some time before 1560. Garcilaso's statement that the Negroes invented the word could mean that in reality it was an African word, an African version of an indigenous word or an African version of a Spanish word. It is unlikely that it was a total neologism unrelated to any other word.

Support for the African source of the word also came in 1627 from Friar Pedro Simón:

*Criollo es vocablo de negros, y quiere decir persona nacida en la tierra; no venida de otra parte, el qual vocablo se ha ya Españolizado, y significan con ellos nacidos en las Indias, aquien llaman criollos, y al nacido en una o en otra parte, o ciudad llaman criollo de tal, o tal parte.*

[*Criollo* is a word of the Negroes and it means a person born in that land, not coming from elsewhere. This word has already become Spanish and they mean by it those born in the Indies whom they call Creoles, and someone born in one place or another or city they call creole of such and such a place.]

Although similarities in the wording ('*los nacidos en las Indias, aquien llaman criollos*') indicate that Simón followed Garcilaso's explanation and probably also that of Acosta, his focus on the way the word was actually used (first, *criollos* as a general term and, second, to specify the particular place of birth – '*criollo de tal, ó tal parte*') meant that he was familiar with the word and was not merely repeating what those before said. Early instances of the use of the word that accord with the second part of Simón's explanation are given by Teresa de Rojas 1947 (cited above) as well as by Aguirre Beltrán:

*Se dijo entonces: **negro criollo de Campeche, criollo de Oaxaca, criollo de Querétaro**, etc.* ([1946] 1989: 161)

[They say therefore: a Negro Creole of Campeche, Creole of Oaxaca, Creole of Querétaro, and so on.]

Aguirre Beltran saw this use of the word as one stemming from the absence of 'nations' in the New World with which negro natives of the New World could be identified. According to Aguirre Beltrán:

*Hubo ... negros de **nación** Castilla, por haber visto la luz primera en tierras del Cid; negros **naturales de Portugal**, por haber nacido en la nacion referida. Si bien España y Portugal, ya con justicia podíanse considerar naciones y el calicativo daba una exacta connotación, en las tierras nuevas el concepto de **nación**, tal y como modernamente es considerado, no se había fijado aún; hasta el siglo XVII encontraremos negros de **nación Mexicana**, mas por el siglo de la Conquista se daba al negro originario del país la designación de **Criollo**, posponiendo a este vocablo el locativo geográfico de donde había nacido y lo que era más importante, donde se había criado; la expresión **nación** o **tierra** se reservó para los extranjeros. ([1946] 1989: 161)*

[There were ... Negroes of the Castilian nation, since they were born in the land of the Cid. [There were] Negroes who were natives of Portugal, since they were born in that nation. Whereas Spain and Portugal by that time could justifiably be considered nations, the word having the correct connotation, in contrast, in the new territories the concept of 'nation', as it is regarded in modern times, still had not become fixed. In the seveteenth century we meet Negroes of the Mexican nation, but during the century of the Conquest a Negro born in a [New World] country was given the designation 'Creole', with the geographical location of place of birth being appended to this word as well as, more importantly, where he was brought up. The expression 'nation' or 'country' was reserved for foreigners.]

Aguirre Beltrán's last point is somewhat misleading because '*la tierra*' was also added after '*criollo*' or used after '*hijo*' or some other word in cases where the specific country intended was clear. In other words, when a poet in Silvestre de Balboa's *Espejo de paciencia* spoke of '*criollo de la tierra*', he was identifying Cuba without actually naming it. Likewise, '*hijo de la tierra*' used by a creole meant his specific country. The phrase '*de la tierra*' therefore represented national feelings and spirit, and the frequency with which it occurred in the literature testified to strength of national feelings among Creoles.

Though Aguirre Beltrán etymologically associates the word '*criollo*' with the Spanish verb '*criar*' when he says '*y lo que era más importante, donde se había criado*', there is no evidence to support this association in the early explanations of the word '*criollo*' or the idea that the place where they were brought up was an important feature in the identification of black Creoles. In addition, unless Aguirre Beltrán was making a fundamental distinction between the way in which early black Creoles were referred to as opposed to West Africans, the manner of (geographical) identification of early black Creoles by Europeans was more than likely just as functional as identification of West Africans. As Aguirre Beltrán himself notes in reference to West African slaves: '*En tales casos los negros perdían el nombre de su nación de ori-*

gen y tomaban el del **entrepôt** o punto intermediario de donde pro-
cedían' ['In such cases the Negroes lost the name of their nation of
origin and were given that of the factory or intermediate point they left
from] ([1946] 1989: 148). The ethnic identity (nación) that Europeans
conferred on West African slaves was simply a matter of where the
slave boat left from to cross the Atlantic and not where they had been
born and brought up. So, whatever the word 'criollo' meant when it
was first used, Spanish writers when they adopted it certainly did not
mean for it to represent an immutable or essential characteristic of
ethnicity; it was used for contrastive identification to mean born in the
New World rather than in the Old World.

The early explanations were made in the contexts of Mexico, Cuba
and Peru, but Hispaniola was the first colony in the New World and it
may also have been the place that spawned the word 'criollo'. Garcilaso
was probably the first to direct attention to the islands as the source of
words used elsewhere in the mainland colonies: in his explanation of
new ethnic terms, he said that 'cholo' was 'vocablo de las islas de Barlovento' ['a word from the Windward Islands'] and he also said that
'cimarron' was 'vocablo del lenguaje de las islas de Barlovento' ['a word
from the language of the Windward Islands']. Note that at that time the
term 'las islas de Barlovento' meant 'Cuba, la Española, Jamayca, San
Juan' (Acosta 1591: 154). Santo Domingo, specifically, as a source of
'Spanish' New World vocabulary is supported by Simón (1627): 'la isla
de Santo Domingo, que como fue la primera tierra que se descubre,
tomaron alli muchos [vocablos] los Castellanos y los llevaron, y
introduxeron en otras, que se fueron descubriendo' ['the island of Santo
Domingo, which, as it was the first country that was discovered, the
Castilians got many words from there and took them and introduced
them into other places that were being discovered']. The same idea is
found in the words of Rufino Cuervo, in whose view Hispaniola had the
privilege of being 'el campo de aclimatación donde empezó la lengua
castellana a acomodarse a las nuevas necesidades' ['the place of
acclimatisation where the Castilian language began to accommodate
new necessities'] (Rodríguez Demorizi 1944: 8). Rufino Cuervo goes on
to argue that Hispaniola was the language distribution centre.

Specific evidence of the early use of the word 'criollo' in Santo
Domingo comes from a 1598 Memorial in favour of the removal of
settlements in the bande norte, a then notorious area in the northwest of
the island of Hispaniola, which said, among other things: 'y después acá
[1548] ha habido otros tres alzamientos de negros y que han dado cuida-
do, que han sido sus capitanes Lemba, Ambo y Juan Criollo ...' ['and
after that [1548] there were three other Negro uprisings which had

caused concern and their captains were Lemba, Ambo and Juan Criollo ...'] (cited by Franco 1969: 42). The date of the uprising led by Juan Criollo is not identified except that it was between 1548 and 1598. What is important, however, is that the slave was given the generic name 'Criollo', which meant that at that time the term 'criollo' was applicable to enslaved Africans. In addition, for a person to be called Criollo at this time indicates that the word must have been familiar for some time before its first known date of appearance in writing in 1571–4 (López de Velasco). In effect, this use of the word as a name would constitute the first known reference to the term, one that was to a black slave.

A later and somewhat contradictory use of 'Criollo' as a name occurred in Fray Pedro Simón's account of early seventeenth-century events in Venezuela, in which the person so named was an indigenous inhabitant: 'Como también lo andaban otros dos o tres negros, que huyendo de sus amos se metieron en su amparo, y otros dos indios, el uno llamado Cristóbal ... el otro se llamaba Pedro Criollo ...' ['As two or three other Negroes were also going, fleeing from their masters, they entrusted themselves to his protection, and two other Indians, one named 'Cristóbal' ... the other named Pedro Criollo ...'] (Simón [1627] 1987: 638). There is no immediate explanation why the name 'Criollo' was applied to an indigenous inhabitant; it may have been that it seemed appropriate for an indigenous inhabitant who had become domesticated in European service, who had become more of a ladino, in which case the emphasis was not on place of birth but on cultural difference, that is acclimation to European behaviour.

The belief that (in Santo Domingo) the word 'criollo' was first used to refer to enslaved Africans is implicit in the words of Vicioso (1979: 160), who, in commenting on the use of the phrase 'la gente criolla' by the Archbishop of Santo Domingo, Nicolás de Ramos, in a letter in 1593/4, says: 'este párrafo tiene el valor histórico de que en él aparece, por la primera vez, la palabra **criolla** aplicada no sólo a los negros sino a todas las personas nacidas en la isla ...' ['this paragraph has historic significance for it was here that the word 'creole' appeared for the first time applied not only to Negroes but to everyone born in the island ...']. It would seem as if in the island of Hispaniola generally the notion of creole was never really separated from the black population.

The word 'criollo' was obviously used by the people in the colonies to identify a reality that had become meaningful, that is a noticeable difference between those persons who had come to the New World from the Old and those who were born there (excluding from the comparison the original indigenous inhabitants). These new natives of the Americas had grown up in a world quite different from that of their parents but at the same time they were not of the same culture as the indigenous inhabitants. The first consciousness of this new group of persons within the individual colonies must have developed during the second quarter of the sixteenth century so that by the middle of the century the name '*criollo*' must have crystallised and become a generic term. The first major factor in the difference, a formative one, was believed to be the climate. The effect of the climate on people in the New World was a theme that came to reverberate in the writings on the islands over the centuries. This aspect of the Americanisation process was first identified and explained by López de Velasco writing in the third quarter of the sixteenth century:

> *Los españoles que pasan á aquellas partes y están en ellas mucho tiempo, con la mutación del cielo y del temperamento de las regiones aun no dejan de recibir alguna diferencia en la color y calidad de su personas; pero los que nacen dellos, que llaman criollos, y en todo son tenidos y habidos por españoles, conocidamente salen ya diferenciados en la color y tamaño, porque todos son grandes y la color algo baja declinando á la disposición de la tierra; de donde se toma argumento, que en muchos años, aunque los españoles no se hubiesen mezclado con las naturales, volverían á ser como son ellos: y no solamente en las calidades corporales se mudan, pero en las del ánimo suelen seguir las del cuerpo ... (1899: 37–8)*

> [The Spaniards who went to those places and were there for some time, with the change of sky and temperament of the regions continued to experience some difference in colour and quality of their persons; but those who were born there, called Creoles, and in everything are regarded as Spanish, it is well known that already they have turned out to be different in colour and size, because they are all big and their colour darker inclining towards the look of the soil; as a result, there is an argument that eventually, even though the Spaniards had not mixed with the natives, they would become like them; and it is not only in their physical features that they would change but the mental features would follow suit.]

López de Velasco believed that the climate had an effect on skin colour, physical shape and mentality. These changes were not presented as minor or temporary, as one would conceive of a skin tan, but were seen to be more abiding changes.

Simón in 1627, speaking about certain indigenous inhabitants, stated what became one of the most common beliefs among Europeans about the behaviour of people in fertile, tropical countries: '*porq como la tierra ... es tan fertil y abundante, que con poco trabajo les dà lo que han menester para sus pobres comidillas; casi todo el tiempo gastan en estar con ociosidad echados, y assi son pereçocissimos*' ['because, since the soil ... is so fertile and bountiful, with little effort it gives them what they need for their poor meals, they waste almost all the time lying at ease, and so they are very lazy'] (p. 41). Such beliefs about the effect of the climate (for the Spanish *la tierra*) on the body, mind and habits of people therefore provided an easy avenue for the explanation of all types of differences.

The other factor that no doubt distinguished the new natives from their parents as well as from the indigenous inhabitants was language. The *criollos* or children born in the Spanish colonies in the first half of the sixteenth century were surrounded by a kind of linguistic diversity out of which they had to fashion their native language and behaviour. The kind of native/foreign language duality in everyday interaction that perforce characterised the linguistic behaviour of their parents as well as that of the original inhabitants was less pronounced in their own native language competence since they unconsciously integrated into a wholesome entity the culture of their parents and their own experience of the flora, fauna and society in which they were brought up. So, while these children, especially the white ones, may have regarded themselves as buds or shoots of the old plant, their appearance, language and behaviour were seen to represent the beginnings of the creolisation process.

Yet, even though by the end of the sixteenth century the creole element (native-born whites) had become significant in the Spanish colonies, had become Americanised and had Americanised the Spanish spoken in most areas, there is no doubt that the societies were politically dominated by persons born in Spain, who in their religious mission increased their Latinisation of the colonies through the printed word and through the imposition of institutions and legal structures introduced from Spain. So, in spite of the actual Americanisation of people and language, written material (religious, historical and narrative) maintained the supreme position of Latin and the Spanish of Europe, and in it Americanisms were no more than explanatory addenda.

The word '*criollo*' was an Americanism in the sense that what it referred to was a product of post-Columbus American society and, in addition, the first appearance of the word in print in Spanish was after 1492. Many of those who have made suggestions about the etymology of the word, especially in the twentieth century, have linked it to a European source[9] rather than to any indigenous language or African language or any kind of American neologism, but it very well may have come from the Canary Islands. The sugar cane plant was introduced into the New World from the Canary Islands by Columbus on his sec-

---

9. Two of the most influential of these have been Arrom (1971: 14–15) and Hymes (1971: 15).

ond voyage in 1493. The Canary Islands were the first stop after ships left Seville on their way to the Antillean islands, and *canarios* or *isleños* were among those who came over because they knew about growing the cane. They continued to come in numbers up until the nineteenth century to form an essential element of the agricultural base of the Spanish islands. Originally, migrants from the Canary Islands were not ethnically uniform and it is of interest to note that in 1589 Juan de Castellanos mentioned *Joan Canario negro*, about whom Alvarez Nazario makes the following remark: *'señala hacia la indudable presencia de un isleño de color entre los primitivos españoles de las Antillas'* ['it points to the almost certain presence of a coloured Canarian among the early Spaniards in the Antillean islands'] (1972: 41)

The word *'cri'* is attested today[10] as a word of address for a young child in Canary Spanish and may well have been the sixteenth-century source of the word in the New World, where it was also used to refer to children. In the context of planting, in which the word *'criollo'* developed, it is also likely that the ending of the word had something to do with the commonly occurring words *'pimpollo'* (Acosta 1591: 161) and especially *'cogollo'*. Contextual evidence to support this etymology comes from Oviedo, who used *'cogollo'* several times toward the end of the *Sumario* and then in an explanation of reproduction in the plantain he said:

> *Este tronco (o cogollo que se puede dezir mas cierto) que dio el dicho razimo tarda un año en llevar o hazer esta fruta y en este tiempo ha echado en torno de si diez o doze y mas o menos cogollos o hijos tales como el principal que hazen lo mismo que el padre hizo assi en el dar sendos razimos desta fruta a su tiempo como en procrear y engendrar otros tantos hijos segun es dicho.*
>
> *Estos platanos los ay en todo tiempo del año: pero no son por su origen naturales de aquellas partes porque de España fueron llevados los primeros[11] y han se multiplicado tanto que es cosa de maravilla ver la abundancia que ay dellos en las yslas y en la tierra firme donde ay poblaciones de christianos y son muy mayores que mijores y de mejor sabor en aquellas partes que en aquestas.* (1526, fo. xliij)

> [This stem (or shoot as can more certainly be said) which produced the same bunch takes a year to bear or make this fruit and during this time it has produced around itself more or less ten or twelve shoots or little ones just like the main one which do the same as the parent did each producing a bunch of this fruit in time as well as sending up so many other little ones, as one might say.

10. *Tesoro lexicográfico del español de Canarias* (1992: 312).
11. Wentworth (1834 1: 264) notes that the plantain 'was introduced into Hispaniola in 1516, by Thomas de Berlanga, a Dominican, who carried it from the Canaries'.

These plantains grow throughout the year; but they are not native to these parts because the first were brought from Spain and have multiplied so much that it is amazing to see the abundance of them in the islands and the mainland where there are populations of Christians and they are usually bigger and better tasting in those parts than in these.]

This 1526 meaning of *'cogollos'* ['local shoots or *hijos* of things brought from Spain'] was exactly the same as that given for *'criollos'* later in the century. Historical variation between /g/ and /h/ in Spanish spelling is attested for *'cogollo'*; Monardes (1565: cii) records the word as *'cohollos'*, which indicates that the consonant was probably as weak in this word as it was in *'agora/ahora'* ['now'].

Another possible source for *'criollo'*, or a possible influence in its development, comes out of North Africa. Blofeld (1844) and D'Estry (1845), in their respective (English and French) descriptions of Algeria, identified one of the ethnic groups living there as *'Cologlies/Koulouglis'*. Knight ([1640] 1747) gave the name as *'Collolies'*. This name could have been written in Spanish as *'*Colollos'*, which is extraordinarily close to *'cohollos/cogollos'*. Blofeld (1844: 113) said that the *Cologlies* were 'the offspring of Turks and native women', and it should be noted that the Turks were in Algeria from 1516, which was about the same time that the Spanish began to establish their New World colonies.

In addition to the pronunciation and meaning of the word, the comments made about the people called *Cologlies* were very much in keeping both with the negative attitude of the parents towards the first Creoles and with comments that were to become typical of the Spanish and Spanish Creoles in the New World colonies. In the incidents of which Knight spoke, which took place in the 1630s, the *Collolies* were regarded as reprehensible, and attempts were made by the Turks to wipe out 'so monstrous a lineage' (1767: 467). In Blofeld (1844), the following comments were made:

> they pass the greatest part of the day in extreme idleness. Rich in possessions inherited from their fathers, they seldom follow any occupation. The grounds which surround their country houses are cultivated by slaves, whom they greatly ill-treat, if not perfectly satisfied with their services.
> An excessive vanity and profound ignorance distinguishes them from the other people of Algeria... .
> The Cologlies seem born to pass their life in that state of tranquil and luxurious idleness, which allows the time to pass, without even an effort to enjoy any other pleasure. (p. 145)

There was also a remarkable closeness between the comments made about the *Cologlies* (of mixed ethnicity) and comments that came to be characteristically made about mulattos in the New World:

> Their character is a mixture of these two natures; they have the apathy of the Turks, and the lymphatic temperament of the Moorish women ...

> ... they hardly possess one virtue, and seem to have inherited the evil qualities only, of their parents. (Blofeld 1844: 144–5)

The similarities in form and meaning, the centrality of *Argel* as a geographical area of conceptual influence on the New World and the closeness in time of emergence make it very difficult for *Cologlies* and *creoles* to be totally unrelated developments. In addition, Haedo (1612: 116) claimed that the Turks and Moors regarded *Argel* as their *Indias*. It is quite possible then that the milieu of emergence of some elements in the notion of '*criollo*' was *Berbería* and its *hablar franco*.

One of the remarkable facts about the word '*criollo*', which says something about its origin, was its early widespread use. In 1571–4 López de Velasco mentioned the word with no particular geographical reference; in 1589 Juan de Castellanos mentioned it when he was writing about Puerto Rico (i.e. the Caribbean islands); in 1590 Acosta mentioned it when he was writing about Peru (i.e. South America), though it should be noted that he spent a year in Mexico before he went back to Spain where he wrote his history; and Garcilaso was writing about Peru when he gave his explanation of the origin of the word. It is unlikely that the word came into being independently in different places. In other words, in the same way that it is clear that Acosta repeated what López de Velasco said, Juan de Castellanos had to have been aware of López de Velasco's use of the word when he used it in relation to Puerto Rico. So, as was generally the case, the first mention of the word in print most likely served as the source for its subsequent diffusion among European writers, especially since López de Velasco stated that that was what the people were called generally '*en las Indias*' ['in the Indies']. At the same time, the word must have been spread informally among the people themselves through migration and travel, seeing that the Spanish used their island colonies as bases in their exploration of the mainland.

No specific characterisation of *criollos* in the Spanish island colonies emerges from the literature of the sixteenth century. The decline of the islands as a result of the surge in riches in the mainland colonies meant that for those who remained in the islands life was miserable. In describing Hispaniola, Cuba, Puerto Rico and Jamaica, López de Velasco said that they were:

*pueblos de españoles en que había vecinos, y que cada dia van siendo
menos, así porque despues que se descubrieron las otras provincías más
ricas, de mala gana las gentes quieren parar en éstas, como porque à causa
de no sacarse el oro, por falta de los indios, las mercaderías todas las pasan
adelante, y así ellos viven pobres y miserablemente.* ([1571–4] 1971: 49)

[settlements of Spaniards in which there were residents and which every
day dwindled because after the discovery of other richer provinces the
people were unwilling to stay there, because no gold was being mined,
there being a shortage of Indians, the businesses all left and so they lived
poorly and miserably.]

Even when Cuba and Puerto Rico recovered some importance, it was
because of their strategic geographical position and not because of generation of wealth internally. Consequently, *criollos* in the islands were
not closely identified with industry and hard work of the kind associated with agriculture and mining. The one *criollo* mentioned by Juan de
Castellanos in 1589 was a white Puerto Rican who led a force to the island of Trinidad in 1570, but, unlike most of the other *varones ilustres
de Indias* [famous men in the Indies], Juan Ponce de Leon did not seem
to have impressed the poet that he had done anything *digna de memoria* [worthy of being remembered]. In fact, he was mentioned principally because of his well-known grandfather of the same name.

Further afield, Garcilaso noted the vicious attitude and behaviour
that was characteristic of the Spanish in their use of racist terminology
when dealing with non-whites. The same attitude was demonstrated in
Guamán Poma de Ayala's account in which black Creoles were verbally
abused and pictorially lampooned (1980: 129, 133). The picture given
was that black Creoles were thieves and liars, which would suggest that
in Peru *criollo* had negative connotations, even if only non-white
Creoles were intended. Yet, if the acts of black Creoles described by
Guamán Poma de Ayala toward the end of the sixteenth century were
generally true, it could explain partially the negative attitude of
Africans to their offspring, as recorded by Garcilaso. This would
definitely support the idea that *criollo* started as a negative term.

If, following López de Velasco, the word *criollo* is interpreted to
have been applied first to whites, then it quickly was extended to
blacks. If, according to Garcilaso do la Vega the Inca, the word was
first applied to blacks born in the New World and started out as a term
of social discrimination to label them as inferior (and consequently
made them the lowliest group in the early New World colonies), then it
quickly lost its social stigma and came to refer to the exactly opposite
group in the society. In fact, as a term of geographical differentiation, it
became respectable enough for literature, particularly verse, and it was

familiar enough to readers to be used without any kind of elaboration. There was no trace of negativity attached to being a *criollo*; the term did not seem to carry any strong social connotations and must have been at the time a very useful one.

In European writings the term *criollo* was not used in the early years to identify children born to parents of different races. López de Velasco listed *mestizos* as a group different from *criollos* and Garcilaso de la Vega the Inca in his explanation of *criollo* was careful to say *los hijos de español y de española* [children of Spaniards by Spanish women]. The *criollo* was presented as an Old World person regenerated in the New World, rather than the product of a fusion of races. Thus, *criollo* was associated with natural modification rather than with racial or cultural syncretism. This specific conception of *criollo* continued for much longer in, and was more characteristic of, the Spanish mainland colonies than the French or the English island colonies.

In order to identify mixtures (*mestizaje*), a whole new terminology developed. Garcilaso de la Vega, the Inca, for example, specifically called one of his chapters *Nombres nuevos para nombrar diversas generaciones* [New names for various kinds of offspring]. In this chapter he gave the following names and explanations:

> *Al hijo de negro y de india, o de indio y de negra, dicen mulato y mulata. A los hijos de éstos llaman cholo; es vocablo de las islas de Barlovento; que quiere decir perro, no de los castizos, sino de los muy bellacos gozcones; y los españoles usan de él por infamia y vituperio. A los hijos de español y de india, o de indio y española, nos llaman mestizos, por decir que somos mezclados de ambas naciones; fué impuesto por los primeros españoles que tuvieron hijos en Indias ... Aunque en Indias si a uno de ellos le dicen sois un mestizo o es un mestizo, lo toman por menos precio. De donde nació que hayan abrazado con grandísimo gusto el nombre de montañés, que entre otras afrentas y menosprecios que de ellos hizo un poderoso, les impuso en lugar del nombre mestizo....*
>
> *A los hijos de español y de mestiza, o de mestizo y española, llaman cuatralvos, por decir que tienen cuarta parte de indio y tres de español. A los hijos de mestizo y de india, o de indio y de mestizo, llaman tres alvos, por decir que tienen tres partes de indio y una de español. Todos estos nombres y otros, que por excusar hastio dejamos de decir, se han inventado en mi tierra para nombrar las generaciones que ha habido despues que los españoles fueron a ella ...* (BAE 1963 133: 373–4)

[They call the child of a Negro man and an Indian woman, or an Indian man and a Negro woman a mulatto. They call the child of the latter a 'cholo'. It is a word from the Windward Islands which means 'dog', not of the nobles but of the very vicious type, and the Spanish use it to curse and insult. The child of a Spanish man and an Indian woman or an Indian man

and a Spanish woman they call us 'mestizos', to say that we are a mixture
of both nations; it was given by the first Spaniards who had children in the
Indies.... Although in the Indies if they say to one of them 'you are a
mestizo' or 'he is a mestizo', it is regarded as an insult. As a result, they
have embraced with the greatest pleasure the name 'montañés' which of
the other affronts and insults that a powerful person addresses to them
impressed them in place of the name 'mestizo'...

They call the children of a Spanish man and a mestizo woman, or a
mestizo man and a Spanish woman 'cuatralvos', to mean that they have a
quarter Indian and three quarters Spanish. They call the children of a
mestizo man and an Indian woman or an Indian man and a mestizo
woman 'tres alvos', to mean that they have three parts of Indian and one
of Spanish. All these names and others, which we will not mention in
order not to overdo it, have been invented in my country to name the
generations that were born after the Spaniards went there ...]

It is significant that Garcilaso did not comment on the word *mulato*, as
he did in all the other cases. It suggests that by that time *mulato* must
have been a normal and common word. The other prominent terms
which he identified (*cholo, mestizo, montañés; cuatralvos, tresalvos*)
testified to a double system of valuation – one subjective and the other
more objective. While words with extended meanings like *cholo, mestizo* and *montañés* had strong valuations attached to them, 'new' words
such as *cuatralvos* and *tresalvos* attempted to give a mathematical precision to mixtures.

Even though they were excluded from the category of *criollo*, children of mixed race were the essence of the creolisation process. The
native language of children of mixed race born in the islands most likely varied according to what their mother's language was and according
to the circumstances of their upbringing. The language of these children
differed in nature from all others because it was acquired under local
circumstances in which there was a varying input from different languages and a need to interact with different ethnic groups. Such children therefore added another dialectal variety to the mix already
present in the island societies and were through necessity consciously
and unconsciously creative in their language.

# The mapping of colonial identity

The reality of life in the early Spanish colonies, as it can be described
today, in a way stands in stark contrast to the picture of it created in
the literature of the time. The year 1492 is today portrayed as having a
certain aura of a dramatic event that it did not have at the time when
Columbus returned from his voyages and even throughout the sixteenth

century. Visions and tales of lands and people across the Ocean predat-
ed Columbus, and there was also a rich mix of mythology and reality in
contemporary accounts of voyages to exotic lands. In the literature of
the day, and presumably in the minds of people who were aware of
them, Columbus's voyages were an addition to the mix and did not
present an image of a new reality in which there were lands a great distance away with a completely different climate and people with different cultures. Even the purely economic preoccupation with gold, fine
cloths and spices created for Europeans a vision of exoticism, which
contrasted sharply with the heat, bush, mosquitoes, hurricanes and unfamiliar people that were the reality of the New World. Geographical
beliefs about the shape and size of the earth, religious
bi-polar concepts of heaven and hell, Christians and heathens, good
and bad, and beliefs about race, colour and intellect – all of these fundamental props of European life were not easily or dramatically
changed by the news of lands and people across the Ocean or by the
formation of societies there.

The new 'discoveries' were made to fit old beliefs, which gave
ground grudgingly. Note, for example, that the 'Island of Seven Cities'
(or Antilia) was still historical reality in the seventeenth century:

> When the Mores assaulted Spaine, and brought it into miserable servitude;
> those who survived the slaughter, retyred themselves to the steepie
> Mountaines of Biscay, and Aragon; other imbarking themselves, fled to
> the Iland of Seven Cities; so called, because seven Bishops, with their
> people, there remained. (Botero 1635: 8)

References to claims in the Bible, in classical Greek and Latin literature,
and arguments made by philosophers of antiquity were for a long time
part of the introduction of many works on the New World, which
sought to situate these new 'discoveries' within the tradition and scholarship of Europe, especially Western Europe, to explain the people as
descendants of people from the known world, and to explain differences in living things in the new lands as climatic modifications of familiar
things. It was a tradition that clearly established the point of departure
and the point of reference as Europe and European scholarship respectively. This was a matter of adapting the known (words and concepts)
to suit the unknown in an attempt to keep the Latin 'diaspora' tied to
its source.

In European 'scholarship' wild visions of exotic people, animals and
places in the East Indies appeared together with more factual knowledge of the people of North Africa. *Berbería* or *Argel* at the beginning
of the sixteenth century was seen as a place of great ferment with

various nationalities of Africans coming together there from all directions, with Moors returning from Spain, and with Turks, Jews and other Europeans coming from the east and north. It was portrayed as a situation of great linguistic diversity that led to the emergence of a common language of communication. The situation in the New World was presented in the literature in much the same way, with a diversity of native peoples interacting with the various European nations and therefore requiring a common language of communication. The new world of *ladinos* was clearly a replication of the old world of *ladinos*.

In time, however, the challenge a new world of people presented to the traditional beliefs of Europeans brought about fundamental changes. A theology that stressed the inherent weakness of all human beings because of original sin, and therefore equality, was satisfactory for Europeans as long as the world was Europe, but with additional human beings, who were different, coming on to the scene, the notion of superiority became more important to French priests in the colonies than equality. This superiority was established in a sense by divine right and in a sense by reason that was epitomised in thought and language written down. 'Civilised' and 'savage' were two polar opposites. The establishment of a tradition of literature, the promotion of cultural refinement and the emergence of printing as the cornerstone of European society established the intellectual superiority of Europeans over other peoples, especially those in the emerging world. Conquest by the sword and dominance by the whip in the field of battle (i.e. the colonies) were ably supported at the court by an army of writers/artists who waged war on the intellect of all combatants and won by getting them to accept a hierarchical society in which order was determined by reason and refinement.

With reference to language and literature specifically, at the time of Columbus's voyages and throughout the sixteenth century, European grammarians kept looking back to Latin for validation of their emerging standards of language (e.g. spelling), and artists revived and embellished Greek and Roman topics to give their work credibility. In time, however, the creolisation process in the New World began moving the European vernaculars even further away from classical models and the New World provided artists in Europe with new themes and topics. In reaction, Latinisation or the (re)creation of the nobility of Rome and Greece was more vigorously pursued as a cult of a pure European past. It was a response in a way fostered to confront and invalidate novelty and Americanisation from the New World and in another way to sanitise them to make them conform. In this context, the use of the unsophisticated indigenous inhabitant to show up the falseness of

European manners and to promote the notion of unspoiled Nature was ahead of its time, for it was not until the eighteenth century that such a philosophical approach became normal. For the sixteenth-century European therefore, western Europe had to its immediate east its presumed cultural founts, Greece and Rome, but to its south were *Argel*, *Berbería* and dark Africa, whose influence on Europe itself and the New World it preferred to suppress in its quest for supremacy, cultural purity and religious uniformity (i.e. Christianity).

In the New World itself, European sailors, soldiers and colonists fought the indigenous inhabitants for the riches of the land, which they repatriated to strengthen their own native land. Children of all types in the New World created societies and languages out of the contrasting and sometimes contradictory realities. So, while Latinisation in one sense (with all kinds of *ladinos*) may be seen as a continuous process moving geographically westwards but continuing to draw its lifeblood from Latin Europe, it may also be seen as a contrast to and a bulwark against Americanisation, creolisation and *mestizaje* [miscegenation]. Paradoxically, as the confrontation between western Europe and the New World became even sharper in the centuries that followed, the modifications in the continuous process became more subtle and complex.

One of the results of the Latinisation process in the New World is that Spanish, the first of the European languages and the longest one to be exposed to creolisation in the island colonies, is today the one most uniformly preserved in its structure across a variety of ethnic groups. Put another way, Spanish turned out to be the European language most fully acquired in the New World, since in most New World dialects it has a syntactic structure not very different from that in the dialects of Spain. This virtually complete acquisition may have been the result of the dynamics of Spanish–American society, the efforts of the Spanish orders of the Catholic Church, or the nature of the Spanish language itself. In considering the nature of the Spanish language and the facility with which it could have been acquired, the following comment by Howell (1737) is revealing as an affective factor:

> The French Nation is quick and spriteful, so is his Pronunciation; the Spaniard is slow and grave, so is his pronunciation: For the **Spanish** and **French** languages being but branches of the **Latin** Tree, the one may be call'd Latin shorten'd, and the other Latin drawn out at length ... . Besides, the **Spaniard** doth use to pause so in his pronunciation, that his **Tongue** seldom foreruns his wit, and his brain may very well raise and superfoete a second thought before the first be utter'd. (pp. 472–3)

It is not only that the concept of *drawn-out* speech will be seen to be noted by Europeans as a characteristic of creole speech but also that since the English and French believed the Spanish to be racially tainted and, as the citation claims, their speech to be slow, the idea that Spanish could be more easily acquired by 'lower' peoples would logically follow.

The Romance language family, which developed from the original Latinisation process, is regarded as the most uniform of language families. Yet, its component languages (Spanish, Portuguese, French, Italian, Roumanian) are today quite different from each other. The Latinisation process in the Americas, though admittedly shorter in duration, in contrast has exhibited more maintenance than divergence of structure, with its New World dialects not being much different from each other. It is possible that the formal bilingual approach adopted from the beginning among the indigenous inhabitants, backed up from early on by locally printed material and printing, facilitated the acquisition of Spanish by giving it an appearance of homogeneity. It is possible also that the long, pre-1500 history of ladinisation and influence from North Africa made it better able to absorb acquisition by foreign speakers without significant structural adjustment, or, in other words, the language had acquired elements that made it easier to learn by a great diversity of people.

The Spanish islands led the way historically in the acculturation of the New World towards Europe and as such are important for an understanding of the development of the other islands later, even though Latinisation as a linguistic and cultural process in a sense brought them closer to Europe than any of the other Caribbean colonies subsequently. It was in reference to a *varon ilustre*, a white man, in one of these Spanish islands, Puerto Rico, that *criollo*, one of the most important New World words, was first elevated in verse. Almost 20 years later the word was again used in verse in three contexts of honour and distinction, one of them being in reference to a black Cuban. The very early role of Hispaniola and the pivotal geographical position of Cuba and Puerto Rico as bases and transshipment points for Spanish conquest facilitated the spread of the word *criollo* throughout the Spanish New World, and if the spread of *criollo* is taken as a single example of a more general process, then it is clear that Latinisation was a vehicle for creolisation. Nevertheless, however ideal or apt *criollo* was as a name to identify the people and things of the New World, it was too much of an Americanism to supplant those names that the Europeans had earlier projected from their lore on to the region.

The dominance and maintenance of the Spanish language in the Spanish colonies and the spread of a word like *criollo* should not be allowed to conjure up a vision of linguistic uniformity in the initial encounters and the early colonies. The language situation that emerged during the first encounters of Europeans and Tainos was no doubt characterised by instability and idiosyncrasy; the additional presence of Africans increased this instability. It was only over the subsequent decades that the potential influence of the languages of the indigenous inhabitants swiftly declined. According to many accounts, the indigenous population of the bigger islands was virtually wiped out within the span of a hundred years. So dramatic was the decline of the indigenous inhabitants that their language is generally regarded as having completely disappeared. This is not really so, because the linguistic legacy of the indigenous inhabitants of the bigger islands proceeded essentially from the initial years of contact with the Spanish when these latter had to come to terms with the flora and fauna of unfamiliar lands, which their language had no terms to deal with. Features of the languages of the indigenous inhabitants also survived because the Spanish had to develop terms for activities, elements of interaction and social phenomena that life in the islands created, and, it should be remembered, it was for the newcomers life with few European women and many scantily clad native ones, a circumstance that led to direct and intimate exchanges. Clearly also the contact 'languages' fashioned by the indigenous inhabitants to communicate with Europeans left their mark on the colonies that came into being.

Early evidence from those who actually came into contact with the indigenous inhabitants is contradictory and does not allow for simple conclusions about the nature of the indigenous languages of the different islands, the relatedness of these languages or the kind of language used in the early encounters. It was only long after, when Europeans had become more familiar with some of the languages of the indigenous inhabitants and had recorded them, mostly in the form of isolated words or word lists with meanings, that comments became less impressionistic. Such citations of words showed similarities across these languages, which were interpreted by some as evidence that all the people had descended from the same origins and by others that the people in the islands had originated from the southern mainland. However, during the sixteenth century there was no clear idea about the diversity or commonness of the languages of the indigenous inhabitants; the only distinctions that mattered were those created by the names given by Europeans to people in various geographical groups. The reactions conjured up by the names *Taino* (also referred to as *guatiao*) and *Carib* coincided with communication and lack of communication respectively. As a result of this, those identified as *Taino*, because they interacted more consistently with the Spanish, have left behind in the Spanish language a greater legacy than those identified as Caribs.

The Spanish created the model of colonisation in the West Indies, which the French, English and others adopted. It was a model of colonisation that was linked historically and in essence to the Roman model from which Spain itself had developed. It is not accidental therefore that the words *ladino* and *latino* persisted in this process of colonisation over a long period of time, perpetuating fundamental characteristics of the culture of Europe in the bodies of 'other' people. The Caribs, who survived the initial encounters with the Europeans for much longer than their neighbours in the bigger islands, unwittingly gave their name to the region because they were not receptive to the advances of the Europeans, as a result of which they were vilified and labelled as cannibals and savages. They were seen as the antithesis of civilised society and their small island homes deliberately labelled as such. Paradoxically, their neighbours in the bigger islands were also portrayed as the antithesis of so-called civilised society and provided for writers an image of the Garden of Eden. *Criollos* were regarded as remodelled people, partially determined by the elements of Nature and, through such modification, constituting a synthesis of the Old World and the New. Diversity in identity was thereby established in a way that facilitated colonial rule according to the old Roman principle of *divide et impera* [divide and rule], and in a way that subsequently made it difficult for homogeneity in identity to emerge.

# Societies in the raw: from diversity to corruption

## Early 'settlements' and 'colonies' and the beginnings of local identities

According to the treaty of Tordesillas (1494), Spain was entitled to all the lands west of an imaginary line drawn around the world from north to south. At the time there was little knowledge of what was west of that line. So, profiting from the treaty meant first of all finding out what was there, then dispossessing the indigenous inhabitants of it, making some use of what was discovered, and, later, warding off the challenges of other European nations. The Spanish term *la conquista* [the conquest] symbolised the mentality that was fundamental to this undertaking.

There was no possibility of Spain establishing *de facto* ownership of the vast extent of the lands that it discovered by having a presence of its own people everywhere. There were not enough Spanish soldiers and colonists for that. After first establishing themselves in the bigger Caribbean islands, the Spanish soon expanded to the main land masses beyond the islands, but they had to put in place a defence system around each settlement to prevent it from being overrun by 'others'. There was therefore a very pragmatic element to the various 'settlements' that were established, which had little to do with local, pre-existing organisations, except to the extent that the 'settlements' were established in such a way as to obviate or overcome local hostility. Initially, the 'settlements' in their extent did not correspond to natural geographical entities, such as whole islands, and, with the expansion to the main land masses, the limits of settlements were determined by how well they could be secured, though natural barriers such as mountains and rivers constituted limits in themselves.

At the same time, starting with Columbus, European adventurers developed the habit of claiming whole islands and vast areas of land in the name of their sovereign. Columbus was convinced of the practicality of his claim to all the islands because of the presumed guilelessness of the indigenous inhabitants there, as well as the superiority of his weapons. Columbus's claim was mostly based on ignorance of what was there, but later claims, like Harcourt's ('all that tract of Land, and part of Guiana, betweene the river of Amazones, & Dessequebe, scituate in America, under the Equinoctiall Line' – Harcourt 1613: *Epistle Dedicatory*) were determined in a more 'literate' way, that is by the use of maps, and were essentially creatures of a mentality of acquisition and expansion and colonial rivalry.

The disparity between settlement and colony as well as the European mentality of possession, and continual increase in possessions, for some time militated against the association of local identities with specific geographical areas. Furthermore, already existing ethnic and geographical identities among the indigenous inhabitants figured to varying degrees in the formation of European colonies. In the case of the Spanish, who initially incorporated themselves into existing indigenous societies (as symbolised in the *guatiao* relationship) and then began to reshape them in their own image and to their own purpose, indigenous cultural influence and pre-existing localities were relatively important in the shaping of their colonies. In the case of later Europeans in the smaller Caribbean islands, several colonies were established in geographical opposition to indigenous communities or in places where these were absent, in which cases direct indigenous influence was much less.

As to the first islands claimed by the Spaniards, their identity was also determined by the political structure that Spain put in place to govern them. At the wider regional level, Hispaniola, Cuba and Puerto Rico were attached to the viceroyalty of New Spain, with its headquarters in Mexico City, but the viceroyalty had within it a more immediate administrative structure, the *audiencia*, the oldest of which was that of Santo Domingo, and this one was responsible for the administration of Hispaniola, Cuba and Puerto Rico. The three islands were therefore linked together at the subregional level, while at the same time they belonged to the larger colonial unit, the viceroyalty. Even if these administrative structures were not apparent to the individual on a daily basis as a result of slowness of communication, over time they established a structure and bonds of identity that were like the walls around Spanish forts and which preserved bonds of association long after the decline of Spain as a world power. Santo Domingo, the first city in the New World, functioned as the administrative and at first cultural centre for the three islands in the seventeenth century, and the Spanish language facilitated and fostered cultural identity among all, even those who were born in the islands.

# Seventeenth-century *criollos* of various shades and distinctions

In the Spanish Caribbean of the seventeenth century there were Spanish administrators, soldiers and priests 'protecting' the settlers and civilised society, while there were also maroons and buccaneers who operated outside the law and at the negative pole of society. Cervantes' negative view of the former (1916: 172), however, did not suggest that they were any different from the latter. In these raw societies, rebelliousness and slavery went hand in hand. From the early sixteenth century Oviedo highlighted the notion of 'fugitives', speaking first of cats that had become *bravos o cimarrones* (BAE 1959, vol. 1, p. 38) and then of an *indio cimarron y bravo* [wild and fierce Indian] (BAE 1959, vol. 1, p.221). From the beginning, the word *criollo* (> English 'creole') appeared in reference to specific individuals in the islands in society within and outside the law, which meant that the early notion of *criollo* had both a 'patriotic' or civilised and a 'wild' or uncivilised flavour to it.

In Puerto Rico in 1589 it had been used in a work that was celebrating *varones ilustres* [great and famous men] in reference to Juan Ponce de Leon, the leader of a naval expedition. In Hispaniola in the 1598 *Memorial* cited by Franco (1969: 42) the leader of the slave uprising was called *Juan Criollo*. In Cuban literature of the seventeenth century the earliest examples of *criollos* occurred, three of them, in Balboa's *Espejo de paciencia* [mirror of patience], a kind of epic poem written in 1608 dealing with the struggles that the Spanish in Cuba had with pirates. The first mention of the word was in a sonnet (preceding the poem) written by a *Criollo* in praise of the author (Balboa) in which he described his sonnet as *Este soneto criollo de la tierra* [This native creole sonnet] (Balboa [1608] 1941: 57). In this first case the word was being used by a locally born person with a certain amount of pride; even more so was its third occurrence, when it was used in high praise of a black man – *Oh, Salvador criollo, negro honrado!* [O creole saviour, honourable negro] (Balboa [1608] 1941: 103). These early instances of the word in poetic works began to create the concept of *criollo* as a courageous product of the native land and gave positive connotations to the word in a context where *criollos* were not even socially, economically or demographically dominant.

In Hispaniola at the beginning of the seventeenth century, the *banda norte* [northern region] was a law unto itself and a source of problems for the Spanish government. It was not only that it was populated by wild cows (*ganado ... bravo y cimarrón*), which attracted undesirables, but also by people who in the eyes of the Spanish officials, according to a 1604 document (cited by Franco 1969: 41), were probably just as wild: *la mayor parte de los vecinos y habitantes en la isla por aquella parte es gente común y mestizos, mulatos y negros ...* ("the majority of the natives and residents in that part of the island are common people, mestizos, mulattos and negroes ...") The document went on to say that the area also harboured maroons, and, even if not all the maroons were creole-born, the *mestizos* and *mulatos* certainly were, and therefore constituted a kind of wild *criollo* society in the north of Hispaniola.

In the more settled south of the island, there was also a visible presence of mulattos, according to Oexmelin: *'Le Bourg de S. Jean de Gouave n'est habité que de **Mulatos** ... qui sont la pluspart esclaves des marchands de S. Domingue'* ['town of St John de Gouave is inhabited by mulattos alone ... they are for the most part slaves of the merchants of Santo Domingo'] (1686 1: 77, 78). The *criollo* in this area therefore already had a strong flavour of *mestizaje* [miscegenation], which no doubt was caused by the fact that the slaves were house slaves or urban. It is likely too that there were those females among them who were like the *ganadoras* [prostitutes] described by the archdeacon of Santo Domingo, Alvaro de Castro, in 1542 (Franco [1961] 1979: 39). In contrast to this prostitute image in Santo Domingo, there was in Puerto Rico, according to what Diego de Torres Vargas wrote in 1647, the image of very beautiful, virtuous women (Fernández Méndez 1983: 50). These were contrasting images of women in the two islands that proved to be very abiding.

The contrast between 'good' and 'bad' *criollos* extended to a contrast between *tierra* [land] and *mar* [sea]. Marrero captured this in his account of early colonial Cuba:

> *Muchos criollos se distinguieron en la defensa de **la tierra** contra corsarios, filibusteros y piratas, pero algunos, movidos por resentimientos o víctimas de injusticias, se sumaron a quienes hacían del **mar** vasto escenario de la lucha internacional contra España.* (1973 3: 118)

> [Many Creoles distinguished themselves in the defence of the native land against corsairs, freebooters and pirates, but some, moved by resentment or because they were victims of injustice, joined with those who made the sea a vast scene of international struggle against Spain.]

The most notorious *criollo* in the seventeenth century was Diego Martin, who was referred to by López de Cogolludo in 1633 as *Diego el mulato, corsario tan conocido, criollo de la Habana donde fue bautizado* [Diego the Mulatto, well-known corsair, Creole of Havana where he was baptised] ([1633] 1971: 419). He was at first a slave but later became a well-known pirate who operated under the Dutch flag and who joined with the English, French and Portuguese to attack the Spanish in and around his own homeland, Cuba. As a mulatto, he was said to be driven by hatred for the Spanish resulting from 'some wrongs which had been offered unto him from some commanding Spaniards in Havana' (Gage [1648] 1929: 350). The actions of Diego Martin and others, because they were caused by social and racial factors, demonstrated the deep rifts in Cuba at the time and more precisely the differences between *criollos* themselves.

As to *criollos* who were part indigenous (*mestizos*), there is little to indicate that they were influential in the Spanish islands (Hispaniola, Cuba, Puerto Rico) through their numbers or because of their deeds. In fact, Coll y Toste (1914a: 77) cited a 1582 Spanish, colonial survey, *Memoria de Melgarejo*, which sought to find out information on each region where there were Spaniards, and which said about Puerto Rico: '*el dia de hoy no hay de los naturales ninguno, salvo unos poquitos que proceden de yndios de Tierra-firme traidos aqui ...*' ['today there are no indigenous inhabitants, except a few brought from Terra Firma ...']. This in itself implied that in seventeenth-century Puerto Rico there was little if any direct indigenous influence on *criollos* and that *criollos* there were by and large products of foreign (Old World) parents.

Mention, however, is made in Chapter 7 of Cervantes' 1613 work, *Viage del Parnaso* (see Cervantes 1922) of a *mestizo* and *criollos*, presumably as the enemy in some military activity ('*dos criollos mató, hirio un mestizo*' ['he killed two Creoles, he wounded a mestizo']), but there is no indication that he was referring specifically to the Caribbean islands. This was the first occurrence of *criollo* in European literature proper, in the sense that Cervantes was not giving an account of the Spanish colonies. In this context *criollos* were not heroes and were not being glorified, but the very mention of them meant that they had achieved some fame or notoriety in the world of the day, and the association of the *criollo* with verse, with literary language, with formal written Spanish projected a vision of an ethnic type worthy of note.

In the seventeenth century, then, there were *criollos* of various shades and distinctions in the Spanish islands, reflecting the nature, composition and dynamics of these societies, but there was neither any specific stigma nor any specific element of glory attached to the term itself. It was a non-distinctive term that was applied to 'off-shoots' of Old World people born in the New World – it was a term that writers found useful to identify a colonial type, which had come into being in the Spanish New World even before a notion of a new geographical distinctiveness developed. It illustrated the basic habit of identifying a person with place of birth. Yet, there was a certain cultural distinctiveness that was already being ascribed to the *criollo* by the end of the seventeenth century by Labat ([1724] 1931 2: 51, 52), in his explanation of the Spaniards' love of the *calenda*, the favourite dance of the African slaves. There was no doubt in this French writer's mind that two hundred years of interaction in the colonies had produced a culturally distinctive *criollo* who had become 'Africanised'. However, such an interpretation on the part of a Frenchman was consistent, as will be seen, with a general French view of the Spanish.

## Savages, Moors and Christians and the attraction of the wild in the small islands

When the French and English sought to challenge the exclusive dominance of the Spanish in the islands, French writers, in the form of priests, were in the forefront of the expeditions that set out to establish settlements. As a result, sixteenth- and, more so, seventeenth-century religious accounts in French played an important role in the shaping of the European vision of the indigenous inhabitants and their intellectual and moral capacity. It was a vision with two sides, originally projected by the Calvinist, De Lery in the sixteenth century and eventually epitomised in Rousseau's 'noble savage' in the eighteenth century. This idea arose out of accounts that were different in attitude toward the indigenous inhabitants, which in some cases contradicted each other and which were driven by contrasting levels and types of religious fervour as well as approaches to conversion. The account of Brazil given by Thevet (1558, 1575) was ridiculed in some of its details by De Lery (1578). In the middle of the seventeenth century (1654) Du Tertre hurried to get out his account before another one, either that of Pelleprat (1655) or that of De Rochefort (1658), and within a few years of these appeared Breton's dictionary of Island-Carib words and expressions (1665) and Du Tertre's second version (1667–71) of the history of the region. In between these periods Claude d'Abbeville (1614) had written his biographical accounts of evangelical work, and a French translation of De Laet's 1633 work had appeared. However antagonistic these authors were to each other, what they all had in common was a fascination with the indigenous inhabitants and the use of the word *Sauvages* [Savages] to refer to them. This word was so consistently used that in French accounts it came to be used as a regular name and not just a common descriptive noun.

In contrast to the word *Sauvages*, there was no single name for black Africans that dominated French or English accounts of early colonisation. The term *African* was not synonymous with *black* because Africa was best known to Europeans from its Mediterranean coast southwards. The several words that were used to identify black Africans were either colour adjectives or names of geographical areas in the continent. The only partial exception to this was the name *Moors*. In an account of two voyages made to Africa in 1553, Eden identified sub-Saharan blacks as Moors who were sometimes called by other names (1555: 344). Barbot also used the term repeatedly in his accounts of the west coast of Africa. It is not surprising then that French writers of the seventeenth century – Fleury (1987: 152); Bouton (1640: 98); Du Tertre (1654: 473) – used the name *Mores* to refer to the black slaves brought to the New World.

A Dutch derivative of the basic word *moro/more*, which had an even stronger connection with slaves, is pointed out by Marees in 1602:

> They make a clear-cut distinction between the words *Moriaen* and Negro and do not want to be called *Moriaen*, but rather negro or Pretto (meaning black people); for they say that *Moriaen* means Slave or captive, or also idiot or half-wit. So they do not want to be called *Moriaen* but Negroes or Pretto, because if one calls them *Moriaen*, they will take it amiss and will not want to answer[1] ([1602, f. 89] 1987: 176).

It is interesting to note that, while *Moriaen* [Moor-like] had a pejorative connotation along the Gold Coast, its Portuguese–Spanish equivalent had the opposite connotation.

In spite of the ethnic and climatic diversity of the continent, in the eyes of Europeans it was uniform because it was conceived of as having an extremely hot climate. The original European conception of the continent, as presented by Haedo (1612: fol. 125.3), made it seem very much like the Christian vision of Hell. Difference in climate, difference in race and difference in religion meant that Africa was in polar contrast to the picture of western Europe its scholars created. Moors stood in contrast to 'pure' Europeans, and as one went further southward in Africa, where the people were blacker and European knowledge of them was vague, the contrast increased. Paradoxically, however, in the real commercial and military world of the Mediterranean, all ethnic groups from surrounding countries came into constant contact with each other and Moors were central in this complexity.

In their earliest accounts of the New World the Spanish referred to themselves just as often as a religious force (*christianos*) as an ethnic group (*Castellanos*[2]). Other Europeans coming after did likewise. It

---

1. This same passage was copied from Marees by Samuel Purchas in 1625 (1905 vi: 341).

was when racial stratification became a prerequisite in the preservation of dominance and privilege that European writers' use of the religious term declined and the use of racial distinctions increased. Accordingly, Black, White and Red/Yellow became more appropriate in the minds of Europeans than Moor, Christian and *Sauvage* respectively. Shoemaker, in reference to North America, points out, however, that 'the English continued with "Christian" until about the 1730s, when the term "white people" began to appear with more frequency' (1997: 631).

The missionary accounts of the interrelationship between the so-called Christians, Savages and Moors in the small islands were situated within, and adopted the perspective of, European colonisation. The role of the French missionaries who accompanied the first colonists was partly to civilise the *Sauvages* of the smaller islands (St Kitts, Guade-loupe, Martinique) through religion but more so to provide moral and spiritual support for the 'Christians' and to keep them within the faith. The presence of 'Moors' in this situation only served to triple the task of the missionaries, for, as far as they were concerned, these 'Moors' were just as uncivilised as the *Sauvages*. However, it is clear from the French accounts that harsh reality in an 'uncivilised' new world drove many of the newcomers towards the indigenous inhabitants and so towards 'savagery' and 'corruption'. In other words, the early experi-ence for many Christians was far different from that of white overlord, for in a sense they behaved no differently from the *Sauvages*. Similarly, the 'Moors' sought to escape their bondage by fleeing from the Christian settlements.

The island that best illustrates the interrelationship between the three groups (Christians, *Sauvages* and Moors) is the one that both the English and the French happened to pick to start their colonisation of the islands, and it was the Carmelite, missionary priest, Maurile de S. Michel, who gave the best picture of the society there after the first twenty years. Maurile de S. Michel, who was in St Kitts in 1646, made a general point about the social structure in a way that is slightly differ-ent from the way in which it came to be normally presented later. He said: '*Nos François maistres de caze, ont trois sortes de serviteurs; François, Negres, & Sauvages*' ['Our French heads of household have three kinds of servant: French, Negroes and Savages'] (1652: 78). While he pointed out the legal differences between the three types of servants, he also showed, quite ingenuously, that there was similarity in their

---

2. The name used to identify early explorers and colonists coming out of the non-Portu-guese part of the Iberian Peninsula to the New World because the kingdom of Castile was at the time the most influential area in what was to become Spain. The same name was also used to refer to their language. The word is usually translated into English as 'Spanish'.

actual experience in St Kitts. First, he explained that: *'quelques François ont espousé des Sauvagesses ... . J'ay veu d'autres François mariez avec des Negresses ...'* ['Some French men have married female Savages ... I have seen other Frenchmen married to Negresses ...'] (1652: 36). He then continued:

> *Il y a icy de nos François qui deviennent sauvages, se cachants dans les bois, vivants des fruicts d'iceux, & comme ces Hiboux & oyseaux nuictiers, n'en sortants que la nuict pour aller picorer; Je sçay quelques-uns de nos passagers, qui ont plustost choisy cette vie, que de supporter les peines des pauvres serviteurs, & de vivre privément avec ceux qui avoient payé leur passage ...* (1652: 38)

> [There are some of our French people here who become Savages, hiding in the woods, living off the fruits there, and like owls and night birds, coming out only at night to go pilfering. I know a few of those who came over with us who have taken to this life rather than put up with the problems of the poor servants and live a destitute life with those who paid their passage.]

The matter of Frenchmen running away into the bush and becoming *Sauvages* was also a problem in Martinique because it had earlier been alluded to by Bouton, who, because his primary role was to save souls, said: *'il n'est pas moins necessaire, & agreable à Dieu, d'empescher que les anciens Chrestiens ne deviennent Sauvages, que d'attirer les Sauvages à se faire Chrestiens'* ['It is no less necessary and pleasing to God to prevent old Christians from becoming Savages than it is to attract the Savages to become Christians'] (1640: 132–3). Maurile de S. Michel's further comment on the point (p. 38) showed that the situation with the English in St Kitts was no different. This was a picture of the harsh reality of the early colony where religion certainly did not have absolute sway and where 'civilization' was not overwhelmingly appealing to all, especially in the face of a shortage of white women. This was a situation where the barriers set up were broken down by sex, and where language barriers had to disappear to facilitate child-rearing, escape and survival.

While there was this picture of community among those who opted for the 'savage' life, there was also a different picture of language communication, which complemented the situation in St Kitts. The French and English colonists found it necessary to communicate for their own security and for this purpose the leaders of the two colonies each had their own interpreter. Apparently then, in general, the French and English did not understand each other's language. This lack of mutual intelligibility is further illustrated in an incident given by Maurile de S. Michel. He tells of going over to the English side of St Kitts where there were some Catholics, among whom there was a gentleman who spoke

Latin well. Maurile de S. Michel, having been invited to say mass and deliver the sermon, did this in Latin, which the gentleman then translated into English for the small congregation. The 'savage' informality at the one extreme was therefore offset by sophisticated erudition at the other.

The contrast within St Kitts was part of an even bigger network of communication. The English kept in contact with the English on other islands; so did the French, and so did the indigenous inhabitants. In relation to inter-island communication among the indigenous inhabitants, Maurile de S. Michel pointed out that:

> de nostre temps il y avoit si grande communication des Sauvages de cette Isle [Martinique] avec ceux des voisines, qu'ils se visitoient journellement par des batteaux ... & tel avoit une femme en une Isle pour faire son jardin, qui en avoit d'autres és Isles voisines. (1652: 36)

> [in our time there has been such a great deal of communication between the Savages of this island and those of the neighbouring ones that they would visit each other daily by boat ... and a man might have a woman in this island to work his land as well as others in neighbouring ones.]

This eye-witness view directly contradicts the generalised claim in Barbot 1752 that 'the **Caribbees** of **Dominica** and those of **St. Vincent** and **Santa Lucia** scarce understand one another's language ...' (p. 643). The conjugal and economic ties among the indigenous inhabitants of different islands created an effective network of communication and presumably a language of wider communication among them. In comparison, the conjugal ties between the indigenous inhabitants and the Europeans had to have been sustained by varying kinds of language, although the children of such unions no doubt came to understand both parents and so had a more complex linguistic competence than either of their parents. It seems quite reasonable to assume then that the languages of the indigenous inhabitants had some influence on the kind of language that became native to these mixed communities in the wild.

Maurile de S. Michel's realistic picture of commonality and mixing contrasted with that of Du Tertre, who preferred to cultivate an image of a European apart and superior. For instance, Du Tertre (1654: 473), speaking of the same islands at the same time as Maurile de S. Michel, did not mention the French servants in the same category as the 'Moors' and the *Sauvages*. Du Tertre also created the impression that the indigenous inhabitants used by the Europeans were imported and different from the local ones, even though Bouton (1640: 104), whose work Du Tertre read, had earlier said that there were few of the South American natives left. What was important here was the idea that the

South Americans were more amenable than the Island-Caribs. Du Tertre thus perpetuated hostility towards the island people (the so-called 'cannibals'), who chose to defend their homes.

Du Tertre's general attitude made it easier for the prejudices that were provoked by the indigenous inhabitants to be transferred to the Africans who were to replace them. Moreover, even though some indigenous inhabitants, based on Las Casas' arguments, were seen to be vulnerable and physically weak when they began to die out, it did not stop Du Tertre from being even more inhumane towards the Africans in his suggested treatment of them. Indeed, instead of learning from the experience of the one in the treatment of the other, he justified the treatment by setting up a contrast – '*c'est un Proverbe dans le pays, battre un Negre c'est le nourrir; mais au contraire, crier un Sauvage c'est le battre, & le battre c'est le faire mourir*' ['There is a proverb in this country which says that to beat a Negro is to feed him; but on the contrary, to shout at a Savage is to beat him and to beat him is to cause him to die'] (1654: 481). What this 'proverb' shows is that by the mid-seventeenth century Du Tertre had already encapsulated a characterisation of the two non-European groups in a contrast of them. Actually, Du Tertre had distorted the words of his predecessor, Bouton (1640: 121), who had said about the indigenous inhabitants that they became hostile at the slightest word and that, if you offended one, you offended all.

That it was necessary to beat the Africans for their own well-being, according to Du Tertre, must have been an emerging view as their numbers grew. It is true that from the beginning of the sixteenth century the Africans had been contrasted with the indigenous inhabitants in the Spanish islands and had been believed to be a bad influence on them, but the idea that beating was beneficial to them seems like a rationalisation of European practice. A more common view of the Africans at the time was given by the same Du Tertre – '*Ils font de toute terre leur patrie, pourveu qu'ils y trouvent à boire & à manger ...*' ['They make every land their homeland, as long as they find drink and food ...'] (1654: 476). This was a view repeated shortly thereafter by Ligon in 1657 – 'They are happy people, whom so little contents' (p. 44). In Du Tertre, therefore, there was an expression of views that persisted in the centuries following and reflected a warped attitude towards the non-European groups in the societies.

It is true that when he was discussing social structure among the white population in the French islands, Du Tertre conceded fleetingly that some debt was owed to the indigenous inhabitants: '*Il y a deux sortes de familles dans les Isles; les premiers sont composées de person-*

*nes mariées, les autres de certaines garçons qui vivent en société, qu'ils appellent matelotage aux termes du pays ...*' ['There are two sorts of family in the islands; the first is made up of married persons, the other is made up of young men who live together in an arrangement which locally is called *matelotage*'] (1667 2: 452). The practice of *matelotage*, which was adopted from the indigenous inhabitants by the white servants, was used by them after their period of indenture and this communal relationship proved to be a source of strength among them as well as a general model for survival in unfamiliar and difficult conditions.

Hospitality was another social characteristic that Du Tertre pointed out among the indigenous inhabitants (1667 2: 390), in the slave population (1667 2: 527–8) and in the white population (1667 2: 473). However, he made no overt link between the three and certainly did not credit the indigenous inhabitants with being the model that the others followed. This is a characteristic that will be seen to be highlighted subsequently in various Caribbean islands.

Early English views of indigenous inhabitants preceded the English founding of colonies in the small islands. English involvement with them started in the sixteenth century with the pirates, Hawkins and Drake, and then English colonists, led by Sir Walter Raleigh at the start of the seventeenth century, concentrated their efforts on the east coast of South America. The English managed to maintain some kind of settlement in different parts of the vast expanse of land called 'Guiana' and also succeeded in befriending some of the indigenous inhabitants there, whom they used subsequently for their own purposes.

On his trip from England to Suriname, Harcourt took with him two Surinamese who had been taken to England previously. On his arrival in Suriname one of the South Americans to meet him was one who had lived for some time in England and who 'was a great helpe unto us, because hee spake our language much better than either of those that I brought with mee' (1613: 6). From this varying competence in English among the Surinamese it would seem as if relations between them and the English were cordial and, in fact, the importation of 30 South American *Arawacos* for the start of English colonisation of Barbados was reported without comment by Smith (1630: 55). However, the French interpretation of the actions of the English and of the Island-Caribs' view of the English was quite the opposite of what Harcourt, Smith and Gardyner (1651: 160) suggested. De Rochefort did not note any friendly contact between the English and the indigenous inhabitants in English colonisation of the islands and in fact presented a picture of the English as pirates when he wrote:

*Nos Caraibes n'ont ... aucuns plus grands ennemis que les Anglois ...*
*Cette inimitié a pris son origine de ce que les Anglois, sous le pavillon des*
*autres Nations, ayant attiré plusieurs des Caraibes dans leurs vaisseaus ...*
*ils levérent l'ancre, & porterent les Caraibes, hommes, femmes, & enfans,*
*en leurs terres, où jusqu'à present ils les tiennent esclaves.* (1658: 476–7)

[Our Caribs have no enemy greater than the English ... This enmity had its
origin in the fact that the English, under the flag of other nations, having
enticed several of the Caribs onto their vessels ... raised anchor and took
the Caribs, men, women and children, to their lands where up to the
present they hold them as slaves.]

Although rivalry between the European nations led to comments of this
type, it seems quite likely that there was some truth in what De Rochefort said and that the Englishmen had not simply asked some of the
South American natives to accompany them to establish a settlement in
Barbados.

In fact, the best-known story coming out of the early colonisation
attempts by the English is that of Inkle and Yarico, first told by Ligon
(1657: 55, 65) but later cited as an example of English treachery by the
Frenchman Raynal in the eighteenth century. Yarico (< Yaio), the young
indigenous Guianese maiden, having nourished the lost and dying Inkle
(< Ingles/Anglais 'English') back to health, begged to be taken to Barbados with him. Inkle agreed, but when he got there he repaid her kindness and love by selling her into slavery. A consequence of this pointed
out by Raynal was that: '*Les Indiens, qui n'étoient pas assez hardis*
*pour entreprendre de se venger, communiquerent leur ressentiment aux*
*negres, qui avoient encore plus de motifs, s'il est possible, de haïr les*
*Anglois.*' ['The Indians, who were not bold enough to undertake to
avenge themselves, communicated their resentment to the Negroes, who
had even greater reason, if that is possible, to hate the English'] (1773,
Bk. xiv, 199–200). This story was in keeping with De Rochefort's views
of the behaviour and mentality of the English.

Even though the English settlement in Barbados itself did not spawn
any maroon community of runaway servants, slaves and indigenous
inhabitants, English actions were said to have partially contributed to
the emergence of such communities in St Vincent and Dominica.
According to De Rochefort, it was in reprisal for English attacks and
the theft of their people that the Island-Caribs raided English settlements and took back with them to Dominica and St Vincent not only
their own men, women and children but also Africans whom the English were using as slaves. In 1698 Froger described St Vincent as having
on its eastern side '*12. à 1500. Negres fugitifs des Isles voisines, & sur*
*tout de la Barbade, d'où ils viennent vent arriere avec les Canots de*

*leurs Maîtres'* ['1200–1500 fugitive Negroes from neighbouring islands and especially Barbados, from which they came with the wind behind them in the boats of their masters'] (1698: 195). Barbot (1752) spoke of an abortive slave revolt in Barbados, from which 'some hundreds made their escape to the island of **St. Vincent** ... where they continue to this day among the **Indian** inhabitants' (p. 644). Africans were thus said to have become mixed into the populations of St Vincent and Dominica.

To supplement this explanation of the Black Carib population of these two latter islands and also to put down the other Europeans (i.e. the Spanish), De Rochefort related that, again in retaliation for bad treatment, the Island-Caribs had been forced to attack and capture Spanish ships, which they looted. From these ships they got Africans, about whom De Rochefort said:

> *Il est vray qu'ils pardonnoient aus esclaves Négres qu'ils rencontroient, & qu'ils les conduysoient à terre, pour les faire travailler en leurs habitations. Et c'est de là que sont venus les Négres qu'ils ont à present en l'île de Saint Vincent, & en quelques autres.* (p. 478)

> [It is true that they spared the Negro slaves that they came across and took them ashore to make them work on their plantations. And from there came the Negroes who are at present in the island of St. Vincent and in others.]

This explanation of the presence of Africans and Island-Caribs together in St Vincent was really three-fold, for, in addition to those taken from the English in the Leeward Islands and those from Spanish ships, there was also the other more common explanation (p. 439) that they got there from a shipwreck off the island. In spite of the obvious biases of De Rochefort's account, there is no doubt that in the seventeenth century alternative communities developed in St Vincent and Dominica, the last of the *Caribbee Islands*.

In these two islands the relationship between Island-Caribs (as masters) and Africans (as slaves) was reported by De Rochefort to have become stable and civilised over a period of time:

> *Dans les iles de Saint Vincent, & de la Dominique, il y a des Caraïbes qui ont plusieurs négres pour esclaves, à la façon des Espagnols & de quelques autres nations. Ils les ont en partie, pour les avoir enlevez de quelques terres des Anglois: ou de quelques navires espagnoles, qui se sont autrefois échovez à leur costes. Et ils les nomment **Tamons**, c'est à dire Esclaves. Au reste, ils se sont servir par eus, en toutes les choses où ils les employent avec autant d'obeissance, de promtitude, & de respect, que le pourroient faire les peuples les plus civilisez.* (1658: 439)

[In the islands of Saint Vincent and Dominica there are Caribs who have several Negroes as slaves, like the Spanish and some other nations. They got them in part by taking them from some English territories or from some Spanish ships which once ran aground on their shores. And they call them 'tamons', that is to say, slaves. Besides, they are served by them in everything and they use them with as much obedience, promptitude and respect as the most civilised people could.]

An indication of the linguistic consequences of the contact is seen in some of the words and meanings that developed. Breton's dictionary of the Island-Carib language in Dominica, which was compiled from experiences in the 1640s, contains the entry 'Chibárali, cachíonna yaboúloupou, sont les enfans engendrez des Sauvages & des Negresses, qui sont nommez ainsi' ['"Chibárali, cachíonna yaboúloupou" are the children born of Savages and Negresses, who are so called'] (pp. 12–13). The development of a specific term to refer to the offspring of Africans and indigenous inhabitants suggests that interaction between the two ethnic groups was fairly common. Furthermore, the identification of *chibárali* as the product of indigenous men and Negro women specifically makes suspect the idea put forward by Du Tertre and later repeated by Raynal and others that the Black Caribs were the product of African men and indigenous women solely, as was stated by Raynal: 'the proprietors of it [St. Vincent] gave them [blacks whom the Island-Caribs took from the Spaniards] their daughters in marriage; and the race that sprang up from this mixture were called Black Caribs' ([1798] 1969 v: 810). It is also not consistent with De Rochefort's claim that the Island-Caribs used the Africans as slaves.

The word *chibárali* had a common word ending *-li*, which, when removed, left *chibára*, and, if it is borne in mind that Taylor said that 'In early borrowings from Spanish, orthographic s, x, j, and ch of the model are alike replaced by Breton's IC ch [š] ...' (1977: 39–40), this means that *chibára* could otherwise be spelt *jibára* or *xibára*. The Island-Carib word and its meaning appear to be the source for the word *gibaro/ivaro/givero/jibaro* that became so common later in the Spanish islands. Of further interest is the origin of the word within the language of the Island-Caribs. The meaning – child of a Island-Carib man and a Negro woman – was apparently a metaphorical extension; the basic meaning given was:

*Chibárali, fleche qui a pour pointe une queuë de raye, c'est la plus dangereuse, parce qu'elle est pointuë par le bout & élargit en montant, outre qu'elle est dentelée comme une scie, & venimeuse de soy.* (Breton 1665: 90)

['Chibárali', an arrow that has as its point the tail of a ray; it is the most dangerous because it is pointed at the end and increases in size as it goes up; besides it is jagged like a saw and poisonous in itself.]

The child born of a relationship between a Island-Carib man and a Negro woman may have been seen metaphorically within the Carib community as a dangerous, poisoned arrow, but what is more important is the shape of the arrow head described by Breton – *pointuë par le bout & élargit en montant*. It is the same as the head of the Black Carib with the flattened forehead. Most likely because of the circumstances of his birth as well as the deformation of the head the *chibárali* gained a negative reputation. It was a reputation that was to stay with them as they developed into a significant cultural and racial group, an independent alternative society. It is not surprising that it was the English in the next century who vilified them and were responsible for their decline even as the word *chibárali* gained greater currency and was modified to *xibara* in the bigger Spanish islands.

In summary, then, one of the off-shoots from early French and English colonisation of the smaller 'Antillean' islands was the emergence of alternative communities outside European control. They were made up of three different races and were the result of a preference for freedom rather than servitude and slavery. These alternative communities were not as marginal and insignificant as it may at first appear because, in reference to the situation in Martinique in the 1660s, DuTertre (1671 3: 201) spoke of a parallel increase in the number of maroons or fugitives as the number of slaves imported increased. Obviously there was a problem with 'civilised' life in these colonies if you were not the master or the priest, and for those who were not, the 'savage' life was attractive.

This attractiveness of the 'savage' life was not inconsistent with the ambiguous image of the indigenous inhabitant (i.e. the noble savage) that had sneaked into French literature in the sixteenth century and which had begun to counteract the vision of the cruel, man-eating native that was typical of the earlier Spanish accounts. In the seventeenth century therefore, since there were no French missionary reports of servants or Africans being eaten in the smaller islands, almost imperceptibly the vision of the horrible, man-eating Carib, which the Spanish had created, was disappearing and, in the French accounts, was being replaced by a picture of a more human, though uncivilised, *Sauvage*. In fact, a fundamental similarity between the native inhabitants of the bigger islands and those of the smaller ones is evident in a comparison of the notion of *guatiao*, said to be characteristic of the bigger islands, and that of *banare/compère*, characteristic of the smaller ones. The bond of friendship, which Martyr (Anghiera 1628: 11) first spoke of and which Las Casas later elaborated on, is mirrored in the description of the Caribs in the account of the Capitaine Fleury 1618/20 voyage (Fleury 1987: 96) The French (or at least this account) thus converted the indigenous inhabitants into as pleasant a group as the Spanish had done a century before in the bigger islands. In any case, the element of reciprocity spoken of in these initial encounters between the French and the indigenous inhabitants of the smaller islands confirmed an attraction towards the communal and hospitable way of life of the indigenous inhabitants. The social contribution of the indigenous inhabitants to the colonies was therefore as important in the smaller islands as it was in the bigger ones, whether European writers acknowledged it or not.

# Buccaneers, mulattos and maroons: origins of people outside the pale of polite society

The activities of the French in the smaller Caribbean islands were parallelled by their activities to the north, in Tortuga and Hispaniola, which were different in nature, but which also resulted in the emergence of communal more so than hierarchical societies. Moreau de Saint Méry mentioned that among the slaves in Saint Domingue[3] at the end of the eighteenth century there were '*la descendance de quelques Caraïbes, de quelques Indiens de la Guyane, de Sauvages Renards du Canada, de Natchez de la Louisiane ...*' ['the descendants of some Caribs, some Indians from Guiana, Renard Savages from Canada, Natchez from Louisana ...'] ([1797] 1958: 83). The presence of these various indigenous groups in colonial communities in Hispaniola stretched back to the sixteenth century when the Spaniards used them to mine gold and silver. The pre-Columbus links between the South American mainland, the smaller Caribbean islands and the bigger ones were therefore reinforced by colonial enslavement practices.

In addition to the historical link and the colonial link established between the smaller islands and Hispaniola by the capture and enslavement of Island-Caribs in Hispaniola, the first French migrants to

---

3. The name *Santo Domingo* was commonly used to mean the whole island of Hispaniola but, when the French established themselves in the western part, the French equivalent *Saint Domingue* was used to refer to that part specifically. However, since the French from early on used *Saint Domingue* as the French translation of *Santo Domingo* (the whole island), Moreau de Saint Méry distinguished between the two by calling one *la partie française de l'isle Saint-Domingue* ['the French part of the island of ...'] and the other *la partie espagnole de l'isle Saint-Domingue* ['the Spanish part of the island of ...'].

Tortuga and the Spanish islands also came from the smaller islands. Oexmelin (1686 1: 62) identified the time the settlements started as the time during which Ogeron was governor (i.e. from 1664 onward). Moreau de Saint Méry, writing a hundred years later, not only said that the migration started earlier but also focused on a religious reason for the start of the colony:

> Des François, chassés de l'isle de Saint-Christophe, où ils se croyoient à l'abri de l'intolérance religieuse, dirigerent, en 1636, la barque qui étoit désormais leur unique ressource, vers la petite isle de la Tortue, qui touche, pour ainsi dire, à celle de Saint-Domingue.[4] (Moreau de Saint Méry 1788: 1–2)

> [Some French, chased from the island of Saint Christopher, where they thought they were protected from religious intolerance, in 1636 directed their bark, which was from then their only resource, toward the small island of Tortuga which touches, so to speak, the island of Santo Domingo.]

Even if his dates were incorrect, Oexmelin was an active participant, had a keen knowledge of the hunter's life and thus appreciated the desire on the part of migrants to shift to the ease of the planter's life. The difference between hunter and planter, however, was one that persisted in the island.

Many of the migrants involved in the emergence of a French colony in Hispaniola were in no sense colonists: they were French adventurers who came to be known as *boucaniers*. Ogeron, who was sent by the French government to bring French activities in Hispaniola within a colonial framework, is usually given credit for having domesticated the *boucaniers* by bringing women from France to satisfy their needs, but Brutus (1948: 10) argues that black slave women satisfied their needs long before the French women came. Indeed, the kind of picture given for social life in the early stages of French presence in Hispaniola in the buccaneer period appeared to be very much like that given for St Kitts, Martinique and Guadeloupe – Frenchmen marrying and having children with indigenous women and negro slave women, such children (mulattos) often being freed from slavery or fleeing into the bush, together with the servants and slaves, to escape hard work and harsh treatment. As a result of such intercourse between black and white, *buccaneer* was closely associated with *maroon* and *mulatto*. In fact, the current meanings of these words have become so distinct from each other that they conceal the interrelationship and close history of the

---

4. This French version of 'Santo Domingo' was the name used for the western third of the island of Hispaniola, which became a French colony. The name was replaced immediately after the 1791–1804 revolution by the indigenous name 'Haiti'.

**Figure 3.1** A buccaneer in Hispaniola. *Histoire générale des voyages.* Published by Pieter d'Hondt 1747–9. New York Public Library.

three concepts, the first two of which symbolised the adventurous life-style that preceded the Saint Domingue colony (what became Haiti).

The early meaning of the French–American word *boucanier* in the Hispaniola context was a person who killed and skinned cows for their hides. The root of the word occurred in French writings on the New

World from as early as the last quarter of the sixteenth century – it was first cited in the context of South America (Brazil) and said to be an indigenous word '... *en leur langage ils appelent* **Boucan** ...' ['... in their language they call "boucan" ...'] (De Lery 1578: 153). The adjectival form of the word was the one used most often by the early French writers and it was translated as 'roasted', e.g.:

> *cuicte & boquonee á leur façon* ... [cooked or 'boucanned' in their way]. (Thevet 1575: 913)

> *Boucanez, c'est à dire, rostiz* ... ['Boucanned', that is to say, roasted ...]. (De Lery 1578: 46)

> *du poisson boucanné (c'est à dire rosti)* ['boucanned' fish (that is to say roasted)]. (d'Abbeville 1614: 96)

It was De Laet who gave the clearest definition of the early use of *boucan*:

> *Ils vivent en outre de poisson, d'oiseaux, de toutes sortes d'animaux, qu'ils rotissent ou grillent à la flamme du feu sur une grille de bois, qu'ils nomment* **Boucan** ... (1640: 559)

> [Besides, they live on fish, birds and all sorts of animals, which they roast or grill by the flames of fire on a wood grill, which they call 'boucan'].

The word *boucan* gave rise to *boucan-ier*, the meaning of which shifted from the person who did the roasting to the way in which the person got the meat that was roasted, and eventually the idea of roasting disappeared. However, while the root of the word was indigenous, the development of the concept of *boucanier*, as a person and a function was the work of French writers.

The practice of running down, killing and skinning cows had been typical within the mestizo and mulatto cattle rearing community that had come to dominate the *banda norte* ['northern zone'] of Hispaniola in the late sixteenth century, so that when the French came on the scene in the following century they were really reviving a practice. At the time when the first French adventurers came, that is years after the Spanish devastation of the *banda norte*, which had taken place in 1605–6, the cows, because they had been left to roam wild, were regarded as fair game by those who intended to set themselves up in Tortuga. In spite of what may have seemed like a windfall, the situation of those who had left St Kitts to settle in Tortuga was one that required a tri-partite division of labour and partnership for survival. Oexmelin explained it thus:

> *Voilà comme le petit nombre de ces Avanturiers fut divisé en trois bandes, dont les uns s'appliquerent à la chasse, & prirent le nom de* **Boucaniers**, *les*

*autres à faire des courses, & prirent le nom de **Flibustiers** ... ; les derniers s'adonnerent au travail de la terre, & on les nomma **Habitans**.* (1686 1: 29–30)

[See how the small number of adventurers was divided into three bands: one dedicated themselves to hunting and took the name of 'boucaniers'; another to running routes and they got the name 'freebooters' ... and the last devoted their time to working the land and they were called 'habitants']

This kind of partnership and interdependence was not new because the system of *matelotage* had been introduced among young, unmarried Frenchmen in St Kitts for their mutual support and was maintained among those who migrated to Tortuga. About it Oexmelin said: '*ils se joignent toûjours deux ensemble, & se nomment l'un & l'autre **Matelot**. Ils mettent tout ce qu'ils possedent en communauté ...*' ['they always joined together in twos and called each other "matelot". They pooled everything they owned ...] (1686 1: 151).

While the confusion of or lack of distinction between *boucanier* and *flibustier* was almost inevitable in this kind of relationship, the survival of the former word at the expense of the latter came about when the notion of the pursuit of free booty on land (i.e. wild cows) was transferred to the pursuit of 'free' booty at sea (i.e. Spanish ships). This latter practice (piracy) had become a part of French and English policy against the Spanish much earlier, from the late sixteenth century, but the name *buccaneer* was only given to the French and English pirates after the word was used in the Hispaniola context in the early seventeenth century. The eighteenth-century explanation of William Russell shows some understanding of this relationship or confusion in the extension of the meaning of *boucanier*:

This is the true origin of those pirates, formerly distinguished in England by the appellation of Freebooters, and in France by that of Flibustiers. But as they are now generally known in this country by the name of Buccaneers, which probably began to be applied to them soon after their junction with the hunters of wild cattle, properly so called, we shall continue that name, as more expressive than any other of their ferocious character, whether men or animals were the objects of pursuit. (1778 1: 531–2)

Eventually, as the notoriety and business of pirates (now *Buccaneers*) bloomed, the original meaning of *boucanier* was forgotten.

Among the early French adventurers, those who had the job of *boucaniers* were not always the ones who actually ran down the cows and caught them – this was also done by servants or slaves, who went on foot in contrast to their masters who went on horseback. In Oexmelin's account, even though there is little said about the involvement of persons other than the French in the events and activities of the day, there is mention of a skilful and speedy mulatto named Vincent des Rosiers as an example of a buccaneer, and Oexmelin said further: '*On voit des boucaniers ... qui courent avec tant de vitesse, qu'ils lassent souvent les Boeufs, les attrapent à la course, & leur coupe le jaret*' ['"Boucaniers" are seen running so fast that they make the cows tired, catch them on the run and hamstring them'] (1686 1: 158). This recalls a similar description by Acosta, in reference to the period before the devastation: '*Aprovechan se deste ganado para cueros: salen negros, o blancos en sus cavallos con desjarretaderas al campo, y corren los toros, o vacas, y la ves que hieren, y caen ...*' ['They use this cattle for hides. Negroes, or Whites on horseback with hamstringers on foot, and they run after the bulls or cows and as soon as they are wounded and fall ...'] (1591: 180). Apparently the practice of using negro slaves to do the capture and kill had not changed much in the intervening years and it is unlikely that the Spanish devastation of the area had removed the mulattos. In fact, referring elsewhere in his account to some Spanish colonists in Hispaniola, Oexmelin said: '*Ils sont à cheval & n'ont que quelques Mulastres à pied*' ['They are on horseback and have only a few mulattos on foot'] (1: 170). There is therefore a strong indication that during the early presence of the French on the north coast of the island there was a continuing practice of whites on horseback having mulattos in attendance on foot and that it was normal for these mulattos to do the work of *boucaniers*. Moreover, though information about Africans is sparse in the account of Oexmelin, who was actually a servant, he did point out, in his explanation of the term *mulastre* that even the Frenchmen had a liking for the black slave women, the consequence of which was an increase in the number of mulatto servants/slaves.

In the early works on the New World mixed persons were identified as *mulatos*, apparently by biological analogy with the mule. This analogy has been commonly accepted as the source of the word *mulato*, but there are at least three other strands of meaning identifiable in the historical use of the word, and it can be argued that the analogy with mule is a more modern argument, contemporary with African slavery in the New World, intended to rationalise a separation of races.[5] The issue of two races, which came to be termed *mulatto* or some similar word, must have become quite evident, if not before, at least with the Moorish occupation of the Iberian peninsula from the eighth century

---

5. Young (1995 pp. 8 et seq.) makes a case that there was a belief that mulattos had diminished fertility, which was adduced to prove that white and black humans were different species.

onward. Later, the same kind of mixed-race person became even more evident when the Portuguese established settlements on the west coast of Africa. Then, with the colonisation of the New World from the sixteenth century onward *mulatto* not only became more evident but also problematic as a result of the hierarchical racial structure implemented to preserve privilege for whites. The amalgam of characteristics associated with the issue of black and white therefore has a long history.

An explanation put forward in the nineteenth century by Lafuente (1850) related the word back to the contact situation between the Spaniards and the Moors around 850 in Córdoba. Lafuente said first: '*El hijo de mahometana y de cristiano ó vice-versa, el **mulado** ó **muzlita**, era reputado por mahometano tambien ...*' ['The child of a Muslim woman and a Christian man or vice versa, a 'mulado' or 'muzlita', was deemed to be a Muslim also ...'] (3: 300). He then went on to explain:

> *Estos **mulados** (de donde vino nuestra voz **mulato**), **muzlitas, mozlemitas** ó **mauludines**, eran hijos ó nietos de musulmanes no puros, sino que habian sido cristianos renegados, ó hijos de cristiana y musulman, ó de mahometana y cristiano. Como el número de españoles era infinitamente mayor que el de las familias árabes, y se fueron haciendo matrimonios mixtos, al cabo de algunas generaciones eran ya mas los **mulados** que los árabes puros ...* (3: 300)

> [These 'mulados' (from which came our word 'mulatto'), 'muzlitas', 'mozlemitas' or 'mauludines', were the children or grandchildren of Muslims who were not pure but had been renegade Christians or children of Christian women and Muslim men or Muslim women and Christian men. Since the number of Spaniards was infinitely more than that of Arab families, and they were getting into mixed marriages, at the end of a few generations the 'mulados' were already more than the pure Arabs ...]

Though it was later repeated authoritatively by Saco (1879), Lafuente's religious/racial mixture explanation of the origin of *mulatto* was not generally corroborated, probably because the mule analogy was much more attractive and powerful.

In addition to this Moorish connection between *mulado* and *mulato*, Forbes refers to Santa Rosa de Viterbo (1798), who, in a discussion of archaic Portuguese terms, identified a twelfth-century use of the word *malado* to mean a kind of servant:

> *Malado: O que vive em terras de Senhorio ... Tambem no Seculo XII. Se chamárão **malados, mancebos**, ou **criados de servir**, os filhos, que ainda estavão de baixo de Patrio Poder ... No Foral de Thomar de 1174 onde diz no Latin: Pro suo malado, o Tradutor verteo: Por seu mancebo ... No Foral de Pena – cova de 1192 se diz = **Miles, e sui maladi.*** (Forbes 1993: 142–3)

[Malado: A person who lives on the landlord's property ... also in the
twelfth century. Those called malados would be young manservants, or
servants, or sons still under parental control... In the Foral de Thomar of
1174 where the Latin says: pro suo malado, the translator translated 'for
your young manservant' ... In the Foral de Pena – grave of 1192 it says
'miles, e sui maladi'.]

The putative connection between *malado* and *mulato* through the
meaning 'servant' corresponds neatly with the role of the mulatto in
Hispaniola, as in Oexmelin's '*Ils sont à cheval & n'ont que quelques
Mulastres à pied*' ['They are on horseback with only a few mulattos on
foot'] (1: 170). It also corresponds with the more general role of the
mulatto throughout the colonial period as attendant or servant to
whites.

Another thread of meaning relates *mulatto* to a colour. The French
word *mulastre*, as does the word *olivastre*, has the ending -*astre*, which
is normally attached to colour terms (= English '-ish' e.g. reddish)
though the first part *mule* is not a colour. The French term, therefore, is
a peculiar formation with a colour focus. In English and Portuguese
historical references, variant spellings of the word in which the first
vowel is not [u] suggest that such variants were not conceptually connected to 'mule'. Marees, giving the Portuguese word in his 1602
account, says:

> This picture shows the condition and appearance of the women-folk ... **A**
> is a Portuguese woman living in the Castle d'Mina, half black, half white
> and yellowish: such women called **Melato** ... (1987: 36)

Phillips gives the variant *Malatto* in his 1693–4 account ('... then came
Mrs **Rawlisson**, the factor's wife, who was a pretty young **Malatto**'
(1732: 201). The Portuguese word given by Marees is connected to
Spanish *melado,* which means 'honey-coloured' or 'cane syrup'. Both of
these meanings are relatable to the colour of mulattos. It should be noted, however, that in Old World usage, *mulato* was not the only (one
word) colour term used to identify products of racial mixture. There is
bibliographic evidence of the competing term *loro* in at least one pre-
1492[6] document and the same word was also used by Oviedo ([[1535]
1851, Lib XV1, Cap 7) to refer to a mulatto who accompanied Ovando
to Hispaniola (*mancebo de color loro*)[7] ('a young man of a chestnut
colour').

---

6.  See Ortiz (1986: X1), which refers to a Spanish cédula of 1474.
7.  Forbes (1993: 156) gives a comparative picture of the use of 'intermediate terms'
    between the years 1514 and 1600.

Figure 3.2 A Portuguese–West African *melato* woman. Pieter de Marees ([1602] 1987). *Description and Historical Account of the Gold Kingdom of Guinea* (1602). Translated from the Dutch and edited by Albert van Dantzig and Adam Jones. Oxford: Oxford University Press.

These explanations gave the origin of *mulatto* three basic elements – a religious one, one of racial mixture and one of service; the other element (colour) can be said to derive from racial mixture. These basic elements link the word straight back to the Old World and show it to be complex in its development. Bearing in mind that the early New World usage of the word had no other basic meaning than a genetic mixture of two races, the religious element must have disappeared completely before the word was transferred to the New World. The element of service, within the New World social and racial context, would have to be seen as a derivative of the element of racial mixture.

At first, writers in the New World used the word *mulato* to mean any mixture of races. For example, in his explanation of the term López de Velasco said: '*Demás de éstos hay muchos mulatos, hijos de negros y de indias ... mulatos hijos de españoles y de negras no hay tantos ...*' ['Besides these there are many mulattos, children of Negro men and Indian women ... there were not many mulattos who were the children of Spaniards and black women ...'] ([1571–74] 1894: 37). Garcilaso de la Vega, the Inca, at the beginning of the seventeenth century defined the word in a similar way: 'The child of a Negro by an Indian woman or of an Indian and a Negro woman is called **mulato** or **mulata**' ([1609] 1966: 607). Maurile de S. Michel, commenting on social relations in St Kitts in the middle of the century, said: '*quelques François ont espousé des Sauvagesses ... J'ay veu d'autres François mariez avec des Negresses; les enfans des uns & des autres s'appellent **Mulastres**, estant de couleur olivastre ...*' ['Some French men have married female Savages ... I have seen other Frenchmen married to Negresses; the children of the former and the latter are called mulattos, being olive in colour ...'] (1652: 36).

The primary meaning of *mulato* as a mixture of any two races continued into the latter half of the seventeenth century, for Oexmelin's initial definition of the word was *gens de sang meslé* (1686 1: 77). However, it is likely that as the numbers of indigenous inhabitants declined dramatically in the islands, *mulato* was less commonly used for the combination of white and 'Indian'. Oexmelin himself added an element of confusion to the meaning of the term when, after identifying the names of specific mixtures of white and black, he went on to make the following general statement: '*L'on void aujourd'huy plusieurs endroits dans l'Amerique, qui ne sont peuplez que de ces gens-là, que les Espagnols & les Portugais ont produits, parce qu'ils sont fort adonnez aux femmes noires Indiennes*' ['One sees today several places in America which are populated only by those people that the Spanish and Portuguese produced, because they are very fond of black Indian women'] (1686 1: 77–8). It is not clear here whether *Indiennes* meant 'in the Indies' or 'born in the Indies' or even 'Indian'.

From the start, miscegenation was regarded, especially by French and English writers, as degradation. Oexmelin made this quite clear in the following comment: '*Ils* [mulattos] *ont le fond des yeux jaune, sont hideux à voir, de mauvaise humeur, traistres, & capables des plus grands crimes*' ['The pit of their eyes is yellow, they are hideous to see, bad-tempered, treacherous and capable of the greatest crimes'] (1686 1: 77). This comment embraced supposed visual, psychological and behavioural characteristics of mulattos. What seemed to have triggered it was the burgeoning number of mulattos in Hispaniola as well as the growing spectrum between races, which led to variations in terms and inconsistency in use across languages (Oexmelin 1686 1: 77). Among those living outside the pale of polite society it is likely that racial mixture was as significant as it was for those inside. It certainly had been for the Cuban mulatto, Diego Martin, who seemed to have found the life of *corsario* (= hired mercenary/privateer) an appropriate one to seek revenge on those who had wronged him. The mulatto as *boucanier* in the society of the French adventurers on the northeast part of Hispaniola may have gained the same sense of satisfaction living and hunting outside 'civilised' society. In other words, among hunters (on land and sea) physical prowess may have counterbalanced miscegenation in a way that it did not among planters in a hierarchical plantation system.

The connection between not only *mulatto* and *buccaneer* but also *maroon* is substantiated by the editor of the anonymous text of the 1618/1620 Capitaine Fleury voyage, who interpreted one of the meanings of *marron* (p. 216), in reference to a *masteur* (= *boucanier*), as '*mulâtre*' (Fleury 1987: 238). Identity in the reference of *buccaneer* and *maroon* is also substantiated by Hickeringill's (1661: 67) pairing of the words '... French Buckaneers, or Hunting Marownaes ... who live by killing the wild Beeves for their Hides ...'. Indeed, the two words (*boucanier* and *maron*) overlapped in their early use in Hispaniola as a result of extension of meaning from people to animals. According to Oexmelin (1686 1: 162–3), *maron* was a word that '*ces gens ont entr'eux, pour dire que leurs serviteurs ou leurs chiens se sauvent: Ce mot est Espagnol, qui signifie beste fauve ou sauvage*' ['these people have among them to mean that their servants or dogs have run away. The word is Spanish and it means "wild animal"']. So, *maron* meant not only the cows that roamed wild and were the targets of the *boucaniers*, but also the servants and slaves who pursued them, when they (the servants and slaves) deserted their jobs.

As was the case in some of his description of the buccaneers, Oexmelin most likely got the word *maron* from Du Tertre, in whose account it was used repeatedly to refer to the Africans who ran away into the bush. In reference to the problem of fugitives in Martinique in the 1660s, Du Tertre wrote:

> Mais à mesure que le nombre de ces esclaves s'accroissoit dans la Martinique, pour la consolation & la richesse des Habitans, celuy des marons, c'est à dire fugitifs, s'augmentoit tous les jours pour les affliger. (1671 3: 201)

[But as the number of slaves in Martinique increased, to the comfort and wealth of the residents, the number of maroons, that is to say, fugitives, increased everyday to their consternation.]

The seriousness of the problem of *negres marons* in the smaller French colonies put the focus on Africans, but the word at that time was not synonymous with escaped African slave. Escape from early European colonial domination was the experience of indigenous inhabitants, white servants, African slaves and mulattos, and the same word was used to refer to all of them.

The first citation of the original of *maron* (i.e. Spanish–American *cimarrón*) is given by Zayas y Alfonso (1931: 202) and Arrom (2000: 122) as '*la ysla al presente está muy pacifica de yndios cimarrones ...*' ['the island is very peaceable with wild Indians'], which appeared in a letter by Gonzalo de Guzmán in Santiago de Cuba dated September 1530. In some of its early occurrences in Spanish (and even in English later e.g. *Symerons*; Drake 1653: 7) the word was spelled with an initial 's', which, together with the doubtfulness of the idea that the top of a mountain could be a safe haven (see Arrom 2000: 126), has discredited the claim that the Spanish word *cima* is the origin of *cimarron*. It is more likely that the word came from the indigenous inhabitants of the islands, as Garcilaso de la Vega, the Inca (lib. VIII, cap. 3) said ('*vocablo del lenguaje de las islas de barlovento*' ['a word from the language of the windward islands']), that it meant 'fugitive' or 'wild', that it was applicable to plant, animal or human being, and that it was Americanisation in word and deed.

Notwithstanding the Gonzalo de Guzmán reference, several lexicographical works have credited the first appearance of *cimarron* in print to the better-known Oviedo (1535), in whose work it was indeed used to refer to people and animals: *indio cimarron o bravo* was used to refer to indigenous inhabitants who had escaped from under the control of Spaniards and *puercos cimarrones o salvajes* to refer to wild hogs. Simón (1627) in his glossary pointed out that *cimarron* was used to refer to both people and animals, but the latter were identified as cattle ('*... se entiende assi de ganados, como de personas, especialmente esclavas*' ['it means both cattle and persons, especially slaves']). Long, more than a century later, tried to make something of the distinction between hogs and cattle when he said, in an explanation of *Marons*, that it was 'Probably derived from the Spanish **Marráno**, a porker, or hog of one year old. The name was first given to the hunters of wild hogs, to distinguish them from the bucaniers, or hunters of wild cattle and horses' (Long [1774] 1970 2: 338 [note]) This attempt to separate *maron* (hog hunter) from *boucanier* (cow or horse hunter) is of course contradicted by Hickeringill's words (1661: 67) cited above and it clearly showed Long's ignorance of the Spanish source word *cimarrón*.

Oviedo's alternatives, *cimarron* and *salvaje*, developed to some extent as geographical or language variants in the islands, for although Du Tertre used the word *maron*, the early French writers, Bouton (1640) and Maurile de S. Michel, in their reference to servants and slaves escaping into the bush in Martinique and St Kitts respectively, regarded these fugitives as becoming *Sauvages*, and so too did Oexmelin (*'il s'etoit rendu Sauvage'* ['he became a Savage']; 1: 165). The preference in the bigger islands and with English writers was for *cimarron*. In the case of the English, evidence of a preference comes firstly from the [1626] 1653 Drake account in which the form *Symerons* occurred, in reference to a people in the area of Panama. Also in the last quarter of the sixteenth century Robert Duddeley, in reference to an island near to Trinidad, said: 'The Simerones of the yland traded with me stil in like sort ...' (Hakluyt 1598–1600 3: 575). It is not surprising then that in Jamaica a variant of *cimarron* was preserved even when Jamaica ceased being a Spanish colony. Long's ([1774] 1970 2: 445) shortened form of the word (i.e. *maroons*), occurring over a century after the Drake account and used in reference to Jamaica, had the same meaning ('the free descendants of the aboriginal Spanish Negroes'). Dallas also used the shortened form of the word, even in his historical explanation of it (1803 1: 33). Patterson (1970: 297) claims that *Maroon* was used from around 1670 in English accounts to identify the Spanish Negroes in Jamaica who refused to be subject to the English after the English take-over, but this did not seem to be intended as a specific statement about the shortened form of the word.

Later (mis)interpretations of its etymology in the eighteenth century added more colour to the meaning of *maroon*. Yet, the threads of meaning from the original context persisted. Edwards, having read Long's ([1774] 1970 2: 338 [note]) explanation of the history of the word, converted suggestion into fact and added further to the image of the word:

[Maroons] The word signifies, among the Spanish Americans, according to Mr. Long, **Hog-hunters**: the woods abounding with the wild boar, and the pursuit of them constituting the chief employment of fugitive negroes. **Marráno** is the Spanish word for a young pig. The following is the derivation, however, given in the *Encyclopédie*, article **Maron**: '*On appelle **marons**, dans les isles Françoises les nègres fugitifs. Ce terme vient du mot Espagnol **Simaran** qui signifie un Singe. Les Espagnols crurent ne devoir pas faire plus d'honneur à leurs malheureux esclaves fugitifs, que*

*de les appeller **singes**, parcequ'ils se retiroient comme ces animaux aux*
*fonds des bois et n'en sortoient que pour cueillir des fruits qui se*
*trouvoient dans les lieux les plus voisins de leur retrait.'* ['In the French
islands fugitive Negroes are called maroons. The term comes from the
Spanish word "simaran" which means "a monkey". The Spanish believed
that they should do no greater honour to their wretched fugitive slaves
than to call them monkeys, because they retreated like animals to the
depths of the woods and only came out to pick fruits near the closest
places to their place of retreat.'] (1796: iii [note])

The latter part of the explanation recalls the words of Maurile de S.
Michel:

*nos François qui deviennent sauvages, se cachants dans les bois, vivants*
*des fruicts d'iceux, & comme ces Hiboux & oyseaux nuictiers, n'en*
*sortant que la nuict pour aller picorer.* (1652: 38)

[our French people who become Savages, hiding in the woods, living off
the fruits there, and like owls and night birds, coming out only at night to
go pilfering.]

Whereas in the original it had been the French who had become *sau-*
*vages*, it was now black slaves who had been converted into fugitives
and were being likened to monkeys. The two other concepts in the
Edwards citation, the association of maroons with pigs (*marrano*[8]) and
fugitive black slaves with monkeys, fitted in well with eighteenth-
century racist philosophy.

It is quite evident, however, that the original link between wild live-
stock, buccaneers, maroons and mulattos declined as the buccaneering
lifestyle was replaced in reality and imagination by that of the plantation and slavery. In addition, the virtual disappearance of the indigenous inhabitants, the decline in the numbers of white servants and the
increase in the numbers of Africans eventually caused *cimarron/maron/*
*maroon* to be almost exclusively associated with escaped Africans.

In summary, in the early French period in Hispaniola, *boucanier,*
*mulâtre* and *maron* could, in numerous instances, have referred to the
same individual. The three words represented prismatic features of the
reality that was the beginnings of the French colony, Saint Domingue,
which in itself was a challenge to the formal colonial (Spanish) structure. The maroon, a by-product of tight European control, was the
embodiment of the fugitive and the love of freedom. The mulatto, a
product of racial mixing, sought escape from discrimination and dominance. The *boucanier,* a product of independent island life, was one element in a practice of division of labour and roles within an agreement

---

8. The relationship between *marrano*/Jew and maroon/Black has already been commented on in the discussion of the concept of *ladino* (Chapter 2).

of interdependence. The word *matelot*, first mentioned with a special-ised meaning by Du Tertre (1667 2: 452), gave a maritime flavour to this agreement. Evidently, *matelot* was a reinterpretation of Montaigne's *moitié*,[9] which was a reinterpretation of Las Casas's *guatiao*. After Du Tertre, *matelot* next appeared in Oexmelin 1686 (1: 151), where it referred to the same kind of relationship, but between the buccaneers in and around Hispaniola. This relationship, which was continued from the first French settlements in the smaller islands and was related to the *banare/compère* concept among the indigenous inhabitants there, provided a communal fabric, outside the established hierarchical one, for the growing French colony of Saint Domingue. Undoubtedly it had an effect on levels of formality in communication and the general use of a single, 'locally grown' variety of language among all groups in the colony.

# European distortions of language communication with and among the indigenous inhabitants

'Savages', 'Moors', buccaneers and maroons were the people integrally involved in shaping the identity of seventeenth-century Caribbean island societies. There is a notion of rough and uncultured that seems characteristic of all these groups. It was a notion that was to be re-inforced by an image of language 'corruption', which Spanish writers had previously developed to characterize the ladino and which French writers now used to describe the language communication of and with the indigenous inhabitants. The fact is that, though the French settle-ments in the smaller islands and in Hispaniola were established in a context of confrontation with Spain, French concepts of the languages and the linguistic ability of the indigenous inhabitants of the islands were substantially formed by the early literature, which came out of Spanish and other sixteenth-century expeditions.

In the accounts of the earliest explorers the comments about lan-guage were contradictory. For instance, Vespucci's version was that 'Many are the varieties of tongues ... so that they are not understanda-ble each to the other' (Young 1893: 121). Columbus, on the other hand, in his 1493 letter said: '*entodas estas islas no vide mucha diuer-sidad dela fechura dela gente ni en las costumbres ni enla lengua: saluo que todos se entienden ꝗ escosa muy sĩgular ...*' ['In all these islands, I did not see much diversity in the people's features or in customs or lan-

9. See Chapter 2, p. 76.

guage, except they all understand each other which is a remarkable thing ...'] (Obregón 1991: 60). In Martyr's account it was Columbus's view of language uniformity rather than Vespucci's view of diversity that was put forward, for the notion of 'the universall languages of those countries' in (Hakluyt's 1628 translation of) the *Decades* came directly from Columbus's letter of 1493. Herrera, in giving his account of events in Hispaniola in 1494, came up with a compromise when he said *'avia diversidad en la isla, aunque generalmente todos entendian una que era la Cortesana'* ['there was diversity in the island, although generally all of them understood one which was the Cortesana'] (1601, Decade 1, p. 88). This compromise – diversity with a common language – was repeated by subsequent writers in reference to indigenous inhabitants in various places in the New World.

In this matter of the language communication among the indigenous inhabitants in the islands the extent to which seventeenth- and early eighteenth-century writers drew upon the works of those preceding, even by only a few years and across languages, is quite evident. For example, De Rochefort's wording *'les Caraïbes de toutes les Iles s'entendent tous universellement entr'eus'* ['the Caribs of all the islands all understand each other everywhere'] (1658: 393), which was very close to the wording of Hakluyt's 1628 English version of the *Decades*, is repeated by Oldmixon – 'the Charibbeans of all the Islands do generally understand one another' (1708 2: 230). Oldmixon's additional comment: 'yet there is in several of them some dialect different from that of the others' (1708 2: 230) is also a straightforward repetition of De Rochefort's *'ce n'est pas à dire pourtant, qu'il ne se trouve en quelque une quelque dialecte different de celuy d'une autre'* ['that is not to say, however, that there is not in some one of them a dialect different from that of another one'] (1658: 393). This in turn is partially based on a comment in the account of the Capitaine Fleury 1618/20 voyage:

> *parce que leur langue ne s'entend pas plus loin que leurs îles, et même d'un bout à l'autre ils ne s'entendent presque point, et ont différents accents qui est cause qu'ils se moquent les uns des autres soit pour cela ou pour le langage, qui se prononce presque tout du gosier ...* (Fleury 1987: 103)

> [because their language is not understood beyond their islands, and even from one end to another they almost do not understand each other and they have different accents which is the cause of them making fun of each other either because of this or because of the language which is pronounced almost completely in the throat ...]

So, as a consequence of such repetition among writers, the idea of language diversity, together with a common language, emerged in the literature as the reality of the linguistic situation among the indigenous inhabitants in the Caribbean islands.

The early, honest views about the linguistic ability of the people of Guanahani, who were the first to meet Columbus, as recorded by the Spaniard Herrera at the beginning of the seventeenth century, were quite positive: '*Parecian de buena lengua, e ingenio, porque facilmente bolvian a pronunciar las palabras que una vez se les dezian ...*' ['They seemed to be well spoken and intelligent because words that were said to them once they could remember how to pronounce them ...'] (1601, Decade 1, p. 27). The reactions of the English (Harcourt, 1613) in their earliest encounters with natives of South America at the beginning of the seventeenth century also did not reveal any general, negative view of their language-learning competence. Admittedly, the comments of both Herrera and Harcourt referred to individuals, but they were made with the knowledge that a great amount of 'successful' work had been done by the Spanish priests among natives of Mexico in the sixteenth century. As far as the French are concerned, one of the earliest comments about influence in language contact showed that the indigenous inhabitants were influencing the French. According to De Lery (1578: 243), the French began, among themselves, to imitate the *façon de parler des Barbares* [the barbarians' way of speaking] and instead of saying *crever*, they said *casser la teste* [burst someone's head].

This initially positive, or at least neutral, outlook illustrated by these writers soon changed as actual experience with inhabitants of the islands increased. By the middle of the seventeenth century a different picture was beginning to be presented. Signs of this changing attitude came from the French, some of whom had had a rough experience with the natives of the islands. A clear indication of this changed attitude is given by the Catholic priest, Raymond Breton, whose early work on the Island-Carib language of Dominica drew the following reaction from the famous Du Tertre:

> *J'ay donné aux pressantes importunités du R.P. du Tertre ... une parcelle de mes traductions de Sauvage en Latin, mais il ne les agrea pas, il voulut quelque chose en langue vulgaire qui fit connoistre l'imperfection de la langue Caraibe, ce qui m'obligea de changer la traduction Latine, en construction Françoise qu'il arrangea à la fin de son livre comme une traduction.* (Breton 1665: In 'Aux Reverends Peres Missionaires')

[I gave Father Du Tertre, after his importunate demands, a parcel of my translations from Savage to Latin, but he did not approve of them; he wanted something in the vernacular that showed up the imperfections of the Carib language, which forced me to change the Latin translation to French which he set out at the end of his book like a translation.]

To equate the Island-Carib language with Latin was obviously, for Du Tertre, to raise it to too high a level and thereby give it validity, which was not the intention of the Church when it endeavoured to reach the indigenous inhabitants in their own languages. In fact, Du Tertre's assessment of the Island-Caribs' language was totally negative:

> *Tu verras dans ce peu de lignes combien cette langue est ingrate & indigente, & les grands travaux que ce bon Pere* [Breton] *a pris pour s'y rendre parfait.* (1654: Advis au lecteur)

> [You will see in these few lines how barren and deficient this language is and the great effort which the good Father took to make it perfect.]

This negative characterisation of the Island-Caribs' language was simply a part of what developed out of a general negative attitude to the Island-Caribs.

It is evident that in spite of the early and widespread practice by Spanish religious persons in Hispaniola and on the mainland to learn the languages of the indigenous inhabitants and to translate religious material into them, the normal practice that developed in the smaller islands among the French, as De Rochefort indicated, was for the Island-Caribs to learn the European language rather than the reverse: '*neantmoins ils aprénent plus facilement la nôtre, que nous n'aprenons la leur, comme il se reconnoit par l'experience*' ['nevertheless they learn ours more easily than we learn theirs, as is known from experience'] (1658: 394). In other words, the indigenous inhabitants, in their own islands, were becoming acculturated into the French way of life, moreso than the reverse. This was said to be a result of their natural curiosity – '*Ils sont fort curieux d'apprendre les langues et moeurs des étrangers*' ['They are very eager to learn the languages and ways of strangers'] (Fleury 1987: 96) and their obviously unsuccessful attempts at reciprocity: '*et ils nous disaient aussi comment ils nommaient en caraïbe, nous exhortant d'apprendre leur langue ...*' ['and they tell us also what it is in Carib, encouraging us to learn their language ...'] (Fleury 1987: 96).

In such a context and process, the accounts of earlier writers, which referred to specific events and people, provided material that was used to show the difficulties the indigenous inhabitants in general had in making themselves understood. The earliest island account itself (i.e. the Capitaine Fleury account) was fairly objective:

> *Au commencement de notre arrivée chez eux, ils nous faisaient entendre ce qu'ils voulaient dire de deux façons. La première par quelque mot espagnol ou français, et l'autre par signes, et souvent il fallait deviner, et ne pûmes rien comprendre qu'après être demeuré longtemps avec eux.* (Fleury 1987: 96–7)

[At the beginning of our arrival among them they made us understand what they wanted to say in two ways. The first by some Spanish or French word, and the other by signs, and often we had to guess and we could not understand anything until after having stayed with them for a long time.]

However, it was easy for De Rochefort later to change the exact words of his predecessors and cause his readers to believe that Island-Carib language was deficient and that, for example, they did not have any verbal way of indicating numbers in their own languages, and therefore, in communication with foreigners, had to supplement their speech with non-verbal communication:

> *Ils ne peuvent exprimer un plus grand nombre que* **vint***: Et encore l'expriment ils plaisamment, étant obligez comme nous avons dit, à montrer tous les doits de leurs mains, & tous les orteils de leur pieds.* (De Rochefort 1658: 399)

> [They cannot express a number greater than twenty. And even that they express quaintly, being obliged, as we said, to show all the fingers on their hands and all the toes on their feet.]

Such a statement was a distortion of the texts from which it was taken. De Rochefort's account of the Island-Caribs' method of communicating numbers was taken from De Laet ([1633] 1640), from Bouton (1640: 128) and from Breton's (at that time) unpublished work, all of which were merely repetitions. De Laet's, which was the earliest of these versions, said:

> *Or encore qu'ils monstrent presque toujours les nombres par les doigts, & quand ils veulent dire dix, ils dressent tous les doigts des deux mains, & pour signifier vingt, ils conjoignent les doigts des deux mains avec ceux des pies; neantmoins les Yaios (car je ne scai rien des autres) ont leurs propres noms des nombres ...* ([1633] 1640: 582)

> [Now even though they almost always show numbers by their fingers, and when they want to say ten, they put up all the fingers on their hands and to signify twenty, they join the fingers on their two hands with the toes on their feet, yet the Yaios (for I know nothing of the others) have their own words for numbers ...]

De Laet, in turn, took this from Harcourt:

> And to shew their meaning more certainely, they will hold up one, two, three, or more of their fingers, expressing the numbers, still making signs as they speake, the better to declare their meaning: when they will reckon twenty, they will hold downe both their hands to their feete, shewing all their fingers and toes, and as the number is greater, so will they double the signe. (1613: 25)

What is remarkable about the very first version (De Lery) of this so often repeated notion of an inadequate numbering system is that it seems as if it was the Europeans and not the indigenous inhabitants who were making the gestures:

> *Mais s'ils ont passé le nombre de cinq il faut que tu montres per tes doigts & par les doigts de ceux qui sont aupres de toy, pour accomplir le nombre que tu leur voudras donner à entendre. Et de toute autre chose semblablement. Car ils n'ont autre maniere de conter.* (1578: 342)

> [But if they pass the number five, you have to show by your fingers and the fingers of those who are near to you, in order to make the number that you want them to understand. And everything else in the same way. Because they have no other way of counting.]

This, more so than any other example, captures the reifying power of repetition and the conversion that can happen in the process, for, by the time the story reached Labat (1722 2: 63), the indigenous inhabitants could not count to more than ten.

In addition to De Rochefort's last statement above, he also said on the same topic:

> *Lors qu'ils veulent signifier un grand nombre, où leur conte ne peut atteindre, ou bien ils montrent leurs cheveus, ou le sable de la mer ...* (1658: 399)

> [When they want to signify a big number that their counting cannot achieve, either they show their hair or the sand from the sea ...]

This was taken from Bouton (1640: 128) and Breton (1665: 78), the latter of whom said:

> *quand ils veullent compter d'avantage, ils disent* **tamigati cachi nitibouti-bali**, *or,* **saccao bali**, *il y en a autant que de cheveux en teste, ou que de grains de sable au rivage de la mer.*

> [when they want to count higher, they say 'tamigati cachi nitibouti-bali' or 'saccao bali', there are as many as the hair on your head or the grains of sand on the seashore.]

The distortion in each version up to the De Rochefort versions was at the expense of the indigenous inhabitants, who were converted into inarticulate, underdeveloped and even comical persons.

To complement their gestural language, the indigenous inhabitants, because of their extended period of contact with Europeans speaking different languages and also because of their own experience communicating among their own different ethnic groups, were credited with having developed a rough and ready contact language for use with the

Europeans, which they modified as the situation demanded. It was Bouton who first announced this in the context of the smaller Caribbean islands: '*ils ont un certain baragouïn meslé de François, Espagnol, Anglois, & Flament ...*' ['they have a certain mixed dialect of French, Spanish, English and Flemish ...'] (1640: 130). Du Tertre then repeated what Bouton said:

> *Ils ont composé eux-mesmes une sorte de langue, dans laquelle il s'y rencontre de l'Espagnol, du François & du Flamand, depuis que ces nations ont eu commerce avec eux; mais ils ne s'en servent que lors qu'ils negotient.* (1654: 463)

> [They themselves made up a sort of language in which is to be found Spanish, French and Flemish, since these nations did business with them, but they only use it when they are negotiating.]

A few years later Du Tertre's words were repeated by De Rochefort:

> *Mais lors qu'ils conversent, ou qu'ils négocient avecque les Chrétiens, ils employent leur langage corrompu.* (1658: 392)

> [But when they converse or do business with the Christians, they use their corrupt language.]

De Rochefort also proceeded to illustrate this *langage corrompu* by bringing together words from Claude d'Abbeville (1614), the Capitaine Fleury account, Bouton (1640), Maurile de S. Michel, Breton's unpublished Carib dictionary and, by adding some from other works, even from South America. In doing this, De Rochefort can be said to have consolidated the *langage corrompu*.

In the use of this *langage corrompu* reversals in 'credit' also occur. For instance, the Island-Caribs were said by De Rochefort to have had a liking for the word *Compere* (first mentioned by Claude d'Abbeville 1614: 74) in addressing their friends, but what the Capitaine Fleury account said was:

> *Mais beaucoup de ceux qui avaient des hôtes aimaient mieux qu'ils les nommassent <<banare>>, c'est-à-dire compere ou ami, que de quelque autre nom d'alliance, disant que ce nom était plus beau que celui de père ou enfant, et ordinairement nous les nommions compères ou banare en leur langue.* (Fleury 1987: 96)

> [But many of them that had guests preferred that they call them 'banare', that is, 'compere' or 'friend', rather than any other name of friendship, saying that this name was more beautiful than that of father or child, and ordinarily we call them 'comperes' or 'banare' in their language.]

In other words, it was the French who called the Caribs *compère*. In comparison, Breton's Carib dictionary was divided in the assignment of 'credit':

> *Ce mot de compere est en usage en toutes les Isles où il y a des Sauvages, tant parmy les François, lors qu'ils traittent avec leur amis Sauvages, que parmy les Sauvages quand ils parlent au François ...* (1666: 320)

> [This word 'compere' is used in all the islands where there are Savages, as much among the French when they are dealing with their Savage friends as among the Savages when they are speaking to the French ...]

In like fashion, where De Rochefort put *bon France* in the mouth of a Carib, in the Capitaine Fleury text the words *'France bon, France bon'* (meaning 'we, the French, are good and to be trusted') were used by hungry Frenchmen begging the diffident Island-Caribs for help.

Generally, citations of speech given by De Rochefort (1658), which were said to have been produced by the Island-Caribs, in no way resembled the phonological, morphological, syntactic or semantic structures of the language of the Island-Caribs as presented by Breton. The only feature that was Island-Carib was the name of the deity *Maboya*, a word used freely in the Capitaine Fleury account. In fact, it was really only vocabulary items that conferred any validity to the notion of a contact language as used by the Island-Caribs themselves, since at least the Spanish words confirmed a historical connection with the Spanish. Spanish words identified by De Rochefort (1658) were:

*Pisket* – [French] poisson      (< Spanish *pescado* [fish])
*mouche* – [French] beaucoup      (< Spanish *mucho* [a lot])
*maron* – [French] sauvage      (< Spanish *cimarron*).

Other Spanish words given by Breton (1665) were:

*boulatta*      (< *plata* [silver])
*kirissiani*      (< *cristiano* [Christian])
*méguerou*      (< *negro* [Negro])
*camicha*      (< *camisa* [shirt])
*carta*      (< *carta* [letter]).

About the last two Breton said:

> *camicha, c'est un mot qui leur sert universellement pour toute sorte d'habits, toile, mouchoirs, mesme pour leurs voiles, il y a apparence qu'ils empruntent ce mot, aussi bien que carta des Espagnols.* (p. 107)

> [*camicha* is a word which they use universally for all sorts of clothes, cloth, handkerchief, even for their veils. Apparently they borrowed this word, as well as *carta* from the Spaniards.]

*cárta, les Sauvages n'ont que ce mot pour dire papier, lettre, parchemin, &*
*quelque livres que ce soit, encore je crois qu'ils l'ont appris des Espagnols.*
(p. 110)

[*carta*: this is the only word that the Savages have for paper, letter,
parchment and whatever books there may be. Again, I believe they learnt
it from the Spaniards.]

Breton also gave the following words, which were said to be of French
origin but modified phonetically according to Carib pronunciation:

| | |
|---|---|
| *apourietoutoni* | (*prières* [prayers]) |
| *choucre* | (*sucre* [sugar]) |
| *baïna* | (*peigne* [comb]) |
| *kaloon* | (*canon* [cannon]) |
| *kanabire* | (*navire* [ship]). |

De Rochefort chose to disregard Breton's words that it was the French,
more so than the Caribs, who might have produced the kind of speech
he illustrated. So, in addition to the kind of image of the indigenous inhabitant he created in the specific examples of non-verbal communication that he had earlier given, the details of the *langage corrompu*
increasingly became a European conception of simplification of a European language. The contact language presented by de Rochefort did not
start from the Island-Caribs' own language, as it inevitably would have
had to do in actual reality, but from a Frenchman's simplification of
French.

De Rochefort, himself, and Du Tertre before him both credited
Breton (that is, the dictionary in manuscript form) with having given
them their information on Carib language. They did not mention
Bouton (1640) or the anonymous Capitaine Fleury account, which both
of them clearly used. However, Breton, who respected the Caribs and
their language, denied having given De Rochefort any of the words that
he presented as part of the *langage corrompu* and furthermore seemed
to have little knowledge of it – '*ceux qui les luy ont donné les peuvent
bien avoir dire aux Sauvages, & aux François, mais comme un jargon
pour se faire entendre, & non pas pour un veritable langage Caraibe.*
['whoever gave him may very well have said them to the Savages and
the French, but as a way of being understood and not as a true Carib
language'] (Breton 1665: In *Aux Reverends Peres Missionaires*). It is
again significant, and indicative of its source, that Breton here again
regarded De Rochefort's jargon as what could have been spoken *to* the
Caribs, that is, by Europeans. Of course, De Rochefort was only giving
substance to what the eyewitnesses, Bouton and Du Tertre, said existed,
and his illustration was consistent, from his point of view, with what
Du Tertre later said:

*Davantage nos Sauvages, au moins une bonne partie commencent desja à baragoiner François.* (1654: 465)

[Moreover, our Savages, at least a great number of them already have begun to 'jargonise' French.]

Yet, one may of course agree with Labat's harsh assessment of De Rochefort and regard what he did as the result of imagination:

*Le Ministre Rochefort, qui n'a jamais vû les isles de l'Amerique que par les yeux d'autrui ... puisqu'il a copié le P. du Tertre; mais il a entierement gâté sa narration par ses descriptions tout-à-fait éloignées de la verité, dans la vue de rendre les choses plus agréables, & de mieux cacher son larcin ...* (Labat 1724: Preface 11)

[Minister Rochefort, who never saw the islands of America except through the eyes of others ... since he copied Father Du Tertre, but he completely spoilt his narrative by descriptions totally removed from the truth, in attempting to make things more pleasant and to hide his thievery ...]

However, De Rochefort's words in themselves were not fabrications and, presented as a bastard language, they appealed to French feelings of superiority and were soon transformed into a reality.

The concept of an Island-Carib contact language was transmitted to the English directly from De Rochefort by Ogilby:

The Carribeans which converse with the Europeans, speak two sorts of languages, the oldest of which is smooth, acceptable ... But their mix'd language hath many of the European words, especially of the Spanish, which they speak whensoe're they converse with the Europeans. (1671: 355)

The same notion and words were repeated a little later by Oldmixon:

The Charibbeans have an ancient and natural language, and a kind of bastard Speech; in which they have intermix'd several European Words, especially Spanish: The last they speak among the Christians, and the first among themselves. (1708 2: 230)

For the English, then, bastardisation or corruption was also the overriding characteristic of the contact language.

The only creditable, detailed, lexicographical work on the language of the Caribs of Dominica was Raymond Breton's *Dictionaire Francois-Caraibe*. It was a work that the author was initially virtually ordered to do, but, because of his long stay (1641–1653) among the people, his respect for them and his commitment to his job as a missionary, he remained faithful to his source. His work was used and praised by others even before it was published. It is not surprising therefore that the only person at the time who seemed to have been concerned about linguistic accuracy was Breton and it is from his comments that there appears an early indication of the confusion of sources of words then current. In trying to defend himself against falseness or for having conveyed false information about the Carib language, Breton said the following about the list of words given by De Rochefort (1658):

> je luy fis écrire & ponctuer en ma presence le vocabulaire, & je le confesse mien, à la reserve des mots de **banaré, manigat, carebet, aioupa, Amac, coüi, mouchache, cacone, coincoin, maron, piknine, boucan, Tortille, pisquet & canari**, qui ne sont point mots Sauvages, & qui ne viennent point de moy ... (1665: In Aux Reverends Peres Missionaires)

> [I made him write and punctuate the vocabulary in my presence and I admit that it is mine except for the words banare, manigat, carebet, ajoupa, amac, coüi, mouchache, cacone, coincoin, maron, piknine, boucan, Tortille, pisquet and canari, which are not words of the Savages and which did not come from me ...]

It would have been more accurate if Breton had said that the words were not from the language of those Island-Caribs he knew in the small French islands, because some of the words were indeed from the native languages of the region. In any case, this disclaimer by the person most familiar with the varieties of language of the Island-Caribs clearly undermined the notion of *langage corrompu* that De Rochefort sought to promote.

The syntax of the *langage corrompu* was a Frenchman's stereotypical reproduction of simplified speech more so than a faithful recording of what was heard. The closeness of the structure of the examples given for the supposed Carib language to the structure of the French language is quite evident. These sentences contain:

- utterances with subject–verb–object word order made up of corresponding isolatable words

  e.g. *toy trompe Caraïbe*     [you trick Carib]
  *si toy bon pour Caraïbe ...*     [if you good to Carib ...]

- reduction by omission of the verb 'to be' e.g. *moy bonne Caraïbe* [me good Carib]

- marking of time by an adverb only e.g. *magnane* [tomorrow]

- No words or endings for marking of tense/aspect e.g. *trompe*

- omission of clause connectors

- repetition of nouns instead of use of pronouns

- unlikely retention of French function words/prepositions (e.g. *à, pour, contre, de*).

All these features, which seemed natural to the European, were inconsistent with the structure of the indigenous languages, which repeatedly were admitted to be difficult for Europeans to learn. The difficulty lay not only in the differences in syntax but also in tonal distinctions that determined meanings, as is pointed out by Warren (1667: 25) in reference to 'Guiana' ethnic groups, some of whom had been taken to the early island colonies. In sum, therefore, the structures presented as evidence of language corruption were not much more than a revelation of the attitudes and preconceptions of one or two French writers.

## Distortion and reality in the shift from the Island-Carib language to that of the Africans

There is no doubt that the early writers were ignorant about the many languages in use across the islands and about the sources of many words in them. In reality, it is quite probable that one or more contact or common languages were used by the Tainos and Island-Caribs before the advent of the Europeans. It is also probable that the common or contact language(s) provided some basis for communication with Europeans. What is quite clear, however, is that in this context of confusion and ignorance neither the French nor the English were particularly concerned, as the African presence in the islands increased, to make any distinction between indigenous versions of European languages and African versions of them. In fact, not much was said by most writers in the middle of the seventeenth century about the structure of the utterances of the Africans and few gave examples of their speech. What was said and given differed little from what was said and given for the indigenous inhabitants.

For example, Bouton, commenting on the linguistic ability of the Africans, said:

> *Ils entendent desia pour la plupart aucunement le François, & en disent quelques mots sans les articles, & autres particules que nous y adjoustons.* (1640: 100)

> [For the most part up to now they understand no French and say a few French words without the articles and other particles that we add.]

Only brief illustrations of their speech came from Maurile de S. Michel:

> *car il y en a qui vous entendent, & parlent asses bon François ... c'est pourquoy levant la teste, il dist à toute peine,* **Moy non mort**

[because there are some who understand you and speak quite good French ... that is why, raising his head, he said with great difficulty 'Me not dead'] (1652: 80)

*Ils apprennent aussi-tost le François pour se faire entendre, en disant les mots sans autre article que Moy & Toy.* (1652: 86)

[They learn French quite quickly to be understood, saying the words without any article other than 'Me' and 'you'.]

Du Tertre (1667 2: 510) made a comment about language learning among the black slave children at a time when the ratio between black and white was not so disproportionate and when black children were more consistently and directly exposed to white children and other whites speaking French. Du Tertre claimed that as a result of their association with whites the black slave children could not speak their parents' African languages, that they learnt to speak French and that they also spoke the *baragouin*. This suggests that at the time the *baragouin* was still nobody's first language, and could still have been, as Du Tertre would have it, *un jargon composé de mots François, Espagnols, Anglois, & Holandois* [a jargon made up of French, Spanish, English and Dutch words].

There was one writer who concentrated his comments on the language of the black slaves and that was Pelleprat (1655). Like the other priests, Pelleprat's goal was the conversion of the 'infidels', specifically those in St Kitts and, in order to do this, he had to get them to understand his language, for, as he said, one would have had to have a special aptitude for languages to teach them in their own native languages. He said that their way of speaking French was ordinarily by the infinitive of the verb and to illustrate this he gave the following example – '*moy prier Dieu, moy aller à l'Eglise, moy point manger*' ['Me pray God, me go church, me no eat'] (p. 53). He went on to say that in order to indicate future or past time (i.e. tense) they added a word – '*demain moy manger, hier moy prier Dieu*' ['tomorrow me eat, yesterday me pray God'] (p.54). To further illustrate the speech of the Africans, he related incidents that involved Africans speaking. The following are excerpts from those incidents:

*Seigneur, toy bien sçavé que mon frere luy point mentir, point luy jurer, point dérober, point aller luy à femme d'autre, point luy méchant, pourquoy toy le voulé faire mourir? ...*
　　*Mon frere, toy te confesser, toy dire comme moy: Seigneur, si moy mentir, moy demander à toy pardon, si moy dérober, si moy jurer, si moy faire autre mal à toy, moy bien fâché, moy demander pardon.* (p. 63)

[Lord, you know well that my brother he no lie, he no swear, he no steal, he no go with another's woman, he not wicked, why you want to kill him?
...

My brother, you confess, you say like me: Lord, if me lie, me ask you pardon, if me steal, if me swear, if me do other bad, me very upset, me beg pardon.]

*pourquoy toy jeusner la veille de S. Ignace? ...*

*Nous jeûner aussi aujourd'huy, pource que demain festes des Rois, & Roy Negre, luy Patron à nous.* (p. 64)

[why you fast on the eve of St Ignatius? ...

We fast also today because tomorrow [is] festival of the kings and Negro King he is we patron.]

*Moy Chrestien ... moy donc souffrir cela.* (p.66)

[Me Christian ... so me will take that.]

These excerpts, as well as those of Maurile de S. Michel, contain the same features as those given for the *langage corrompu* of the Island-Caribs. Such excerpts, seemingly innocent, were really being used to illustrate the grammatical points specified and so they showed a decided preference for utterances involving the pronouns in French that have distinctive, strong forms, to the point of overuse or misuse. For example, *toy* was not the most used second person pronoun and it was not *toy*, but *vous* or *tu* that actually provided the creole form. Such examples of speech were really cited to satisfy the expectations of readers and were easy for them to understand.

Even a cursory reading of what Pelleprat said shows that this kind of speech was less a reflection of what the Africans said and more an illustration of the kind of language the Europeans thought that it was easiest for them to understand in the initial stages:

*On leur fait comprendre par cette maniere de parler tout ce qu'on leur enseigne: Et c'est la methode que nous gardons au commencement de leurs instructions.* (Pelleprat 1655: 54)

[They are made to understand by this way of speaking all that they are taught. And it is the method that we maintain when we start to teach them.]

To what extent this kind of speech affected what the Africans actually produced is difficult to determine, but such ad hoc speech would have varied from one European to another and would have been too inconsistent in itself to provide a model for learning. While it is likely that European writers' concept of simplification obscured any distinctions between the Caribs and the Africans in their production of European languages, it is also likely, especially in a situation like that in St Kitts, that the Africans both learnt and adopted a language of general communication. They learnt through direct contact with Europeans and adopted that which the Island-Caribs were using. Some of the recorded similarities between the speech of the Island-Caribs and that of the Africans were therefore more than figments of the imagination of French writers.

De Rochefort's list of words contained *maron*, *piknine*, and *canari*, words that linked Island-Caribs with Africans. The word *canari* became a common word among the black slaves for earthenware containers used for holding edible items. It is generally regarded as an indigenous word, but the origin of the use of word *canari* among them was explained by Du Tertre thus:

> *La boisson ordinaire que l'on appelle Oüycou, se fait dans de grand vaisseaux de terre, faits en façon de cloches, qui tiennent environ un demy poinçon. Les Sauvages les font eux-mesmes, & les appellent à l'imitation des Espagnols, Cannary.* (1654: 184)

> [The ordinary drink called 'ouicou' is made in big earthenware vessels made in the shape of bells which hold about half a puncheon. The Savages make them themselves and call them, following the Spanish, 'Cannary'.]

In support of Du Tertre's claim is the fact that the word is not mentioned in the early descriptions of the pots in Thevet (1575) and De Lery (1578), and a little later Harcourt (1613: 66) talks of the need for the importation of *Canary wine*. Breton, who specifically said that *canari* was not a Carib word, provided further support for Du Tertre's claim that *canálli* were *grands vaisseaux de terre dans lequels les Sauvages font leurs vins* [big earthenware vessels in which the Savages made their wines] (1665: 107).[10] In any case, the Africans adopted the utensil and its name from the natives of the region and not from the Europeans, which suggests a close, early association between the two ethnic groups, and an overall movement from European to indigenous inhabitant to African.

Another word in the area of drink, which started with the indigenous inhabitants and later was more closely associated with the black

---

10. Labat, 40 years after Du Tertre, believed that the Europeans got the name from the Caribs: 'On se sert pour cela [Ouycou] *de grands vases de terre grise que l'on fait dans le pays. Les Sauvages, & à leur imitation les Européens les appellent Canaris; nom générique qui s'étend à tous les vaisseaux de terre grands & petits, & à quelque usage qu'ils soient destinez.'* ['For this one uses big, gray earthenware vases which are made locally. The Savages and, following them, the Europeans call them "canaries", a generic name used for all earthenware vessels, large or small and whatever they are used for'] (1742 1: 412).

part of the population, is *mabi*. The drink *mabi* was identified with the native population from the early accounts of the region, at which time it was a drink said to have been made from potatoes, logically because *mabi* was an alternative word for 'potato'. Later, confusion with another drink occurred. Breton included as an item in his dictionary the word *comáti* with the gloss:

> *c'est un arbre, dont les Sauvages levent & gratte la seconde ecorce, dont ils expriment le jus qu'ils meslent avec un peu d'eau afin de le detremper & de s'en pouvoir servir, ils l'appellent (ainsi preparé)* **noucoumátiri**. (1665: 175)

> [it is a tree from which the Savages take and grate the second skin/bark from which they squeeze the juice which they mix with a little water in order to soak it to be able to use it. They call it (thus prepared) 'noucoumátiri'.]

While the black part of the population adopted the word *mabi*, the actual drink that they called by this name, in spite of variations, more closely resembled what Breton identified as *comáti*.[11] Yet, it can be said that, in spite of the change, both the name and the drink were adopted by the black slaves.

Two other words that have come to be associated with the Caribbean region, although not with the indigenous inhabitants, are the modern variants of *chicke* and *masseta*. The one was the bane of all who lived in the region and was mentioned in almost every descriptive work, while the other was a most useful implement. De Laet (1633) 1640: 583 gave the word *masseta* with the French meaning *sarpe*[12] as a word from the language of the Yaios. This Yaio word was the source of the almost indispensable *machete*.[13] Breton gave *chicke* as a Carib word and it is without question the source of the word *chigger* and its variants. Unlike *mabi*, the words *chicke* and *masseta* did not undergo any semantic shifts or confusion. All three words passed from the languages of the indigenous inhabitants to the Africans and were for a long time symbolic of the region in the literature of the Europeans.

Another less general, but no less interesting word is *Cai*, given in the Capitaine Fleury account (Fleury 1987: 99) as an exclamation of

---

11. Alvarez Nazario 1974: 278–9 [note 72] gives a varying explanation of the development of the term, even though it involves both Native Americans and Africans.

12. Larousse (1865) gives the word *sarper* with the meaning *couper avec une petite faux à manche cintré* [to cut with a curved-handle sickle], e.g *sarper du blé* [to cut wheat]. *sarpe* is a variant of *serpe*.

13. Simon 1627 gives a different origin for this word: '*Machete, es vocablo Vizcayno, muy usado en estas tierras, aunque en otras no se entiende, es lo mismo que puñal, o cuchillo de monte*' ['Machete is a Viscayan word common in these parts, although elsewhere it is not understood; it is the same as "dagger" or "hunting knife"'].

astonishment used by Carib men. The word *Kai* is given much later by Dickson (1789) as a word used by black slaves and interpreted by Turner (1949: 196) (for the Gullah dialect) as an African exclamation. Even though it is not unequivocal that the two spellings represent the same pronunciation, it is not inconceivable that the two words may have had one source or that the later African word may have been reinforced by the earlier Island-Carib word.

There were also other language features identified in the literature which connected Island-Caribs with Africans. For instance, the addition of *-ee* at the end of words was given as a characteristic of the Island-Caribs from very early in the history of the colonial West Indies by Hakluyt (1589), a reference later cited by both Dillard (1972: 136–7) and Taylor (1977: 25–6). What is quite clear from the Hakluyt account is that the Island-Carib addition of a final vowel to English words pre-dated a similar feature, later given as typical of the pronunciation of Africans.

Also in relation to the articulation of European syllable structure by the Island-Caribs, Breton says in explanation of the pronunciation *boulatta*:

> *ils ont emprunté le mot **plata** de l'Espagnol, & parce qu'ils ne prononcent pas aysement deux consonantes, ils inserent entre deux une voyelle, & prononcent souvent le p. comme un b. quand ils parloient de Mr. du Plessis ils le nommoient du Boulessi.* (1665: 420–1)

> [they borrowed the word 'plata' from Spanish and because they do not pronounce two consonants easily, they inserted a vowel between the two and pronounce the 'p' as 'b'. When they speak of 'Mr Du Plessis', they call him 'Boulessi'.]

Three other examples of the same process given by Breton are *apourietoutoni* (1665: 433), *kirissiani* (1666: 16) and *méguerou* (1666: 252). The French word *prieres*, after the insertion of a vowel and the addition of an Island-Carib prefix and suffixes, became *apourietoutoni*, while *kirissiani* derived from *Christian*. The word *méguerou* was based on *nègre* [negro, Moor]. These examples of the restructuring of the consonant clusters of European languages closely parallel one of the most prominent features later given as typical of the pronunciation of Africans in the Caribbean.

As to the articulation of specific consonants by the Island-Caribs, the changes from [p] to [b] (*plata* > *boulatta*), from [n] to [l] (*canon* > *kaloon*) and from [v] to [b] (*navire* > *kanabire*) are also later given as characteristic of the speech of Africans. In relation to the Galibi of South America, who were thought to be the ancestors of the Island-

Caribs, Pelleprat (1655: part 3: 10–11) said that they made one consonant of [p] and [b], and they did the same with [l] and [r]. Interchange between [l] and [r] is also later pointed out in the speech of the Africans and their descendants.

It was not only in words and sounds that similarities between the indigenous inhabitants and Africans are to be found, but also in the word formation. De Lery (1578: 364), in explaining the distribution of the first person pronoun *ché*, claimed that it had the same form whether it was subject, oblique, possessive or otherwise, and that the same held for the other pronouns. While this claim was being made about a specific language in South America, it was probably interpreted as a general characteristic of the indigenous languages of the region, for the strong form of the pronoun was always used in the representation of their contact speech. Later the same kind of uniformity in the pronoun was given as typical in the speech of the African slaves.

Undoubtedly the Island-Caribs had some influence on the Africans and undoubtedly also there were similarities in the languages of the two, but those features that were transferred from their native languages respectively and formed part of a more general lingua franca would not only have been difficult for most of the European writers to record, but would also have posed a problem for European readers. In other words, the expectations and capabilities of the European reader partly determined what writers said about language and the examples they gave. What is clear then is that a certain view and explanation of everyday language in the Caribbean became fixed in the minds of the French after being consistently repeated by French writers. It was a portrait painted by people most of whom had a gullible audience because of their profession, for the seventeenth-century French accounts of their early settlements were written mostly by priests (Claude d'Abbeville, Bouton, Maurile de S. Michel, Pelleprat, Breton, Du Tertre).

The accounts of the English (Gardyner, Ligon, Ogilby) were not by priests. The state of Protestantism, the Church of England and English politics did not at that time allow for the same kind of missionary zeal displayed by the different elements of the Roman Catholic Church, which had from the very beginning with Columbus established its claim to the *Savages* of the New World. The English writers were not primarily interested in philosophical and moral questions; they were not giving accounts of the flock or of new converts to the flock. Their tradition of writing about nature of the early settlements did not really demonstrate a major interest in the humanity of the non-Europeans, of which language was the basic thread. Early English accounts of the settlements did not therefore say as much about details of language as the French accounts.

In the Ligon account of Barbados, which described roughly the same period as the accounts of Maurile de S. Michel and Pelleprat, there were no examples of the speech of Africans or Caribs. What Ligon's comments about Africans suggested was that they were not very competent in English:

> The substance of this, **in such language as they had,** they delivered, and poor Sambo was the Orator [the person chosen to speak] ...
>     And this is all I can remember concerning the Negroes, except of their games, which I could never learn, **because they wanted** [lacked] **language to teach me.** (1657: 54)

A little later, the 1676 report *Great Newes from the Barbados* presented the speech of two Coromantee slaves with no comment about their competence in English, as if they were normal English speakers. The only word used by them that was not English was *baccararoes* (< *mbakara* [ruler] Efik and Ibibio), which was an already familiar word and one which was to have a long life in the speech of the Africans and their descendants.

The absence of dialectal or simplified features for the slaves in the early English texts may have been not only because English writers were not particularly interested in the linguistic details of the speech of the black slaves but also because citation of such features was not seen as necessary to the understanding of incidents. In addition, unlike in the French islands, there were no immediately preceding examples or model of *langage corrompu* for English writers to repeat. It is also likely that within the first 30 or 40 years of the English settlements in the islands no general or typical form of English emerged among the slaves, since at that time they did not constitute a coherent element within the total population.

In neither the French nor the English settlements were there any communities of speakers of any single African language, even though certain African ethnic groups were said to have been preferred by the colonists. The only reference to what seemed to be a concentration of any one African ethnic group in one place is by Maurile de S. Michel: '*Il* [Major Auger, Normand] *appelle les cazes de ses Negres & Negresses, qui estoient plus de cent, sa ville d'Angole, à cause qu'ils sont venus d'Angola d'Afrique ...*' ['call the huts of his Negroes and Negresses, which were more than a hundred, his village of Angola, because they came from Angola in Africa ...'] (1652: 45). This was said about a specific slave owner in St Kitts, but, besides this, the comments suggest diversity. In a general comment Du Tertre said: '*Pour ce qui regarde les Negres, ils sont amenez dans toutes les Indes des costes d'Angole, de Guynee, ou du Cap vert ...*' ['As to the Negroes, they are brought to the Indies from the coasts of Angola, Guinea or Cape Verde ...'] (1654: 474). Pelleprat was even more specific in his identification of the diversity: '*On compte dans les Isles jusqu'à treize nations de ces Infidels, qui parlent toutes de differentes langues ...*' ['In the islands there are up to thirteen nations of these unbelievers, all of whom speak different languages ...'] (1655: 52–3). It was, however, Ligon's explanation of the importance of diversity that captured the attention of Europeans: 'They are fetch'd from several parts of Africa, who speak several languages, and by that means, one of them understands not another ...' (1657: 46). The policy of diversity was then generally seen and said to have been pursued as a basis for European security and the quick loss of African culture.

Of the three racial groups the Africans were the only one whose native languages were not represented in the literature. The language of the African, after his/her capture, suddenly became useless in practice and non-existent formally. Formal non-existence of African languages, except for the occasional mention of a word for some cultural item, by a continuous process of omission caused all those who read and eventually all those whom reading affected to believe that the Africans did not have languages. On the other hand, dictionaries, word-lists and discussions of the languages of the indigenous inhabitants confirmed that they did have languages, even though such material was part of an immediate process of re-education. However, as Pelleprat said (1655: 53), the Europeans had to wait until the Africans acquired some competence in the European language before any attempt could be made to re-educate them. This hiatus lasted a long time for Africans as a group.

In time, contact between Europeans and Africans in most places superseded all other types of contact and stabilised ad hoc speech into a recognisable, general form of speech. Without doubt, this was influenced by the early methods of communication involving the indigenous inhabitants. The community language that the Africans and their offspring acquired or was transferred to them had, from the beginning, been projected as a 'bastard' form of language and a mixture of languages from different groups. The concept of 'bastard' language was also linked to the results of miscegenation (e.g. mulattos), with the degree of 'bastardization' parallelled by a scale of miscegenation. In this conception, the European was always represented as the norm and the African as deviant and dependent. It was not exactly the same in the European–indigenous inhabitant relationship, if only because, as Bouton said: '*Ils disent que c'est nous qui avons besoin d'eux, puis que nous venons en leurs terres, qu'ils se sont bien passez de nous, & s'en passeront bien encore*' ['They say that it is we who need them since we came into their country, and that they did well without us in the past and they will do so again in the future'] (1640: 135–6). The 'bastard' language was regarded as the creation of indigenous inhabitants for practical purposes, but in the mouths of the Africans and their descendants it became a symbol of inferiority. So, while the indigenous inhabitants were said to have consciously created a useful method of communicating using the words of the Spanish and Portuguese, the Africans were seen as passively adopting and corrupting the languages of the Europeans.

# The historical background to the notion of *langage corrompu*

## Spanish sources

The notion of *langage corrompu* developed by French writers did not do so in a vacuum; it developed within a context of expanding French classicism in a century that is often regarded as the greatest for French culture, literature and language. It was a century in which France superseded Spain and culturally dominated the rest of Europe. It was a time when the rise of French influence in Europe and the growth of French self-confidence and image were being contemporaneously expressed on the other side of the Atlantic by initial excursions and settlements in the islands of the Antilles. However, the centre of the French world always remained the king and the court of scholars who surrounded him. There was no New World challenge or alternative to the centralised uniformity of French letters because there was no printing of French material in the New World. The court of Louis XIV became the sole star of French culture.

By the second half of the century, the society of Louis XIV regarded itself as civilised and disciplined and the French language was being promoted as a superior medium and a reflection of the highest culture. In the whole of France and especially in the king's court, there was respect for the social hierarchy and the king, who, with absolute power, was the shining sun of civilisation. Earlier, in the first half of the century, in 1635, the *Académie française* [the French Academy] had been established to develop, among other things, a dictionary and a formal grammar of the French language. It was made up of writers whose job it was to make decisions about the language. So, in 1647 Vaugelas published his influential *Remarques sur la langue française* [Comments on the French language] in which he advocated that correct usage should be based on that of courtly society as well as that of refined Parisian society. Following the foundation of the *Académie française*, there was a keen consciousness about language and, throughout the century, writers and scholars passionately discussed points of grammar and the appropriateness of specific words. French literature, then, in remaking the classical works of Greece and Rome was intensely taken up with linguistic form.

De Rochefort, who was no traveller but a writer conscious of his social and cultural milieu, naturally regarded the contact language of the Island-Caribs from the point of view of a sophisticated Frenchman. Consequently, for him, it could have been nothing else but a corruption of language. The notion of corruption of language was also prominent in the minds of the French for another reason, for, with regard to their own language, there was a consciousness of it having evolved from Latin, thus being in a sense inferior to it, and that there had been a need for the French language to be codified. There was an awareness or belief that it was through discussion in refined society and use by good writers that standards were established. Judged within this framework, language in Antillean society, since it arose out of natural communication, was in the distant depths because there was no refinement of nature through reasoning and written literature to lift it.

To a great extent, French notions of the superiority of the culture of France that developed in the seventeenth century were a reaction against Spanish behaviour and dominance in the sixteenth century and were a contrast to them. The relative laxity of racial barriers in intercourse that from the beginning characterised Spanish New World colonies was discouraged by law and Church when it began to show itself in the early French island colonies. The concerns of the French missionary priests about maroonage among the French servants were as much a concern about social and cultural hierarchy as they were about salvation of souls. While the Spanish outlook facilitated the spread of Spanish, the French view of a more rigid social hierarchy militated against the easy spread of French and tended toward linguistic separation. Yet, the notion of *langage corrompu* that the French developed as a New World phenomenon was substantially an extension of *bastarda lengua*, which for the Spanish was an Old World reality, part of an experience of travel and culture contact with North Africa that had begun to become familiar to the French and English through Spanish literature in the first half of the seventeenth century.

It was Fray Diego de Haedo who, in his description of Algeria of that day (i.e. *Argel*), identified the Old World precursor of the *langage corrompu*:

> *La tercera lengua que en Argel se usa, es la que los moros y turcos llaman franca, o hablar franco, llamando ansi a la lengua y manera de hablar de christiano,[n]o porque este hablar (aquellos llaman franco) sea de alguna particular nacion christiana, que lo use, mas porque mediante este modo de hablar que esta entre ellos en uso, se entiendien con los christianos, siendo todo el, una mezcla de varias lenguas christianas, y de bocablos, que por la mayor parte son Italianos, y Españoles, y algunos Portugueses ...* (Haedo, 1612: 24)

> [The third language used in Argel is that which the Moors and Turks call 'franca' or 'hablar franco', thus referring to the Christians' language and way of speaking, not because this speech (they call 'franco') is that of any specific Christian nation that uses it but because this mode of speaking which is in use among them is understood by the Christians, being a mixture of various Christian languages and words which for the most part are Italian, Spanish and some Portuguese ...]

It was not so much this notion of mixture of European languages that foreshadowed what was said about the language used by the natives of the New World to communicate with the Europeans but the more precise characteristics of 'corruption' that were identified:

> *Y juntando a esta confusion y mezcla de tan diversos bocablos y maneras de hablar, de diversos Reynos, provincias y naciones christianas, la mala pronunciacion de los moros y turcos, y no saben ellos variar los modos, tiempos y casos, como los christianos (cuyos son propios) aquellos bocablos y modos de hablar, viene aser el hablar franco de Argel ...* (1612: 24)

> [And joining to this confusion and mixture of such diverse words and manner of speaking, from diverse kingdoms, provinces and Christian nations, the bad pronunciation of the Moors and Turks, and they do not know how to change moods, tenses and cases, like the Christians (whose languages do) those words and manners of speaking come to make the 'hablar franco' of Argel ...]

Then, as if to make sure that the connection between the Old World and the New World was clear, Haedo made a significant comparison: '*casi una gerigonça, o a lo menos un hablar de negro boçal, traydo a España de nuevo*' ['almost a jargon, or at least the speech of a bozal negro, brought back to Spain'] (1612: 24). It was this identification of the *negro boçal* that constituted a clear link between speech in Argel, speech in Spain and speech in the New World colonies.

Haedo further said that the *hablar franco* was in widespread use:

*Este hablar franco, es tan general, que no ay cosa do no se use, y porque
tampoco no ay ninguna do no tengan christiano y chistianos, y muchas
que no ay turco ni moro grande ni pequeño, hombre o muger, hasta los
niños, que poco o mucho y los mas dellos muy bien no le hablan, y por el
no entiendan los christianos: los quales se acomodan al momento a aquel
hablar* (1612: 4)

[This 'hablar franco' is so general that there is nothing for which it is not
used, and also because there is no place where there are no Christians and
... and many where there are no Turks or Moors, big or small, man or
woman, even children, that little or much and the most of them do not
speak it very well and through it they do not understand the Christians
who become accustomed at the moment to this speech.]

This not only led to the belief that the common language of communication in North Africa and that in the Caribbean islands were of similar
types, but to a more recent belief that the former was more or less the
parent of the latter.

Quite clearly one of the determining factors in the emergence of the
*hablar franco* was the diversity of languages in the city of *Argel*.
According to Haedo:

*Las Gentes habitadores desta ciudad, se dividen generalmente en tres
generos o maneras de personas, es a saber, moros, turcos, u judios: no
hablamos de christianos, aunque ay una infinidad dellos de toda suerte y
nación ... Los moros son tambien de quatro maneras ...* (1612: 8)

[The groups who live in this city divide themselves into three types or
kinds of persons, that is, Moors, Turks, or Jews; we are not speaking of
Christians, although there are an infinity of them of all sorts and from all
countries ... The Moors are also of four types ...]

It was in order for communication to take place between such a diversity of people speaking so many different languages that the *hablar
franco* developed.

It was not only Fray Diego de Haedo who provided evidence of the
*hablar franco* of *Argel*, but also the more illustrious Cervantes, who
was a prisoner there. The matter of divergent language occurs in *Don
Quixote* in relation to Moors and it is here that the notions of 'bastard'
language and 'mixed' language are developed:

*el cual me dijo en lengua que en toda la Berbería, y aun en Constantinopla,
se habla entre cautivos y moros, que ni es morisca, ni castellana, ni de otra
nación alguna, sino una mezcla de todas las lenguas, con la cual todos nos
entendemos, digo, pues, que en esta manera de lenguaje me preguntó que
qué buscaba en aquel su jardín, y de quién era.*

*... Ella tomó la mano, y en aquella mezcla de lenguas que tengo dicho
me preguntó si era caballero ...*

> *Servianos de intérprete a las más destas palabras y razones el padre de
> Zoraida, como más ladino; que aunque ella hablaba la bastarda lengua
> que, como he dicho, allí se usa, más declaraba su intención por señas que
> por palabras.* (1 Chap XLI)

[who told me in a language which is spoken in the whole of Berbería and
even in Constantinople by captives and Moors that is neither Moorish
nor Castilian nor that of any other nation, but a mixture of all languages,
by which we all understand each other. I say then that it is by means of
this dialect he asked me what I was looking for in his garden and whose it
was.

... She took my hand and in that mixture of language that I have spoken
about she asked me if I was a nobleman ...

Zoraida's father, being the most ladino, functioned as our interpreter
for most of these words and information, for although she spoke the
bastard language which, as I said, is used there, she declared her intention
more by signs than by words.]

For Cervantes, then, this 'bastard' and 'mixed' language was a widespread reality. Cervantes' view of it, and presumably that of Spaniards
more generally, was one that acknowledged the historical reality of ethnic diversity in Spanish experience. However, from the lack of clear
communication that characterized the interchanges between the Spaniard and the Moorish girl, her resorting to gestures and the need for a
ladino interpreter, the 'bastard' language, as Haedo said, was not uniform. Spaniards' versions and Moors' versions were not the same, but
the Moors, by experience, had the advantage because they knew more
of the Europeans' languages than the reverse.

If this was a language used for communication between the Moors,
their captives and others through all of North Africa and Turkey, no
doubt it would also have been known in West Africa, which formed
part of the trading area of the Moors and which was an area to which
the Moors had spread their religion. This language, which, according to
Cervantes, was a product of Moorish trade, would clearly have been
influenced by the languages of the Moors. Whether or not it was taken
to the Caribbean islands by the earliest slaves, those from the Sene-
gambia area, which was within the ambit of Moorish influence, is
speculation. If it was and it came into contact with that used between
the indigenous inhabitants and Europeans, then it would have been a
case of double 'bastardization'. In *Don Quixote* there are no direct
references to New World people, except that the *zagalejas* are described
in their dress as well as in their social and moral knowledge just as the
indigenous inhabitants were in the histories of the time (evoking a
*golden world* and *noble savage* image):

*Entonces sí que andaban las simples y hermosas zagalejas de valle en valle y de otero en otero, en tranza y en cabello, sin más vestidos de aquellos que eran menester para cubrir honestamente lo que la honestidad quiere y ha querido siempre que se cubra ... No habia la fraude, el engaño ni la malicia mezclándose con la verdad y llaneza.* (Cervantes 1949 1 chapter X1)

[Then the simple and beautiful zagalejas went from valley to valley and hill to hill, in ... and bareheaded, with only the amount of clothes they needed to cover properly what propriety wants and always wanted to be covered ... There was no fraud, trickery or malice mixing with truth and simplicity.]

Besides the possibility of the spread of a Moorish 'bastard' language through trade and travel, what is quite evident is that the representation of 'bastard' language became a general part of the reality of early seventeenth-century Spanish literature. There was, for instance, a 1615 letrilla of Góngora (*Al nacimiento de Cristo Nuestro Señor* [On the birth of Our Lord Jesus Christ];[14] Henriquez Ureña 1939: 250), which was obviously meant to represent the speech of a Muslim/Arab. There were also two short plays (*Entremés primero: de Melisendra* (1609)[15] – Cotarelo y Mori 1911: 109 – and *Entremés octavo: del Indiano* (1609)[16] – Cotarelo y Mori 1911: 139) containing the speech of Moors, which appeared around the very time that Cervantes mentioned the *bastarda lengua*. The linguistic features presented in the letrilla and the two sketches overall convey the sense of a variety different from that given for ladino Africans, but they contain some interesting features. For example, the use of an uninflected/infinitive form in a finite context (*tener yo; Yo estar* [have I; I be]), the articulation [e] instead of [i], the variant articulation of -ll-, the lack of gender concord between the article and the noun, and the loss of the initial unstressed syllable in *nemego* (< enemigo), all these were given, even if in a slightly different way, in the speech of ladino Africans (see Alvarez Nazario 1974: 106–130). Moreover, the use of the fully written infinitive form of the verbs in the speech of the Moors was remarkably similar to those given by Pelleprat (1655) for the black slaves in St Kitts.

Besides the speech of *moros* and *moriscos,* the speech of other ladino Africans in Spain was common enough for it to have become a source of amusement in Spanish Golden Age literature,[17] which indicates that the comic features highlighted were based on reality, although most likely exaggerated for effect. Such features, which were

---

14. See P. Henriquez Ureña (1939: 250).
15. See Cotarelo y Mori (1911: 109).
16. See Cotarelo y Mori (1911: 139).

to become common in the speech of natives of the Spanish Caribbean island colonies, were:

- the non-articulation (loss of) of [s] at the end of words – *habemo, casamo, vimo, vamo, alcoholemo, bailemo*

- using [l] for [r] or the reverse– *crara* < clara, *neglo* < negro, *branca* < blanca

- articulation of -ll- as [y]– *beya, estreya, eya*

- loss of unstressed syllable at the beginning of words – *mana* < hermana, *legante* < elegante, *panta* < espantar

- metathesis (change of order of consonants within a word)– *presona* < persona

- making the words end in the same way – *Christa, trista, denta, fustana*.

These features were not regarded by Golden Age Spanish writers as retentions of dialectal variants of Spanish but as corruption of the Spanish language by black speakers. It was clear to them at the time that Africans in Spain had a characteristic variety of speech of their own and so, when later writers were to present the variety used by black speakers in the New World, the same features appeared, as, for example, in the piece called *Porto Rico* written by Sor Juana Inés de la Cruz in 1677. Alvarez Nazario (1974: 134) regards these verses as having the same characteristics as those found in Spanish Golden Age literature and as pointing to general characteristics in slave speech in Spanish America – '*lengua española corrupta, como communmente la hablan todos los negros*' ['corrupt Spanish, as is commonly spoken by all Negroes'].

In summary, the early definition of *ladino* as a 'Spanish-speaking Moor'(Kany 1960: 22) of medieval times and the later use of *ladino* to refer to black Africans in Spain speaking a variety of Spanish establish a direct link between the *bastarda lengua* and ladino African speech. Moreover, the Spanish works with comically speaking *ladinos* were popular enough that the notion of 'corrupted' language would have been familiar in France next door as well as in other parts of Europe.

---

17. Instances of ladino speech in Spanish literature occur in *Entremés de los negros* by Simón Aguado 1602 (see Cotarelo y Mori 1911: 231–35); *El capellán de la Virgen* by Lope de Vega (See Albornoz and Rodríguez Luis 1980: 45); *En la fiesta del Santísimo Sacramento* by Góngora 1609 (see P. Henriquez Ureña 1939: 235–6); *Al Nacimiento de Cristo Nuestro Señor* by Góngora 1615 (See P. Henriquez Ureña 1939: 253); *A la <Jerusalem Conquistada>, que compuso Lope de Vega* 1609, attributed to Góngora (see P. Henriquez Ureña 1940: 85).
A full and detailed discussion of the features of black peninsular speech presented in Spanish Golden Age literature is given in Alvarez Nazario 1974: 106–30.

So, there is little doubt that the concept of *bastarda lengua* occurring in Spanish literature at the beginning of the seventeenth century was converted into the concept of *langage corrompu* by French writers in the middle of the century, since Africans were involved in both situations and since the French initially used the word *more* [Moor] to refer to black Africans. The fact that there was no continuity and no uniformity in structure across so many different ethnic groups was not important, but it was important that, as was the case with the concept of *ladino*, other people, who were not white European, were being lumped together and characterised as having corrupted European language.

## Portuguese sources

Though it is quite evident that the designation *langage corrompu* was the adoption of a notion from Spanish literature and its perpetuation by French writers in their accounts of the French colonies in the Caribbean, a view developed in the last half of the twentieth century among linguists that the variety/ies of language previously called 'corrupt' but now renamed 'pidgin' came from Portuguese sources. In this view early varieties of language spoken by Africans in the islands were seen as resulting from a real variety of language born in the Old World and transmitted to the New. The *lengua franca* of the Mediterranean was linked, in this twentieth-century reinterpretation, directly with the 'pidgin' languages of the African slaves in the islands (without any specific mention of the *langage corrompu* of the indigenous inhabitants), and Portuguese sailors and traders were given credit for having transmitted it there and in fact all over the world. The historical literature, however, tells a different story.

In his history of Brazil, the French writer, De Laet, provided an early account of Portuguese involvement with the indigenous inhabitants there and the powerful role played by language and language learning in the process of subjugation:

> *Les nations qui habitent la Continente du Brasil, sont pour la plupart differents de langage: toutefois ils en ont un commun entr'eux, duquel se servent ordinairement dix nations d'iceux, qui demeurent proche du rivage de la mer & mesme au dedans du païs: presque tous les Portugais l'entendent, car il est aisé, copieux & assés agreable: Or les enfans des Portugais nés ou eslevés de jeunesse dans ces Provinces, le sçavent comme le leur propre ... par le moyen de cette langue les Peres de la Societé ont aussi coustume de traicter avec ces nations, car ces Sauvages sont les plus humains & familiers de tous, & ont de long temps paix & amitié avec les Portugais: de sorte que par leur moyen & armes, ils ont en partie subjuguees les autres nations du Brasil, & les ont renduës tributaires, ou*

*les ont du tout destruictes, ou contrainct de quiter leurs maisons & de s'enfuir au dedans du païs.* ([1633] 1640: 478)

[The nations which live in the continent of Brazil are for the most part different in language. However, they have one language that is common among them which is normally used by ten of them who live near the seaboard and even inside the country. Almost all the Portuguese understand it, because it is easy, copious and quite pleasant. The children of the Portuguese born or raised from early in these provinces know it as their own ... using this language the Fathers of the Society are accustomed to deal with these nations, for these Savages are the most humane and familiar of all and for a long time now have lived in peace and friendship with the Portuguese so that by their help and arms, they have partially subjugated the other nations of Brazil and have made them dependent or have totally destroyed them or forced them to leave their homes and flee into the interior of the country.]

This was a story of the Portuguese learning the common language of the indigenous inhabitants, quite different from the idea of them bringing a language and spreading it among the local community. Here there is not even mention of 'bastard' Portuguese as there was on the west coast of Africa where Portuguese settlements developed.

There is little support in the historical literature for this claim of a 'Portuguese pidgin'; it was really based on odd bits of evidence. An early but isolated case of evidence in the New World islands was provided by the French Catholic priest, Antoine Biet, who sought temporary shelter in Barbados in 1654. He was accosted by a man there and tried to deny that he was a priest because he was afraid of Barbadians' reputation for being hostile to priests. Biet recounted the incident as follows:

*Il me parla un certain langage corrompu entremeslé d'Italien, de Portugais, de Provençal ou pour mieux dire d'un certain langage corrompu, que ceux qui voguent sur la mer Mediterranée entendent tous fort bien: **Seignor Padre**, dit-il, **io so servitore à vostra Signoria**. Je le regarday comme un homme fasché, & luy répondis en ce mesme langage, que voille dire, **mi no so Padre**.* (1664: 275)

[He spoke a certain corrupt language to me with a mixture of Italian, Portuguese and Provençal, which people who sail in the Mediterranean all understand very well: '*Seignor Padre*,' he said, 'io *so servitore à vostra Signoria*.' I considered him an angry man and I replied to him in the same language, '*voille dire, mi no so Padre*'.)

This was a chance encounter between two Europeans who did not share a common language and had to resort to the Mediterranean *hablar franco*. This incident was apparently generalised and taken to mean that the *lingua franca* of the Mediterranean played a role in the development of language in the islands because the different ethnic groups did not share a common language. In addition, since the Portuguese were the earliest European traders down the coast of West Africa, where the slaves in the islands came from, they were the ones to be credited with the spread of the *lingua franca* for general communication and the Portuguese language was assumed to be the main component of it.

In addition to the Biet reference, there is evidence of a link, via the West African coast, between *hablar franco* and contact language in the New World. Firstly, Jean Barbot, speaking of his visit to Sierra Leone in 1678, said: 'Almost all the blacks of this district speak either Portuguese or **langue franque** ...' (Hair *et al.* 1992 1: 220). Barbot did not explain here or elsewhere what he meant by *langue franque* or *lingua franca*, but his emphasis was probably on the idea of a language that was used for communicating with Europeans. A few years later, more evidence occurs in the 1685 account of a Frenchman named La Courbe: '*Ces gens la, outre la langue du pays, parlent encore un certain jargon qui n'a que tres peu de ressemblance a la langue portugaise, et qu'on nomme langue créole, comme dans la mer Mediterranée la langue franque* ...' ['Those people, in addition to their own language, speak a certain jargon/dialect which has only a slight resemblance to Portuguese; it is called creole language, as in the Mediterranean Sea the lingua franca ...'] (La Courbe 1913: 192). La Courbe was referring to the mixed descendants of the Portuguese in an area named Barre, just north of the River Gambia. The Portuguese had established themselves in the general area since 1446 so that by the time that La Courbe went there, the descendants of the Portuguese, who had mixed with the local population, were speaking a variant of Portuguese.

La Courbe's association of *la langue franque* with *la langue créole* highlighted its 'mixed' composition and the labelling of a local variant as a *jargon* by French writers paralleled their characterisation of *langage corrompu*, the language varieties used by the indigenous inhabitants and African slaves in the Caribbean in the seventeenth century. The identification of an African variety of a European language as a creole language not only linked two focal areas on different sides of the Atlantic but actually preceded any such documented designation of a language in the Caribbean.

Thomas Phillips, an English captain of a slaving vessel which visited the Gold Coast in 1693–4, did not use any variation of the term *hablar franco/langue franque*, but referred to 'the bastard Portugueze the negroes use upon this coast'. Even though the area referred to by

Phillips was quite a distance from that La Courbe was talking about and even though La Courbe was talking about mulatto descendants of the Portuguese whereas Phillips was talking about Africans, the notion of 'bastard' Portuguese was being identified as common to both.

About 50 years after La Courbe's linking of *langue franque* to *langue créole*, the English traveller Francis Moore identified the language of the descendants of the Portuguese in the same way but without using the term *langue franque*:

> When this country was conquer'd by the Portuguese, which was about the Year 1420, some of that Nation settled in it, who have cohabited with the Mundingoes, till they are now very near as Black as they are; but as they still retain a sort of bastard Portuguese language call'd Creole ... (1738: 29)

Smith, a few years after Moore, also commented on the same people, though he did not label their language as either *franque* or *creole*:

> This Place was first discover'd and settled by the Portuguese, whose Progeny are still pretty numerous up in the Inland Country ... Tho', to speak Truth, there is but little of the Portuguese to be found in them, beside the Language, being quite degenerated into Negroes ... (1744: 25)

Even though Smith did not label the language itself, one would have to assume that his view of it did not differ from that of his predecessors, Phillips and Moore.

It is instructive to note the comments on the racial mixture of black and white made about people in the area of the Gold Coast by the same Smith:

> Upon this Coast are a Sort of People call'd Mullatoes, a Race begotten by the Europeans upon Negroe Women. This Bastard Brood is a Parcel of the most profligate Villains, neither true to the Negroes, nor to one another, yet they assume the Name of Christians ... In short, whatever is bad among the Europeans, or Negroes, is united in them; so that they are the Sink of both. (1744: 213)

This is the same comment, cited above, made about the *Cologlies* by Blofeld – 'they hardly possess one virtue, and seem to have inherited the evil qualities only, of their parents' (Blofeld 1844: 145). The word *creole* then, in its West African references, parallelled its use in the Americas, but with a greater focus on mixture and corruption. Here there was a direct relationship drawn between 'bastard' people and 'bastard' language.

In the middle of the nineteenth century the notion of a Portuguese creole resurfaced when Bertrand-Bocandé, a French writer, in his 1849 description of Guinea, gave what amounted to a definition of a creole language. This definition included two elements critical to the development of the twentieth-century 'Portuguese pidgin' theory – structural simplification and local modification. Starting with the view typical of his day that Africans' level of intelligence did not allow them to manipulate a European language and that it had therefore to be reduced to its bare essentials, Bertrand-Bocandé, in an explanation of what he called *créole portugais*, said:

> *Il se fit un retranchement graduel de toutes ces modifications qui servent à exprimer les diverses nuances de la pensée, et quand il ne fut plus possible de rien retrancher pour conserver le discours encore intelligible, l'idiome fut fixé dans sa grammaire particulière, devenue aussi simple que peuvent le permettre les règles de la grammaire générale de toute langue. Il exista alors ce que l'on appelle la langue **créole** portugaise... Le créole portugais n'est donc qu'une altération de la langue portugaise ...* (1849: 74–6)

> [There is a gradual cutting off of all the endings which express the diverse nuances of thought, and when it is no longer possible to cut off more in order to keep the discourse intelligible, the dialect was fixed in its peculiar grammar, having become as simple as the rules of general grammar of language permit. Then that is what is called Portuguese creole... So creole Portuguese is nothing but a distortion of the Portuguese language...]

This simplification explanation was clearly adopted by Bertrand-Bocandé from previous explanations of local languages in the Caribbean and was seen to be an obvious and logical one.

As to the idea of local variation, which was in the twentieth century built up into a theory of 'relexification' (i.e. replacement of the original Portuguese words with local words), Bertrand-Bocandé explained as follows:

> *Ce créole varie dans chaque lieu: il a des mots, des expressions, une accentuation et même quelquefois un ordre grammatical plus ou moins différents, suivant la langue qui a dominé pour faire subir ses modifications à la langue portugaise, qui est toujours partout le fondement du créole.* (1849: 73–5)

> [This creole has variants in each locality: it has words, expressions, an accent and even sometimes a grammatical order more or less different, according to the predominant language that inflicted its grammatical features on the Portuguese language, which is always the base of the creole everywhere.]

Paradoxically, while Bertrand-Bocandé's *créole portugais* was a duplication and extension from the Americas to West Africa of the concept of a creole language, his generalisations about it, occurring in a learned journal, provided a theoretical and geographical source for the

'Portuguese pidgin' theory that sought to explain Caribbean creole languages.

It was therefore odd incidents and observations such as those by Biet, La Courbe and Bertrand-Bocandé, which, taken together and changed around, gave Portuguese a central role in the development of what had previously been called *langage corrompu*. It was French writers who were principally responsible for propagating links on the west coast of Africa between *langue franque* and *langue créole portugaise*. However, what was fascinating about the 'Portuguese pidgin' theory was the way in which the matter of diversity of language, common language and language learning among the indigenous inhabitants, which started out in the words of Columbus and Vespucci, was repeated in various accounts, was transformed and given substance in the account of De Rochefort, was affected by a definition of Portuguese creole, where creole was used in an inadvertently revolutionary way to refer to language varieties in West Africa, and was recaptured with modifications and additions in a twentieth-century theory explaining the languages of the descendants of Africans in the Caribbean.

The fact is that claims were made from the earliest times by those Europeans who followed the Iberians into the New World that Portuguese and Spanish were the European languages that the indigenous inhabitants used in their *langage corrompu*. In a sense, it was reasonable for European writers to assume that contact experience with the Portuguese and Spanish had left the indigenous inhabitants with a number of words and a sufficient knowledge of European languages to make themselves understood, as well as to name those objects and people they had no previous knowledge of. Even in cases where objects and people were already known, it seemed reasonable to assume that the normal borrowing that results from contact, a kind of borrowing that is documented for languages in all parts of the world, was responsible for European words displacing native ones in the speech of the indigenous inhabitants. It was therefore a logical step, by analogy, to propose a general theory that gave the Portuguese a central role in the development of language among African slaves in their transfer from the West African coast to the New World. The labelling that took place fell squarely within the tradition of projecting concepts or notions from European literature and lore on to the Americas.

There were very few, if any, inconvertible facts of language that were put forward to support the 'Portuguese pidgin' theory; it was simply a case of circumstantial evidence at best and confusion brought about through literature at worst. Though the theory has now declined, two of the words adduced to support it (*pickaninny* and *creole*) have still remained largely uncontested, and it so happens that these words not only now further illustrate confused theorising, but, more importantly, provide evidence of substantive and linguistic interrelationships between the three racial groups (Europeans, indigenous inhabitants and Africans) in the islands. Curiously enough, both words were first used by European writers at about the same time, that is at the end of the third quarter of the sixteenth century.

The word *pickaninny*, in that form, was first used by Ligon in 1657 in his history of Barbados. It is a word that actually has at least three threads of history and the one work in which two of them, quite innocently, occur is Herlein (1718). It is in this work that the two threads, which are linked to two different languages, two different peoples and authors writing in two different contexts, occur without connection. The thread that occurred first historically in the literature began with the mention of a word in a French account of South American exploration: '*Les Sauvages l'apellent **Picanene**.*' ['The Savages call it "picanene"'] (Thevet 1575: 976b). Thevet was referring to the smaller of two kinds of flying fish. Thevet clearly thought that the word *picanene* was from the native language of the indigenous inhabitants, in the same way that he thought that the more general word for flying fish was from their language, when he said at the beginning of his long description of the fish: '*vous pouvez voir une infinité de poissons volans ... Les Barbares Sauvages les appellent **Bulanpech**, du nom d'un oyseau qui luy ressemble en sa longueur ...*' [' you can see an infinite number of flying fish ... The barbarous Savages call them *bulanpech*, from the name of a bird which resembles it in length ...'] (1575: 976a). The fact that Thevet did not realise that *bulanpech* was really *bulan* (*volans*) *pech* (*pescado/pêche*) = flying + fish and was from the so-called *langage corrompu* of the indigenous inhabitants meant that it was not a recent adoption. Equally, Thevet did not seem to realise that the words *picanene* was from the so-called *langage corrompu*, that it was actually two words *pica nene* and represented exactly what he was saying 'small/baby fish' – *pica* < *pescado* [fish] + *nene* [baby].

What emerges from Thevet's substantial description of the flying fish is that they were preyed upon by bigger fish and so had to leap out of the water to escape from their pursuers. The flying fish and the image it conjured up were very appealing to subsequent French priests–writers and were repeated and expanded in their works (Claude d'Abbeville 1614: 30–3; Maurile de S. Michel 1652: 23, 316). The fish was seen as jumping out of the water to avoid its bigger pursuers and into the air to be snatched by birds; Maurile de S. Michel linked it directly to Adam:

**Figure 3.3**  The predicament of the flying fish. After page 774, volume 2,
P.E.H. Hair, Adam Jones and Robin Law. (eds.) (1992. ) *Barbot on
Guinea: The Writings of Jean Barbot on West Africa 1678–1712.*
Volumes I and II. London: The Hakluyt Society.

'*ces pauvres poissons volans nous representent l'homme depuis le premier peché, lequel a des ennemis par tout, qui ne taschent qu'à le sur-
prendre*' ['these poor flying fish represent for us Man after the original
sin – he has enemies everywhere who are always trying to surprise
him.'] (1652: 23). The image of innocence and vulnerability became
quite closely associated with the fish. Since the fish itself, as Thevet
(1575: 976b) said, was a favourite meal of the indigenous inhabitants,
the word *picanene* became a fairly prominent word.

In his vocabulary list for the Island-Caribs, De Rochefort (1658:
519) cited the word as *pikenine* with the meaning '*chétif*', which originally meant 'captive'. De Rochefort identified the word as one of those
that the indigenous inhabitants had learned from the Spanish and
which they used in their *langage corrompu*. However, the meaning he
gave it almost exclusively highlighted the literary image of the vulnerability of the flying fish.

The second thread of the word emerged in English writing, the first
citation of the word in an English text being that by Ligon talking
about a (slave) mother in Barbados: 'but when the child is born (which
she calls her Pickaninny) ...' (1657: 48). Pickaninny here had the same
meaning as the Spanish/Portuguese word *nene*, and in the context the

mother was using a term of endearment to refer to her baby rather than as a plain descriptive word. Interestingly enough, the word *picanene* was quite similar to an English dialectal word (now obsolete), which the *Oxford English Dictionary* describes as follows:

> **Pinkeny, pinkany** ... [orig. child's language, fondly imitated by nurses, and so became an expression of endearment] ...
>
> 2. transf. Applied to a person, usually as a term of endearment: Darling, pet ...

The *Oxford English Dictionary*'s illustrative examples of the word start in the sixteenth century, that is before Ligon's use of the word. Ligon himself did not give any indication of the history or source of *pickaninny*, and he has, to some extent, to be blamed/credited for the subsequent development of the word, for in saying that *pickaninny* was the word the mother used to refer to her child, he, probably unintentionally, created the impression that it was a new and specialised word. Furthermore, Ligon's use of the word five times in the same paragraph really made it stand out in that context and caused it to be remembered in that way by subsequent writers. For example, Godwyn (1680: 84), only 30 years later, without comment equated *pickaninnies* with *young negro children*.

Although De Rochefort's citation of *pikenine* (1658) was the year following Ligon's *pickaninny* (1657), De Rochefort seemed to be unaware of Ligon's word. De Rochefort was merely repeating a word that had a history emerging out of contact between the Spanish and the indigenous inhabitants, and, moreover, his words were all being cited from preceding works. It is just coincidental, then, that the geographical context of the early use of both words seemed to have a common factor – Barbados. The flying fish, which was once very widespread, has now come to be associated with the waters around Barbados, which was the same island Ligon was describing. There is nothing else in the literature of the time to associate Barbados with the development of *pickaninny*.

It was probably Sir Hans Sloane who brought these two threads (the one in French and the other in English) together as a result of his own personal experience. Sloane was a well-read and educated man, who had been to France to do his medical training. He was familiar with De Rochefort's work and it appears as if it is he who first linked the word in De Rochefort's *langage corrompu* and the word in Ligon. The English dialectal word *pinkeny* was most likely unfamiliar to Sloane, who was born and raised in Ireland where his father had been put in charge of a Scottish colony. Sloane's (1707) work was based on a 15-month experience mainly in Jamaica in 1687 to 1688, and in it he gave the version *piganinny* in two different contexts with a Spanish source:

> Their children call'd piganinnies or rather pequenos ninnos, go naked until they are fit to cleans the paths (p. lii)
> They have Saturdays in the afternoon, and Sundays, with Christmas holidays, Easter call'd little or pigganinny, Christmas ... (p. lii)

It is quite likely that Sloane's persuasion towards Spanish *pequeño* as the source word for *piganinny* had been occasioned by the critical presence of formerly Spanish Africans and Creoles in Jamaica. Direct connection between them and the English at that time is noted by Hickeringill ([1661] 1705: 23–4), who pointed out that the *Molottoes* and *Negroes* that the Spanish left in Jamaica were taken over by the English. Sloane's work was very influential and in particular his use of *piganinny* with its presumed Spanish source became just as scientific and objective as the plants he described.

Neither thread of the word was strongly evident in Barbados or in the French territories after their early mention. In Surinam the word survived with a meaning not restricted to 'child' and this was because of English dialectal and Portuguese influence. In Herlein (1718: 121) the word that occurred in the illustration of *Neger-Engels* was *pinkinine*, clearly a term of endearment, said by a young man to a young lady whom he was trying to seduce. Herlein made no connection between this word and *pikenine*, which is included at the end of the work in the word list of indigenous words taken directly from De Rochefort, with the same meaning given by De Rochefort and the same comment that it belonged to the 'bastard talk' of the indigenous inhabitants.

The Spanish origin of the word, linked by De Rochefort to the *langage corrompu* of the indigenous inhabitants and apparently linked by Sloane to Spanish presence in Jamaica, remained uncontested up until a little after the middle of the twentieth century when linguists identified its origin as Portuguese in a bold attempt to substantiate the 'Portuguese pidgin' theory of origin for creole languages – see, for example, Cassidy and LePage (1980: 348). This Portuguese origin has remained unchallenged in spite of the fact that the 'Portuguese pidgin' theory has declined into virtual oblivion. In the face of the literary evidence of two threads of development, the Portuguese origin must now be seen as erroneous.

The third thread of the word *pickaninny* is a word given by Boyer in his 1654 account of the Galibi of northeastern South America. Bear in mind that the Island-Caribs in the lesser Antilles are said to be descendants of these continental Galibi. In his 'dictionary' of the

Galibi language Boyer (1654: 413) lists the word *pitani* with the meaning *jeune enfant* [young child]. It has to be admitted immediately that Boyer's *pitani* could have been a version of what Thevet recorded as *picanene*, because in Boyer's list of 'Galibi' words there are a number that are Portuguese/Spanish in origin (e.g. *caza* [house], *camisa* [shirt], *sombraire* [hat], *cavaye* [horse], *canou* [cane], *siccarou* [sugar], *chamboura* [tambour]). One has to assume that these were words in what the Europeans referred to as the *bastarda lengua* used by the indigenous inhabitants to communicate with Europeans. If *pitani* was a genuine Galibi word and not a borrowed one, then it could have been a source of *pickaninny* or it could have provided some reinforcement for the other two threads of *pickaninny* as it developed. Evidence suggests that it was a genuine word, for Taylor (1977: 145) lists the word *pita:ni (-ko?)* as a Karina/Carib word with the meaning 'children'.[18]

The other word (*creole*) whose etymology was adduced to support the 'Portuguese pidgin' theory did so in a pivotal way. In essence, an argument leading to the belief that the word *creole* was Portuguese in origin led those who believed so to claim that creole languages were Portuguese in origin. This constituted a break with history, for it seems as if up to the beginning of the twentieth century the word *criollo* was assumed to have come directly out of Spanish contact in the New World. However, it was the lack of definitive evidence about the origin of the word that allowed twentieth-century lexicographers to speculate about and re-create other possible origins for it. When José Arrom reproduced his 1951 article 'Criollo: Definition y matices de un concepto' ['Creole: Definition and variations of a concept'] in 1971, he gave additional information about the history of the word in print and also said:

> pues el filólogo Leite de Vasconcellos, en su **Antroponimia portuguesa**, afirma categóricamente que '*crioulo vem de criadouro*, deformado en bôca de pretos', y Cornu establece, en su **Grammatik der Portugiesischen Sprache**, la siguiente evolución: **crioulo** < \***criooiro** < \***criaoiro** < **criadoiro**. (1971a: 14–15)

> [then the philologist Leite de Vasconcellos, in his work *Portuguese Anthroponymy*, categorically asserts that 'crioulo' comes from 'criadouro', deformed in the mouths of Blacks', and Cornu, in his *Grammar of the Portuguese Language*, establishes the following evolution: *crioulo* < *criooiro* < *criaoiro* < *criadoiro*.]

No historical, social or bibliographical evidence was put forward by these late nineteenth/early twentieth-century writers (Vasconcellos and

---

18. Taylor also credits Berend Hoff for supplying the word.

Cornu) to support a development that had taken place in the sixteenth century, though one has to assume that they were accepting the explanation of Garcilaso de la Vega , the Inca, that it was the Africans who were initially responsible for the development of the word. It is very unlikely that either Vasconcellos or Cornu knew anything about the earliest citation of the Spanish form of the word from as early as 1571. Furthermore, the suggested etymology by Cornu (with posited intermediate forms) presupposed the same kind of extended time span and type of evolution that characterised the evolution of modern European languages. The short time between 1492 and 1571 (at most) did not allow for this.

In the same year that Arrom's expanded article appeared, the influential book *Pidginization and creolization of languages* appeared, in which David DeCamp repeated the notion of a Portuguese origin for the word *creole* (Hymes 1971: 15). What evidence this explanation as a whole is based on is not clear. In addition, the evidence for a Portuguese word **crioulo** (in this definition and the ones given by Arrom 1971) as a predecessor of Spanish *criollo* has never been provided and there is no explanation of where and how the development from Portuguese to Spanish took place. Yet, the definition by DeCamp, together with the fact that a few articles in the same book attempted to substantiate a Portuguese pidgin origin for creole languages, gave the Portuguese etymology of *creole* status and acceptance.

The view that the Portuguese were responsible for the transmission of a special language from the Old World to the islands was a bold attempt to link odd historical facts about the Portuguese together to provide a coherent explanation for varieties of language spoken by New World Africans and their descendants. It is now clear not only that the Portuguese did not have a central role in propagating any lingua franca in the islands but also that a Portuguese origin for two of the words used to support this theory is unsupported by evidence. Island varieties of language regarded as 'bastard' and 'corrupt' came directly out of Spanish experiences and views, with projections from the Old World on to the New, and later through French and English modifications of these notions.

# Colonial society formation and corruption

The seventeenth century witnessed the start of a massive introduction of disparate peoples into the Caribbean islands after their initial population and depopulation by the Spaniards in the previous century. As a result, maritime activity and movement into and across the islands had a critical effect on social organisation and language. In the smaller islands, for example, the image of the Island-Carib was a person in a canoe sailing from island to island to plunder. The type that emerged as the most glamorous or notorious among Europeans operating in the Spanish Caribbean world was the buccaneer or hunter type, taking its very name from an indigenous word. Ironically, it was the need to defend against this type that led to the strongest characterisations of ethnic identity among the indigenous inhabitants ('Carib' and 'Taino') and 'national' identity among *criollos* in Cuba. As to the effect of maritime activity on language, for the French, the term *matelot* [sailor] was used to describe a communal relationship between unmarried, young Frenchmen in their fledgling societies. For African slaves, *batiment* [ship] and its equivalents came to represent an unforgettable bond between fellow voyagers across the Atlantic. Among some of the indigenous inhabitants *guatiao*, a symbolic exchange of names, was a bond of solidarity and friendship with visitors from elsewhere.

As with maritime activity, early colonisation was numerically domi-nated by men, which meant that male bonding and fleeting, visceral relationships with women became the pillars of men's view of sexuality. Part of the initial succour provided for European men in the *guatiao* 'exchange' was the companionship of indigenous women, for women were in short supply among the Spaniards and also later among the French. Women themselves seemed to have had little power, for general evidence suggests that they worked harder than men and that they were treated as subservient. Not only did women seem to be regarded as part of the spoils of victory but the acquisition of wives also seemed to be the reason for hunting parties (Rouse 1992: 22). Among the indigenous inhabitants, *caciques* were said to have several wives and men generally were said to have women and families in different islands. The contra-diction in this subject role of women, however, was that women brought together the disparate elements. Indigenous women became the wives and concubines of Spaniards[19] and also later of the French in the small islands. Whether formally sanctioned or not, unions between indigenous women and European men became commonplace and, to a lesser extent, those between African women and European men.

Not all the women involved in relationships with men were seen as innocent victims, for the view first expressed by Cieza de Leon in 1553 ([1553] 1945: 145) about women from one ethnic group in Ecuador

---

19. Sauer (1966: 199) presents figures from a 1514 census in Hispaniola showing that 54 of 392 (i.e. about 14 per cent) of Spanish vecinos had indigenous wives. However, Konetzke (1946: 235), citing the same census, concludes that only 64 of the 689 Spaniards (i.e. less than 10 per cent) had Amerindian wives.

(*'Las mujeres son ... no poco ardientes en lujuria, amigas de españoles'* ['The women are very much given to lechery and are girlfriends of the Spaniards']) was later repeated about local women elsewhere and their love of foreign men. There is no record that indigenous women generally exploited the power that they had to their own advantage, but the emergence of prostitution in Santo Domingo in the sixteenth century as a viable livelihood among slave women no doubt created a heightened awareness of the benefits to be derived from sexual exploitation, which in turn led to visions of moral decay in these societies. In addition, it is not clear whether the *matelot* relationship between young Frenchmen was simply for economic survival and male bonding or whether, from the fact that it seemed necessary to introduce women to regularize the society, it was seen as homosexual. In any case, whether it engendered prostitution or it involved homosexuality, gender imbalance was seen as a primary cause of corruption.

Indigenous inhabitants who escaped from the clutches of the Spaniards were the first to be identified as maroons and all who did thereafter were regarded in the same negative way, as if living outside European 'civilisation' was a return to savagery. The egocentric perspective of Europeans and of the kind of racial evaluation inherent in it was not unconscious. Morgan Godwyn, for example, expressed this relativity of sentiment across ethnic groups in the Caribbean:

> For it is well known, that the Negro's in their native country, and perhaps here also, if they durst speak their inward sentiments, do entertain as high thoughts of themselves and of their complexion, as our Europeans do; and at the same time holding the contrary in an equal disdain ... whereby the missionaries of the Roman Church (who to facilitate their conversion, do condescend to humour them in divers things) are said to represent our Blessed Saviour in the Negro's complexion; themselves also describing the evil spirit in ours. (1680: 21)

The notion of corruption that French priests propagated from the start of French colonisation of the islands was therefore part of an overall intention to make these colonial societies appendages to European society and, as such, to establish a vision of European civilisation as the norm and the model against which the islands were to be judged.

The notion of corruption of Europeans themselves was evident in the writings about the early French colonies in the Caribbean from the very start. This had not been a major worry or theme in Spanish accounts of early Spanish settlements because there was no heavy dependence on indentured white labour. In the eyes of French priests, corruption was a move away from what were regarded as the acceptable norms of European morality and behaviour, either by being attracted away by

171

'uncivilised' people or by indulging in 'uncivilised' practices. Ironically, it was factors and practices in colonial society that departed from those of normal European society which caused the corruption. It was the lack of European women that drove European men to seek pleasure in the arms of indigenous and African women; it was harsh servitude that drove them to join the indigenous society; and it was the love of freedom and adventure that led them to become buccaneers, thus indulging in a wild and rough life, which had no model in Europe. The 'corrupt', however, were by no means the majority in seventeenth-century *Caribbean societies* but, because difference, as it always is, was exciting to the imagination and basic in the construction of identity, they gave Caribbean societies their identifying features in the eyes of Europeans at home.

The notion of corruption of language went hand in hand with the notion of corruption of people. The notion of corruption of language seems at first like a direct and straightforward reaction to language heard and seen, but the *langage corrompu* that was identified for the early Antillean islands was much more than that. As was the case with the notion of *ladino* earlier and more generally the colonial culture of 'otherness', the notion of *langage corrompu* was developed to characterise the language of non-Europeans, non-whites or, in other words, people seen as different from and inferior to western Europeans. As such, it embodied and lumped together value judgements about the linguistic competence of different ethnic groups. The application of the designation *langage corrompu* reflected the western European's consciousness of contemporary dialectal variation, both social and regional, within western Europe and the consciousness of a need to establish a social, political and linguistic hierarchy in which European characteristics were incontestably dominant. There was also a need to create uniformity and clarity where diversity and ignorance existed, and this was done by using a single simple term (i.e. *langage corrompu*).

The *langage corrompu* of the native inhabitants of the islands presented by the French writers, Bouton, Du Tertre and De Rochefort, did not differ in its structure from the *langage corrompu* of the Africans presented by Pelleprat. The difference in attitude to the two, where one was regarded as a useful means of communicating while the other was regarded as a bastardisation of the language of Europeans, can be attributed to the fact that while the indigenous inhabitants were new to the Europeans, Africans had a long history of being regarded as comical and corrupt in Spanish literature, an attitude that was accepted and generalised by the French and English writers. Even so, French views of the contact language of the indigenous inhabitants and subsequently the Africans stressed social and linguistic hierarchy by underlining the notion of corruption and the notion of a European model or norm from which the contact language diverged. This so-called *langage corrompu* was in reality stereotypical modification or corruption of their own languages by Europeans and what this kind of writing did was to provide for its readers an uncomplicated, comforting view of non-Europeans in the island colonies of the New World through their most salient human characteristic, language.

Both France and England were Europe-centred countries with little experience of and appreciation for ethnic groups other than those in their own countries. Spain in the sixteenth century had been a country of independent-minded pueblos, which its ethnic diversity served to accentuate. The people of the Iberian peninsula, after the voyages of Columbus and others, became even more self-centred and the country with its face towards the west turned its back on Europe as its empire stretched further and further across the Americas. The independent-mindedness of the people of Spain, their loyalty to localities rather than to Spain as a whole and their long experience with non-white peoples allowed them to expand outwards and establish communities in the New World without any strong and abiding allegiance to a cultural centre in Spain. In this sense they differed from the French and the English who were to follow them across the Atlantic and whose ties to a metropolitan headquarters in the home country were paramount.

> By a kind of Magnetick Force England draws to it all that is good in the Plantations. It is the Center to which all things tend. Nothing but England can we relish or fancy: our Hearts are here, where ever our Bodies be. If we get a little Money, we remit it to England. They that are able, breed up their Children in England. When we are a little easy, we desire to live and spend what we have in England. And all that we can rap and rend is brought to England.

This was the English view, as expressed by Edward Littleton (1689: 34). When he said 'They that are able, breed up their Children in England', this was part of a desire to erase any creole characteristics from their creole children and make them English. It was no different with the French.

In the seventeenth century, except for Herrera's work at the very beginning of the century, there was hardly any detailed literate view of the islands other than those presented by French and English writers. Through a relatively new institution, the printing press, the French and the English created a world for themselves with their native lands respectively as the centre of and as if bigger than the rest of the world. It was done essentially by presenting their own characteristics and actions as normal or superior and those of others as a deviation from these. Even those of the Spanish and Portuguese were said to deviate. Yet, the very medium that was of paramount importance in the propagation of the French and English vision, that is written literature, was one that was dependent on previous historical literature. So, though they detested the Spanish, the French and the English perpetuated a view of colonial language and culture that substantially emerged out of peninsular and North African Spanish experience. The identification of all the different ethnic groups born in the New World with a single term was a projection into the New World of a Spanish way of seeing others: it was a likening and linking of *criollo* to *moro*. Shakespeare's Caliban was a Creole, born of a Moor from *Argel* who had been taken to a New World island. The notion of corruption, embodied and constantly repeated in *langage corrompu* and associated with *Sauvages*, 'Moors' and Maroons, was a characterisation that came out of North Africa; it was one that came to define Creoles in French and English colonies and was one upon which identity of creole society was substantially built.

Declarations about *la tierra* demonstrated inklings of national feeling in Cuba, but the word *criollo*, to which it was intimately tied, in a sense did little to consolidate them, because it was a non-specific term and was not intended to make any distinction between one island or set of islands and another in terms of place of birth, race or language. What the word reflected was that there were European settlements dotted about the Americas, in different parts of Caribbean islands and the main land masses, and both in these settlements and outside of them there were people who had directly resulted from them. The dominant perspective behind the term was European, one which saw the settlements as assemblages of various ethnic groups issuing either from the Old World or the New or a combination of the two. The dominant category used to label New World people was not geography, which carried with it the element of territoriality, but experience or behaviour.

# Creolisation, nativisation and enlightenment

## Changes in *criollo* identity in the Spanish islands in the eighteenth century

In the first half of the eighteenth century the western world did not experience any momentous events. The second half of the eighteenth century, however, witnessed revolutions that had economic, political and social consequences. Industrial evolution, the American War of Independence, the French Revolution and the beginnings of the Haitian Revolution were to a great extent the consequences of intellectual ferment, criticism and questioning that were already evident by the middle of the century in Europe in the satirical works of Swift, Voltaire and others. Extended prose had come into its own as a genre. Newspapers, political essays and even novels began to be aimed at the general public, more of whom now could read as a result of more general education. The actual, popularly spoken languages of European nations (i.e. Romance languages) had, with their speakers, assumed a position of importance, and had become more powerful than Greek and Latin because they were directed at a larger public and dealt with contemporary issues. The publication of Dr Johnson's dictionary of the English language in 1755 validated English as a language of scholarship, but this was no match, as a dictionary, for the Spanish so-called *Diccionario de autoridades* (1726–1739) [Dictionary of authorities] coming out of the French inspired *Real Academia Española* [Royal Spanish Academy]. Authoritative written literature had become an important tool in outreach, in empire building and in subversion of empire; it proved to be a powerful catalyst in the dialectic on enrichment through slavery.

Spain was no longer a leader in intellectual or any other kind of ferment, for throughout the eighteenth century this former world power, which had declined economically and politically, was ruled by Bourbon kings from France. As a consequence, there was a general attempt to gallicise the country and in literature French ideas ruled. There was no Spanish literary or other figure of note in the eighteenth century and so the Spanish island colonies in the Caribbean had little or no inspiration from the mother country. In Santo Domingo and Puerto Rico especially, there were no great signs of social change or of progress in agriculture or commerce during the eighteenth century. The following *décima* [ten line poem] from Luis Peguero, a Dominican, can be used as an apt indication of the state of life in those islands:

| | |
|---|---|
| *Puerto Rico lugar chico* | [Puerto Rico, a little place |
| *Cathedral grande, clerigos pocos,* | a big cathedral, few priests |
| *abundante de platanos, y cocos,* | an abundance of plantains and coconuts |
| *pocos cavalleros, ninguno Rico,* | few gentlemen, none rich |
| *Es de estima el Cangrejo y el borrico,* | precious are crabs and donkeys |
| *no tiene Monjas ni letrados,* | there are no nuns or learned people |
| *ni calles llanas, ni terrados,* | or flat roads or terraces |
| *no rueda coche, ni moneda,* | no coaches moving about and no money |
| *y el chocolate anda a la queda,* | and chocolate is under curfew |
| *pero sobro de carnes, frutas y pescados.* | but there is a glut of meats, fruits and fish.] |

([1762, 1763] 1975:53)

Among the people, there was what writers at the time referred to as 'indolence' and later writers regarded as *dolce far niente*, said to be typical of the Spanish colonial way of life. The decline of these islands was very apparent, so much so that the Abbé Raynal said of Puerto Rico: '... *cette isle est inconnue à la plupart des peuples*' ['this island is unknown by most people'] (1773, Bk. 12, p. 220), and about Santo Domingo he mentioned: '... *des habitations ... depuis très-longtemps entierement perdu de vue*' ['dwelling places ... totally lost from view long ago'] (1773, Bk. 12, p. 226). Not even in Cuba, which was moving towards a predominantly sugar economy, were there any major signs of progress. In short, there were few indications of an age of enlightenment in the Spanish islands.

Toward the last quarter of the sixteenth century, when the term *criollo* emerged in the written literature it referred to the New World offspring of Old World people. It was a term that had validity beyond identification of place of birth as the New World: it differentiated whites from whites, blacks from blacks, and whites and blacks from indigenous inhabitants. In effect, there was a need to identify whites and blacks born in the Americas from other whites and blacks because there were perceptible differences that made the former recognisable. Such perceptible differences were socio-cultural in a context where both the enslaved and the freed blacks were linked to the whites and where relationships with indigenous inhabitants varied from very close to distant.

Ironically, since *criollo* did not inherently specify racial and political identity, it suggested the absence of a specific nationality. It was a term that fitted a stage in political evolution as well as a specific historical period. When it first emerged in the sixteenth century, the Spanish settlements were economically and administratively linked to Spain, but they were geographically distant. Sixteenth-century *criollos*, that is persons *nacidos de Españoles en Indias* [born of Spaniards in the Indies], were within a disparate set of settlements of people who were co-existing with each other. Then, within a few decades and for almost 200 years thereafter the fierce fighting for territory among the Europeans denied the islands political stability and a coherent identity. Throughout the seventeenth and eighteeenth centuries therefore *criollo/creole* was a convenient term for those born 'over there' (Acosta's *alla*) of varying and variable nationalities. Their status as appendages to Europeans did not change as long as the Europeans fought among themselves.

Indeed, it was the fighting for colonial territory and the partisan literature that accompanied it that had accentuated differences in national identity among Europeans. The sixteenth-century Spanish chroniclers just as often referred to their own people by national identity as by religious identity, but when the French and English began to challenge the Spanish, the term *Christianos*, as a synonym for the Spanish, was no longer appropriate. French and English writers maligned the Spanish, the French maligned the English and the English maligned the French. As a result, political national identity began to dominate the terminology in the literature. The dependence of people born in the colonies, the temporary political affiliation of this dependence, and the great number of these dependencies favoured the maintenance of the non-nationalistic term *criollo/creole*. Eventually, however, distinctions between Creoles began to be more significant and these differences began to be reflected in the literature.

It was almost inevitable that the undifferentiated term *criollo* would be modified, replaced or supplemented by other terms as people changed from being perceived as children of settlers in the New World to third- and fourth-generation natives of different island countries, with differing racial and cultural characteristics. In fact, the very word *criollo*, because its beginnings were unclear, gave rise to varying claims to its invention as well as to contradictions in its meaning. For instance, the *Diccionario de la lengua Castellana* (1726–1739), the first dictionary of the Royal Academy of Spain, identified the word as Spanish in origin with the following definition and claim: '*El que nace en Indias de Padres Españóles, ù de otra Nación que no sean Indios. Es voz inventada de los Españóles Conquistadóres de las Indias y comunicada por ellos en España*' ['Anyone born in the Indies of Spanish parents, or of any nation other than Indians. It is a word invented by the Spanish conquistadors in the Indies and brought by them to Spain'] ([Real Academia Española] 1729, Vol. 2, p. 661). Thus, by the beginning of the eighteeenth century the word had become established and respectable enough to inspire the Spanish to claim it as a part of their colonial heritage. Though, apparently, the idea of a slave origin for such a literary and widely used word as *criollo* had become officially unacceptable, there is no literary evidence to show that this 'official' claim was generally accepted either.

Implicit in the Spanish claim was the idea that it applied initially to (Spanish) whites, and this is the meaning given to it by the Frenchman, Pierre Ledru, who visited Puerto Rico in 1797. Ledru gave as one of his *'quatre classes d' habitants, bien distinctes'* ['four very distinct classes of inhabitants'] in Puerto Rico *'Les créoles ou blancs nés en Amérique'* 'Creoles or whites born in America'] (p. 161), a classification that retained the wording and narrow definition of Velasco 1571–4. These *créoles* he contrasted with *blancs venus d'Europe, mulâtres* and *nègres* [whites coming from Europe, mulattos and negroes]. On the other hand, Abbad y Lasierra in 1788 said that the majority of the population of Puerto Rico were mulattos (*hijos de blanco y negra* [the children of a white man and black woman]) and that the name *criollos* applied to everybody born in the island *'de cualquiera casta o mezcla de que provengan'* ['whatever class or mixture they come from'] (p. 181). It is quite evident then that even in the last quarter of the eighteeenth century the term had both a broad meaning ('all those born in the country') and a narrow one ('whites born in the country').

Changes in classification were reflected not only by the split in the meaning of *criollo* but also by the emergence of alternative terms. One of the first alternatives to appear in the literature was *gibaro*, a term given in his comprehensive history of the world by Padre Murillo Velarde:

> *y aora no hay Indios en toda la Isla* [Española], *sino Españoles, Mestizos, Negros, y Mulatos ... Los Criollos, y Mestizos de la Española, de Puerto Rico, y otras Islas se llaman Gibaros, y es gente de animo, y brio.* (1752, Bk. ix, Chap. xxi, p. 345)

> [and now there are no Indians anywhere in the island, only Spaniards, mestizos, negroes and mulattos ... The Creoles and mestizos of Hispaniola and Puerto Rico and other islands are called gibaros, and they are a spirited and lively people.]

The term *gibaros* thus provided a common identity for Hispaniola and Puerto Rico and *otras islas*, probably meaning smaller islands like St Thomas and St Croix. The essence of what Murillo Velarde was saying was that the original, 'pure' population had disappeared to be replaced by people whose parents had come from the Old World as well as mixtures of these latter with the original indigenous population. According to this definition, then, *gibaros* were Spanish Creoles, and the term included Negroes, Mulattos, *Mestizos* and Whites who were creole. This meant that Spanish Creoles were being separated from other Creoles and were portraying themselves as culturally and linguistically different.

The separateness of Spanish Creoles was also confirmed by the fact that other Europeans saw them as different from their own Creoles. The French view of Spanish *criollos* was not positive, because the French saw the New World as a place to be fully exploited, and *criollos* in the Spanish islands certainly were not doing that. In 1731 the French priest, Charlevoix, presented a view of *criollos* that was to be repeated in bits and pieces by several writers coming after:

> *Ils ne s'occupent à rien pendant tout le jour; ils n'employent pas même alors leurs Esclaves à aucun travail pénible. Ils passent tout le têms à jouer, ou à se faire bercer dans leurs branles, ou Hamacs. Quand ils sont las de dormir, ils chantent, & ne sortent de leurs lits, que quand la faim les presse. Pour aller chercher de l'eau à la Riviere, ou aux Fontaines, ils montent à Cheval, n'y eût-il que vingt pas à faire; il y a toujours un Cheval bridé au piquet pour cet usage ...*
>
> *Ils ne sçavent rien, & à peine connoissent-ils le nom d'Espagne, avec laquelle ils n'ont presque plus de Commerce. D'ailleurs, comme ils ont extrêmement mêlé leur sang, d'abord avec les Insulaires, ensuite avec les Negres, ils sont aujourd'hui de toutes couleurs, selon qu'ils tiennent plus de l'Européan, de l'Afriquain, ou de l'Ameriquain. Le caractere de leur esprit participe aussi de tous les trois, & ils en ont surtout contracté la plupart des vices. Ils ne laissent pourtant pas de se croire encore les premiers Hommes du monde, & de témoigner un très-grand mépris des François. Quelqu'un demandant un jour à un Espagnol, ce qu'il y avoit donc de si estimable chés eux, pour mépriser ainsi leurs voisins, ay* **Hombres,** *répondit-il.* (1731 2: 479)

[They do nothing for the whole day. Even their slaves they do not use to do any hard work. They spend all their time playing or rocking in swings or hammocks. When they are tired of sleeping, they sing and they do not get out of their beds until hunger forces them to. To fetch water from the river or fountains, they ride, although it is only twenty feet away; there is always a horse tethered for this purpose ... They know nothing and are scarcely familiar with the name 'Spain', with whom they almost no longer transact any business. Moreover, since their blood is extremely mixed, first with the native inhabitants and then with the negroes, today they are of all colours, according to the amount of European, African or American they have in them. The character of their spirit also exhibits all three and they have above all inherited most of their vices. However, this does not stop them from believing themselves to be the most important people in the world and from showing very great disdain for the French. When someone one day asked a Spaniard what was it that they had that was so great that made them have contempt for their neighbours, he replied: 'there are men among us'.]

In 1756 Saintard continued in the same negative vein, saying:

*Le génie noble des Castillans n'a point passé les mers; presque toute la colonie Espagnole est formée par des Mulâtres & des Metifs, que l'oisiveté abâtardit encore, gens qui ne cultivent point, & par-là propres à détruire les cultures.* (p. 229)

[The noble spirit of the Castilians did not traverse the sea; almost the whole of the Spanish colony is made up of mulattos and half-breeds, whom laziness still causes to degenerate, people who do no manner of cultivation and accordingly are ready to destroy it.]

This eighteenth-century lifestyle in the Spanish islands was certainly not French in spirit. The fact is that for *criollos* the islands were their home and life was something to enjoy rather than to spend working for others, as was the case in the French islands.

The English view of Spanish *gibaros* was similar to that of the French – quite the opposite of *gente de animo, y brio* or *hombres*. It was the Englishman, Edward Long, who said: 'The Giveros lie under the imputation of having the worst inclinations and principles; and if the cast is known, they are banished' (1774 2: 261). Long's conception of *giveros* was most likely based on the Spanish 'guerrilla' Negroes who fought against the English after their takeover of Jamaica in 1655. In the table of classification of people given by Long, under the broad classifications, *givero* (equivalent to Murillo Velarde's *gibaro*) was listed under *retrograde* types, as opposed to those that were *ascending* [towards white] or *stationary*. Long's comment reflected a view of *givero* as a type, as well as the general negative opinion the English and the French had of *criollos* in the Spanish islands.

In spite of the fact that repeated racial intermixing was a general colonial practice, English opinions about and attitudes towards the racial composition of the people in the Spanish islands were becoming more extreme and strident. The terminology that was adopted to define its products was credited to the Spanish by Long (1774 2: 260) not only because the Spanish were in the region first but also because racial mixture was seen by the English as a Spanish and Portuguese phenomenon, pre-dating the encounters in the New World. One Englishman was quoted by Sanchez Valverde (1785: 146) as saying that there were hardly any 'pure' whites in the Spanish colonies, so mixed were they with Caribs and Negroes and so mixed were they in any case with *los antiguos Moros* [the Moors of old]. Long was even more vehement in the expression of his opinion: 'Let any man turn his eyes to the Spanish American dominions, and behold what a vicious, brutal and degenerate breed of mongrels has been there produced, between Spaniards, Blacks, Indians, and their mixed progeny ...' ([1774] 1970 2: 327). Moreau de Saint-Méry tried to be more matter-of-fact in the expression of his view: *'la majeure partie des colons espagnols sont des sangs-mêlés, que plus d'un trait africain trahit quelquefois ...'* ['the majority of the Spanish colonists are mixed breeds whom more than one African trait sometimes betrays ...'] (1796 1: 59). Even if it was because his image of Cubans was coloured by their hostility towards the English after the English captured Havana in 1762, Bossu's views maintained the same line of argument: *'la vile populace, qui n'est ici composée que de races de mulâtres, métis, quarterons, jambos, pêtris de tous les vices des différentes Nations dont ils descendent'* ['the low class populace which is made up of nothing but mulattos, mestizos, quadroons and jambos, shaped by all the vices of the different nations from whom they descend.'] (1777: 344). The same negative attitude towards the mixed nature of Iberians and the belief in the degenerate result of mixing were also noted in his diary by Thomas Howard during his military expedition in Hispaniola:

> I prefer the Negroes, to the Mulattoes; the latter mostly partaking of all the vices of white, & Black, from whom they spring, without inheriting their virtues; they stand much the same, in the scale of people of colour, as the Portugaise do amongst other nations of Europe or more particularly the Spaniards, & Moors ... (1796–8, Vol. 3)

The views of English and French writers were of course not related to the intentions of their crown to capture Spanish colonies and rob the Spanish of their wealth, which meant that they saw the Spanish not just as enemies but unworthy of their potentially rich possessions, which they were allowing to go to waste. Racial mixing was one way of accounting for and justifying Spanish unworthiness. What was explicitly said by Howard, Bossu and Long was implicit in such views – racial mixture constituted degeneration in that racially mixed persons inherited the negative characteristics of their parents. Spanish *criollos*, as part of the Spanish empire, were therefore degenerate human beings. To a great extent, then, French and English attitudes to Spanish *criollos* mirrored their attitudes to the Spanish in Europe.

In the minds of the French and the English, repeated intermixing in the Spanish colonies, by breaking down the sharp division between

Black, Indian and White, had made possible a gradient system of classification that degraded whiteness by allowing for an expansion of the category of white itself. Long (1774 2: 260), on the Spanish table that he gave, identified the product of White man + Quinteron as white. Then, about 20 years later, in a table identifying the products of the different mixture of races in Puerto Rico, the Frenchman Ledru [1810] 1863: 165 said that the classification white included the child of a white man and a *tercerona*.

To further underscore the negative elements in Spanish *criollos*, French and English writers highlighted their relationship with the indigenous inhabitants, even though the prevailing opinion at the time was that the indigenous inhabitants had disappeared. *Givero* was the only term identified as a Black + Indian + White mixture by Long in his table of classification: he said specifically that it was the product of *Sambo de Indian* and *Sambo de Mulatta*, which meant that it was 62.5% Black, 25% Indian and 12.5% White.[1] Long's use of the Spanish word *de* (as well as one of his own comments) indicates that these terms (*Sambo de Indian, Sambo de Mulatta*) were borrowings from the Spanish rather than common terms in use in Jamaica at the time.

Writing about the people of Santo Domingo early in the century, Raynal highlighted a tri-partite genetic composition, which included an indigenous element:

> *Leur couleur & leur caractere tenoient plus ou moins de l'Amériquain, de l'Européen & de l'Afriquain ... Ces demi-sauvages plongés dans une fainéantise profonde, vivoient de fruits & de racines, habitoient des cabanes, étoient sans meubles, & la plupart sans vêtemens.* (1773, Bk. 12, p. 225)

> [Their colour and character had something more or less of the American, the European and the African ... These half savages steeped in profound idleness, lived off fruits and roots, lived in cabins without furniture and the majority of them wore no clothes.]

Raynal's picture of half-savages living off fruits and roots consciously or unconsciously recreated Maurile de S. Michel's 1652 description of Frenchmen who became *Sauvages* in St Kitts.

In contrast to these opinions, those of two creole writers discounted the indigenous element. Sanchez Valverde, a native of Santo Domingo,

---

1. Blanchard (1910: 55), referring first (p. 44) to illustrations of *castas* on the large eighteenth-century canvas in the Museum of Mexico, gives *gibaro* as 67.19% White, 12.5% Indian and 20.31% Black. Second (p. 48), referring to J.J. Virey's (1824) classification (which was probably adopted from Long 1774), he gives *gmoiddleivero* as 12.5% White, 25% Indian and 62.5% Black. Third (p. 55), referring to R.E. Cicero's (1895) version of mixtures in Mexico, he gives *gibaro* as 48.44% White, 12.5% Indian and 39.06% Black.

did not give the impression either that there were indigenous inhabitants or that there were mixtures of them in the general population. What he said about one specific area, Monte de Plata, was:

> *han procurado conducirse à aquella parte, despues de la extinction de los Indigenas, algunos otros pobres, que han venido de la Tierra firme con diferentes motivos, que tambien se ha acabado, dexando solo unos veinte cinco, ò treinta **Mestizos**, que gozan los fueros, y privilegios de **Indios**.* (1785: 122)

[After the extinction of the indigenous inhabitants, a few other unfortunate ones managed to reach there; they had come from Terra firma with different motives and they also had disappeared, leaving only twenty five or thirty mestizos to enjoy the rights and privileges of Indians.]

A little later, Moreau de Saint Méry said:

> *je veux parler de certains créols, en très petit nombre, qui ont des cheveux semblables à ceux des Indiens, c'est-à-dire, longs, plats & très-noirs, & qui prétendent être issus des premiers naturels de l'île. Ils attachent une grande importance à cette descendance, néanmoins démentie par tous les faits historiques qui constatent tous que cette race d'hommes a été exterminée. Tout ce qu'on peut leur accorder, c'est qu'ils descendent après un mélange avec la race espagnole ...* (1796 1: 61)

[I want to speak of certain Creoles, very small in number, who have hair similar to that of the Indians, that is to say, long, straight and very black, and who claim to be descendants of the first natives of the island. They attach a great deal of importance to this ancestry which is disproved by all historical facts which all show that this race of people was exterminated. What they can be granted is that they descend from a mixture with the Spanish race.]

These statements by natives of the island lent support to the view that the indigenous inhabitants had virtually disappeared without trace. However, both views persisted without resolution – the one that *criollos/gibaros* were partly indigenous genetically and otherwise, and the other that they were not. To complicate the picture further, when Sanchez Valverde used the term *Indo-Hispanos* to refer to those whom López de Velasco first and others following called *criollos*, that is *hijos de españoles en Indias*, he was rejecting the idea of impurity: he meant Spanish whites born in the New World. Valverde's intention was to establish a line of continuity between the Spanish of Spain and the Spanish of the New World, with the concession that the latter were New World variants. However, even though the term *Indo-Hispanos* did not become popular, it facilitated the (mis)interpretation that *Criollos* were a combination of indigenous inhabitants and Spanish.

The idea that the Spanish *criollo* was in part Indian gained strength not only as a result of the French and English continued challenge to the Spanish empire but also as part of the more general argument of ecological determinism, which sought to make a sharper distinction between Creoles and Europeans. The idea of natural modification became even more popular in the eighteenth century as an explanation of the distinctiveness of the *criollo*: it was seen to affect both the genetic make-up of *criollos* and their attitudes and behaviour. Writing about Puerto Rico, Abbad y Lasierra suggested a link between the influence and legacy of the indigenous inhabitants and the effect of natural conditioning by climate:

> Así como los habitantes de Puerto Rico han adquirido de los antiguos moradores de esta Isla la indolencia, frugalidad, desinterés, hospitalidad y otras circunstancias características de los indios, han conservado igualmente muchos de sus usos y costumbres. La construcción e idea de sus casas, su establecimiento y morada en los bosques, la vida sedentaria, la afición a las bebidas fuertes y espirituosas, la propensión a los bailes y otras inclinaciones, son comunes y propias a estos dos pueblos, bien sean contraídas por el trato y unión mutua o por efectos propios del clima o consecuencias naturales de ambas causas. ([1788] 1966: 185)

> [In the same way that the inhabitants of Puerto Rico have acquired from the former natives of the island the indolence, frugality, unselfishness, hospitality and other circumstances characteristic of the Indians, they have also maintained much of their ways and customs. The construction and conception of their houses, their establishment and dwelling in woods, the sedentary life, the love of strong and spirituous liquors, their propensity for dancing and other inclinations are common and characteristic of these two peoples, they may very well have been condensed by intercourse and mutual union or by effects typical of the climate or natural consequences of both.]

However, even though Abbad y Lasierra underlined the legacy of the indigenous inhabitants of Puerto Rico, unlike Raynal in reference to Santo Domingo, he did not identify any indigenous element in the racial make-up of the people. In fact, what he said was that '*Los mulatos, de que se compone la mayor parte de la población de esta Isla, son los hijos de blanco y negra*' ['The mulattos, who make up the majority of the population of this island, are the children of white men and black women'] (Abbad y Lasierra [1788] 1966: 182). For him, then, there was no genetic inheritance of Indian characteristics, but a suggestion that the climate could change one genetically.

Ledru, who was in Puerto Rico in 1797, repeated part of Abbad y Lasierra's description almost word for word:

*Les habitants de Porto-Ricco ont emprunté des anciens indigènes la frugalité, le désintéressement, l'hospitalité et plusieurs autres vertus qui les distinguent: ils ont aussi conservé le goût pour les liqueurs fortes, et la vie sédentaire de ces Indiens.* (1810: 164)

[The inhabitants of Puerto Rico have acquired from the former natives frugality, unselfishness, hospitality and several other virtues which are distinguishing characteristics. They have also maintained the love of strong drink and the sedentary life of these Indians.]

Ledru, for his part, was vague about the composition of the mixture, preferring to focus on the lack of racial purity among the whites:

*Les blancs purs, sans aucun mélange de sang étranger, sont fort rares: Raynal évalue leur nombre à 28,887, mais il n'en est pas la moitié: les races sont tellement croisées, que l'on ne rencontre, le plus souvent, que des visages basanés.* (1810: 163)

[Pure whites, without any admixture of foreign blood, are rare: Raynal estimated their number to be 28,887, but it is probably less than half of that: the races are so crossed that most of the time one meets only swarthy faces.]

The view of the populations in Santo Domingo and Puerto Rico that emerged by the end of the eighteenth century was therefore that they were without doubt in their culture partly indigenous and in their genetic make-up partly indigenous for some and mulatto for others. It is thus the general notion of the effect of the climate and more specifically the conviction about the indigenous element in the culture and possibly racial make-up of the people that led to the spread of the term *gibaro* as an alternative to *criollo* and as a substitute for *mulato*.

The other determinant of Creoles generally continued to be highlighted in the literature with specific reference to Spanish *criollos*. Slave labour was said to have a negative effect on those who used it; it was no different in the case of the Spanish islands, where in fact *criollos* were seen as the epitome of laziness. In the case of the Spanish islands, even before the African slaves, there had been dependence on the indigenous inhabitants. This policy of dependence on others had disastrous results in Santo Domingo, according to Bossu:

*Cette politique* [dependence on the indigenous inhabitants] *a été peu avantageuse à l'Espagne; la colonie de Saint-Domingue lui a toujours été plus à charge qu'à profit; ses habitans sont, à la vérité, d'une paresse extrême; ils ne retirent aucun fruit du plus beau pays du monde. Leur industrie se borne à élever du bétail qu'ils vendent aux François, & à cultiver quelque peu de vivres, pour le soutien d'une vie singulierement frugale.* (1777: 358)

[This policy has been of little use to Spain; the colony of Santo Domingo has always been more a burden than a profit; its inhabitants are truly extremely lazy; they did not reap any benefits from the most beautiful country in the world. Their industry is limited to raising cattle which they sell to the French and to cultivating a few provisions to sustain a singularly frugal life.]

With the replacement of the indigenous inhabitants by Africans, the argument became louder and harsher. Raynal, for instance, said in reference to Puerto Rico:

*Ils ont environ trois mille negres, plus occupés à nourrir l'indolence du proprietaire qu'à seconder son industrie. Les maîtres et les esclaves rapprochés par la paresse vivent également de mays, de patates & de cassave* ... (1773, Bk. 12, p. 220)

[They have about three thousand negroes, more taken up with nourishing the indolence of the proprietor than in supporting his industry. Masters and slaves, bound by laziness, live alike on maize, potatoes and cassava ...]

A fuller comment was given by Ledru, who, in reference to Puerto Rico, repeated some of the words of his predecessor, Bossu:

*La plupart de ces colons sont, en géneral, d'une paresse et d'une insouciance inconcevables ...*
   *Dans chaque ménage, les gros travaux sont livrés aux esclaves du dehors; les menus détails de la maison à ceux du dedans; en un mot, les maîtres ne font strictement que ce qu'il leur est impossible d'exécuter par d'autres. Couchés dans leurs hamacs, ils s'y bercent une partie du jour, occupés à réciter le rosaire ou à fumer ... Leurs enfants, élevés loin des villes, sans éducation, et vivant avec les jeunes nègres de l'un et de l'autre sexe, dans la plus grande familiarité, contractent trop souvent des habitudes corrompues, et deviennent cruels envers leurs esclaves.* (Ledru 1810: 169)

[The majority of the colonists are generally incredibly lazy and carefree ... In every household the major tasks are given to the outdoor slaves; the small house tasks to the house slaves; in short, masters, strictly speaking, only do what it is impossible for others to do for them. Lying in their hammocks, they rock themselves for a part of the day, taken up with reciting the rosary or smoking ... Their children, raised far from towns, with no education and living with young negroes of both sexes in the most intimate way, too often learn corrupt habits and become cruel towards their slaves.]

Ledru pursued his point, showing the further negative effects on the mentality of whites:

*D'autres obstacles ont retardés jusqu'à ce jour l'amélioration de la colonie, malgré les encouragements qu'elle a reçus du gouvernement.*

*1. Un absurde préjugé flétrit le travail des mains; l'agriculture, le premier, le plus honorable de tous les arts, est en partie abandonnée aux esclaves, comme une occupation vile et déshonorante ...*

*Si un blanc se faisait servir par un blanc, l'un et l'autre seraient déshonorés dans l'opinion publique. L'expression la plus insultante que l'orgueil puisse employer contre un créole, est celle-ci: 'Il a des parents à la cote'.* (1810: 175)

[Other obstacles have retarded the improvement of the colony up to today, in spite of the encouragements it has received from the government.

1. An absurd prejudice against manual labour is a blight; agriculture, the first, the most honourable of all the arts, is in part given over to the slaves, like a vile and dishonourable occupation ...

If a white has himself served by a white, both would be dishonoured by public opinion. The most insulting expression that pride can use against a Creole is this: 'He has relatives on the side.']

To complement his pathetic picture of the *criollo*, Ledru made observations on the poor state of education, another consequence of dependence on slavery:

*Vous chercheriez en vain des manufactures ou des colléges ... Le peuple croupit dans l'ignorance: les moines seuls, et quelques femmes, enseignent à un petit nombre d'enfants les éléments de la religion et ceux de la grammaire: les sept dixièmes des habitantes ne savent pas lire.* (1810: 92)

[You would search in vain for factories or colleges ... the people wallow in ignorance: monks alone and a few women teach a small number of children the rudiments of religion and grammar. Seven tenths of the inhabitants are illiterate.]

In summary, then, racial mixture, climate and dependence on slaves were the powerful arguments of the day that were put forward and repeatedly used to explain *criollos* in the Spanish islands.

The overall characterisation of the people of Puerto Rico by Abbad y Lasierra brought together all these negative elements:

*De esta variedad y mezcla de gentes, resulta un carácter equívoco y difícil de explicar; pero a todos convienen algunas circunstancias que podemos considerar como características de los habitantes de Puerto Rico: el calor del clima los hace indolentes y desidiosos; la fertilidad del país que les facilita los medios de alimentarse, los hace desinteresados y hospitalarios con los forasteros; la soledad en que viven en sus casas de campo, los acostumbra al silencio y cavilación; la organización delicada de su cuerpo auxilia la viveza de su imaginación que los arrebata a los extremos; la misma delicadeza de órganos que los hace tímidos, los hace mirar con*

*desprecio todos los peligros y aún la misma muerte; las diferentes clases
que hay entre ellos infunde vanidad y orgullos en unos, abatimiento y
emulación en otros.* ([1788] 1966: 183–4)

[From this variety and mixture of peoples results a character which is
ambiguous and difficult to explain; but to all of them apply some
circumstances which we can consider as characteristic of the inhabitants
of Puerto Rico: the heat of the climate makes them indolent and listless;
the fertility of the country, which provides them the means of feeding
themselves, makes them unselfish and hospitable to strangers; the solitude
in which they live in their country houses makes them accustomed to
silence and suspicion; the sensitive organisation of their bodies helps the
liveliness of their imagination which pushes them to extremes; the same
sensitivity of organs which makes them timid makes them look at all
dangers and even death itself with scorn; the different classes among them
infuses vanity and pride in some and dejection and emulation in others.]

This really was not a characterisation of Puerto Rico alone: it was one
that presented a European view of tropical life and people, putting an
indelible stamp on the people in the Caribbean islands, especially the
Spanish ones.

In 1785 Sanchez Valverde responded to the charges made against his
countrymen, particularly those made by the French priest, Charlevoix,
in 1730–1 and those made by an Englishman he identified as Gascona-
da de Weuves. In citing the charge made by Weuves, Sanchez Valverde
said:

*La insolencia de Weuves, y de otros Estrangeros, no se ha contentado con
insultarnos sobre la actividad, y génio; sino que ha tenido la habilantez de
abrir nuestras venas, y manchar la sangre, tanto de los **Indo-Hispanos**,
como de sus Progenitores **Europeos**.* (1785: 145)

[The insolence of Weuves and other foreigners is not satisfied to insult us
about activity and spirit but have had the ability to open our veins and
stain our blood, not only that of the Spaniards born in the Indies but also
their European progenitors.]

Sanchez Valverde conceded that Santo Domingo was in a backward
condition but he strongly rejected the idea that Europeans, especially
the French on the other side of the island, were superior to *criollos*. He
argued the opposite:

*Hemos manifestado con pruebas convincentes, como fundados en hechos
sujetos à los sentidos, que la actividad personal de los **Franceses** en la
**América**, lejos de hacerlos superiores à los **Criollos**, que llaman, y suponen
poltrones, es muy inferior à la infatigable taréa, y sobriedad de estos, lo
qual se confirmará mejor quando hablemos de nuestros Pastores; y que*

*ellos son en efecto los verdaderos holaganes, sensuales, que hay en la **Isla**.*
(1785: 147)

[We have put forward convincing evidence, well founded on observable
facts, that the personal activity of the French in America, far from making
them superior to Creoles, whom they call and claim to be lazy, is inferior
to the tireless work and sobriety of the latter and this is even further
confirmed when we speak of our herdsmen; and that they are indeed the
lazy and sensual ones in the island.]

He drew a picture of the life of the *Montero* as a very tough one and of
the *Montero* as a hard-working person in contrast to the picture of an
indolent person needing a horse to travel 20 yards. Ironically, he explained the lack of development in Santo Domingo as the result of a
shortage of slaves to do the necessary work and also a result of the fact
that the few slaves that there were in Santo Domingo were *jornaleros*
[day workers] who did very little work. About them he said: '*El abuso
de tener Esclavos à jornal ... inutiliza una gran parte de los pocos, que
tenemos: porque ésta es una especie de **Negros**, que viven sin disciplina,
ni sujecion*' ['The abuse of having day labour slaves wastes a great part
of the few that we have, because this is a kind of negro that lives without discipline and control'] (1785: 150). Additionally, he identified
female slaves as villains because of their habit of seducing their masters
to get themselves out of slavery, thus reducing the slave population and
then continuing with prostitution as a way of life: '*despues de libres no
tienen otro oficio para subsistir, que el que les sirvió para sacudir la
esclavitud*' ['after becoming free they have no other way of making a
living than that which served them to undermine slavery] (1785: 153–
4). This association of the black women of Santo Domingo with prostitution by Valverde perpetuated a view that had been stated as early as
1542 by the archdeacon of Santo Domingo.

Sanchez Valverde's explanation did little to improve the image of
*criollos*, and apart from Sanchez Valverde there was no abundance of
literature from within these islands or written from a Spanish point of
view to counteract the widely disseminated view of French and English
writers. *Criollos* in the islands, then, who had suffered from the decline
of Spain and their loss of interest in the islands as well as from the
decline in their own populations, were now virtually defenceless in the
face of the onslaught by French and English writers. There seemed to
be no way for them to escape the image of indolence, which, constantly
repeated, they themselves began to believe.

# Changes in creole identity in the French and English islands

In the eighteenth century the economies of France and England were transformed through commerce, especially with the colonies. The considerable wealth that was created in turn transformed their social structure by producing a burgeoning middle class whose children had time and money to become interested in the kind of refinement offered by literature and the fine arts. Consequently, where in the previous century literature and the arts were the province of a small number of high-born, courtly persons, by the early eighteenth century they had become part of the normal aspirations of a more general, expanding middle class. Yet, while for the French and the English the eighteenth century was one of social change, for their island colonies in the New World it was one of structural consolidation. The French and English colonies had started from scratch less than 100 years before the beginning of the eighteenth century and the massive influx of Africans, who were subsequently to form the majority of the populations in most of the islands, had started less than 50 years before, in the last half of the seventeenth century. Consolidation against a backdrop of social ferment in Europe inevitably meant that the colonies, in their cultural and literary characteristics, reflected both the old and the new of Europe, as well as the products of their own dynamic creativity or creolisation.

In these colonies the process of consolidation was not uniform. For example, the plantation, as an economic unit, reached its peak in small Barbados in the seventeenth century, in bigger Jamaica in the mid-eighteenth century and Saint Domingue later in the same century. The dilemma that increasingly dominated the written literature appeared as a debate about the morality of slavery, in which those opposed to it were for the most part in Europe. By this time, Creoles had, by force of numbers, come to assume dominance in these island societies. They were conscious of their own identity and their worth to the mother country. They began to speak for themselves and even write to defend what they saw as their way of life against outside interference. As the white Creole was increasingly portrayed as decadent, ignorant and cruel, that same Creole more fervently sought after the embrace of Europe, which was promoting a picture of itself as the seat of civilisation. The dichotomy for white Creoles was sharpened by their practice of sending their children to Europe to be educated – a process that in most cases made the children ill-adapted to fit back into creole society by converting them into 'sophisticated' foreigners in their own homeland. This kind of dislocation and ambivalence eventually filtered down through all classes in the French and English colonies especially, and was epitomised in the concept Creoles had of their own native languages as well as their attitudes to them.

While economically the French and English colonies provided great wealth for their mother countries, culturally, specifically in the area of literature, they also provided them with themes, points of focus and subject matter. In the general curiosity about the New World, books on travel and discovery became popular and, because of European confrontation with indigenous inhabitants of the Americas and black Africans, the supposed contrasts between 'civilised' and 'uncivilised' man and between the 'cultured' and the 'uncultured' began to dominate European literature.

From the middle of the eighteenth century, English writers consistently referred to the English islands as the *West-India Islands* (Belgrove 1755, Hillary 1759) or the *West Indian Islands* (Singleton 1767) or the *West Indies* (Singleton 1777, *West India Merchant* 1778). As a result, Creoles from the English islands came to be known in England as *West Indians*. As is evident from the words of *Toby* in Defoe (1718: 304) ('There [Barbados] lives white Mans, white Womans, Negro Mans, Negro Womans, **just so as live here**[2] [England]'), there were many West Indians and their 'slaves' in England in the eighteenth century. It is not surprising, therefore, that a play titled *The West Indian* (by Richard Cumberland) was staged in England in 1771. It had a very successful run then and enjoyed a fair measure of popularity until the first few years of the nineteenth century. The West Indian in the play was a white Creole from Jamaica, coming to England for the first time. He was given no special creole characteristics, suggesting that there were no sharply distinguished features uniquely associated with whites from the islands. The events in the play did not indicate either that there was any disrespect for or prejudice against those who had been born in the islands. All this, however, was deliberately done by Cumberland, whose explicit intention was to give a positive image to those characters that had been presented on stage as butts of ridicule and abuse, as, for example, the wealthy West Indian. Cumberland was probably responding to Samuel Foote's *The Patron* (1764) in which there was a caricature of the wealthy West Indian in the character, Sir Peter Pepperpot. The success of *The West Indian* as a play suggests that English audiences had no problem with Cumberland's more positive version of the West Indian.

Evidently the concept of the West Indian in the English mind was based on a degree of ignorance, judging from the following comment by Long: 'The many Mulatto, Quateron, and other illegitimate children

2. My highlighting.

191

sent over to England for education ... pass under the general name of West-Indians; and the bronze of their complexion is ignorantly ascribed to the fervour of the sun in the torrid zone' ([1774] 1970 2: 274). Yet, the following remark by Mathews, a Creole writing in defence of Creoles – 'Three West Indians ... went to a public house in London, to dine, they resolved to keep their being Creoles a secret ...' (1793: 132) – gave no indication of colour difference and contradicted the notion that West Indians in Britain were recognisable by their colour. In fact, anecdotes given by Mathews suggest that it was only in their choice of words that West Indians were identifiable, which meant that language had become a defining characteristic of English Creoles, one that was considered socially unsophisticated.

As to the idea that Creoles felt self-conscious and inferior, and tried to hide their identity when in England, Mathews himself expressed some scepticism about the truth of this, saying: 'Now this is well enough for a story; but in one particular it does not sound well, for a Creole is never ashamed of his country' (1793: 132). What emerges from the various, opposing comments is that even though West Indians, for economic reasons, had become a more prominent feature in British life, the characteristics of West Indians, as a people, were generally not familiar to the English, thereby indicating that West Indians had not yet acquired any striking racial or political separateness in their identity. The term *West Indian* was therefore only geographical and economic in significance in the eighteenth century.

As far as Creoles at home were concerned, there were a number of characteristics ascribed to them, mostly having to do with the presumed effects of a tropical climate as well as with the institution of slavery. One observation appearing early in the eighteenth century accounted for a different coloration in some whites in terms of the food they ate, in addition to the climate:

> They [turtles] infect the blood of those feeding on them, whence their shirts are yellow, their skin and face of the same colour, and their shirts under the armpits stained prodigiously. This I believe may be one of the reasons of the complexion of our European inhabitants, which is changed, in some time, from white to that of a yellowish colour, and which proceeds from this, as well as the jaundices, which is common, sea air, &. (Sloane 1707: xviii)

Sloane was a medical doctor who left a detailed record of Jamaican flora and fauna of the last quarter of the seventeenth century. His view was therefore an influential one at the time, but even so there was no subsequent repeated assertion and belief that eating turtle meat would change your skin colour if you were white.

Another claim, by Edwards, was that white Creole children experienced more rapid mental development than did European children:

> Perhaps, the circumstance most distinguishable in the character of the Natives [whites] to which the climate seems to contribute, is the early display of the mental powers in young children; whose quick perception, and rapid advances in knowledge, exceed those of European infants of the same age, in a degree that is perfectly unaccountable and astonishing. This circumstance is indeed too striking to have escaped the notice of any one writer who has visited the tropical parts of America; and the fact being too well established to be denied, the philosophers of Europe have consoled themselves with an idea that, as the genius of the young West Indians attains sooner to maturity, it declines more rapidly than that of Europeans. (1794 2: 12)

Whether or not this claim arose out of some observable behaviour, as Edwards suggested, it was not consistent with the general belief about indolence and the slow pace in the tropics. However, it was only white Creoles who were said to have developed this characteristic, and it was one which, in the final analysis, was counterbalanced negatively.

A much better-known aspect of Creole behaviour that consistently attracted the attention of British writers was the sexual preferences of white Creole men. Browne voiced his concern thus: 'What I mean by vicious habits, are their great attachments to Negroe-women; there being but few gentlemen but what have several of those ladies very early in keeping' (1756: 23). Accordingly, a major concern for social improvement that was repeatedly put forward was the need for a change in these preferences. Singleton expressed his view in verse:

> Attend, ye sons of Carribean lands ...
> Shun the false lure of Ethiopic charms. (1767: 151)

Edward Long expressed this more strongly in prose, saying: 'it might be much better for Britain, and Jamaica too, if the white men in that colony would abate of their infatuated attachments to black women ...' (1774 2:3 27). It is interesting to note that, while in the case of Santo Domingo Valverde blamed the female slaves for seducing the men, in Jamaica Long charged the white men with unbecoming behaviour. In Barbados, in the case of free coloured women, the historian Poyer (1808: 639–640) saw them in 1797 as dangerous and even as subverters of the justice system through their influence.

The sexual preferences of white men were of course facilitated by the institution of slavery, and it is this that some European writers chose to attack. Hughes, for example, claimed that what really prevented Creoles from achieving much was slavery itself, which caused them from infancy to 'acquire (unless happily prevented, or corrected, by the

good Examples of Parents, or Education) an overfond and self-suffi-
cient Opinion of their own Abilities, and so become impatient, as well
as regardless, of the Advice of others' (1750: 13). This was the same
comment that was made by Raynal and Ledru in reference to Puerto
Rico, thereby indicating that the negative effects of slavery must have
been strikingly evident in the behaviour of white Creoles.

British writers' concerns about the decline of whiteness were obviously provoked by the fact that the plantation and slavery, operating
with a system of wet nurses, nannies, domestic servants and kept wom-
en, bound the people together intimately from birth to death. The attitudes that permitted white men to have 'liaisons' were formed from
early in life when white plantation children used young slave children
as playthings. Native-born whites and the native-born blacks had intertwined experiences, which made the term *creole* intimately applicable
to both and which frustrated the desires of English 'puritans'. Yet,
because white racial purity and civilised society were synonymous for
the English, and, because the need for racial separateness was reinforced by the political dominance of the mother country over the colony, there was little chance of any comprehensive racial identity for the
entire creole population emerging in the English islands. So, unlike in
Santo Domingo and Puerto Rico, there was no alternative racial term
like *gibaro* that emerged as a term of national identity. In any case,
Creoles of mixed race did not numerically dominate in the English
islands and there were no pretensions among Creoles themselves about
a common indigenous element in the population mixtures in these
islands. For English writers, *West Indians* were primarily whites associated with commerce and the English colonies; Creoles more generally
continued to be viewed in separate racial categories.

One of the earliest expressions of self-awareness among white
Creoles in the English islands occurred in a letter in the *Barbados
Gazette* of 15 May 1734, in which unmarried Barbadian men were
chastising their lady folk for their preference for foreign men. The letter
writer used the terms *Barbadian, countrymen,* and *Creoles* to refer to
himself and his colleagues, and the terms *Creolia's* and *country-women*
to refer to locally born women. The main point of the letter – to formally establish the comparatively good qualities of locally born men –
made it clear that, a hundred years after the start of the English settlement of Barbados, there was not only an observable distinction
between Creole and British persons but also that the Creole men of
Barbados felt themselves better (i.e. less showy and more solid) than the
English, Scottish and Irish men who came there. On the other hand,
judging from the men's complaint, the women did not share the men's

high view of themselves. Half a century later, Abbad y Lasierra made the same point about the women in Puerto Rico – '*Las mujeres aman a los españoles con preferencia a los criollos*' ['Women love the Spaniards rather than Creoles'] ([1788] 1966: 182). Quite obviously, discrimination in such a fundamental and sensitive area of social life would have led to an escalation, among the locally born men in the islands, of the need to establish their identity and superiority. It also would have added fuel to the politically and economically inspired animosities between the men of the colonies and those from Europe.

There is no indication of comparable tensions in love in the French islands involving a female preference for foreigners rather than locals. There was a common belief, however, one that applied equally to the French and English colonies, about the negative effects of European experience on young creole men – that European 'education' transformed them into debauchees. Thibault de Chanvalon stated this as follows:

> *Dans les premiers tems de leur education, ils donnent les plus grandes espérances pour l'avenir; dès qu'ils reviennent dans nos Isles, dès qu'ils ont même atteint l'âge bouillant de la jeunesse, ils perdent le fruit de leurs études; ils renoncent pour les plaisirs à l'amour des sciences & des belles lettres.* (1763: 35–6)

> [In the first years of their education, they give the greatest hopes for the future; as soon as they return to our islands, even as soon as they reach the tumultuous years of youth, they lose the fruit of their studies; they renounce the love of the sciences and the fine arts for pleasure.]

In this sense European was seen as negative – going to Europe was seen as an introduction to a life of immorality, which was continued on returning home, where coloured and black women were easily available. It caused white creole women embarrassment and suffering, because in most cases they were not sent overseas to be educated, and when they stayed at home they got very little education. In fact, Moreau de Saint Méry ([1796] 1958 1: 42) said that Creole women in St Domingue got no formal education. White creole women were consequently no match intellectually for the men returning home and, probably because they were constrained by strict moral codes, were not exciting to them sexually.

In spite of this negative view of European 'education' and its effects on young men, French Creoles seemed to be in unison socially with their mother country and to have acquired a good social reputation by the middle of the century. Bossu, for example, was impressed with his treatment in Saint Domingue when he stopped there in 1751:

*C'est avec juste raison qu'on accorde en France le titre de noblesse aux Créoles; ils y répondent parfaitement par leurs sentiments distingués soit dans la profession des armes, soit dans d'autres arts qu'ils exercent avec succès.* (1768: 18)

[It is with reason that in France Creoles are accorded the title of nobility; they respond to it perfectly by their refined sentiments either in the military or in other arts which they pursue successfully.]

This positive view of the social refinement of Creoles had already been apparent 20 years earlier, judging from Charlevoix's view:

*Le caractere d'esprit des Créols François commence aussi à se débarasser du mélange de Provinces, d'où sont sortis les premiers Fondateurs de cette colonie. Bientôt même il n'y restera plus aucun vestige du génie de ces anciens Avanturiers, ausquels le plus grand nombre des Habitans doivent leur naissance ... L'héritage, qu'ils ont conservé le plus entier de leurs Peres, c'est l'hospitalité; il semble qu'on respire cette belle vertu avec l'air de S. Domingue.* (1730 2: 483)

[The character of the spirit of French Creoles is also beginning to rid itself of the mixture of the provinces, whence came the first founders of this colony. Very shortly, there will no longer be any trace left of the spirit of these old adventurers to whom the majority of the inhabitants owe their birth ... The heritage which they have most completely retained from their parents is hospitality. It seems as if one breathes this beautiful virtue in the air of Saint Domingue.]

White Creoles were seen to be achieving a measure of uniformity in an orderly colony, somewhat removed from its wild beginnings and in sharp contrast to their 'indolent' neighbours on the other side of the island. Charlevoix viewed refinement in Saint Domingue as, on the one hand, the decline in French provincial characteristics and in what one could conceive of as the adventurous frontier spirit, but, on the other, as the retention of hospitality, a characteristic needed for civilised society.

In the middle of the century, Saintard's picture of French Creoles was more mundane and typical, in its linking of them to their habitat, as had been done for the indigenous inhabitants: '*Ceux qui y naissent, recevant de la nature une Patrie fertile, s'annoncent par des moeurs franches & naturelles plus convenables aux lieux; avec moins d'intrigues ont plus d'attachement au sol; sont plus Colons, & par-là plus sujets*' ['Those who are born there, receiving from Nature a fertile fatherland, are characterised by candid and natural manners better suited to those places; with less intrigue they have a greater attachment to the land; they are truer colonists and consequently better subjects'] (1754: 136). As good colonists and subjects, they were different from

Frenchmen in the colonies, who he claimed no longer had any strong feelings for France.

Even though Saintard presented a contrast between French colonists and Creoles in behaviour and feelings, there was no sense of conflict between them and in fact he went on (1754: 136–7) to bring them together in their reactions to the mother country. Differences though there were, there was no sharp political division between white Creoles and the mother country, signalling the beginnings of a separation into an independent identity. In fact, near the end of the century, Moreau de S. Méry said (1788: 12–13) that whites in Saint Domingue had no feelings of patriotism towards Saint Domingue.

Paradoxically, feelings of patriotism toward colonies generally could have been partially determined by economic success. James puts forward the following economic argument to show the leanings of the white proprietors towards France rather than the colony:

> With the growth of trade and of profits, the number of planters who could afford to leave their estates in charge of managers grew, and by 1789, in addition to the maritime bourgeoisie, there was a large group of absentee proprietors in France linked to the aristocracy by marriage, for whom San Domingo was nothing else but a source of revenue to be spent in the luxurious living of aristocratic France. So far had these parasites penetrated into the French aristocracy that a memoir from San Domingo to the King could say: 'Sire, your court is creole,' without too much stretching of the truth. (1963: 56–7)

It is not surprising therefore that, despite the growth of social refinement in the French colonies, there was no sense of Antillean identity, there was no parallel to the term *West Indian*.

# General characteristics of the creolisation process

## A racial structure of identity

In the eighteenth century, the process of creolisation involved an experience of and in part a validation of a social hierarchy in which there was a gradient system of identity setting out degrees of race and colour. When Garcilaso de la Vega the Inca gave his terms of racial classification in 1609 (*mulato, mestizo, cuatralvos, tresalvos*), he said that they identified the percentage of Indian blood and the percentage of white blood in a mixed person. A later and more developed hierarchical system with three divisions was set out and explained by Edward Long

(1774 2: 260–1). The basic (top part of the) gradient system identified the percentage of black/African that mixed individuals had. This was seen to be necessary in the gradient system, for becoming white, that is getting rid of non-white blood, was the purpose of *lineal ascent*. As Long explained, 'In the Spanish colonies, it is accounted most creditable to mend the breed by ascending or growing whiter ...' (1774 2: 261). In this system persons who had less than one fifth of black or one quarter of Indian were considered white in the Spanish colonies. The fluid concept of white in the Spanish colonies was the result of a long history of racial mixing in Spain itself and the colonies. López Cantos, after citing cases of successful claims for whiteness, concluded that '*La pureza no era total, pero la sociedad y las instituciones así lo consideraban*' ['The purity was not total, but the society and the institutions considered it to be so'] (Sued Badillo and López Cantos 1986: 261). White therefore became a practical, achievable classification in the Spanish colonies as if to make up for the rarity of 'pure' whiteness but really in part to provide Spanish men with an avenue to recognise their offspring legally.

In contrast to what Long identified as 'Direct lineal Ascent or growing whiter', there were the types that were 'Mediate or Stationary, neither advancing nor receding' and others still that were 'Retrograde', meaning that the offspring was going towards black. The terminology used to identify the *mediate* and *retrograde* types was not of the mathematical type but common words with extended meanings. The system as a whole therefore combined precise and imprecise terms along a continuum, and, in the words of Long, was a 'kind of science' among the Spaniards.

The French adopted the gradient system of colour and race, together with most of the actual terms, from the Spanish and eventually the gradient system and the values attached to it became pervasive throughout all the European colonies in the Caribbean. The Dutch were, according to Long (1774 2: 261), the most extreme/exclusive of all the colonial nations in the matter of *lineal ascent* or the attainment of whiteness. Long gave no details of the Dutch system of classification but it was probably little different from the French colonial system given by Moreau de Saint Méry, which did not allow for recovery of whiteness through *lineal ascent*. Therefore, though white was not a practical, achievable classification in the French and Dutch colonies, it was in the Spanish, and concerted attempts were made at the personal and the national levels to achieve it. The English, with strong notions of purity, had a resistance to a continuum between races and preferred to deal with notions of black and white.

Moreau de Saint Méry's classification ([1796] 1958 1: 86–9) was probably the most elaborate attempt to set out the detailed results of mixing. He introduced it in the following way:

> *Résultat De toutes les nuances, produites par les diverses combinaisons du mélange des Blancs avec les Nègres, et des Nègres avec les Caraïbes ou Sauvages ou Indiens Occidentaux, et avec les Indiens Orientaux.*
>
> [Results of all the nuances produced by the diverse combinations of mixing of white with negro, and negro with Carib or Savage or West Indians and with East Indians.]

Thirteen[3] distinct classes of skin colours, each with subdivisions, were listed and discussed. The major continuum was of course from black to white and this was constructed as: *sacatra – griffe – marabou – mulâtre – quarteron – métis – mamelouque – quarteronné – sang-mêlé*. While it might have been practically impossible to keep separate such a great number of different shades of colour, this kind of listing, which occurred less elaborately in many descriptions, attests to the importance of shade of colour, over and above the concept of race, in the identification of people across the society, especially the non-slave part of the society. Yet, even this listing was not exhaustive since it did not contain terms used by all French writers.

The fact that this gradient or hierarchical type of classification was mentioned in most descriptions attests to its abiding importance in the identification of people for practical social purposes. The sliding scale of privileges, which was the essential reason for the hierarchical classification, was topped by whiteness, which carried the greatest privileges. This is evident in the following subjective, nationalistic remark by Long, who was trying to show that the English were less exclusive than the other European nations: 'the laws permit all, that are above three degrees removed in lineal descent from the Negro ancestor, to vote at elections, and enjoy all the privileges and immunities of his majesty's white subjects of the island' (1774 2: 261). As is clear from this remark, the gradient system was not just a social value system; it was part of the legal system. Racial distinctions and terminology became so entrenched in the minds and actions of people in the colonies that they became almost immutable and were bulwarks against social mobility and pillars of the walls which retained privilege.[4]

---

3. Blanchard (1910: 39), using nineteenth-century illustrations from Mexico, makes a case for 16 set types (*castas*): '*Le chiffre de seize types, qui se retrouve dans nos trois exemples, prouve bien qu'il y a là quelque chose de défini, d'officiel, qui échappe à la fantaisie d'un artiste*' ['Sixteen as the number of types found in our three examples proves beyond doubt that it was something definite, official and beyond the whims and fancies of any artist'].

In the Spanish colonies, as well as in the others, since entitlements of succession and privileges of association were accorded legally and officially according to the percentage of white blood that an individual had, it was the pseudo-mathematical terms (e.g. *tresalvo, cuatralvo, terceron, quateron, quinteron*) (*tres* = 3; *cuatro* = 4; *tercer* = a third; *quater* = a quarter; *quinto* = a fifth) that were used to facilitate this. In Saint Domingue, where the *gens de couleur* increasingly resented the attempts of white to restrict their privileges, the legislation of privileges (e.g. the forbidding of non-whites to use white names, and the prosecution of those who maliciously or innocently labelled or referred to a white as non-white) in large measure led to the Haitian Revolution. Yet, in everyday social interaction between people, the pseudo-mathematical terms were not the more common of the two types – it was the more subjective terms that were. Among the subjective terms were those used to identify Indian mixtures (e.g. *mestizo, caboclo, marabou, mameluco, cafuso, givero, grifo*), which, because of ambivalent values accorded to the indigenous inhabitants, were transferred across racial types, thereby acquiring double and extended values.

Terms were adopted and extended beyond their presumed 'accurate' reference to identify the increasing diversity of mixtures. As a result, Indian terms of identity came to be used to refer to individuals who only looked like Indians. Colour was the major feature of most of the terms as they became more stylised and especially when the terms were applied in areas where the population composition differed from that where the term had its source. The White + Black mixtures were equated with the colour of the White + Indian mixtures so that the colour was kept more or less constant. For example, *mestizo* became the French term *moustiche*, which, especially in the context of French Louisiana, became *mustee*. Thus, *métis*, which was White + Indian, was assessed to be the same colour as *mustee*, White + Quateron.

The terms that identified Indian mixtures were Old World (e.g. *mestizo* < Latin *mixtus* 'mixed') or Brazilian/South American Indian, non-specific, non-technical terms, which obviously acquired specific references. The terms that came from South America most likely travelled with sugar technology in the seventeenth century and with the slave trade from Brazil to the islands. So, while the Spanish were identified as the ones who had developed the gradient system of

---

4. Blanchard (1910:38) surmises that '*dans une société aussi hiérarchisée que l'était alors la société espagnole, il fallait pouvoir désigner par un terme technique ces divers degrés de métissage et, au besoin, en cas de contestation, pouvoir les définir par un moyen sûr et indiscutable*' ['in a society as hierarchical as Spanish society was, it was necessary to identify, using technical terms, the various degrees of mixture and, in case it was needed where there were disputes, to be able to define them by reference to an established and accepted measure'].

classification, some of the terms came from Brazil, for it was in the Portuguese, French and Dutch settlements there that these words were used to identify mixtures and colours of people.

Repeatedly setting out a colour classification in written literature gave the terms in it an objective validity that practically defied reality. More than anything else it reflected time-depth, that is a limited number of generations of miscegenation. While sixteenth- and seventeenth-century works mostly identified one generation of miscegenation (i.e. *mulatos*), by the eighteenth century a greater number of generations were evident and could be identified, but, evidently, it had not yet reached the stage where the mixtures were too many and too minutely differentiated to make naming them useless for normal practical purposes. Even so, it was already evident in Moreau de Saint Méry's classification that the term *mulâtre* was no longer simply the product of a white person and a black person but the result of several possible combinations. Yet, a colour classification, which was objectified/reified in writing, established for a relatively new society a hierarchy that gave those at the top a greater feeling of superiority and those somewhere in the middle a better sense of security than a simple division into two or three levels would have done. Written literature therefore went further than everyday communication to provide a detailed racial structure for the slave society: this gradient colour system established the social structure of creolisation and a racial identity for everyone.

## The stigmatisation of African-ness as well as means of obliterating it

Among the slaves the term *criollo* was, according to Garcilaso de la Vega, the Inca, initially a negative term because it contrasted the off-shoot New World 'African' with the true African. In like manner, it contrasted the off-shoot European with the real European from Europe. However, within the plantation system in the islands, *creole* became a positive term, no doubt as the number of Creoles increased, and, in the case of the black slaves, it established a superiority of native-born slave over African slave. In 1731 Charlevoix gave a comparison of the image of the creole slave with that of the African that had emerged in Saint Domingue:

> *Enfin les Negres **Creols**, de quelque nation qu'ils tirent leur origine, ne tiennent de leurs Peres que l'esprit de servitude, & la couleur. Ils ont pourtant un peu plus d'amour pour la liberté, quoique nés dans l'esclavage; ils sont aussi plus spirituels, plus raisonnables, plus adroits,*

*mais plus fainéants, plus fanfarons, plus libertins que las Dandas, c'est le nom commun de tous ceux, qui sont venus d'Afrique.* (2: 498)

[So, Creole negroes, from whatever nation they originate, retain of their parents only the spirit of slavery and their colour. However, they have a little greater love of liberty, even though they were born into slavery; they are also more spiritual, more reasonable, more skilful, lazier, more boastful, more licentious than the 'Dandas', that is the name of all those coming from Africa.]

The superiority of the creole slave over the African was established by the social structure in a number of other ways, principally as a result of the fact that 'informed' views on race during this century established a scale of human-ness with the white European at one end and the black African at the other, with enlightenment at one end and darkness at the other. Black was the negative pole in the colonial structure, with creole black being regarded as better than African on this scale and features of creole-ness being regarded as real standards of attainment for Africans.

The seasoning process that the newly arrived Africans had to undergo in itself put the 'pure' African at the bottom of the social and economic scale. Bearing in mind that a considerable percentage of the newly arriving slaves were young and thus socially unsophisticated in any case, the designation of them as *bozal* in the Spanish islands incorporated negative valuations of race, maturation and intellect, although in the English islands the designation of them as *saltwater* negroes initially highlighted the element of time. In general, because the arriving slaves were in a terrible condition, both in terms of physical health and appearance, there was a vision of them as sickly and deficient. From a psychological point of view, because they were traumatised and did not seem lively, this added to the negative view of them. In sum then, they seemed to fit perfectly at the bottom of the ladder.

In the Caribbean islands[5] Africans generally could be differentiated from creole slaves by physical marks and appearance. Generally, African slaves were branded with the names or initials of their owners, whereas creole slaves were not. Fermin was forthright in his explanation of the need for branding:

---

5. The practice of branding also took place in the factories on the African coast, as noted by Svalesen (2000: 99): 'The branding of slaves usually takes place in the Company's courtyard. The Guinea Company marks its slaves on the front of the right thigh. The Company's mark consist of an "S" inscribed in a heart. The "S" stands for "Slave" ...' See also Marees ([1602] 1987: 177]: 'They are also branded and marked, so that if they were to run away they could again be found because they would be recognised by their marks or signs.'

*il faut scavoir que chacun marque ses Esclaves, pour pouvoir les reconnoitre ... Sans cette precaution, personne ne pouroit distinguer ceux qui lui appartient dans le nombre prodigieux qu'il y en a dans la Colonie ... d'autant plus que ce peuple n'a point de physionomie assez distinctive a nos yeux.* (1769 1: 117–18)

[it is necessary to know that everyone marks his slaves in order to identify them ... Without this precaution, no one would be able to identify one's own among the huge number of them in the colony ... the more so since, to us, the facial features of these people are not distinctive enough.]

Bossu in the same vein referred to branding as '*cette méthode ... pour reconnoître leurs Esclaves*' ['this method ... to recognise their slaves'] (1777: 372–3). Yet, there were those who later managed to interpret branding otherwise – M'Queen (1825: 255) claimed that branding was done for the safety and protection of the African slave, who, if left un-marked, not being familiar with the surroundings and not being able to speak the language of the country, might get lost and even die. The mark, on the other hand, would allow a finder to return the African to the rightful owner. Barclay (1826: 218) made a similar kind of argu-ment.

Branding in the New World, however, did not start with and was not restricted to African slaves; in fact, no ethnic group was excluded from this practice. Early sixteenth-century evidence is given to show that the Spanish were the first Europeans to start this practice. La Rosa Corzo refers to '*una autorización que el 25 de julio de 1511 se le envió por parte a Diego Colón para que marcara con hierro candente a los indios de la Española que por haberse rebelado fueron sometidos a condiciones de esclavitud*' ['authorisation given on the 25 July, 1511 by Diego Columbus to brand with a hot iron the Indians of Hispaniola who for having rebelled were submitted to conditions of slavery'] (2003: 36). Fernández Mendez (1976: 33) also quotes from a letter from Ponce de León [in Puerto Rico] to the king of Spain which in part said: '*lo de haber herrado con una F en la frente a los indios tomados en guerra, haciéndolos esclavos ...* ['having branded with an F on their forehead the Indians taken in war, making them slaves ...'] (cited in La Rosa Corzo 2003: 36).

Further afield, the Spanish branded indigenous boys and women when they subjugated Mexico in a campaign that started in 1517. It was a method of identifying new and desirable property gained through war: '*Y así se volvió con buena presa de mujeres y muchachos, que les echaron el hierro por esclavos*' ['And so it returned with a good capture of women and boys whom they branded as slaves'] (Díaz del Castillo 1955 1: 425). Branding was not reported by Diaz del Castillo, one of

Cortes's soldiers, as unusual, and indeed he explained more than once that the brand, which was administered in the name of the King of Spain, was a **G**: '*Pues ya juntas todas las piezas y echado el hierro, que era una G como ésta, que quería decir guerra ...*' ['Then with all the 'pieces' already together and branded with an iron, which was a G like this, which meant "war" ...'] (p. 428).

Two of the earliest references to branding in the English and French colonies respectively were cited as punishment for incorrigible Europeans. Referring to an incident in Barbados in the 1630s, an early (Trinity College, Dublin) manuscript said: 'Captain Stronge saith, he saw Wiborne stigmatised, and heard the iron hisse on his cheeke.' Du Tertre noted that white *Commandeurs* (i.e. persons put in charge of servants and slaves) were threatened with branding for sleeping with the black slave women:

> *Deffenses sont faites à tous Commandeurs d'Engagez, de Negres, de débaucher les Negresses, à peine de vingt coups de Lienes par le Maistre des hautes oeuvres, pour la premiere fois, de 40. pour la seconde, & de 50. coups, & la fleur de Lys marquée à la jouë pour la troisiéme fois ...* (1671 3: 72)

> [It is forbidden for all commanders of servants and negroes to seduce the negresses, the penalty for this being 20 lashes with the whip by the master of the high works for the first offence, 40 for the second, and 50 lashes and the fleur de lis marked on the cheek for the third offence ...]

The use of symbols or letters to publicly identify a person as bad was a European practice that was transferred to the colonies and used among whites themselves (see Hawthorne's *The Scarlet Letter*, 1850).

Physical marking was also a symbolic punishment for stealing among the Island-Caribs, according to De Rochefort, who said: '*Si les Caraïbes subçonnent quelcun de leur avoir derobé quelquechose, ils taschent de l'attraper, & de luy faire des taillades, ou de couteau ou de dent d'Agouty pour marque de son crime & de leur vengeance*' ['If the Caribs suspect someone of having stolen something, they try to trap him and to cut him either with a knife or agouti teeth as a mark of his crime and their punishment'] (1658: 468).

Branding became commonplace in the French and English colonies with the increasing importation of Africans and their fast turnover, a reality that made recognition of human and mobile property difficult. African slaves routinely escaped soon after they were bought and joined up with servants and indigenous inhabitants in quasi-maroon communities, making identification on recapture difficult. Branding was in part, thus, a punishment for running away and in part a deterrent to

**Figure 4.1** When slaves are purchased by the dealers they are generally marked on the breast with a red hot iron. National Library of Jamaica

those contemplating running away. In 1685 the French *Code Noir* speci-fied the following punishment for running away: '*Le nègre, marron pendant un mois, aura les oreilles coupées et sera marqué d'une fleur de lys sur l'épaule gauche; s'il récidive, il aura le jarret coupé et sera marqué sur l'autre épaule ...*' ['The negro who escaped for a month will have his ears cut and will be marked with a fleur de lis on the left shoul-der; if he escapes again, he will have his leg cut off and will be marked on the other shoulder ...'] (Schoelcher 1948: 64). In time, probably because marronage became so widespread despite branding, branding was used simply as a means of identification in the recovery procedure.

Europeans were no doubt encouraged to brand slaves when they saw that they already put marks on themselves for ethnic identity, marks which Du Tertre called '*une espece de broderie au visage, au sein, aux bras & aux épaules*' ['a kind of embroidery on their faces, chests, arms and shoulders'] (1667 2: 523). Branding of slaves for iden-tification was, in its development, a response to a problem that arose on the plantation, and, as was the case with creolisation generally, it combined features from different sources – it was an extension of the European practice of permanent public condemnation, which coincided with practice among the indigenous inhabitants; it was an extension of the European practice of marking property for identification, and it was in the eyes of Europeans an extension of African ethnic/tribal marking, which itself had an element of belonging, though not a negative one.

In time, branding not only became widespread but as a method of identification it moved beyond the notion of the individual trying to identify property to make it secure and beyond the punitive. In the 1780s in Saint Domingue, for example, African slaves bought at concessionary prices specifically for the population of the south had to be branded for purposes of recognition and control. Beyond this, in 1777 Hilliard d'Auberteuil mused that the only way that the problem of the Negroes going to get baptised more than once could be solved was if they were branded at the first baptism. Bearing in mind that in the Christian tradition baptism is intimately associated with naming (i.e. giving a Christian name) and with babies, what this meant was that, confronted with a vast number of adults in a non-traditional situation, the Christian rite of passage was being associated with symbolic, physical marks, which were seen as more meaningful to the African and easier for the European for record purposes. In contrast, higher up the hierarchical system the creole slaves did not need to be branded because their owners and others had been seeing them from the time they were born and could recognise them on sight.

In spite of its value in creole society as a marker of property or of punishment, the brand posed problems. As Fermin (1769 1: 116) explained, if a buyer (from a ship's captain) put his brand on a slave, he could not after that decide to return the slave. Then, as Moreau de Saint Méry ([1796] 1958: 83) pointed out, when the slave changed status, i.e. became free, the brand raised doubts whether the slave was really free. A third problem presumably was in cases where branded slaves changed from one owner to another. So, because the permanence of branding was inconsistent with a society in which human chattel and manumission were constant and related realities, the practice of branding declined by the end of the eighteeenth century. Additionally, by that time European recognition of their human property had become easier.[6]

---

6. Dickson (1789: 122) said: 'I believe you and the public are yet to be informed, that the negroes in Jamaica are BRANDED with their owner's marks and initials of their names, and, in one instance before me, with the owner's sirname at full length on four parts of the body. This is quite a new discovery to me; for the practice of branding slaves does not disgrace the island of Barbados.' Edwards (1794 2: 126 note (d)) said: 'It is the custom among some of the planters in Jamaica, to mark the initials of their name on the shoulder or breast of each newly-purchased Negro, by means of a small silver brand heated in the flame of spirits, as described in a former cha pter; but it is growing into disuse, and I believe in the Windward Islands thought altogether unnecessary.'

The ethnic markings of the African slaves also constituted a permanent, hierarchical distinction among Africans and between Africans and creole slaves. In relation to the latter distinction, Moreau de Saint Méry noted that '*on pourrait en prendre plusieurs pour des mulâstres, si des marques plus ou moins multipliées ... ne montraient qu'ils sont Africains et nègres*' ['you could mistake several of them for mulattos, if marks more or less multiplied ... did not show that they are Africans and negroes'] ([1796] 1958 1: 50). In relation to the former distinction, the same writer observed that '*On a même vu des nègres Mines reconnaissant des princes de leur pays, à ces signes bisarres, se prosterner à leurs pieds et leur rendre des hommages ...*' ['Mine negroes have even been seen recognising the princes of their country by these bizarre marks and prostrating themselves at their feet and rendering them hommage'] ([1796] 1958 1: 51). A similar observation about African slaves in Jamaica was made two years earlier by Edwards:

> some of the newly imported Negroes display these marks with a mixture of ostentation and pleasure, either considering them as highly ornamental, or appealing to them as testimonies of distinction in Africa; where, in some cases, they are said to indicate free birth and honourable parentage. (1794 2: 124–5)

**Figure 4.2** 'Tattooed' faces. From R. Bridgens (1837). *West India Scenery, with illustrations of negro Character, the process of making sugar, from Sketches taken during a voyage to, and residence of seven years in the island of Trinidad.* London: Robert Jennings. British Library.

Clearly then, it was not simply a matter of African slaves being identified as one ethnic group or another according to the African port from which their ship came, as some have claimed, but according to the knowledge and behaviour of the Africans themselves.

While African slaves attached social valuation to ethnic marks, Europeans attached economic value to them, having come to conclusions about the strengths and weaknesses of each ethnic group in terms of how they fitted into the system of slavery. Thus, the Africans' recognition of their own different ethnicities played into the hands of the Europeans, who used them for economic purposes. This is evident in the words of Fermin:

> On me demandera peut-être à quoi l'on reconnoît ces différentes sortes de
> Negres, & s'il ne serait pas possible au Capitaine d'en imposer, & de
> vendre un **Louango** aussi cher qu'un **Cormantin** ou un **Papa**? Je réponds à
> cette question, qu'on ne doute nullement que tous ces peuples ne se
> connoissent, & qu'à peine l'Esclave serait acheté, qu'il serait déclaré pour
> ce qu'il est, par quelqu'un des autres, ce qui feroit perdre le crédit au
> vendeur, & l'exposeroit à reprendre son esclave, & à se détruire dans le
> pays ... (1769 1: 116)

> [Someone perhaps will ask how these different sorts of negroes can be
> identified and if it is not possible for the captain to impose an identity and
> to sell a Louango for as much as a Coromantine or a Papa. My answer to
> this question is that there is no doubt that all these peoples know each
> other and that as soon as the slave was sold he would be revealed for what
> he is by one of the others, which would make the seller lose credit and he
> would have to take back the slave and he would be destroyed in the
> country ...]

After a time the markings of specific ethnic groups were recognised by Europeans themselves. Bridgens (1837), for example, gave illustrations of tribal marks, which he referred to as *tatoos*.[7] Elsewhere, reports in newspapers said, for instance, that Nagos had tribal markings on the face and Congos had them on the belly. On the other hand, as the search for slaves went further and further inland in Africa among ethnic groups unfamiliar to Europeans, it made recognition of the

---

7. 'We have here given representations of two of the various modes of tatooing in use
among the native tribes of the west coast of Africa. The operation is performed in
infancy. The desired figure having been pricked on the skin with a sharp instrument, a
dyeing liquid is rubbed into the wound. The impression is indelible, and expands with
the growth of the features. The male head in the centre row is marked with parallel
lines of small protuberances. There are the distinctive badge of the Mocha tribe, who
are reproached with the horrid crime of cannibalism.* [*The different tribes have each
their peculiar pattern, which serves to identify the subjects of the different petty
chiefs.]'

slaves' ethnic identity more difficult, as Hilliard d'Auberteuil (1777: 61–2) pointed out.

All physical marks on slaves were important to slave owners, who kept a record of them, and listed them in newspaper advertisements for runaway slaves. Slave owners had no appreciation for ethnic markings, except to the extent that they aided identification. Thus, they were included together with brandings and deformities as part of the identity of the slave, all of which negatively influenced the price paid for the slave in the case of resale. In the European mind, therefore, identification of Africans was strongly associated with mutilation and disfigurement.[8] However, mention of physical markings in a negative way in written literature must have contributed to their reduction in the case of branding and in the case of ethnic markings to them becoming features of embarrassment for Africans. In the later condemnation of the practice of branding, for example, the French claimed to be following the practice of the English (Moreau de Saint Méry [1796] 1958 1: 83), while Mrs Carmichael (British) mentioned (1833 2: 300) that her husband had seen a branded slave '*the* property of a Frenchman in St. Lucia'. In the case of ethnic markings, writers like Moreau de Saint Méry openly expressed disapproval in comments such as '*signes bizarres*' ['bizarre marks'] and '*plus ou moins ridicules pour l'oeil qui n'y a pas été accoutumé dès l'enfance*'['ridiculous, more or less, for the eye that has not been accustomed to them from childhood']. In time, Creole slaves adopted the attitudes of their masters to physical markings and that is why, as Moreau de Saint Méry [1796] 1958 1: 83 pointed out, branding was even used as a method of humiliating creole slaves.

The intent to obliterate African-ness was evident in naming practices. It is not only that names of adult slaves had to be intelligible to Europeans and to fit into a Christian tradition but that slaves did not have absolute authority to name their babies. Many adult slaves were renamed and identified by a single European 'Christian' name, which meant that they were summarily deprived of overt kinship ties through names. This served the purposes of slavery in that it facilitated the movement of slaves from one owner to another without any reminder of family. It also destroyed the slaves' own knowledge of lineage beyond recent generations, thereby removing from them the strength and comfort of traditions, creating a virtual diaspora. In the early years

---

8. Sloane (1707: liv): 'There are few **Negros** on whom one may not see a great many Cicatrices or Scars, the remains of these Scarifications, for Diseases or Ornament. On all their faces and Bodies, and these Scarifications are common to them in their own Countries, and the **Cicatrices** thought to add beauty to them. The **Negros** called **Papas** have most of these Scarifications.'

it fostered a kind of slave isolation that was comforting to slave owners, whose eternal fear was the slaves' banding together in rebellion. In time, it effectively separated creole slaves from their African forebears. However, Barbot, in a very nationalistic spirit, claimed that there was preservation and fostering of the family in the French colonies, in contrast to the English:

> the master buys the woman his man slave likes best, allowing them full liberty to match to their own liking; insomuch that it is an established law in the **French** islands, that when one person's male slave has a mind to marry another inhabitant's woman slave, and she approves of it, one of the two owners is obliged to dispose of his slave to the other ... that they may cohabit in the same house.
>
> This care of marrying and settling them together in a family, allowing them some little parcels of ground to till and make gardens, endears them to their masters, and makes them add to their ordinary labour ... (1752: 649)

Whether or not this practice was as general as Barbot claimed, there was no intention to preserve African ways – this slave *family* was meant to be nuclear in its structure, Christian in its religion and creole in its culture, which contrasted sharply with the family structures of the Africans in their own countries. Furthermore, the benefits of this family unit that Barbot highlighted, even though they may have decreased the incidence of nuclear family separation, were for the masters in particular and the creole society in general.

In the case of the naming of babies born to slaves, Fermin outlined customary 'christening' practice in the eighteenth century: '*Trois ou quatre jours après ses couches, la mere vient elle-même présenter l'enfant pour être nommé, observant que, si c'est un garçon, elle s'adresse au maître, & que si c'est une fille, c'est à la maîtresse*' ['Three or four days after giving birth, the mother comes herself to present the child to be named, noting that if it is a boy, she speaks to the master and if it is a girl the mistress'] (1769 1: 132). In addition, as noted much earlier by Du Tertre (1667 2: 528), the god-parents of the child were normally friends of the slave owner. All this meant that the owner and his friends would know the creole slave and be assured of the slave's existence and name from birth.

African-ness was also obliterated progressively by the Church through the sacrament of baptism, in the case of slaves said to be converted to Christianity. While most writers wondered how intelligible conversion was to the African slaves, Ramsay noted in one case a farcical element resulting from the mercenary intent of the person doing the conversion:

He came to the plantation on a Sunday afternoon, and desired the manager to collect eight or ten slaves to be baptized. They were brought before him. He began to repeat the office of baptism. When he had read as far as that part of the service where he was to sprinkle them with water, if their former name pleased him he baptized them by it, but if he thought it not fit to call a Christian by, as was his opinion of Quamina, Bungee, and the like, he gave them the first Christian name which occurred to his memory. This name the bearer, perhaps, could not repeat, and scarcely ever remembered afterwards; so that he continued to be distinguished among his fellows by his old heathen name. (1784: 158–9)

Yet, even though *old heathen names* continued to be used, the new 'Christian' names were official and in a few generations supplanted the former names, further denuding the slaves of African-ness.

Stigmatisation and acculturation indelibly linked two types of slaves in the public consciousness – Creoles and Africans. Long portrayed the link between them as follows:

The Creole Blacks differ much from the Africans, not only in manners, but in beauty of shape, feature, and complexion. They hold the Africans in utmost contempt, stiling them, 'salt-water Negroes,' and 'Guiney birds;' but value themselves on their own pedigree, which is reckoned the more honourable, the further it removes from an African, or transmarine ancestor. ([1774] 1970 2: 410)

## Accommodation within the creole racial structure and an attempt to breach it

Beyond physical markings and names, African slaves were further known and tagged according to their experience. The Spanish used the word *bozal* to describe the new African brought directly from Africa. Moreau de S. Méry ([1796] 1958: 55) gave *bossal* as the French equivalent, even though he seemed to regard it as a Spanish term. The English did not generally adopt the Spanish word but used the terms *saltwater negro* and *Guinea bird/Guinea negro*, which had the same function and effect. Discriminated against by black and white Creoles, the Africans sought to bolster their own security in their new homes by establishing alliances among themselves based on common experience. The one most significant to them was what is now known as 'the middle passage', and those who went through it together on the same ship regarded themselves as inseparably bound. In Cuba, according to Pichardo (1875), the word *carabela* was used by *bozal* slaves in the process of becoming *ladino* to mean '*Mi paisano, que vino conmigo de Africa*' ['My countryman, who came with me from Africa'] (p. 140). The word *carabela* in its basic meaning was one that occurred in the very early literature of discovery and declined in use thereafter – it referred to the kind of light sailing-ship used by Columbus and the early explorers. (*Carabela* itself derived from Greek *karabos*, which meant a small ship of sails and oars used by the Moors.) Apparently, then, the word was adopted by the slaves in the Spanish colonies with the specific meaning of 'fellow sufferer' from as early as the sixteenth century. It is interesting to note, however, that *carabela* was not far removed in its form from *carabi* and *caribe*, supposedly indigenous words also used to refer to persons in the early literature on the Americas.

According to Moreau de S. Méry ([1796] 1958: 58), the word the slaves in Saint Domingue used among themselves to identify those who came across in the same ship was *bâtimens*, thereby highlighting the idea of the (same) ship. The English version of this had been identified two years earlier by Edwards, who, in discussing the customs of the slaves in Jamaica, said:

> and accordingly we find that the Negroes in general are strongly attached to their countrymen, but, above all, to such of their companions as came in the same ship with them from Africa. This is a striking circumstance: the term **shipmate** is understood among them as signifying a relationship of the most endearing nature; perhaps as recalling the time when the sufferers were cut off together from their common country and kindred, and awakening reciprocal sympathy, from the remembrance of mutual affliction. (1794 2: 78)

The significance of the term *shipmate* is evident in the fact that it continued to appear in the literature through the following century. Renny stated that:

> A **shipmate** is one of their most enduring appellations; and they who have been wafted across the Atlantic ocean in the same vessel, ever after look upon each other as brethren: so natural is it for partners in misfortune, to become dear to each other! (1807: 172)

Some years later Waddell said:

> He came early every Sunday to the services of God's house in his clean white Osnaburgh frock and trousers that a 'shipmate',* good as a daughter to him, had always ready.
> *She was called his daughter. The attachment of the slaves who had come in the same vessel, to each other was like that of blood relations. ([1863] 1970: 107)

The terms *carabela*, *bâtimens* and *shipmate* showed a commonness in the concept across Spanish, French and English slaves,[9] but the word

---

9. A number of variants based on the same concept are given for different Caribbean societies by Mintz and Price (1976/1992: 43–4).

*shipmate* contained two concepts (ship + mate), the second of which also appeared separately in the French accounts.

In other words, in addition to the concept of 'ship', a notion embraced in the word 'sailor' also came to indicate a special kind of relationship among the slaves. In this case, the development of the word in the Caribbean did not start with the slaves. The word *matelot* was first mentioned by Du Tertre (1667 2: 452), then by De Rochefort (1658: 461)[10] and then by Oexmelin (1686 1: 151). It appeared about a hundred years later in Girod-Chantrans' version of Haitian speech – '*Li dit moi que li mériter mieux amour à moi que matelot à li*' ['He told me that he more deserved my love than his rival'] (1785: 189–90). Here, although the basic idea of having something in common remained, the meaning changed to one where the persons involved were seen as rivals rather than partners. The word then occurred a little later in Moreau de S. Méry's account:

> *Cependant, en général, les Africaines accoutumées à des maris polygames, n'ont pas une jalousie furieuse, et il est même assez commun d'en voir plusieurs qui vivent dans une sorte d'harmonie quoiqu'elles aiment le même objet. Elles se nomment alors entr'elles **matelotes**; mot tiré d'un ancien usage des Flibustiers qui formaient des sociétés dont les membres s'appelaient réciproquement matelot.* ([1796] 1958: 57)

> [However, in general, African women, accustomed to polygamous husbands, do not have any mad jealousy and it is quite common to see several of them living in a sort of harmony although they love the same person. Among them they are called 'matelotes', a word taken from the old usage of the freebooters who formed societies whose members called each other 'matelot'.]

The meaning here was in a sense the same as that given by Girod-Chantrans, but with women as 'rivals' rather than men. It was basically the same meaning that occurred in Descourtilz ([1809] 1935: 125) a few years later, and it is also the meaning that is current today. What is striking in Moreau de S. Méry's explanation, however, is the association of the early and essentially Caribbean colonial practice of *matelotage* with the traditional practice of polygamy among those ethnic groups of Africans that provided slaves for the Caribbean.

The word *matelot* and the concept that it embodied went back to Montaigne (or those of his time), who had reinterpreted the indigenous word *guatiaos* as *moitié*. The indigenous word was also later

10. '*Ce mot de **Matelot**, est commun aussi entre les François habitans des Iles, pour signifier un Associé. Et lors que deus habitans ont acheté, ou defriché une habitation ensemble on dit qu'ils se sont **enmatelotez**'* ['This word *matelot* is commonly used among the French colonists to mean a partner. And when two colonists have bought or cleared a plantation, you say that they have become partners'].

interpreted as *matelot* among the white servants in the French Caribbean colonies. The slaves of the French used the words *matelot* and *batimen*, which was either a bifurcation of the indigenous word (through folk etymology) or a misinterpretation by European writers of what the slaves were saying, that is a word that they had acquired directly from the indigenous inhabitants. The various forms of the word used by the slaves of the English did not develop independently of those used by the slaves of the French. Among the English slaves there was also a bifurcation, for while some used the combination *shipmate*, others used *mati* on the one hand or *sipi* on the other. In addition, while *matelot* (with its variants) was used by the African slaves as a term of address to establish solidarity of experience, *morrogou* and later *countryman* were used by slaves from the same area in Africa to express a bond of kinship and ethnicity.

The violent consequences of the refusal to abide by and accept the creole structure of identity and its terminology are nowhere better illustrated than in Saint Domingue. The Frenchman's tenacious maintenance of a sharp distinction between white and all others had been intended primarily to safeguard privilege and wealth for whites in the face of growing numbers of sophisticated people of mixed race. Since, for the French, the notion of pure race increased in measure as the evidence of mixed race grew and since whites had conceived for themselves an essence of superiority, maintenance of a sharp distinction was converted into a means of preservation of species. The idea of a species apart meant simply that, in the social sphere, whites were not to be confused with any other group. The effect of this was pointed out by Satineau:

> *du milieu du XVIIIe siècle à l'abolition définitive de l'esclavage (1848) l'appelation de sang-mêlé est considérée à Guadeloupe et dans toutes les îles françaises comme l'insulte la plus grossière qu'on peut proférer à une personne de race blanche. L'appellation de sang-mêlé est un outrage passible des peines les plus sévères.* (1928: 358)

> [from the middle of the eighteenth century to the final abolition of slavery (1848) the term *sang-mêlé* in Guadeloupe and all the French islands was considered the grossest insult that one could address to a white person. To call someone *sang-mêlé* is an insult liable to the most severe penalties.]

Satineau then went on (1928: 359) to give court cases in Saint Domingue in the middle of the eighteenth century where people were punished severely for having called whites *sang-mêlés*. The important factor here is that *sang-mêlé* was in physical appearance virtually indistinguishable from white – note Moreau de Saint Méry's definition,

'*Sang-mêlé, qui s'approche continuellement du Blanc*' ['Sang-mêlé: one who is getting nearer and nearer to white'] ([1796] 1958 1: 86). It was therefore the conception of mixture that was totally unacceptable to whites.

In 1785, the number of freed people in Saint Domingue was about 30,000, the same number as the whites (Lespinasse 1882 1: 274) and they had come to be identified as *gens de couleur* (Lespinasse 1882 1: 232). These *gens de couleur* represented all distinctions on the colour scale between black and white, but it was those who were nearest to white, in Saint Domingue as well as in other French colonies of the eighteenth century, who became an insistent social and political force because they wanted to be accorded the privileges of whites. In 1767, however, the French King, Louis XV, in a letter to the Conseil Supérieur in Port-au-Prince, denied Negroes and all their descendants the privileges of whites. Furthermore, according to Lespinasse (1882), there was a comprehensive attempt made to keep *gens de couleur* separate from whites. In an ordinance of 1773, *gens de couleur* were forbidden from giving their children the same surnames as whites and those who had done so had to discontinue using those names and find new ones. They were advised to give their children '*un surnom tiré de l'idiome africain ou de leur métier et couleur*' ['a surname taken from an African language or from their job and colour']. In an ordinance of 1779 *gens de couleur* were required to accord full respect to their former masters and mistresses and whites in general. They were also forbidden from wearing clothes that white people wore and items of value – they were expressly required to dress simply, as they had done before they were freed. These stipulations only served to foment resentment, especially since in practice many white fathers had treated their mulatto offspring well, sending them to Europe to be educated and providing them with ample means to live well. In addition, since the keeping of women of all colours by men in positions of privilege was an integral part of the social fabric of the day, distinctions and scale of privilege, which were important to coloured women in their valuation as prospective wives or mistresses, were fundamentally threatened by such ordinances.

For the *gens de couleur*, then, the attempt to deny them privilege was not just insulting but a call to arms. The *gens de couleur*, while they aspired to whiteness, no doubt felt that Saint Domingue was more their home than anybody else's. It is not surprising therefore that before the end of the century whites were being forced to flee from Saint Domingue. In other words, it was the French desire to maintain a strict distinction between black and white rather than to expand the category of white, as the Spaniards had done, which helped to bring about their greatest humiliation in the colonies. More generally, the intent to re-africanise *gens de couleur* by making them use African names was a direct and flagrant attack on, and a contradiction of, the values of creole society. It demonstrated, ironically in a French world dominated by the watchwords *liberté, égalité, fraternité* and in a colonial situation where everyone tried to escape from being labelled *Sauvage*, an underestimation of how strongly European values had gripped colonial societies.

## The changing identity of those outside colonial society – *Sauvages africains* or Black Caribs

The eighteenth century saw the emergence of an ethnic phenomenon in the island of St Vincent that was the result of a process which stood in contrast to European creolisation: it developed outside the pale of European acculturation and produced what Raynal (1773, Bk. xiv, p. 263) called *'cette race doublement sauvage'* ['this doubly savage race']. It was the result of a process of nativisation – the acquisition of traits of the Island-Caribs by Africans to produce a people African in appearance and Island-Carib in culture – *Sauvages africains*[11] or Black Caribs.

From the days of the earliest French and English settlements in St Kitts there were runaways, including white indentured servants who escaped into the bush and joined with indigenous inhabitants in their own societies (Maurile de S. Michel 1652: 38). As the literature indicates (De Rochefort 1658: 439, 478), the indigenous inhabitants also carried off Europeans and Africans from the European settlements in reprisal for European misdeeds as well as to bolster their own communities. The process of nativisation of Europeans and Africans, which began in the first half of the seventeenth century, evidently continued throughout the entire century, but it changed from being intra-island to regional and became restricted to Africans. Thus, as the total area of St Kitts and some of the smaller islands came increasingly under the control of the Europeans, runaways fled to those islands that the Europeans formally in 1660 conceded to the Island-Caribs, St Vincent especially. Increasingly, it was the African slaves who made up the greatest number of escapees in the Carib islands. So began a different process of acculturation for these Africans, one that continued for over a hundred years. They assumed the ethnic identity of their hosts and, as these latter decreased drastically in number, the newcomers increased and took over their lands and communities.

---

11. This is a term used by Schoelcher, the French abolitionist.

**Figure 4.3** Chatoyer the Chief of the Black Charaibes in St Vincent with his Five Wives. Bryan Edwards *The history, civil and commercial, of the British Colonies in the West Indies.* 1801; engraved from a painting by Agostino Brunias. British Library.

In addition to Raynal's, a few short accounts of the Island-Caribs were published in the last quarter of the eighteenth century. One of these was by George Davidson, who prided himself on knowing the Island-Caribs because of his 'residence near the boundary for two years, some turn for observation, and the footing on which I stand with them' (Coke 1788: 7). Davidson's intent, like that of Thomas Coke, who published his description in an appeal for funds to provide a school and teachers for the children of the Island-Caribs, was religious and colonial. Another account was by Alexander Anderson, who also lived in St Vincent in the last quarter of the eighteenth century.

In commenting on the matter of the safety of Europeans among 'uncivilised' Africans who greatly outnumbered them, early European writers reassured their readers and themselves with the notion that the bitter hatred among various groups of Africans militated against their working together in unison to remove their masters. In the case of the relationships between the Black Caribs and the slaves of the English planters who were introduced into St Vincent in the first quarter of the eighteenth century the same notion is restated by Anderson (1797): 'A fortunate thing for the preservation of the colony was the constant and almost mortal antipathy that existed between the English slaves and black Carribs' (pp. 68–9). Presumably, by the last half of the eighteenth century the Black Caribs had emerged into a clearly differentiated ethnic group, distinct from the African slave and from the creole slave, and, according to Anderson, 'The only appellation they wished to go by was Indian and much hurt if any other was used to them ...' (p. 63).

In highlighting the critical visible feature of the Black Caribs, Anderson said: 'The men in structure of their bodies and color of the skin differ but little from their other African brethren in the islands. The only external difference was the flattening of the head, which custom they adopted from the yellow Carrib' (p. 63). It is on this point that there is some difference of opinion in the accounts. Another explanation given was that the free, much longer resident blacks of St Vincent wanted to differentiate themselves from the black slaves when the latter were introduced there by the French in 1719:

> When the French came to St. Vincent, they brought slaves along with them, to clear and till the ground. The black Caribs, shocked at the thoughts of resembling men who were degraded by slavery, and fearing that some time or other their colour, which betrayed their origin, might be made a pretence for enslaving them, took refuge in the thickest parts of the forest. In this situation, in order to imprint an indelible mark of distinction upon their tribe, that might be a perpetual token of their independence, they flattened the foreheads of all their children as soon as they were born. The men and women, whose heads could not bend to this strange shape, dared no longer appear in public without this visible sign of freedom. The next generation appeared as a new race. (Raynal [1798] 1969 v: 82–3)

Davidson (Coke 1788: 10) and later Coke himself ([1810] 1971 2: 180–1) repeated the same. Anderson's explanation seems more logical if one assumes that the acculturation process of those who became part of Carib societies would have started in the seventeenth century. In other words, flattening of children's foreheads was not an ad hoc response but the perpetuation of a cultural feature among an ethnic group (the Island-Caribs) that had long practised it.

Indeed, because of the stated antipathy between Caribs and slaves, the creole slaves caused this mark of difference to disappear, according to Anderson, who said: 'the flattening the head was left off among some years previous to the insurrection, owing to the negroes by way of ridicule calling them flatheads' (p. 63). So, like the African slaves, whose ethnic markings were ridiculed by creole slaves, the Black Caribs were also ridiculed into modifying their cultural practices with the result that in physical appearance they began to come under the dominant European aesthetic system. Yet, in contemporary accounts, there seemed to be no confusion of Black Carib with slave.

Raynal made the speech of the Black Caribs match what he regarded as their doubly savage state when he said: *'cette race doublement sauvage parle avec une véhémence qui semble tenir de la* colere'[12] ['this doubly savage race speaks with a vehemence that seems to hold anger'] (1773, Bk. xiv, pp. 263–4). Anderson gave his own negative impression of the results of nativisation by referring to the language of the Black Caribs as 'a mixture of their native tongue [African] and that of the aborigines, caribbean, much more guttural and harsh than that of the aborigines' (p. 63). Davidson made a similar kind of assessment of their language, saying: 'Their language, like their natures, is harsh and dissonant. They speak with the utmost impetuosity, and as if they were constantly in a passion ...' (Coke 1788: 20). While these remarks were impressionistic, they were made by people who presumably heard the speech of both the indigenous and the Black Caribs, and indicated that the Black Caribs had a linguistic identity of their own that was clearly non-European and clashed with European sensibilities.

In the period 1719 to 1763 the language of the Black Caribs had undergone changes as a result of their economic, military and everyday association with the French and their slaves. During this period, the French had managed to establish themselves as de facto rulers of part of the island and their language as the general form of communication. This is evident from the words of Young, who, in referring to British actions in 1768, that is five years after they had assumed rule, said:

These instructions were proclaimed and published in English and French, throughout the island of St. Vincent's. It is to be observed, that from the long intercourse with the French all chiefs of the Charaibs speak that language. (1795: 37)

In the same vein, Anderson noted that:

During six years the French were masters of the island [actually the four years 1779–1783] the Carribs had constructed a number of large canoes, many of them capable of carrying from 30 to 40 men. With these they had constant and regular traffic with the French islands, and even as far as

---

12. '[T]his doubly savage race speak with a vehemence that seems to resemble anger' (Raynal [1798] 1969 v: 81).

Trinidad, bartering their commercial articles for arms, ammunition and trifling bagatelles of dress. (p. 57)

They adopted a number of French words, so much so that (with) strict attention (to) their conversation the general tenor of the subject might be known. Most of them spoke a jargon of French, their chiefs tolerably well. (pp. 63–4)

Davidson, who saw himself as giving first-hand information, also observed that 'the French language is almost generally spoken by them' (Coke 1788: 20). In fact, the linguistic repertoire of the Black Caribs at the end of the eighteenth century had a triple layering – part of it was what the Black Caribs had inherited from their hosts, part was what had developed from contact in the nativisation process and part was what they acquired through contact with European colonial communities.

In addition to this layering there was a gender distinction that was identified repeatedly in early historical accounts of the indigenous inhabitants. Anderson also mentioned it:

It is very remarkable that the men had a dialect not understood by the women which they made use of in their councils of war and other matters the females had no concern in. As the women were not originally of their own country, they probably adopted this lingo for want of confidence in them and fear of their communicating their plans to their countrymen. (p. 63)

In contrast, the slave-like status of women in eighteenth-century Island-Carib society was pointed out by Tobin (1785: 121–2) and Davidson (Coke 1788: 11) and no mention is made of linguistic difference between men and women. Indeed, the idea of betrayal seems inconsistent with Davidson's presentation of the women and their total subjection to the men – 'So entirely ... are the wives devoted to the despotism of their husbands, that quarrels among them [the various wives] are never known.' So, with the rise of the Black Caribs and with the increasing European control of the islands, there was a decrease in acquisition of wives from external sources, which in time brought about more gender uniformity in language.

There was a long-held belief about the Caribs' aversion to the English and their preference for the French. It was a belief that dominated the minds of the British and influenced their actions. When in 1658 De Rochefort had said that the Caribs had an aversion to the English and their language and a preference for the French and the French language, he was referring to the indigenous inhabitants (the 'Red' Caribs) and his words were probably no more than nationalistic prejudice and jingoism in a context of colonial conflict between his own country and the English. However, when the British took over St Vincent in 1763, it had become far too French in disposition and philosophy for their liking. According to Young: 'The island was at that time inhabited by about 3000 Black Charaibs, or free Negroes, by 4000 French (their Negroes included), and by about 100 Red Charaibs, or Indians; so reduced were that aboriginal people!' (1795: 18). Toward the end of the eighteenth century the same English belief and attitude about the Caribs' preference for the French persisted, as is evident in a comment by Anderson (1797):

> These traffic and connection with the English inhabitants were trifling in respect to what they had with the French islands, which daily increased. Seldom any of their chiefs or principal men came to Kingstown, to such a height had their aversion to the English become. Although they knew the language so as to speak it equally well as the French, they would use only the last to the English. (p. 57)

The Black Caribs may have come to have some preference for the French over the British, but they were not always pleasant and subservient to the French, for, as Raynal claimed, in their early land dealings with French planters in St Vincent they used a policy of might over written agreement in the same way that many so-called civilised nations did:

> The black Caribs, conquerors and masters of all the leeward coast, required of the Europeans that they should again buy the lands they had already purchased. A Frenchman attempted to show the deed of his purchase of some land which he had bought of a red Carib; **I know not,** says a black Carib, **what thy paper says; but read what is written on my arrow. There you may see, in characters which do not lie, that if you do not give me what I demand, I will go and burn your house to night.** In this manner did a people, who had not learnt to read, argue with those who derived such consequence from knowing how to write. They made use of the right of force, with as much assurance, and as little remorse, as if they had been acquainted with divine, political, and civil right. (1798, vol. v, p. 84)

By the end of the eighteenth century, however, not only because of the presence of French planters in the island for over 50 years, not only because the Black Caribs had adopted French names, but also because of the unmistakable intention of the British to dispossess the Caribs, the Caribs' (both groups) preference for the French over the English had reached a flash point, judging from the ominous words of the British governor of the island, Sir William Young – 'Sad and fatal experience has shown that the combination of barbarous and of national enmity is not to be broken, and that the Charaib will ever be French' (1795: 122). Confirmed in their view of the Black Carib, the British colonial government in St Vincent came to the following conclusion in 1795:

> Under all these circumstances and considerations, the Council and Assembly of St. Vincent's, in the instructions to their agent in London, declare the sole alternative to be – 'That the British planters, or the Black Charaibs, must be removed from off the island of St. Vincent's.' (Young 1795: 124–5)

Of course, it was not the British planters who left but the Caribs who were removed.

The expulsion of the Caribs to Central America in 1797 was really the implementation of an idea that had been contemplated by the British soon after their assumption of rule in St Vincent in 1763 but it became increasingly necessary as a result of events in France and Haiti and in order to halt the spread of the hated republicanism among the black population. It was not only that the Caribs occupied some of the most fertile lands in the island, but that these free and independent people, who were black in colour and considered to have captured the island from the original inhabitants, were too close in character to the revolutionary negroes in Saint Domingue. In the early 1790s the British thought that Victor Hugues, the leader of the French Revolution in the Caribbean, was fostering a regional conspiracy to further republican ideals through revolution by using, among others, Fédon in Grenada, Delgrès in Guadeloupe and the leaders of the Black Caribs in St Vincent. For British ideals to prosper in the islands generally and for the British to gain possession of the richest land in St Vincent specifically, the Black Caribs, who had never ceased to be negroes in the British mind,[13] had to be removed. For the British, there was a certain parallelism between the Black Caribs in St Vincent and Maroons in Jamaica – both had chosen to live outside European 'civilisation'; the beginnings of both were located in the 1650s; both were seen as Africans; both were said to treat their wives worse than slaves; both were banished from their island homes to the north in the 1790s. In short, the British were not prepared to tolerate any competing group and image in their midst.

In contrast to this attitude to the Black Caribs, a kind of pathos developed in the English writing towards the 'original inhabitants', expressed in the words of Anderson:

---

13. Note, for instance, that Tobin (1785: 62) said: 'the free negroes in St. Vincent, (very improperly called Cairibs)' and Young (1795: 6) said: 'The Negroes, or Black Charaibs (as they have been termed of late years) ...'

As has been before observed, the original inhabitants, or Carribs properly so-called, were a numerous people in St. Vincent before the arrival of the blacks from Africa, but so small is the remains of them at present that they deserve hardly to be mentioned as constituting part of the inhabitants. But as they were by God and nature the true proprietors of the country, they deserve some notice; nor can the mind susceptible to humanity refrain from the melancholy reflection as to the destiny of human beings in an offence harmless but brave race of men totally extirpated from their native land, which was the last but most favorite asylum for them when Europeans dispossessed them of all the other islands. (p. 69)

In time this concept of the Caribs as a 'brave race of men' developed into mythological status. At the time, however, a reaction of this type was either a matter of glorifying a vanquished enemy in order to glorify oneself, or (in reference to the Spanish islands) a matter of expressing sympathy for a victim in order to highlight the barbarity of his con-queror. In reality, the 'Yellow' Caribs' demise was substantially a result of their inability to withstand the diseases, deception and might of those who had come into their islands.

While the 'Yellow' Caribs in their demise were being accorded the status of a 'brave race of men', the Black Caribs, who had replaced them, had grown in number and were fighting the English, were being seen as the opposite – 'The St Vincent Carribs were a cowardly race, but what they wanted in courage they made up by cunning' (Anderson 1797: 68). Presumably, the English wanted the Black Caribs to stand in the open, to be shot down and to die gloriously, whereas the Black Caribs preferred to be more intelligent and to use guerilla warfare to frustrate them. So, while eventually they may have gained some conces-sion for their military skill, they could not achieve in the British mind the status of a 'brave race of men' because they were negroes and the perfect example of what the slaves would become if they were freed – 'totally averse to the least civilization' (Tobin 1785: 122) and with their lands 'still nearly in a state of nature' (Tobin 1785: 121). They were a century later even deprived of their exploits and branded as cowards by Ober: 'It may have been that the innate cowardice of the Black Caribs, born of their negro blood, prevented them from taking an active part in the war, and may have induced them to seek the protection of the English' (1880: 217). Ober, who was supposed to be a naturalist and an academic, had by the end of the nineteenth century completely turned around history and recreated a fictional native of the islands.

The Black Caribs were also saddled with a negative image, which dogged them subsequently in their new homes in Central America; their name was corrupted into *crab* because of what Raynal said: 'The flat-headed Caribs, who were nearly of the same age, tall, proper men, hardy and fierce, came and erected huts by the sea-side' ([1798] 1969 v: 83). The Black Caribs were thus likened to crabs because of their appearance and their preferred place of abode. It was an image that apparently was spread across the region, for the awareness of Black Caribs, the negative attitude towards them and the insulting association of them with crabs were quite evident in the boastful remark of a black Barbadian slave – 'Me neder Chrab, nor Creole, Massa! – me troo Barbadian born' (Pinckard 1806 1: 133). The linguistic development of the Black Carib, from African to Carib with an addition of French and English in a situation where the man's language was believed by some to be different from the woman's and where the English known was deliberately not used, must have constituted the zenith of corruption and deviance in the minds of the British. On top of this, the image of the crab and its way of foraging for food made the Black Carib even more detestable.

In the eighteenth century, the islands presented a picture of varying cultural identity in which the central process was a graduated form of creolisation, stretching from black Creoles at the lower end through mulattos to white Creoles at the higher end. In this process whitening was accepted as the mode of *ascent* and was thus associated with enlightenment or emergence from darkness. The islands themselves were to be seen within the bigger picture of the New World in which a new white world was establishing itself but where the encounter between Europeans and the indigenous inhabitants had not been fully resolved. In this continuing encounter Europeans were more forcefully portraying themselves as enlightened, while they were portraying the indigenous inhabitants as savages. Within the islands, however, the indigenous inhabitants, who for the most part were outside the central process of creolisation, were increasingly being regarded in a sympathetic and idealised way as they declined and disappeared. In contrast, the relatively new phenomenon of the Black Carib was tarnished with the original 'hostile savage' image of the Carib in the smaller islands. The newly developing image of the free, brave and unspoiled (Yellow) Carib, together with the 'doubly savage' Black Carib, enhanced the identity of exoticism that the islands had. It was an image that was further amplified by visions of roving buccaneers at sea and roaming Maroons on land, both of whom were also outside the central process.

Exoticism was not automatically seen as negative: there was ambivalence in European presentations of those people living outside the central process of creolisation as well as those living within it. For instance, there was some admiration for pirates and buccaneers in the bigger islands who lived outside colonial law and institutions and whose actions led to considerable loss of life and destruction of property. While there was admiration for the disappearing 'Yellow' Carib, there was much less for the Black Carib in St Vincent especially when the Black Carib became a colonial foe. There was none for the unfettered or unconquered Maroon except when the Maroon was seen as ally against runaway slaves. There was little – except when it affected profits of European proprietors – for the white Creole when the white Creole defended his habits and characteristics and when the white Creole expressed his desire to be out from under the domination of European colonial power. In other words, though the same yearning for freedom characterised all these groups, it was seen as positive and the people enlightened only when it was in accord with colonial policy. The various creole groups were in no position to trumpet their own values and achievements through written literature. Creole literate output was no match in volume and intensity for European literate output. European literature put forward the political, economic and cultural practices of Europeans as models and norms against which everything was to be measured and with which 'enlightenment' had to be aligned.

# European satirical literature and its relation to creole love of banter and ridicule

The eighteenth century in literary Europe was characterised as the age of satire and this was reflected in the literature on the West Indies right from the beginning of the century. In 1679 Thomas Trapham, a doctor, published a book called *A Discourse of the state of health in the island of Jamaica With a provision therefore calculated from the air, the place, and the water; The customs and manners of living &c.* In 1705 the third edition of Captain Hickeringill's 1661 *Jamaica viewed* was published and in 1707 Dr Sloane's *A voyage to the islands Madera, Barbadoes, Nieves, S. Christophers and Jamaica* appeared. These standard descriptions of Jamaica occurring in quick succession were separated by one that was quite different in nature, Edward Ward's *A trip to Jamaica: with a true character of the people and island* (1700). In his introductory statement *To the Reader*, Ward stated his intention: 'I only Entertain you with what I intend for your Diversion, not Instruction, Digested into such a Stile as might move your **Laughter**, not merit your Esteem.' He then went on to make light of his own profession in a cleverly worded comparison:

> The Condition of an **Author**, is much like that of a **Strumpet**, both exposing our **Reputations** to supply our **Necessities**, till at last we contract

such an ill habit, thro' our Practices, that we are equally troubl'd with an **Itch** to be always Doing; and if the **Reason** be requir'd, why we betake our selves to so Scandalous a Profession as **Whoring** or Pamphleteering, the same excusive Answer will serve us both, viz. That the unhappy circumstances of a Narrow Fortune, hath forc'd us to do that for our Subsistance, which we are much asham'd of.

The chiefest and most commendable Talent, admir'd in either, is the Knack of Pleasing; and He or She amongst us that happily arives to a Perfection in that sort of Witchcraft, may in a little time (to their great Honour) enjoy the Pleasure of being Celebrated by all the **Coxcombs** in the Nation.

The only difference between us is, in this particular, wherein the Jilt has the Advantage, We do our Business First, and stand to the Courtesie of our Benefactors to Reward us After; whilst the other, for her Security, makes her **Rider** pay for his **Journey**, before he mounts the **Saddle**. (p. 3)

Ward intended to enjoy himself and he did this by poking fun at Jamaica. Since he obviously spent a lot of his time in Jamaica at Port Royal with a nefarious crowd, his account reflected the kind of banter that would have characterised such a crowd.

His portrayal of Jamaica was done in the kind of spirit of ridicule that came to characterise writing in England and France in the eighteenth century:

### A Character of Jamaica

The Dunghill of the Universe, the Refuse of the whole Creation, the Clippings of the Elements, a shapeless Pile of Rubbish confusd'ly jumbl'd into an Emblem of the <u>Chaos</u>, neglected by Omnipotence when he form'd the World into its admirable Order. The Nursery of Heavens Judgments, where the Malignant Seeds of all Pestilence were first gather'd and scatter'd thro' the Regions of the Earth, to Punish Mankind for their Offences. The Place where <u>Pandora</u> fill'd her Box, where <u>Vulcan</u> Forg'd <u>Joves</u> Thunder-bolts, and that <u>Phaeton</u>, by his rash misguidance of the Sun, scorched into a cinder. The Receptacle of Vagabonds, the Sanctuary of Bankrupts, and a Close-stool for the Purges of our Prisons. As sickly as an Hospital, as Dangerous as the Plague, as Hot as Hell, and as Wicked as the Devil. Subject to Turnadoes, Hurricanes and Earthquakes, as if the Island, like the People, were troubled with the <u>Dry Belly-Ach</u>. (p. 13)

Ward's satirical characterisation of Jamaica, written less than 50 years after the English had taken Jamaica from the Spanish, contrasted both in its tone and substance with one of Barbados, written by Keimer (1741: iv), in which the older British colony continued to be associated with the matter-of-fact, business side of the British. Jamaica, 'an Emblem of the *Chaos*', and Barbados, 'a single family', were in a sense portrayed as opposites, but the former came over as a much more exciting place, in keeping with the effervescent spirit of the age. Evidently, the kind of buccaneering spirit that was characteristic of the northwest of Hispaniola had extended to Jamaica.

Ward's book on Jamaica was satirical, but the focus of it was still the people of the New World. The next book, purporting to be about Jamaica, was different. People had become tired of the many accounts of travel to the New World repeating the same things and written by doctors. The accounts and their writers became the subject of satire, an early example of which written in 1709 was William King's *Useful transactions for the months of May, June, July, August and September, 1709. Containing a Voyage to the island of Cajamai in America.* This was a satire on authors on Jamaica (> Cajamai), specifically Sir Hans Sloane and Thomas Trapham, both of whom were doctors. In relation to Sloane, King was suggesting that what he wrote was not worthy of being in a book because it was obvious, common knowledge, and that he was not qualified to write about botany, that he was not a good doctor and that the illnesses mentioned were not serious. In the case of Trapham, King ridiculed his introductory pompous reference to a 'great' authority, his long discourse on water as well as uninteresting treatments for illnesses. King was suggesting that travellers' accounts were trivial in a world where political and philosophical matters were of great importance.

King was known to Jonathan Swift, who in 1726 published one of the most famous books of the eighteenth century – *Travels into several remote nations of the world. In four parts. By Lemuel Gulliver, first a Surgeon, and then a Captain of Several Ships.* Although Gulliver spanned a whole concocted world, Swift's references to the West Indies (vol. 2, p. 157) showed that he was familiar with the deeds of pirates and buccaneers, which had formed the basis of Ward's satirical description of Jamaica. Swift's book, written in the style that had become familiar for travel accounts, completed the process of turning inward. For, while the seventeenth-century European reader of travel accounts was allowed to feel superior to the un-Christian, unintelligent, uncivilised Savage and African over there, this book called into question behaviour, morals, intelligence, self-centredness and, more specifically, the dominance of one set of people over another, as well as indulgence in trivialities by writers, scientists and philosophers, in what Moore suggests was 'an extravagant burlesque on voyages' (1941: 216). More generally, Swift was seen as a champion of the Irish in their struggle with the English and it is this kind of colonial theme that obviously appealed to persons in the Caribbean.

The vision of the plantation as a rural village would suggest that it was in a different world from Swift, outside the reach of literary satire. The plantation was, however, not devoid of cultural activities, and the general process of acculturation that was taking place there meant that European features, introduced by local whites, were providing frameworks or the parameters of operation for the cultural expressions of the whole society, including slaves and servants. In Barbados in the 1730s there was a prominent person, a Dr Towne, who was said[14] to know Voltaire and to have translated part of his *Henriade*. In addition, the local contributions in the *Gazette*, the earliest newspaper in Barbados, as well as in other newspapers, were characterised by repartee in the sections that could be regarded as cultural. Somewhat later, in the 1780s in St Kitts, a response to an attack on Creoles by Rev. Ramsay started off with a reference to Dean Swift, and another reference in a newspaper (*Barbados Mercury* 14 August 1787) mentioned him with the same kind of admiration.

No doubt the enslaved African demonstrated a variety of cultural features in their conversations and social intercourse, but certain ones may have been highlighted by writers and may have outlived others precisely because they coincided with European preferences. In 1667 Du Tertre made the following observation about the African slaves:

> *Comme ils sont grands railleurs, ils relevent les moindres défauts de nos François, & ils ne sçauroient leur voir faire rien de reprehensible, qu'ils n'en fassent entre eux le sujet de leur divertissement, & de leur entretien.*
>
> *Je ne sçay si les chansons qu'ils marmotent en travaillant, procedent de la gayeté de leur temperament, ou s'ils les disent pour charmer leurs fatigues; mais ils paroissent d'une humeur assez enjoüée, & chantent ordinairement chacun en son particulier quand ils travaillent une chanson, dans laquelle il repetent tout ce que leurs Maistres ou leurs Commandeurs leur font de bien ou de mal.* (p. 497)

> [As they like to make fun of people, they pick out the slightest faults of us French and to them there is nothing that can be made to seem reprehensible that is not made among them a subject of amusement and entertainment.
>
> I do not know if the songs that they hum as they work come from the gaiety of their temperament or if they are used to soothe their tiredness; but they seem to be quite playful and they normally sing, each in his own way, a song in which they repeat all the good or bad things that their masters or commanders have done them.]

It should not be surprising to learn that another French writer made a similar claim some years later. Referring to the year 1698, Labat said about the African slaves:

---

14. *Barbados Gazette* Wed. 8 August 1732.

*ils sont railleurs à l'excès, & que peu de gens s'appliquent avec plus de succès qu'eux à connoître les défauts des personnes, & sur tout des Blancs, pour s'en mocquer entr'eux, & en faire des railleries continuelles. Si-tôt qu'ils ont reconnu un défaut dans quelqu'un, ils ne le nomment plus par son nom, mais par quelque sobriquet, qui a du rapport à ce défaut. Ce sobriquet est parmi eux un mistere, qu'il est bien difficile aux Blancs de pénétrer, à moins que sçachant leur Langue, on ne le découvre en les entendant se divertir des personnes dont ils parlent par des railleries piquantes, & pour l'ordinaire très-justes. J'ai souvent été surpris des défauts qu'ils avoient remarquez, & de la maniere dont ils s'en mocquoient ...* (1742 iv: 480–1)

[they really like to make fun of people and there are few people who have been so successfully committed to discovering the faults of people and especially those of whites in order to make fun of them among themselves and to make them into constant jokes. As soon as they recognise a fault in someone, they do not call the person by name, but by some nickname that is related to the fault. This nickname is a secret among them which is very difficult to understand unless you know their language and come to understand it when you listen to them making fun of persons with stinging jokes which are most times very accurate. I have often been surprised by the faults that they observe and the way they make fun of them ...]

Toward the end of the eighteenth century, another French writer, Moreau de Saint Méry, also made it seem as if a kind of sharp repartee was the most typical characteristic of the slaves in Saint Domingue in their conversations:

*Un grand plaisir pour lui, c'est de causer en mangeant, et s'il se trouvent plusieurs nègres ensemble, chacun a son assiette, ou bien chacun puise à son tour dans un plat commun. C'est le moment des contes, qu'interrompent de grands éclats de rire. La saillie, l'épigramme, car le nègre est railleur, animent les convives et l'hyperbole est admise pourvu qu'elle amuse.* ([1796] 1958 1: 61–2)

[A great pleasure for them is to chat while eating and if many negroes are assembled together, each one has his plate or rather each one takes his turn from a common plate. It is the time for stories, which are interrupted with great peals of laughter. The witticism, the epigram, for the negro likes to make fun of people, enlivens the participants and hyperbole is allowed as long as it amuses.]

Moreau de Saint Méry went on further:

*C'est dans ce langage qui, comme l'on voit, comporte la rime et la mesure, que les Créols [de toutes les couleurs] aiment à s'entretenir, et les nègres n'en ont pas d'autre entr'eux. C'est encore par son moyen, que les nègres expriment et leurs mots sententieux, [que j'ai dit qu'ils aimaient] et leurs traits piquans.* ([1796] 1958 1: 82–3)

[It is in this language which, as can be seen, has rhyme and beat, that Creoles (of all colours) like to amuse themselves; and the negroes have no other among them. It is again by this means that the negroes express both their proverbs and their banter.]

It is quite unlikely that the comments of Du Tertre, Labat and Moreau de Saint Méry were based on independent observations; it is more likely that they were repetitions of comments previously occurring in the literature over a 150-year period.

The love of ridicule and banter and the penchant for the epigrammatic which were reported for the French colonies were also observed by Long in Jamaica:

Instead of choosing panegyric for their subject-matter, they generally prefer one of derision, and not unfrequently at the expence of the overseer, if he happens to be near, and listening: this only serves to add a poignancy to their satire, and heightens the fun. ([1774] 1970 2: 423)

Again the similarity of the comments does not suggest totally independent observation but rather the influence of topical literature. So, while the satirical features among the black section of the population were undoubtedly part of their traditional culture, they may have publicly dominated other non-satirical ones because initially they found encouragement in the general atmosphere of the eighteenth century and also because later writers kept on repeating/featuring them, thereby illustrating the power of the printed word in the formation of identity.

In spite of the fact that derision/satire was highlighted among the slaves by French and English writers, it was more a result of harsh social and economic conditions than an exceptional trait of the people. First, as said before, it was French and English writers who singled out and repeated this characteristic. Second, derision, as an offensive tactic, could have been a part of inter-ethnic rivalry, the kind reported among the slaves in the early years, but which abated with the decline in the slave trade. Used against masters and those in superior social positions, it was also symptomatic of powerlessness, which was an economic reality subject to change. It was thus a human response to circumstances of tension and oppression and it is only in its specific manifestations that its cultural character was notable.

# Creole literature and language and assertion of identity

Critics differ about the intentions and personalities of Swift and Voltaire, as to whether they were just mean people or whether they were interested in the well-being of the societies in which they lived, but there is no question that their satire and that of the eighteenth century generally were accompanied and followed by significant technological, industrial and social changes. The characterisation of the eighteenth century as the age of enlightenment applies primarily to Europe, but the idea of 'enlightenment' can also be seen as applying positively and directly to the eighteenth-century Caribbean, in the way that Creoles in their literate expression became aware of themselves, showed some amount of intellectual achievement and sophistication and displayed some level of artistic creativity. The Americans' assertion of political independence in the last quarter of the century showed that 'Creoles' on the mainland had come to believe in themselves, especially since they had the press for support. Creoles in the islands also had similar sentiments and aspirations.

In spite of all the negative elements of Creole life, Thibault de Chanvalon, a French Creole, was optimistic in his vision of the Caribbean when he said:

> *il commence à regner dans nos Isles quelque emulation pour les lettres; le desir d'acquérir des connaissances paroît s'introduire dans ma patrie. Peut-être qu'enfin cette masse de lumiere qui éclaire l'Europe depuis un siecle, qui a pénétré par-tout successivement, passera les mers un jour, & qu'elle étendra ses rayons & son influence jusqu'à ce nouveau continent.* (1763: 36)

> [in our islands some love for literature is beginning to show itself; the desire to acquire knowledge seems to be emerging in my native land. Perhaps finally the mass of light which has been illuminating Europe for a century and which has penetrated all places, one after another, will traverse the seas one day and will spread its rays and influence as far as our continent.]

On the other hand, Girod-Chantrans, a foreigner, suggested otherwise, saying:

> *Si, à mon tour, je demandois aux blancs de cette colonie pourquoi elle n'a pas encore fourni un seul écrivain célebre depuis son établissement, quoiqu'ils sussent au moins lire & écrire en y arrivant, de même que la plupart de leurs prédécesseurs, qu'auroient-ils à répondre?* (1785: 168)

[If, in turn, I asked the whites of this colony why it has not yet produced a single famous writer since its founding, although they could at least read and write when they came, just like the majority of their predecessors, what would they say?]

In direct response to this question, Dauxion-Lavaysse some years later remarked:

Partial or ignorant writers have said that the American Islands have never produced a man distinguished in literature and the fine arts; but Martinico, for instance, did it not give birth to the late M. du Buc? Could he have been an ordinary person, I allude to Blanchetiere Bellevue, who, never having left that colony before, nor received a literary education, at the age of thirty-six, appeared like a meteor in the constituent assembly, where he was admired for his captivating eloquence, and the variety of his knowledge? The celebrated physician Lamure was a Creole. France, Spain, and Great Britain, reckon among their celebrated existing characters a great number of Creoles ... ([1820] 1969: 180)

This response did not precisely answer Girod-Chantrans' question, but it showed that there were European writers who were willing to challenge the bigotry of their countrymen. Yet, one may ask whether the optimism of the Creole, Thibault de Chanvalon, was any more or less merited than the apparent pessimism of the foreigner, Girod-Chantrans?

Even if the claim that Saint Domingue had not produced a famous writer was true, a similar claim could have been disputed for the whole of the French Caribbean, since Thibault de Chanvalon, a Creole who referred to himself as an *Americain*, had written a respectable work. In addition, within a few years of Girod-Chantrans' question Moreau de Saint Méry, who was born in Martinique, had written his work on Saint Domingue, and by the end of the century other Creoles had written works in response to attacks on Creoles. It is not all of these works that were of historical significance, but in some respects they showed a different level of understanding of things Caribbean. Most significantly, their comments on language showed a finer sense of appreciation of the cultural use of the creole language. Even though the writers born in the Caribbean were not making absolute claims about the capability of the creole language, they clearly had reached a level of self-confidence that they could favourably compare the creole with European languages, pointing to what they regarded as its genius and finesse. For example, Thibault de Chanvalon, without directly mentioning the creole language, implicitly validates it in an explanation of poetic and musical composition among the slaves:

*Ils sont tout-à-la fois poëtes & musiciens. Les regles de leur poësie ne sont pas rigoureuses; elles se plient toujours à la musique. Ils alongent ou*

*abregent au besoin les mots pour les appliquer à l'air sur lequel les paroles
doivent être composées.*

*Leurs compositions nous ramenent à l'idée que nous pouvons avoir de
la naissance de la poësie dans les premiers âges du monde. Un objet, un
événement frappe un Negre, il en fait aussi-tôt le sujet d'une chanson.
Trois ou quatre paroles, qui se répetent alternativement par les assistans,
& par celui qui chante, forment quelquefois tout le poëme; cinq ou six
mesures font toute l'étendue de la chanson.* (1763: 66–7)

[They are all at the same time poets and musicians. The rules of their
poetry are not rigorous; they bend to the music. They lengthen or shorten
the words according to need to make them fit the tune within which the
words must be composed.

Their compositions take us back to the idea that we may have of the
birth of poetry at the beginnings of the world. An object or an event strikes
a negro and he immediately makes it the subject of a song. Three or four
words repeated alternately by the singer and those present sometimes
make up the whole poem; five or six beats make up the full extent of the
song.]

More explicitly, Moreau de Saint Méry directly pointed out the subtlety
and the superiority of the creole intentionally to contradict Girod-
Chantrans's view (1785: 189) that the creole was *'faible, maussade &
embrouillé'* ['weak, crude and muddled']:

*Il [le créol] a aussi son génie ... et un fait très-sur, c'est qu'un Européen,
quelque habitude qu'il en ait, quelque longue qu'ait été sa résidence aux
Isles, n'en possède jamais les finesses... ..*

*Il est mille riens que l'on n'oserait dire en français, mille images
voluptueuses que l'on ne réussirait pas à peindre avec le français, et que le
créol exprime ou rend avec une grace infinie. Il ne dit jamais plus que
quand il employe les sons inarticulés, dont il a fait des phrases entières. Le
Chia, le Bichi même, qu'on a tant voulu ridiculiser, est-il un terme de
dédain qui renferme plus de sens? Et pour qu'on ne prétende pas que je
crée des merveilles imaginaires, je vais rapporter une chanson bien
connue, qui fera voir si le langage créol est un jargon insignifiant et
maussade. Elle a été composée, il y a environ quarante ans, par M.
Duvivier de la Mahautière, mort Conseiller au Conseil du Port-au-Prince.*
(Moreau de Saint Méry [1796] 1958 1: 80–1)

[The creole has its genius ... and one thing that is very certain and that is
that a European, whatever customs he has and however long he has been
living in the islands, never masters its finer points... .

There are a thousand nothings that one would not dare say in French
and a thousand voluptuous images which one could not paint with French
that the creole expresses or renders with infinite grace. It never says
anything better than when it uses inarticulate sounds with which it makes
whole sentences. The very 'chia' and 'bichi', which people wanted to

ridicule so much, is there an expression of disdain which has more force? And so that no one can claim that I believe in imaginary marvels, I am going to recall a well known song which will show if the creole is an insignificant and crude jargon. It was composed about fifty years ago by Mr Duvivier de la Mahautière, late councillor in the Council of Port-au-Prince.]

A point similar to the one made by Moreau de Saint Méry, that complete mastery of the creole language was beyond the ability of the foreigner, was made a few years earlier by Samuel Mathews, a native of St Kitts. What Mathews said (1793: 141–2) was that no Creole white or European took the time or trouble to learn the language of the slaves. Mathews was so proud of his competence in the slaves' language that he cited two songs 'which I wrote in the year 1786, and had the honour of singing both, for His Royal Highness Prince William Henry, the last day of Feb. 1787'.

Such positive expressions by Creoles about the languages of the region, the languages of the slaves and, especially in Saint Domingue, according to Moreau de Saint Méry, the language of white Creoles, demonstrated feelings of pride in their own characteristics and identity. The strongest indication of the strength of the link between creole identity and language came from Saint Domingue: Moreau de S. Mery used the term '*creol*' to refer to both the language and the people and, as Girod-Chantrans (1785: 191) pointed out, the creole language was used by all the people. It was within a few years of each other that some of the earliest verse in French and English Caribbean creoles respectively (Mathews 1793; Moreau de Saint Méry 1796) appeared in print. The pieces were cited in reaction to ridicule by foreigners – just as Mathews derided Moreton's citations of creole, claiming that they were the author's concoctions, so too did Moreau de Saint Méry respond to what Girod-Chantrans claimed to be an example of creole. Both writers, white Creoles, said that their pieces had been composed earlier. Both dealt with man–woman relationships from the man's point of view. Neither poem was intended to be a caricature, seeing that the themes of infidelity and unrequited love were significant for the black, powerless slave in the plantation context. In fact, there is a certain similarity in theme in this kind of poem/song and the early 'blues' songs of African Americans in the southern states of the USA. So, in addition to the fact that the creole languages in the Caribbean had clearly reached a level of stability, there were native speakers of them who were willing and able to illustrate and defend them. European writers, therefore, could no longer trot out the *langage corrompu*, filled with European simplifications of European languages, and expect to have it universally accepted as what Creoles really said.

Why did these white Creoles praise the creole language, illustrate its literary capability and thus associate themselves with their slaves against outsiders? Actually, it was a consequence of the spirit of ridicule and self-examination that Ward, King and, more generally, Swift and Voltaire had spread and the kind of reactions it had engendered. Writers on the Caribbean, through harsh and direct criticism, held up Creoles and creole society to ridicule, as, for example, the blatant attack on Creoles made by Edward Thompson (1770: 126), who was incensed by his reception as a foreigner in Antigua and Barbados, as well as by the poet Ulton's favourable words about the people of these islands. In time, Creoles began to react quickly and strongly in defence of themselves, thereby, no doubt, confirming in the minds of Europeans 'that volatile spirit so peculiar to the Creole' (Thompson 1770: 112). The virulence of the attacks and counter-attacks that came to characterise Caribbean writing right through to Emancipation really first became dramatic and deadly between two Frenchmen who spent some time in Saint Domingue. When Hilliard d'Auberteuil wrote his critique of Saint Domingue in 1776, he probably had no idea of the intensity of the reactions to his book. In the first place, it was suppressed by the Conseil d'Etat. Then, in 1780 Pierre Dubuisson wrote a systematic response to it, going through the book section by section, chapter by chapter attacking almost everything that Hilliard d'Auberteuil had had to say. The attack and counter-attack must have caused some acute personal animosity between the two men because when Hilliard d'Auberteuil was killed in Saint Domingue in 1789, Dubuisson was believed to have been behind it.

The response to Rev. Ramsay's attack on St Kitts in 1784 by Some Gentlemen of St. Christopher the same year, the response to Moreton's (1790) attack on Creoles in Jamaica by Mathews (1793) and, to a lesser extent, the response of Moreau de Saint Méry (1796) to Girod-Chantrans (1785) showed the same kind of systematic response to details in attempts to refute them. In these cases, beside the fact that nobody was murdered because of what they wrote, there was a critical difference. In each case, the counter-attack was made by Creoles, who, by having to defend slavery, their treatment of slaves and their way of life, highlighted what they thought were their own good points and the enormous value of their contribution to European economies. So, by being put in the spotlight, they in a sense became enlightened. On the other hand, those in Europe who were behind the attacks were in part trying to discredit this new phenomenon, the parvenu, who sent his children to live among them in decadence and immorality and who came to retire among them and to display himself with servants. White Creoles were socially and culturally a problem to Europeans because, being 'shoots', they were neither foreign nor native, but their association with other races and their life in hot climates obviously tainted them, making them degenerate, unenlightened Europeans.

In the Spanish colonies, Cuba was somewhat different from Puerto Rico and Santo Domingo as a result of the fact that in the eighteenth century there was a slow but prolonged build-up of *bozal* Africans there, which gradually affected the identity of the island: it became less white and more black. The build-up also affected the language situation in that the speech of these Africans, called *habla bozal*, became one of the characteristic and prominent language varieties in the island. It may not have been as pervasive as *creol* was in Saint Domingue at the end of the century, but it was much more so than in Puerto Rico and Santo Domingo, for in those islands there was no association of specific language characteristics with the people, either by Raynal in the case of Santo Domingo or by Abbad y Lasierra or Ledru in the case of Puerto Rico. In fact, Abbad y Lasierra made no mention of the everyday speech of the people of Puerto Rico. So, evidence of vernacular language from the Spanish islands in the eighteenth century comes only from surviving anonymous songs from Cuba, which Arrom commented on as follows, giving some idea of their cultural context as well as the nature and significance of the *habla bozal* used in them:

> *el canto folklórico africano trasplantado por el esclavo a su nuevo pais. Ahí encontramos el canto de cabildo de la negrada del ingenio, la marcha de comparsa de sus fiestas, o el himno fervoroso a sus dioses, entonados a menudo en ese idioma, mezcla de español y dialectos africanos, que venía a ser como una **lingua franca** entre los infelices extraídos de distintas regiones por los ávidos buques negreros. Esa poesía popular negra, que coexistía con la décima del campesino blanco y la composición culta del poeta letrado, es la que, viviendo vida de obscuridad durante los siglos XVIII y XIX, ha venido a brotar redimida en lo más puro de la música y el verso afrocubanos de hoy.* (1941: 388–9)

> [the African folk song transplanted in his new country by the slave. Here we find the cabildo song of the sugar mill negroes, the parade of his fiestas or the fervent hymn to his gods, often intoned in this language, a mixture of Spanish and African dialects, which came to be like a lingua franca among the unfortunate slaves taken from different regions by greedy slave boats. This popular black poetry, which coexisted with the décima of the white peasant and the sophisticated composition of the educated poet, is the one which, living a life of obscurity during the eighteenth and nineteenth centuries, has come to sprout redemption in the purest Afro-Cuban music and verse today.]

No doubt it was the contrast between, on the one hand, the language and literature of the *peninsular*, aspired to by the upper society *criollo*, and, on the other hand, the *habla bozal* and the *poesía popular negra* that made the latter stand out and caused it to be recorded.

The anonymous eighteenth-century songs recovered by Ramón Guirao ([1938] 1970) illustrate features of the *habla bozal* that link the users not only to varieties of *ladino* speech in Spain but also to characteristics of language in the French and English islands, which were numerically dominated by Africans and people of African descent. Arrom's assessment that these songs were a mixture of Spanish and African languages is based on the fact that, while many of the words in them are Spanish, a fair proportion of them, especially in the choruses/refrains, can be identified as African, specifically Kikongo (e.g. *cutu, mambo, diambo, dinga, muana, cabanga, lucuanda, mayimbe, kangala, baquini*). Lipski regards the language of '*Un canto de cabildo*' as 'pidginised Spanish' and goes on to comment that 'the phonological modifications are those of vernacular Cuban Spanish, and suggest that Africans in eighteenth century Cuba were extending already existing consonantal reduction ...' (1994: 107). However, if 'pidginisation' is accepted to be a process of contraction with influence from first-language features, then it does not seem logical to claim that eighteenth-century Africans (i.e. speaking African languages) in Cuba were, in their 'pidginised' Spanish, *extending* already existing features of vernacular Cuban Spanish. The fact is that these songs were more than mixtures or 'pidginised' Spanish, but what no doubt leads to such erroneous assessments is that they are consistent in their appearance with later (nineteenth-century) stereotypical presentations of *habla bozal*.

Some pronunciation features of the songs resemble those that were typical of creole speakers in the islands where African-derived people were numerically dominant, for example:

| | | |
|---|---|---|
| loss of unstressed first syllable | | *cucha* (< escucha), |
| | | *nimá* (< animal) |
| reduction of consonant clusters | [-sk-] > [-k-] | *roccá* (< ro<u>sc</u>ado) |
| | [-sm] > [-m] | *mimo* (< mi<u>sm</u>o) |
| | [-st-] > [-t-] | *asutta* (< asu<u>st</u>ar) |
| | [-rm-] > [-m-] | *fomma* (< fo<u>rm</u>a) |
| | [-rt-] > [-t-] | *cotta* (< co<u>rt</u>a) |

preference for consonant + vowel syllable structure as evidenced by
a.  the tendency for words to lose their final consonant and end in vowels
    e.g. *nimá, matá, mayorá*
b.  the adding of the final consonant in the article to the beginning of
    noun which starts with a vowel, e.g. *sojo* (< los ojos).

As to word and sentence structure, there are also similarities with creole languages, for example:

| | |
|---|---|
| absence of plural endings on nouns | *diente, sojo* |
| a single form for all persons of the verb | *son/so, ta* |
| inconsistent marking of gender | *Ma Rosario ta malo;* |
| | *cosa malo; la cabayero* |

subject copying, that is, the insertion of a pronoun after a noun subject and before the verb – this is pointed out by Castellanos and Castellanos (1992: 341):

> *En una de las muestras más antiguas (del siglo XVIII) el pronombre "né" es también empleado como copia del sujeto:*
>
> > *Mira sojo d'ese nimá*
> > *Candela **né** parese*
> > *¿Que nimá son ese*
> > *Que **né** parese majá?*
>
> [In one of the oldest pieces of evidence (from the eighteenth century) the pronoun *ne* is also used as a copy of the subject:
>
> > Look at the eyes of this animal
> > It looks like fire
> > What kind of animal is this
> > that looks like a snake?]

In fact, Castellanos and Castellanos (1992: 321–56) give a full analysis of *el habla bozal cubana* in terms of its similarities to creole languages.

In addition to the African-influenced linguistic features, there are in the songs a few words of indigenous origin – *guataca* [hoe], *majá* [snake], *jocuma* (= *arbol silvestre* Pichardo [a tree that grows wild]). However, these are words that were a part of everyday Cuban life and had become a normal part of Cuban Spanish; they do not necessarily indicate any direct connection between Africans and the indigenous inhabitants in Cuba.

On close examination the songs are not as simplistic as they may at first appear to be. As expressions of cultural identity, they mark a stage in the linguistic development of *bozal* Africans in Cuba. The song *Un canto de cabildo* has the nature of a competitive challenge between drummers. The high pitched sounds in the first lines *Piqui, piquimbín, piqui, piquimbín* and the lower pitched sounds in last lines *Pa, pa, pa, práca, Prácata, pra, pa* contrast the sounds made by different drums (that is, *mula* and *tumba* (< *tumbador*). The spirit of competition is also captured in the word *gayo* from *gallo* in the context of cockfighting. From a 'literary' point of view, there is a very clever play on words in the main (second) part of the song. First, between *jacha* [axe], *quiebra-hacha*, literally 'break axe' and thus 'a tropical tree with very hard wood' (therefore a harder tree) and *resquebrajarse* [crack, break, split]; and, secondly, between *tumbador* [kind of drum] and *tumbar* [to fell a tree]. The relationship between tree and drum is that the drum is made from a hollow tree trunk, a hard tree that had to be felled, and the relationship between the two is further heightened by the alternation of *tumba* and *tambó* at the beginning of lines in the song. Such 'literary' features would hardly be typical of 'pidginised' Spanish.

The song *Canto Congo de cabildo* demonstrates the process of accommodation in communication between groups speaking different languages. Castellanos and Castellanos give some idea of this with the notion of *desplazamiento*, that is the displacement of African language features by Spanish ones:

> El **bozal** y, más tarde, el español se convirtieron así en una suerte de **lingua franca** entre el padre Nganga y sus espíritus, ya que éstos, a lo mejor, no comprendían el congo. Los lucumíes, por el contrario, podían y pueden confiar en la competencia lingüística de sus dioses yorubas, quienes prefieren que se les hable en su idioma. Vemos, por tanto, como factores religiosos pueden contribuir a los procesos de retención o desplazamiento lingüísticos. (Castellanos and Castellanos 1992: 317)

> [The bozal and later Spanish thus became a kind of lingua franca between the Papa Nganga and his spirits, now that the latter at best no longer understood Congo. The lucumís, on the contrary, could and can trust the linguistic competence of their Yoruba deities, who prefer to be spoken to in their language. We see therefore how religious factors can contribute to the linguistic processes of retention or displacement.]

However, accommodation was a more subtle process than displacement. It may have actually preceded it and it shows how some words are selected to enhance communication. It is what would in a literary and figurative context be regarded as a play on words, which may well be the reason for and the source of the love of such word play in creole contexts. The examples in this song are *cutu and dinga*, both of which can be interpreted to have a double source:

(1)     *cutu* < *kutu* (Kikongo) [ear]
        *cucha* < *escuchar* (Spanish) < *auscultare* (Latin) >
        *ecouter* (French) [hear]

(2)     *dinga* < *ndinga* (Kikongo) [speak, say]
        *diga* (Spanish) +addition of a nasal sound before sounds
        like /g/ (normally in initial position in the word) common
        in creoles such as Palenquero and Saramaccan [say, tell]

It should be noted that 'ear/hear' and 'speak' are two very basic words and no doubt primary in initial communication between speakers of different languages.

Viewed from the point of view of a Spanish speaker or a monolingual African speaker, the poems may be characterised as mixed in their language, but as items of cultural expression of a group, they represent language use at a higher level than basic communication for functional (economic) purposes: this was a coherent code used for purposes of social cohesion. In any case, it is unlikely that any group of people would have been expressing themselves among themselves in a cultural context in a language that was not near and dear to them or in a language that did not have some currency as a genuine medium of communication of emotions, desires and intentions. In fact, this variety of language was noticeable as a variety within Cuba because it was a social and cultural entity.

The appearance in print of 'cultural' instances of slave language (i.e. poems and songs) around the same time in the eighteenth century in the English and French islands and in Cuba was not purely coincidental. Writers in each colonial context wanted to demonstrate the literary capabilities of their own slaves, and general evidence suggests that the songs were genuine or at least reflected contemporary cultural reality. What is clear in a comparison of the songs from the three linguistic groups is that the Cuban songs contained a greater number of African words and other African features than those from Saint Domingue and St Kitts, indicating therefore that there was a difference between the degree of evolution of the language of Africans (*bozales*) and that of Creoles. On the other hand, the stage of cultural development among the black slaves in the much older colony, Cuba, appeared to be the same as that in Saint Domingue, which itself had a massive influx of African slaves in the eighteenth century.

## Saint Domingue creole – tracing the change from 'corruption' and diversity to autonomy

The historical literature points to what would be a neat account of the development of French Creole language in the islands. There is a clear indication that the starting point was St Kitts, the place of the first French settlement, and that it was the the indigenous inhabitants who modified the contact language previously used with the Spanish to communicate with the French. This frenchified, Indian–Spanish contact language apparently was then taken to the other islands as well as to

Tortuga and Hispaniola by the the indigenous inhabitants, French people, Africans and Creoles. The historical accounts indicate that migration from the smaller French islands, which had started in the seventeenth century, seemed to have continued unabated until 1739 when a royal order prohibited the further transportation of slaves from the smaller islands to Saint Domingue and the reverse. It would seem reasonable to conclude from these historical accounts then that the line of continuity of French Creole started in St Kitts with the first encounters between French settlers, the indigenous inhabitants and Africans, and that constant traffic between the smaller islands followed by traffic to Tortuga and Hispaniola extended the line of continuity to Saint Domingue.

However, this neat account of the initial development of what became Saint Domingue or Haitian Creole has to contend with the eighteenth-century, more complex reality of massive importation of slaves from diverse places into Saint Domingue. Note, for instance, that between 1720 and 1740 the southern part of Saint Domingue was built up by the English from Jamaica and the Dutch from Curaçao. Nicolson remarked that 'La proximité de la Jamaïque facilite le commerce des Nègres avec les Anglois ...' ['The proximity of Jamaica facilitates the trade in negroes with the English ...'] (1776: 107). Such commerce could hardly have been carried on without linguistic consequences. Even before that, according to Raynal (1770: 94), the French stole slaves from the Spanish and more from the English, including 3000 in 1694. These earlier acquisitions of slaves presumably speaking some kind of Spanish and English respectively were made during the formative period of the society before the population mushroomed. These demographic factors could not have produced uniform language development, and more so since many of the African slaves made no direct contribution to the creolisation process because they continued to use their own languages. Hilliard d'Auberteuil (1777: 35) observed that the lack of understanding of the slaves' different languages caused some masters to react with renewed cruelty. Most important of all, however, is that, at the time Raynal was writing (1770), Saint Domingue was really made up of two different colonies – one to the south and west and the other to the north, with no overland connection between them. Rather than uniformity in development, then, Saint Domingue was in the eighteenth century characterised by separation, isolation and uneven development. Its creole language not only arose out of this diversity but also had to cater for it daily.

In spite of its diversity of development, comments of writers at the end of the eighteenth century indicate that the creole language in Saint

Domingue was already an autonomous entity, but the time of the first perception of this autonomy cannot be identified with any certainty. In 1786, Girod-Chantrans, commenting on the language of Saint Domingue, wrote:

> *Quoi qu'il en soit, le langage créole a prévalu. Non-seulement il est celui des gens de couleur, mais même des blancs domiciliés dans la colonie, qui le parlent plus volontiers que le français, soit par habitude, soit parce qu'il leur plait davantage.* (p. 191)

> [Notwithstanding, the creole language has prevailed. Not only is it the language of the 'coloureds' but also the whites who have made the colony their home. They speak it more freely than French either out of habit or because it pleases them more.]

What is significant about this statement is that it indicates that by that time Saint Domingue Creole had become a clearly identifiable linguistic entity and seen as separate from French – it was being referred to else-where in the literature of the time as *créole*, *jargon*, *patois*. In going backwards to establish the time of the first perception of the autonomy of the creole, the creole song *Liset kite laplen* recorded by Moreau de Saint Méry constitutes important evidence – it was said to be from the 1750s, which suggests that what was said about the creole in 1786 equally applied in the 1750s.

What appears to be a contrary indication about the autonomy of the creole in the smaller French islands is found in a comment by Thibault de Chanvalon, who, himself a French Creole, said:

> *Quand ils [les Caraïbes] parlent aux Européans, ils employent, ainsi que font les negres, un françois si corrompu, que c'est presque un langage étranger pour les François nouvellement arrivés aux Isles. Ce langage grossier n'est qu'une fausse imitation de notre langue, dont on a conservé quelques termes, & à laquelle on a donné des inversions & une construction très-informe.* (1763: 57)

> [When the Caribs speak to the Europeans, they use, just like the negroes, a corrupt French, which is almost a foreign language for French people newly arrived in the islands. This crude language is nothing but a false imitation of our language, from which some terms have been preserved and to which have been given inversions and a very informal structure.]

The difference in this case, however, was that this was the creole as spo-ken by the Island-Caribs, for whom it was more of a lingua franca and a second language. It is not surprising therefore that Thibault de Chanvalon did not credit this Island-Carib variety of the creole lan-guage with the same degree of autonomy as Moreau de Saint Méry did the Saint Domingue variety. Yet, even more important in the comment is the idea that the creole language could not be understood by a recently arrived French speaker, for the same point had been made earlier by Labat. Thibault de Chanvalon's comment, then, while it referred to the variety used by the Island-Caribs, can be seen as a repetition and extension of Labat's comment after his initial experience with African slaves in the islands: *'c'étoient des Negres nouveaux qui ne parloient qu'un langage corrompu, que je n'entendois presque point, auquel cependant on est bien-tôt accoûtumé'* ['it was the newly arrived negroes who spoke only a corrupt language, which I almost did not understand, but which one soon gets accustomed to'] (1722 1: 98). At the end of the eighteenth century, the non-comprehension of Africans' speech was again pointed out by Moreau de Saint Méry ([1796] 1958 1: 80–1), who said that the creole language in Saint Domingue was often unintelligible in the mouth of an old African. If this remark by Moreau de Saint Méry is taken to mean an African who had been in Saint Domingue for a long time, it would indicate that the creole, say, of the first half of the eighteenth century was different from (i.e. less French than) what was then spoken at the end of the century. In other words, it could be interpreted as evolution in the language more so than difference between a native speaker's version and a new learner's version. This would then confirm the autonomy of the creole from at least the first half of the eighteenth century.

Labat's remark about his initial non-comprehension of the *langage corrompu* referred to the year 1694, a short time after his arrival in Martinique. What this initial lack of intelligibility indicates is that, contrary to the examples of simplified language given in the French accounts, the speech of the slaves took some time for Europeans to understand. So, at the end of the seventeenth century there was a variety of language that was common to the slaves and it was not a simple variety of French. As a result, it can be said even then to have had a measure of autonomy. It is probably not accidental then that it was at the beginning of the eighteenth century that the word *creole* began to be used more frequently by writers (e.g. Oldmixon 1708, Labat 1722).

While Labat suggested that the specific slaves knew only a *langage corrompu* because they had recently arrived (*des Negres nouveaux*) and had not had time to learn French properly, Thibault de Chanvalon's *langage corrompu* was something much more fixed, stable and mixed. These two concepts – a new learner's version of French and a mixed contact language – were not always seen as different by French writers and in fact were gradually confused. Labat himself did not confuse the two for in reference to a creole slave between 15 and 16 years old whom he was given, he not only says that the slave spoke French but indicated that he also spoke the '*baragouin ordinaire des Negroes*' ['the everyday jargon of the Negroes']. Labat's characterisation of the language situation at the end of the seventeenth century perpetuated the constant notion of contact language as *langage corrompu* but it also now identified a *baragouin ordinaire des Negres*. Thibault de Chanvalon in the middle of the eighteenth century brought together the *baragouin* initially credited to the Caribs and the *baragouin* of the Negroes in his comment cited above ('*ils [les Caraïbes] employent, ainsi que font les negres, un françois corrompu … Ce langage grossier …*').

The confusion was even more apparent in later writers. Comments on Saint Domingue creole by writers in the eighteenth century show that they went back to the seventeenth century for explanations of the early stages of the language and that in doing so they regarded the *langage corrompu* as the starting point of the creole. For example, Girod-Chantrans (1786) said:

> *Le langage créole de cette colonie n'est autre chose que le françois remis en enfance.* (p. 189)

> [The creole language of this colony is nothing but French returned to its infancy.]

> *La tournure insipide du langage créole vient peut-être de la stupidité que les premiers colons supposoient aux negres.* (p. 190)

> [The insipid turns of phrase of the creole language perhaps come from the stupidity that the first colonists presumed the negroes to have.]

> *Cette manière de s'exprimer est en effet plus facile à comprendre, jusqu'à un certain point, que la maniere ordinaire. C'est aussi celle de tous les commençans.* (p. 191)

> [This method of expressing oneself is indeed, up to a certain point, easier to understand than the normal method. It is also one used by all beginners.]

These comments of Girod-Chantrans are a repetition of seventeenth-century comments by Pelleprat:

> *On leur fait comprendre par cette maniere de parler tout ce qu'on leur enseigne: Et c'est la methode que nous gardons au commencement de leurs instructions.* (1655: 54)

> [All that they are taught they are made to understand by this way of speaking. And it is the method that we keep at the beginning of their instruction.]

Girod-Chantrans was therefore associating Saint Domingue creole with what in the smaller islands was described more than a hundred years before by Pelleprat and others as the *langage corrompu*. Moreau de

Saint Méry's belief about the source and time of origin was no different, because he said that it was '*un français corrompu, auquel on a mêlé plusieurs mots espagnols francisés* ... ['a corrupt French with which have been mixed several frenchified Spanish words ...'] ([1796] 1958: 80). This was a repetition of the same kind of statement made about the *langage corrompu* by French writers in the smaller islands in the previous century. So, the historical linkages appeared much clearer to these writers then and with them the perpetuation of the concept of *langage corrompu*.

In the bringing together of the *baragouin* of the Caribs and the *baragouin* of the Africans and in establishing a historical link between the smaller islands (St Kitts, Guadeloupe, Martinique) and Saint Domingue, the notion of *langage corrompu* was the central thread. Both by their comments and by what they cited, writers consolidated the notion that corruption of the French language was the common-sense explanation of the origin of the creole language in Saint Domingue. Girod-Chantrans (1785: 190–1) believed that the corruption was a matter of simplification done deliberately by whites, who believed blacks to be stupid. The purpose was to make the language easier for them to understand. While conceding that for beginners this might have been practical, he thought that after a time it was counter-productive in that it made understanding more difficult and deprived its speakers of capable language. In any case, he thought that simplification was unnecessary because the Negroes learnt French just as easily as any other foreigners, and as proof he identified '*tous ceux que l'on voit en France, & même les domestiques des bonnes maisons de S. Domingue*' ['all those whom one sees in France and even the servants of good houses in Saint Domingue'] (p. 191).

In addition to its development through 'corruption', Moreau de Saint Méry ([1796] 1958 1: 80) also claimed that the creole had elements of sailor language (*termes marins*). Moreover, he repeated the idea given by Pelleprat (1655 part 3: 12) in relation to the Island-Caribs that the vocabulary was assisted by onomatopoeic words:

*Ils aiment surtout à exprimer les sons imitatifs. parlent-ils d'un coup de canon? Ils ajoutent* **boume**; *un coup de fusil,* **poum**; *un soufflet,* **pimme**; *un coup de pied ou de bâton,* **bimme**; *des coups de fouet,* **v'lap v'lap**. *Est-on tombé légèrement?* *c'est* **bap**; *fort, c'est* **boum**; *en dégringolant,* **blou coutoum** ... (p. 56)

[Above all they love to use imitative sounds – if they are talking about a cannon shot, they add *boume*; if it is a gunshot *poum*; a slap *pimme*; a kick or a lash *bimme*; whip lashes *v'lap v'lap*. If one falls lightly, it is *bap* and heavily it is *boum*. If one tumbles down, it is *blou coutoum* ...]

He also said that the language was assisted by gestures and, classifying both the Island-Caribs and the Negroes as uncivilised people, he made the generalisation that gestures were an intrinsic part of the languages of all uncivilised people – '*Chez les nègres, comme chez tous les peuples non-civilisés, les gestes sont très-multipliés et ils forment une partie intrinsèque du langage*' ['Among negroes, as among all uncivilised peoples, gestures are multiplied and form an intrinsic part of language'] (p. 56). This of course went back to the oft repeated story in the early literature about the the indigenous inhabitants that they used their fingers and toes to indicate numbers. Another feature that Moreau de Saint Méry identified was repetition to intensify meaning – '*et toutes les fois qu'on veut rendre un son augmentatif, on le répètent loin, loin, loin, qui exprime une grande distance*' ['and whenever one wants to intensify, one repeats – "far, far, far" means "a great distance"'] (p. 56). This to some extent contrasted with Girod-Chantrans's claim that creole speakers used '*peu d'adjectifs & beaucoup d'adverbes, surtout de ces adverbes amplificateurs comme très, trop, &c.*' ['few adjectives and many adverbs, especially intensifying adverbs like "very", "too", and so on'] (1785: 189). Unlike other writers, Moreau de Saint Méry underlined the meaningful role played by the intonational structure of the language, which made the language seem generally miserly in words – '*Ce jargon est extrêmement mignard, et tel que l'inflexion fait la plus grande partie de l'expression*' ['This dialect is extremely miserly and such that inflexion forms the greatest part of expression'] (p. 81). However, no specific explanation of the intonation or the source of it was given.

Other writers who cited Saint Domingue creole did not focus on 'corruption' but, by citing the creole, reflected attitudes about it and its speakers. For example, Nicolson (1776: 56, 57) used the presumed simplicity of the creole language to demonstrate a kind of philosophical clarity and unemotional straightforwardness in its speakers at moments of extreme distress. The creole was obviously used to strike a note of compassion for the slave and revulsion towards those responsible for the acts of cruelty towards slaves. On the other hand, in the *Journal de Revolutions* (15 November 1793) the presumed crudeness of the creole language was given to show the harsh, direct words of a revolutionary, almost vengeful Negro and so to show clearly the kind of spirit prevailing at the time in Saint Domingue. Another instance of a writer's attitude towards the creole was Hilliard d'Auberteuil's (1777: 68 footnote) use of it where it was meant to be consistent with the Negro's simplemindedness and lack of understanding and appreciation of the concept of sacraments in Christian faith.

The citations do not show any evidence of straightforward African features, even though at the time of the revolution about two thirds of the slaves in Saint Domingue were African, according to Moreau de Saint Méry ([1796] 1958 1: 44). Yet, it would be unwise to come to a conclusion in this respect because the French writers, not knowing any African languages, wrote down what they did in accordance with conventions of their own language, French. It is this which gave rise to the repeated notion of 'corruption' of language, meaning the French language. The absence of African words can be interpreted to mean that on face value Saint Domingue Creole was not an African language. However, some features that were assumed to be French by writers were most likely a combination of features similar in French and some African languages. In short, the sharp distinction between Africans and Creoles that was elsewhere evident was not recorded in the citations and there is therefore no surviving picture of the variation in this area.

As to other social and geographical variation, the citations are not helpful either. Moreau de Saint Méry reprimanded Girod-Chantrans for what he judged to be a false, theatrical version of *creol*, thereby suggesting that the creole had an established and 'proper' structure. This may well have been so, but there had to be differences between racial and social groups. The problem is that each writer, according to his immediate purpose, gave his own version of Saint Domingue Creole and it is virtually impossible in most cases to recover the reality of what the writer heard from what the writer wrote down. As a result, the extent of variation, whether geographical or social, can only be guessed at.

# Creole language in the English colonies – 'bad English larded with the Guiney dialect'

The earliest, negative comments about white Creoles' speech in the English islands were directed at women. In a letter in a 1732 issue of the *Barbados Gazette*, Barbadian men charged Barbadian women thus: 'You are accused of several great foibles, such as valuing yourself too much on your Negroes, lisping their language ...' (Keimer 1741: 56). About 20 years later Poole (1753: 280) identified as one of the faults of the creole women that they 'run too much into the negro brogue in their language'. Almost another 20 years later Thompson (1770: 126) again singled out white creole women when he accused them of 'swearing in a vulgar corrupted dialect at their slaves'. A few years later, in reference to Jamaica, Long recommended 'The utility of a boarding-school for these girls, where their number might admit of employing the ablest teachers, where they might be weaned from the Negroe dialect' ([1774] 1969 2: 250). Such consistently negative reactions to the way women spoke suggested that women of the highest social class sounded, in their pronunciation, like the slaves did and that such speech was thought to be unsuited to their social position and to their role as the fairer sex.

That such speech was not restricted to women, however, is clear from the following comment by Leslie – 'for a boy, till the Age of Seven or Eight, diverts himself with the Negroes, acquires their broken way of talking' (1740: 36). Long also said: 'This sort of gibberish likewise infects many of the white Creoles, who learn it from their nurses in infancy, and meet with much difficulty, as they advance in years, to shake it entirely off, and express themselves with correctness' ([1774] 1970 2: 427). Evidently it was not unbecoming of men to speak like their slaves because in any case they were known to consort with the female slaves from their teenage years.

Besides sounding like their slaves, white Creoles were reported to use peculiar expressions. As to the sources of some of these expressions, Long, following Sloane (1707: lii), said that in Jamaica there was an influence of 'many nautical, or seafaring terms of expression' ([1774] 1970 2: 319). Long specifically mentioned that the use of the word 'cow, to signify all sorts of horned cattle' was a 'bucaneer term', and he also gave the word 'jerk' with the meaning 'to salt meat, and smoak-dry it', as well as words for rum drinks (grog, toddy, kill-devil), all of which were associated with the buccaneering life-style. By the third quarter of the eighteenth century, therefore, the speech of whites born and bred in the English islands was seen as having its own creole identity – partly slave influenced and partly sailor influenced.

The speech of a young, black Barbadian was the first creolised variety of language from an English colony to be illustrated in print more than just by citation of odd words. This appeared in 1718 in Daniel Defoe's *The Family Instructor* (vol. 2, pp. 302–6), which presented a conversation between a father in England and a slave, Toby, who had been taken from Barbados to England to serve as the son's personal servant. It was because Defoe's intention was in part to show up the un-Christian treatment of slaves by slave owners in the colonies that he sympathetically presented the innocent, 14-year-old boy, Toby, speaking in his own words, as first-hand evidence. Toby's 'imperfect English' therefore fitted Defoe's purpose admirably. In fact, 'imperfect English' was even regarded as a surprising shortcoming for a Barbadian slave.

If, in the absence of comments to the contrary, Toby's speech presented no basic problems of communication for his young English master and if Toby was a typical Barbadian creole slave, then Barbadian speech was not far removed from varieties of British English at the time. Nothing unmistakably African is revealed in Toby's speech. Defoe's designation of it as 'imperfect English' was quite apt in that it resembled a learner's imperfect mastery of English. The only non-English feature was the addition of -ee to the verb ending in a consonant sound, a feature that subsequently was cited in a stereotypical manner as a mark of an English creole speaker. However, the tendency to make syllables and words end in vowels had earlier been cited as an indigenous feature. The use of *me* as subject pronoun and uninflected *be* could have resulted from source languages (English and African) as well as from rules that language learners use. In short, what Defoe presented as Barbadian speech was consistent with the image of Toby that he wanted to present and might have been a reasonable representation of speech addressed to a white person.

Citations of words from the English islands by other writers before 1789 do not really give an adequate picture of the slaves' language: they merely identified and explained items that were new to writers, or that writers thought would not be known by readers in Britain and would be interesting to them. Examples of such words are:

rap, is what in Nieves they call cool drink, viz. molossus and water ... (Sloane 1707: lxii)

jumbee (that is to say the Devil...) (Smith 1745: 49 re Nevis)

the Yaws, or a sort of Itch, which the Negroes call in their Language Cro crow (Hillary 1759 re Barbados)

mogoss, a term made use of in the leeward islands, signifying the same as mill-trash in Barbados: It is used to signify the reliques of the cane after it has gone through the mill. (Singleton 1767: 7 [note] re St Kitts)

wowra, the decayed leaves of the cane ... (Ramsay 1784: 75)

The edda, called also the vegetable wash ball, from its apparent soapy quality ... (Luffman 1789: 68 re Antigua)

tenah = a headdress, composed of one or more handkerchiefs, put on in a manner peculiar to these people [negroe women] (Luffman 1789, note p. 74)

This kind of explanation of names and words was a normal part of descriptive writing: such names and words were intended to add exotic colour and a flavour of authenticity.

In 1774 Long did more than cite odd words, he gave his views on the language used by the slaves in Jamaica. After first saying 'The Africans speak their respective dialects, with some mixture of broken English' ([1774] 1970 2: 426), he continued:

> The language of the Creoles is bad English, larded with the Guiney dialect, owing to their adopting the African words, in order to make themselves understood by the imported slaves; which they find much easier than teaching these strangers to learn English. ([1774] 1970 2: 426)

In these two comments Long indicated stages in the process of language learning as well as dynamic factors in the communication.

The idea of conscious accommodation of the 'lower' speaker to ensure proper communication was an explanation favoured by Long to account for 'higher' speakers' use of 'lower' features. This interpretation contrasts with that given by Fermin (1769 1: 22–3) in his explanation of the English creole language in Suriname. Fermin's idea was that the creole resulted from incomplete acquisition of English on the part of Africans and addition of African words not only to make up the deficiencies but also to make the language more elegant. The major difference between Long's idea and Fermin's was that Fermin was giving an explanation for something that had taken place a hundred years previously, whereas Long seemed to be explaining what he saw as current. Nevertheless, Fermin's explanation, coming five years before Long's, shows that a view of the creole other than as corruption was not beyond European writers.

In addition to his idea of conscious accommodation, Long obviously had in mind a specific African language or variety of language when he said that what the slaves mixed with the 'bad English' was 'the Guiney dialect', but he gave no further information. Long's summary view of the slaves' language in Jamaica then was that it was a corruption of English to which a fair number of words from African slave-coast language were added. Long then went on to make more detailed comments about the structure of this language and in so doing showed more than a fleeting familiarity with it:

> The Negroes seem very fond of reduplications, to express a greater or less quantity of any thing: as **walky-walky, talky-talky, washy-washy, nappy-nappy, tie-tie, lilly-lilly, fum-fum**: so **bugs-a-bugs** (wood-ants); **dab-a-dab** (an olio, made with maize, herrings and pepper); **bra-bra** (another of their dishes); **grande-grande** (augmentative size, or grandeur) and so forth. In their conversation, they confound all the moods, tenses, cases, and conjugations, without mercy: for example; **I suprize** (for, I am surprised) **me glad for see you** (pro, I am glad to see you); **how you do** (for, how d'ye do?) **me tank you; me ver well**; &c. ([1774] 1970 2: 427)

Yet, these comments by Long only served to further illustrate his belief that the slaves' language was a corruption and that it depended on simplistic or simplified devices to get over meanings.

Slaves' language in other English islands is practically invisible in the literature up to about the last decade of the eighteenth century. Then, more substantial citations occur: Dickson (1789) gave a few short sentences from Barbados, Mathews (1793) provided extended stretches of language from St Kitts and Ford (1799) is a 'transcription' of the speech of two African women in Barbados.

The citations of speech of different Barbadian slaves in Dickson (1789) were basically features that writers associated with what they regarded as corruption of English. Mathews (1793) was somewhat different – this work contained the earliest and most substantial citations of what Mathews called 'the negro language', that is of the slaves in the smaller English islands. The illustrations of 'negro language' in Mathews (1793) were: a 'typical' conversation between Mathews and a slave (pp. 44–5); a few phrases from a female African slave (p. 68); two songs composed in 1786 by Mathews and two other popular songs (pp. 134–9). Although Mathews identified one of the slaves as African, her speech was not said to be different from that of the creole slaves and there were no features given that pointed to any difference. Mathews's citations, more than any other in the eighteeenth century, gave a more detailed and accurate picture of the structure of the slaves' language: they showed it to be made up of elements from different sources, which came together formally and functionally to serve the purpose of communication. Mathews himself was proud of his ability to speak the 'negro language', and as a Creole his primary intention was to display the language rather than to belittle it.

In comparison, the illustrations of slave speech by John Ford (1799) differ in a number of respects. First, they are handwritten on a sheet of paper and so are without context. In their story content they are very much like emancipation literature, which sought to get from the slaves themselves accounts of their earlier life in Africa and their comparison of this life with life in the West Indies, with the intention of showing that it was much better or much worse. In this case the slaves point to Africa as a better place – one of them describes the pain of the enslavement process, while the other talks positively about customs in her homeland. Both of the transcribed speakers were women who were African born – one is identified as Fantee – and had learnt their 'English' as a second language in Barbados. An author intending to put out these kinds of views in 1799 would hardly have been Barbadian or West Indian or a colonist. More than likely, John Ford was an English visitor who wanted to use the technique of first-hand accounts and actual slave language to get over his own personal views about slavery.

The worth of the 'transcriptions' then from a linguistic point of view is more in the identification of the specific non-English features used by the two slaves rather than as an accurate record of the variety spoken by African slaves. John Ford may have been struck by and remembered specific features, which he inserted into an essentially English account. This does not mean that all the more standard English features in the texts are misrepresentations and could not have been produced by the slaves in question, but that they are more in keeping with the author's own unconscious production and that the accounts were being 'edited' in such a way as to make them intelligible and persuasive to readers who were involved in the national debate on slavery. The features given by Ford were meant to portray a simplistic kind of English and to convey a sympathetic picture of an innocent slave. For instance, when one of the slaves referred to the words of encouragement given to her and others by the black people in the slave factory on the West African coast, Ford changed to a repetitive and simplistic form of language – 'yonder yonder, where we would workee workee picka-nee-nee, and messy messy grandee and no fum-fum [beating]'. This kind of syntax together with the repetition of the _-ee_/_-y_ ending (earlier given in Defoe (1718) with the same intent) came to be the stereotypical simulation of the English of slaves.

African features in the eighteenth-century citations were not categorically identified as African. One would have to interpret Long's phrase 'larded with the Guiney dialect' as a suspicion that features such as those that follow were African:

> *law*/*naw*. The most likely source of this preposition/conjunction and its variant forms is Igbo, which today has an all-purpose preposition which varies according to the specific sounds that surround it – *la* (basic form) and *na* (used in the environment of nasal sounds). Igbo also has a particle *là*/*nà* [and]. It seems as if the two Igbo words (i.e. the preposition and the particle), which are distinguished tonally, collapsed into one word in the creolisation process. However, the alternation between initial [*l*] and [*n*] could also have been a survival from Mende, for Sengova (1994: 180) identifies the change from [*nd*] to [*l*] as an example of consonant mutation in Mende. In addition to *law*, the form *long* also occurs, apparently having a different function – it precedes animate nouns: *tank long Sabina; bex long ee Mosser; pake long Obeshay; play long Nanny Muccoo*. Judging from these examples, *long* was not used as a spatial preposition but as an associative particle meaning 'with'. This is in keeping with the probable source, the Mende word *leŋga* "in unison" given by Turner (1949: 122). (The variants *langa* and *nanga*, used commonly to mean 'with' and 'and'

in Sranan, are attested in Suriname from early in the eightenth century.) Mathews's spelling of the word suggests that in St Kitts it was thought to come from English *along*. If, as the examples above indicate, *law* and *long* complemented each other, this would have been an instance of two separate African items coming together in the creolisation process to blend with an English one.

*edda* comes from Igbo (< *édè* [coco yam]).

*cro crow* seems to be a Mande word (< *korokoro* [syphilitic disease] akin medically to yaws (Bazin 1906)).

*tenah* may have come from Mande (*te-ŋ'-ya* [physical beauty; to become beautiful; embellish]).

*jumbee* is given varying origins, e.g. Banoo *njambe* and Congo *zumbi*, but the Mande languages have a root *dyo/dyu* [divinity, amulet or talisman], which is widespread, to which can be added the intensifying ending *mbe*.

*mogoss* is usually explained as a derivative of the French word *bagasse*, but it has always had an initial [*m*] variant of the first syllable, which could have been caused by Mande slaves *mugu* [powder, dust residue, substance made into small pieces, crushed, crumpled] (Delafosse 1955: 519).

*Kai*, occurring in Dickson (1789), has, since Turner (1949: 196), been regarded as an African exclamation. An equally valid and earlier use of the word was as a form of address, for Delafosse (1929: 294) explains that in Mande *kyè/kè* [man] was used typically in addressing a man whose name you do not know. This interpretation is obscured by the punctuation and glosses given by Dickson specifically.

*fum-fum* [beating] is an Eastern Ijo word.

*back-erah* [white person, boss], one of the oldest cited African words, is from Efik and Ibibio *mbakara*.

*wowra*. While Ramsay gave the meaning of this word as 'the decayed leaves of the cane', it was used in St Kitts to refer to thatched roof of the slaves' huts. In Christaller's *Asante and Fante Dictionary* (1933) the word *àwórám* is given with the meaning 'a certain plant, with which houses are thatched'.

These words reflect a range of West African languages, for the Mande languages themselves (Malinké, Bambara, Dyula) stretch over a wide area – Senegal, Mali, Gambia, Ivory Coast, Guinea, Upper Volta, Ghana, Sierra Leone and Liberia (Alexandre 1972: 10). The words not only seem to confirm Ligon's (1657) view of the widespread sources for the slaves ('Guinny', 'Binny' and 'River of Gambia'), but the apparently greater influence of the Mande languages suggests that slaves from these areas were more important linguistically in the formative period than numbers for slave importations tend to suggest.

Added to the words above are the following used in the Mathews texts with their probable sources:

*bamboa* [loose shirt] also occurs later in Carmichael 1833 (1: 258, 311) where it was used to mean 'woollen dresses'. It was given earlier by Ramsay (1784: 81) with the meaning 'coarse woolen cloth'. Voorhoeve and Lichtveld (1975: 238–9) explain that *bamborita* in Suriname is a 'brightly colored weekend shirt'. The African versions of the word, however, do not have the nasal in the first syllable – *bubu* Wolof (Cole and Scribner 1974: 104), *buba* Yoruba (Turner 1949: 68).

*morrogou* [countryman] comes from Mande *mògò* [man] or *mòrò* [man] + *dugu* [country] and was used as a form of address.

*naungaw* ['proud' and 'pride'] comes from Mende *nyàngá* and seems to have been widely used and long-lasting because Bates (1896: 38) cites it in the Jamaican proverb *Nyanga mek crab go sideways*.

*nawm* [yam] from Wolof *nyam-bi*.

*naunnam* [boiled corn meal] actually identifies a dish and is not simply a variant of the word for 'food/eat'; the word comes from Mande *ŋyõ*, which Delafosse (1955: 583) explains as 'sorghum or coarse millet; food grain in general'. European writers had difficulty transcribing the words that the slaves used to refer to what they were eating and to mean 'food' and 'eat', e.g. Pinckard (1806 1: 116) – *gnhyaam* and *nhyaam*. In the case of Mathews (1793) the first syllable in *naunnam* most likely comes from the same nasal sound [ŋ] which Pinckard writes as *gnh*.

*cobbacon* is used to refer to a kind of bed. The Mande languages have the word *kala-ka*, which Delafosse (1955 2: 328) explains as 'a bed made of stalks of reeds or raffia'.

*sho* occurs in *Brudder Dublin kum home sho von Shatteray nite* and, while it could be a transcription of English 'sure', it may also be the Mande word *so* [home], in which case there is a repetition (i.e. *home sho*), which is common in language contact and change.

*ye*. There is again a possible repetition in the sentence *ee hed tan lek dem dry wedder naum, ee ye yie tan lek granfarrar bobboon* [he head stand like them dry weather yam, he eye stand like grandfather baboon]. Cassidy and LePage (*Dictionary of Jamaican English* 1980: 484) suggest that the word *yai* [eye] in Jamaica may involve 'some concurrent influence from African languages]. It is again a Mande language (Bambara) that provides the African source for 'eye' – *ŋ-yè*.

Although in total there are comparatively few African words in the Mathews citations, they again point to the Mande and neighbouring languages as a major element in the formative period.

Very seldom in the eighteenth century was the slaves' language identified as having links to their native (i.e. African) languages, and almost never was any feature identified as coming from a specific African language. The repeated absence of any named African source removed any valid alternative historical identity, thereby allowing the notion of corrupt language to flourish. Additionally, the repeated absence of any comment on the similarities between features of regional varieties of English and vernacular speech in the English islands allowed the idea of corruption of English to flourish.

# Creole language as a symbol of identity and a medium of enlightenment

European literature of the eighteenth century was dominated by ridicule as European societies tried to relinquish the old order in the face of the reality of a bigger world, greater wealth and more social mobility. European awareness of colonial territory and the wealth it provided was attended by moral questions, some of which had arisen in the transformation of the hierarchical structure of society, with different classes of persons and different levels of privilege, from the old feudal, regional or small-scale system to a national one. In fact, the eighteenth century in the Caribbean in a sense repeated the experience of Europe in that there was a transformation from colonies made up of small, isolated village-like plantations towards more monolithic, centrally governed and defended entities, dominated by a creolised social structure and values that encompassed all persons. The literature on the Caribbean specifically addressed the creolisation process in all its facets and struck many raw nerves in doing so. The arguments in the literature between mother country and colony were like arguments between father and son as the son moved to maturity and acted independently. In the replication of the European social system in the Caribbean, it is the dimensions of colour and race that took some time to be established and accepted, not because there was much question about the validity of what was implemented but because the amount of violence it took to establish the system upset many of those who were not immediately involved.

The conflict between the mother country and colonies was resolved in part by the use of ridicule and reason as techniques for establishing power and dominance. Control of the printing press added another critical factor in overall control and dominance. The introduction of the printing press into the Caribbean islands in the eighteenth century allowed white Creoles to defend themselves to some extent, but, more generally, by adopting the forms and subject matter of European printing, the Caribbean press guaranteed the propagation of the European interpretation of colonial characteristics. Enthusiastic initial attempts to display the creole language and to defend creole values were counteracted by a more sustained and widespread identification of the speech of Creoles as corruption and consequently identification of Creoles themselves as the same. The widespread perceptions of identity were determined by a kind of reasoning that started with prejudiced assumptions and the need to maintain power at every level of a social hierarchy. So, even though Creoles had become or were becoming numerically dominant in most islands and, through child rearing practices that brought members of all classes into intimate contact, were all part of a language spectrum that facilitated mutual intelligibility, white Creoles were being ridiculed into renouncing and dissociating themselves from the most fundamental characteristic of their identity – the features of their language that were peculiarly their own – because they were common to their black slaves.

It was in 1734 that a comment in the *Barbados Gazette* first referred to the 'language' of the Negroes; in the middle of the eighteenth century Poole (1753) spoke of the 'negro brogue' and some time between 1752 and 1754 Thomas Thistlewood in Jamaica mentioned a slave speaking 'in the Negro manner' (Hall 1989: 54). Hillary (1759) talked about the 'Language' of the Negroes; 15 years later Long (1774) talked about the 'Negroe dialect' and, near the end of the century, Mathews talked about the 'negro language'. These references prove that from early in the eighteenth century there was a local, slave variety of language throughout the English possessions and that its existence was quite evident. Moreover, the word 'language' used to characterise it suggests that it was significantly different from other languages in use and as such was not immediately understood by foreigners.

In 1777 Oldendorp, in contrast to the English writers mentioned, used the term *creole* to identify these, for him, versions of European languages spoken in the West Indies:

> By the term Creole language, I mean the language that is spoken by the Negroes on St. Thomas and St. John and to a certain extent by those of St. Croix. The domain of this language extends no further than these islands. It is not the only Creole language because every European language which is spoken in a corrupted manner in the West Indies is called Creole. So it is that Creole English is spoken by the Negroes of English masters who have come to St. Croix with them from other English islands. ([1777] 1987: 251)

In 1785 Girod-Chantrans also used the term *créole* to refer to the vernacular language in Saint Domingue. By that time Saint Domingue had become the richest sugar colony in the New World, the envy of all other European nations. A most dramatic language development had taken place there. At the beginning of the eighteenth century Saint Domingue's population was miniscule and scattered; by the end of the century importation of Africans had brought it to about half a million. Yet, in spite of the great variety of African and non-African languages spoken in the colony, Moreau de Saint Méry could say unhesitatingly 'C'est dans ce langage ... que les Créols [de toutes couleurs] aiment à s'entretenir ...' ['It is in this language ... that Creoles (of all colours) like to amuse themselves ...']. There is no question that the dramatic events of the last years of the eighteenth century and the first years of the nineteenth in Saint Domingue could not have taken place without this langage créole.

The negro language/langage créole was without doubt a medium of enlightenment for the slaves in that it gave them access both directly and indirectly to their immediate source of power – their master. From a situation even in the last quarter of the seventeenth century when, according to contemporary writers, the Africans could not easily under- stand each other or their masters because of differences of language, by the end of the first quarter of the eighteenth century, in the smaller Car- ibbean islands, and later in Saint Domingue they had a language com- mon to themselves and intelligible to their masters. As a result of this, news of immediate concern, local and overseas, oral and written, could more quickly filter down to the lowliest slave on the plantation as long as that slave could understand the negro language/langage créole. Where initially linguistic difficulties among the slave population provid- ed their masters with a modicum of security, the negro language/lan- gage créole removed this and gave the slaves greater capability for concerted action in rebellion as well as greater facility for deception in face-to-face encounters. The negro language/langage créole also provid- ed the slaves with a common medium for spiritual, artistic and other cultural expression; it had developed into a medium capable of express- ing emotion in song and verse.

In addition, there had developed among the Creoles slaves a more 'cultured' register. This was explained by Long as follows:

> The better sort are very fond of improving their language, by catching at any hard word that the Whites happen to let fall in their hearing; and they alter and misapply in a strange manner; but a tolerable collection of them gives an air of knowledge and importance in the eyes of their brethren, which tickles their vanity, and makes them more assiduous in stocking themselves with this unintelligible jargon. (1774 2: 426–7)

Long did not identify this variety as a performance register, but it sounds very much like what became characteristic of cultural events among the black section of the population in several islands in the nineteenth and the early twentieth centuries. Long effectively identified it as a higher-class register among the creole slaves by saying that it was typical of the 'better sort' and the result of imitation of white speech. As such, it might have been more characteristic of domestic slaves than field slaves. However, the notion that it was substantially malapropism gives no consideration to the extent to which it could have been deliberate aping of white pompous behaviour or, on the other hand, a form of playing with words and sounds as a cultural practice.

More common among the slaves generally was the deliberate cultivation of persuasive language, which was addressed to the master or mistress to achieve some specific goal. In referring to this practice among the slaves, Labat said:

> *Ils sont naturellement éloquens, & ils sçavent fort bien se servir de ce talent, quand ils ont quelque chose à demander à leurs Maîtres, ou lorsqu'il s'agit de se deffendre de quelque accusation qu'on fait contre eux, il faut les écouter avec patience, si on veut en être aimé.* (1742 iv: 458–9)

> [They are naturally eloquent and know very well how to use this talent when they want something from their masters or when it is a matter of defending themselves against some accusation made against them; you have to listen to them patiently if you want them to love you.]

Long later said: 'They are excellent dissemblers, and skilful flatterers' ([1774] 1970 2: 407). No doubt, in a situation where punishment was swift and severe and where they were virtually powerless, slaves had quickly to learn to use words to survive and prosper. In 1793 Edwards said of the slaves in Jamaica that 'They are fond of exhibiting set speeches, as orators by profession …' ([1819] 1966 2: 100–1). Edwards was struck by the similarities in the patterns of argumentation among the slaves, who had developed strategies for pleading their own cases. The slaves' imitation of whites resulted in 'cultured' speech, that is, speech seen as 'improving their language' and giving 'an air of knowledge and importance'.

The greatest triumph of a creole language was in Saint Domingue, where it emerged as a symbol of national identity and a medium, as the white colonists had feared right from the beginning, for revolution. Its actual and symbolic significance is captured in the words of Lionel Fraser, who recounts the outbreak of the revolution in Saint Domingue in dramatic words:

Suddenly a negro of the name of Boukman, looked upon his fellows as a Papaloi, or High Priest, advanced into their midst. After chanting some of the mysterious incantations of the Vaudoux worship, he broke out into the following prayer, if such it can be called, which local tradition has preserved: –

'Bon Dié qui fait soleil qui clairé nous d'enhaut,
Qui souleve la mer, qui fait grondé l'orage
Bon Dié la, z'autres tende, li caché dans yon nuage,
Eh lâ li gardé nous. Li voué tout ça blancs fait.
Bon Dié blancs mandé crime et cela nous vlé bienfêts,
Mais Dié qui si bon, ordonné nous vengenace;
Li va condui bras nous, li ba nous assistance.
Jetté portrait Dié blancs qui metté d'lo dans yeux nous,
Couté la liberté qui parlé coeur nous tous.'[15]
...

It may well be supposed that words like these, spoken at such a time and by such a man, took immediate and fearful effect. Before midnight the bright flames shooting upwards to the sky in every direction on the great plain behind Cap Français, showed the terrified inhabitants that the long-dreaded servile revolt had at last broken out. (1891: 47–8)

No one knows what the exact words of Boukman were at that historic moment, but, as recorded here, they recapture the powerful sentiments coming from the heart of a common man who was to be remembered forever.

By the end of the century the *negro language/langage créole* had virtually become a part of the identity of all Creoles, within or on the margin of the central creolisation process. Author after author remarked with some concern that white Creoles were acquiring the language of their slaves. The bond of language was therefore becoming a badge of identity for all Creoles. The social and racial divisions were barriers that controlled the distribution of power across creole societies, but the *negro language/langage créole* surmounted these barriers and sustained the society from its base upward and made it work. It was an identity, however, that had little value beyond its own society because it was seen as corruption, and as such was regarded as something that

---

15. Translation:
  God who made the sun which lights us from on high
  Who raises the sea, who makes the storm groan
  That God, you hear, he is hidden in a cloud
  And there he watches over us. He sees whatever Whites do
  The God of the Whites tells them to commit crimes and Ours wants us to do good
  But God who is so good commands us to seek vengeance
  He will guide our hands, He will assist us
  Throw away the picture of the God of the Whites who put tears in our eyes
  Listen to the freedom which is speaking to the hearts of us all

had to be suppressed. Ironically, then, the more strongly creole identity emerged, the more the Creole had to be self-destructive: the blacker creole identity was, the more the Creole had to deny it. It was only in Saint Domingue that there was some affirmation of creole identity. There the accent on culture within the French self-concept apparently had unintentionally led to an appreciation of the beauty of the creole language, in spite of its repeatedly identified *langage corrompu* source. In the English vision, empire and control were considered as absolute and for them the creole language was good for slaves and children but needed to be eradicated in young whites so that they could assume a proper role in society as adults. For the Spanish *criollos* there was no problem – for them no *negro language* existed or, if it did, it existed only among Africans.

## Creole enlightenment – looking inward and looking outward

The prominence of the words 'reason' and 'enlightenment', words that came to be associated with the eighteenth century, was a direct result of Europeans having to confront their previous ignorance about the geography of the world, about Europe as the centre of the world, about the image of God and man in God's image and about the diversity of human culture. The encounters with other humans and other climates required a reconstruction of views of man, the world and the universe. The dominance of the European in this new cosmology was not diminished. Indeed, the western European, attaching his culture to what was called the classicism of Rome and Greece, put himself at the apex of a pyramid of civilisation.

Accordingly, enlightenment, creolisation and nativisation at first appear to be three processes at three points in descending order on the scale of civilisation. From the viewpoint of European writers the three were plainly in evidence during the eighteenth century in the islands of the Antilles. Nativisation, the 'lowest' of these processes, had at its centre the 'native'. In a narrow sense, 'native' referred to the peoples of the Americas, but on a global scale Africans were probably the prototypical native for European writers. Nativisation in the islands of the Antilles therefore embraced both the indigenous inhabitants and the Africans living outside direct European control. Even Africans living within the European colonies can be said to have been in large measure outside the sphere of (European) enlightenment, for the identity of the slaves was shaped for them in a way that was opposite to the idea of enlightenment – they were prevented from entering the arena of literacy because the slave masters believed that enlightenment of the slaves would pose a danger to them and the good of the society. The intention of the owners was therefore to keep the slaves in darkness, which meant that essentially the identity of a slave was the antithesis of enlightenment and the retention of slavery was equivalent to a struggle against enlightenment. The system of slavery therefore intended to preserve the slave intellectually in the 'native' state.

Creolisation in the islands is generally seen as a synthesis of European and African elements, at least, in a process of acculturation in which there is a gradual movement towards European culture. Creolisation, even more generally, is seen as the experience of being born and raised in the New World. Using this latter view of creolisation, the eighteenth-century Maroon and Black Carib can be regarded as quintessentially Creole, representing alternative models of creolisation. The African Maroon and the 'Yellow' Carib of the previous century had given way to the Creole Maroon and the Black Carib respectively. In the case of the Maroons in Jamaica, by the middle of the eighteenth century when they had gained their 'independence', most of them must have been creole-born. Thus, the Maroon experience in Jamaica had lasted through several generations, which meant that Maroon culture was creole. The same is true of the culture of the Black Caribs, who had established themselves as an independent group in St Vincent for over a century. In both cases, then, 'native'-dominated creolisation presented a contrast to the more widespread European-dominated creolisation. The combination of 'native', Creole and free/independent meant that the Jamaican Maroon and the Vincentian Black Carib were not only attractive models of Creoles for many within the European colonies but also, as such, were really precursors of independent Haitians, who are usually regarded as the first to break out of the European-dominated mould. The Maroon and the Black Carib may therefore be identified as the first examples and the prime symbols of enlightened creolisation.

In Puerto Rico and Santo Domingo the native element in creolisation also came to the fore in the eighteenth century in the person of the *jíbaro*, who was changing from a person of mixed race that was predominantly Black to one that was more indigenous. This change was facilitated by the fact that, like the Black Carib, although not so markedly, the *criollo* was seen to have adopted much of the culture and lifestyle of the previous indigenous inhabitants. With an added change of colour, therefore, the *jíbaro* came nearer to being indigenous. In Cuba, although the *criollo* population was regarded as mixed in race, as in the other two islands, the preoccupation with *bozales* and the fear of them as a creeping majority indicated the development of a polarisation between black and white rather than a simple preference for an indigenous racial type as a national figure. In addition, since Cuba was more important to Spain than the other two, the conflict between *criollo* and *peninsular* was more acute there, and consequently whiteness was more important overall. The indigenous inhabitant remained as part of the lifestyle and the historical consciousness of the population, but was used primarily to assert a *criollo* identity different from both the white foreigners (*peninsulares*) and the black ones (*bozales*).

In the French colonies generally there was from the start intercourse between the French and the indigenous inhabitants, even if it was not said to be as pervasive as it was in Spanish colonies. Indeed, Boucher (1989: 45) argues that Colbert, the French minister for the colonies, tried to promote population growth 'by encouraging French–Indian marriages' in the late seventeenth century. In spite of the fact that in the smaller French islands the Island-Caribs seemed to have developed a preference for the French over the English, there was no promotion among eighteenth-century French Creoles of a mixed racial type involving indigenous inhabitants; there was no nationalistic view of the indigenous inhabitants as the true natives of the region and mulattoes in the French islands portrayed themselves as *gens de couleur*, not as indigenous inhabitants. Economic success dominated the minds of the French and the government of their colonies, while the attainment of French ideals dominated the minds of *Créoles*.

The expulsion of the Caribs from St Vincent was symbolic of colonial thinking in the English colonies – the indigenous inhabitant had no part to play in national identity in the eighteenth century. The real, living indigenous inhabitant was in no sense seen as a positive figure, and it is clear that it was only with his decline or demise that he became a powerful symbol of the region, based on his reported fierceness and courage that embellished the early historical literature on the region. Within English colonial society, the *West Indian* was a fairly innocuous, un-political figure who suffered from vices brought on by slavery and the climate and whose acceptance of Britain as home still persisted.

One of the most powerful of the European's tools, which was also a powerful norm in itself, was the notion of refined (oral and written) language. This normative tool made sure that, in spite of the spirit of, or yearning for, liberty and cultural independence that was variously exhibited in the daily lives of several groups throughout the islands, Creoles were confronted with a contradiction between the vision of unfettered life and enjoyment of the creole way on the one hand and on the other the enticing and overwhelming power of cultural enlightenment through the medium of the colonial language. The unquestioned acceptance of the colonial language as the path to enlightenment automatically conferred an identity of 'corruption' on creole languages and consequently on all creole speakers and their way of life. There could then be little admiration for the cultural mixture and creativity of creole languages, which were on the contrary considered as deviations from the purity of the standard European language forms and structures; there could then be little admiration for the feat of general communication among disparate peoples and the fashioning of a common culture, which these creole varieties of language had achieved. The notion of *langage corrompu*, entrenched as it was in the European vision of things and propagated throughout the colonies in its literature, made sure that this was so.

# The development of national identities in Hispaniola, Cuba and Puerto Rico

## The fate of 'creole' in the face of rising national identities

The whole of the Caribbean was dominated throughout the nineteenth century by the events that took place at the beginning of the century in the island called Hispaniola. Haiti (the western part of the island) had changed its name from Saint Domingue and had emerged, through revolution, to become known to all as the first black republic. The revulsion that this caused in some quarters in itself gave Haiti, at least from outside, a very clearly defined identity and made it a powerful emotional symbol. At the same time, however, Haiti represented successful struggle against colonial domination and this emboldened others in the Caribbean islands and elsewhere to look towards further removal of European domination. The Spanish island colonies specifically were also directly affected by events in Spain at the beginning of the century, that is by Spain's decline in the face of Napoleon's advance. For them, the loss of their 'mother country' (Spain) created a climate of imprecise political affiliation and factional fragmentation. Among the other two major colonial powers, the French world was in upheaval, and the British were trying to recover from their losses in North America by consolidating the gains they had made in the Caribbean islands at the end of the eighteenth century. As to the government of the United States, it was opening its eyes to expanded horizons and in fact was being perceived by many in the Spanish islands as a favourable suitor.

In the New World during the eighteenth century feelings of American identity had begun to crystallise first in the British North American colonies as animosity towards Britain increased. There, revolution, independence and federation followed, and this success provided a model for the French colony of Saint Domingue, which also expelled its colonial master some years later. With the coming into being of new countries, the united states of America and Haiti, at the end of the eighteenth century and the beginning of the nineteenth century respectively, geographical entity (i.e. place) began to be as prominent as race in the establishing of identity among people in the New World. Yet, the united states of America and Haiti, the first two places to become independent in the New World, were extremes in the evolution of national identity. The united states of America together formed a part of a large land mass, so the geographical area of this new entity was not naturally separate; the people were perceived to be the same race as Europeans and their language was English – not distinct from that of the former mother country. On the other hand, Haiti's geography (an island), the black colour of its people, and the new language that had emerged made it stand out very sharply. Furthermore, the term 'Creole', the first term of identity for the new natives of the Americas, was hardly used to refer to the people of the united states of America, whereas it was a prominent term in Haiti. In spite of this lack of separateness, the identity of the people of the united states of America developed fairly strongly in the nineteenth century, thereby indicating that inherent distinctiveness was not a prerequisite for assertion of identity. In the case of Haiti, inherent distinctiveness did not guarantee a national identity of people, but on the other hand, as the first black republic, Haiti became an antithetical point in the construction of identity in the islands, especially in the Spanish ones nearby.

In the North American colonies that became the United States, the designation 'Creole' had long been becoming irrelevant as feelings of regional identity grew; it became more so after the proclamation of independence in 1776 and even more so after assumption of office by the first president in 1789, even though Anderson points out that 'The figure of Benjamin Franklin is indelibly associated with creole nationalism in the northern Americas' (1991: 61). Since this new country did not give itself a specific name, maintaining only the description of an alliance as a name, the term 'American' increasingly was used to identify its citizens. It was a term that was tied exclusively to the name of a place, it established an independent identity for the people of that place and it was the name used by the people themselves.

From its beginning the term *criollo* was one used primarily for differentiation of people rather than for identifying a place or a specific national, cultural or ethnic group. In the case of the black population it was used to distinguish those brought from Africa from those born in the New World; in the case of whites it was used to distinguish those born in the New World from those coming from Europe. It was not applied to mulattos because those born in the New World did not have any sizeable group coming from elsewhere to be distinguished from. With the decrease and final cessation of the importation of slaves into some islands, the number of African slaves diminished to the point where there was no longer a need to distinguish between Africans and black Creoles. Consequently, the term 'creole' became virtually redundant in its application to slaves and blacks generally in those islands, as it was for mulattos. It continued to be needed for whites as long as the relationship between colony and mother country existed, that is, as long as whites kept on coming from the mother country and other places to the colonies. In essence, then, the term 'creole', which symbolised a contrasting and not a self-evident and independent identity, waxed or waned according to the prominence of the contrast between people.

In the Spanish island possessions the term *criollo* became more sharply focused as the contrast between local and foreign heightened, partly as a result of the fact that independence from Spain had been proclaimed throughout continental America, leaving the islands of Puerto Rico and Cuba as the only possessions that Spain had in the Americas in the middle of the nineteenth century. Even on the eastern side of the island of Hispaniola an independent Dominican Republic had been proclaimed in 1844.

In Haiti, where any significant foreign influence was eliminated by the revolution and the first constitution in 1804, the term *créole* sustained itself because it was equally used as a name for both the people and the language of the people, the latter clearly contrasting with the French language and thus acquiring its own name. However, this identity of name of people (Creole) and name of language (creole) did not have any distinguishing effect and it did not confer a sense of separateness on the new nation, because not only was the name (Creole) not unique to Haitians but also the continued perception of the language of Haitians as a corruption of French linked it and its speakers in a subordinate relationship to France and Europe generally. Nevertheless, the influence of events in Haiti on the French islands bolstered the term *créole* throughout the French Caribbean and, with migration, in French America (Louisiana).

In the English island possessions that were won from the French, 'Creole' was a much more regular and characteristic term than it was in the older English islands, where there was no need to distinguish between English- and French-speaking whites. It had never really been a favourite term for English writers to refer principally to the whites born in their territories. It is quite possible that for them the word retained the feeling of being a borrowed word in addition to the fact that the decline of the white population in almost all their territories in the nineteenth century reduced the prominence of local whites as a group.

In the process of change in the structure of identity during the nineteenth century, it was also the apparent contradictions in the major features of identity that came to determine the clarity of characteristic
national identity in the Caribbean islands and the prominence of 'creole' as a term for highlighting national peculiarities. 'Creole' embodied
the major features of national identity (i.e. place, race and language),
but because it did not identify a specific country, since it allowed for
more than one race and since it implied some kind of 'corruption' in
language, it was neither satisfactory to all nor could it really project the
identity of each individual island. Nevertheless, it was a premier term
of identity in the islands. Accordingly, its evolution and respective
meanings reflected the way in which national identity emerged, and the
appearance of alternate as well as supplementary terms to 'Creole' in
the different territories in the Caribbean in a sense revealed the deficiencies of the premier term.

# Race and identity in Haiti

In Saint Domingue from the start of the ninteenth century, as a result of
the sudden and drastic change, there was a need for new terms and
symbols of wider or national identity; there was also a need for a structure to identify people in the post-revolution reality. This structure had
to be different from that which merely reflected the social contrasts of
slavery, that is the contrast between unmarked freedom and marked
property and between European, African and Creole. In other words,
new names were needed to replace 'master', 'slave', 'freedman' and any
others that were a part of slavery. There was, in short, a need for a positive assertion of independence.

Accordingly, with the declaration of independence on 1 January
1804, the former 'European' name 'Saint Domingue' was officially
rejected as the name of the island and replaced by the old, indigenous
name *Haiti*. In addition, the new constitution of 1805 identified the
country as the *Empire d'Haiti*, gave the people themselves a name and
defined them by race – '*les Haitiens ne seront désormais connus que
sous la dénomination générique de noirs*' ['Haitians from now on will
be known only by the generic denomination "blacks"'] (Ardouin
[1853] 1958 vol. 6 p. 34). This act of naming was a new beginning and
was clearly intended to establish the dominance of the majority. The
designation of the people as 'black' clearly indicated that the restoration of the indigenous name 'Haiti' was an act of rejection of the European name and not an attempt to claim to be partly indigenous. The
names *Haiti* and *Haitien* were meant to be synonymous with *noir*, even

though not all Haitian natives were black or wanted to be considered black. The identification of Haiti as an Empire, though grandiose and provocative of images of Rome and the colonial 'mother countries' themselves, was meant to establish it as independent, without superiors and outside the influence of Europe. So began the history of independent Haiti, which within a decade or two became known far and wide as the Black Republic.

The constitution explicitly excluded white people by stipulating that: '*Aucun blanc, quelle que soit sa nation, ne mettra le pied sur ce territoire à titre de maître ou de propriétaire, et ne pourra à l'avenir y acquerir aucune propriété*' ['No white person, no matter from what nation, will ever set foot on this territory as a master or proprietor, and from now on will never be able to acquire any property here'] (Ardouin [1853] 1958 vol. 6 p.33). In an era when possession of property, real and human, was the foundation of white colonisation of the New World, this stipulation must have been a harder slap in the face than the prospect of black leaders. This stipulation was a statement of ownership and right to the island. It was later followed by invitations to free blacks in North America to settle there, a fact that served to strengthen this sentiment and suggested to black people that Haiti was as much their home as Africa. Such sentiments were the consequences of a military victory in a land in which the majority race had not only suffered as slaves but also had had many setbacks during the struggle. They had won it in battles that had lasted over a number of years, in the process of which they had lost thousands of their own. They were tied to their territory no less emotionally than the Americans were tied to theirs – they had both fought the colonial power for it. Consequently, it was not illogical that in this case the state should define and determine the race of the people. It was clearly a reverse of the normal process of emergence – it was a deliberate political act of self-protection and also one intended to reverse the negative concept of black.

However, the clauses of the constitution could not remove the racial divisions in Haiti and the bitter hatreds between the major groups that made up Haiti at the beginning of the nineteenth century. Although the island up until 1844, in the minds of the leaders, who came from the western side, was supposed to be one and indivisible, there were perceptible differences between those on the Spanish side and those on the French. They were all supposed to be Haitians, but they were never really one nation. Although it is made to seem today that those on the Spanish side had been always held against their will by those on the other side, this view was not shared by all at the time. The missionary,

Bird, who spent 30 years in Haiti in the middle of the century, present-ed the following opinion:

> It must, however, be understood that the Spaniards of the eastern part of the island were not, in this case, a conquered people. The movement which terminated in their union with the Haytian Republic originated with the Spaniards, they themselves having wisely seen, that their own interest and those of the island at large, rendered it desirable that the whole population of Hayti, throughout the entire island, should live under one flag, although there doubtless was a strong opposing party among the Spaniards. (1869: 143)

What Bird's view signalled was that on the wide expanse that was the Spanish side there were factions with different interests. There is no question, however, that the union of the eastern and western sides was an uneasy alliance of people of two distinct identities and that those on the Spanish side really wanted to be seen as different.

As a colony, the island was first Spanish and furthermore it had the first Spanish city in the New World. These were facts that those on the eastern side of the island were constantly reminded of and facts that shaped their identity. From that point of view, *dominico/dominicano* was the first nationality in the island and, as a term, it was one that continued to be used consistently to refer to those on the Spanish side. It was certainly used in the 1821 declaration of independence: '*En estas breves y compendiosas clausulas está cifrada la firme resolucion que jura, y proclama en este dia el pueblo Dominicano ...*' ['In these brief and succinct clauses is noted the firm resolution which this day affirms and proclaims the Dominican people ...'] (Mackenzie 1830 2: 216). Yet, the fact that this abortive declaration of independence proposed *Haiti Español* ['Spanish Haiti'], as the name of the eastern part of the island indicates that there was a level of acceptance of, and probably preference for, the indigenous name as opposed to the European name, even if it associated them with the western part of the island.

One major difference between the two sides was language, but it was not the only one. The perception of difference on the two sides was substantially due to the fact that, while the French side of the island had been a plantation-structured, sugar-producing colony, the Spanish side had been seen as an easy-going colony of ranchers. As a result of this, there was a belief and a repeated claim that slavery on the eastern side was 'milder': '*Il résulte de cette opinion une faveur qui s'étend nécessairement sur les esclaves. Ceux-ci sont nourris, en général, comme leurs maîtres, & traités avec une douceur inconnue aux autres peuples qui possèdent des colonies*' ['There results from this view a favourable circumstance which consequently extends to the slaves.

These latter, in general, are fed like their masters and treated with a mildness unknown to the other nations which have colonies'] (Moreau de Saint Méry 1796a 1: 59). Consequently, the slaves on that side were said to be more like their masters, more cultured and less African. In addition, the percentage of whites and mulattos on the eastern side was higher, especially after the revolution. The following comment from Mackenzie gives some indication of the feeling of superiority on the eastern side – 'I was not a little amused with the contemptuous mode in which even the blacks speak of their western neighbours as *aquellos negros* [those negroes]' (1830 1: 215). Consistent comments on the differences between the two sides by contemporary writers indicate that the much bigger population on the smaller western side remained substantially separate culturally from the smaller and scattered population on the much bigger eastern side. In addition, the rhetoric coming out of the eastern side showed that they considered themselves to have been racially different as an entity from the other side.

Within the western side itself, the racial divisions, which had fuelled the revolution, were exacerbated by the bloody conflict of the revolution itself. The exclusionary clauses in the constitution demonstrated the level of hatred of whites, many of whom consequently left the country. The animosity between mulatto and black remained unabated. The crucial first 40 years of independent Haiti were dominated by two blacks and two mulattos – Dessalines and Christophe on the one hand and Pétion and Boyer on the other – and it is during this period that the psychological foundations of the nation were established. The division of the country between Christophe and Pétion was as much political as racial and commentators of the time, foreign and local, showed their preference, as is evident in the remark of Candler – 'Christophe had a strong and invincible prejudice against the coloured class, of whom Pétion was one' (1842: 32). To a great extent the external support or condemnation of one or the other side fanned the flames of hatred, thereby contributing to the economic decline of the country in the nineteenth century.

Unlike most other writers, Lespinasse, in his work of 1882, tried to establish a spiritual union between mulattos and blacks, but still in the end he gave pride of place to mulattos:

> Par la naissance du mulâtre, l'esprit de liberté eut accès dans la population esclave de Saint-Domingue vivant dans l'éloignement de toute lumière. La procréation de l'homme de couleur fut le germe de la destruction de la servitude. Le blanc ayant donné de l'instruction à l'homme de couleur parce qu'il était son fils, l'homme de couleur éclaira le noir parce qu'il était son frère. La révolution de 91 est essentiellement due à l'esprit du mulâtre. (p. 15)

[Through the birth of the mulatto the spirit of freedom became accessible to the slave population of Saint Domingue, living far from all enlightenment. The procreation of the man of colour was the seed of destruction of slavery. The white, having given education to the man of colour because he was his son, the man of colour enlightened the black man because he was his brother. The revolution of 1791 is essentially due to the spirit of the mulatto.]

What Lespinasse said was of course the passionate belief of the mulattos, who maintained the hierarchical system of race, enlightenment and civilisation. The fact that a work as recent as Viau (1955) lists and discusses the leaders of Haiti from Dessalines onward with their primary characteristic and motivation given as 'negro' or 'mulato' indicates how deep and bitter this division has been. In fact, Viau identifies racism as the cancer of Haitian history and society: '*El prejuicio de color en Haití es un sentimiento colectivo, opresivo, sanguinario y monstruoso. Es la causa de todas nuestras desgracias. Es la peste que devora a Haití*' ['Colour prejudice in Haiti is a collective sentiment; it is oppressive, bloodthirsty and monstrous. It is the cause of all our misfortunes. It is the plague that is devouring Haiti'] (1955: 11). It was not simply a skin-colour division, but a division of class and cultural orientation, because from the era of slavery the mulattos had enjoyed privileges of education and money, which the blacks generally had not.

However, the fact that skin colour alone was not the absolute criterion for the distinction between the two groups is reflected in the saying *Neg riche li milat, milat pov li neg* [A rich negro is a mulatto; a poor mulatto is a negro], a saying that was already well known at the time. The privileges of a higher social class as well as cultural orientation towards the 'civilised' (white) world provided the mulatto with his justification for despising the negro, and racial purity and numerical dominance gave the negro his justification for despising the mulatto. A glimpse of the deep division in society still preserved during the second half of the nineteenth century is evident from the following remark made at time: 'Every one who mixes in Haytian society is struck by the paucity of black gentlemen to be met with at balls, concerts, or the theatre, and almost total absence of black ladies' (St. John 1884: 136–7). This was a time when the general population was about nine tenths black and one tenth mulatto.

It is not merely coincidental, then, that the eastern side of the island, which was not dominated by blacks, remained under the union during the time that Boyer, a mulatto, was president, but seceded in 1844, never to return, when a succession of blacks (Guerrier, Pierrot, Riché, Soulouque and Geffrard) took control from 1844 to 1867. The comment of Sir Spenser St. John, British consul-general to Haiti from 1863 to 1875, captured the kind of sentiment that no doubt was typical among the anti-black element: 'The revolution of 1843 that upset President Boyer commenced the era of troubles which have continued to the present day. The country has since been steadily falling to the rear in the race of civilisation' (St. John 1884: iii). St. John's obviously pleasant attitude towards those on the eastern side of the island is quite evident in his remark: 'The Dominicans have few prejudices of colour, and eagerly welcome foreign capitalists who arrive to develop the resources of their country' (1884: xii). The variables in the combination of race, political control and economics therefore precipitated the splitting apart of the island into two identities and the western part itself into two real- ities – one looking outward and the other confined within itself.

The island named 'Haiti' and 'the Black Republic' was certainly not a racially uniform or harmonious place. Yet, in spite of the deep divi- sions, there had been a change in attitudes and behaviour as a result of the Revolution. The imperial constitution of Dessalines was partially revised in 1816, but the deliberate acts of rejection of the French, Euro- peans and whiteness and the assertion of blackness had a long-lasting effect, which can be sensed in the following reaction in an extract in M'Queen: 'The meanest inhabitant of Hayti, considers himself upon a footing not only with his own countrymen, but with any stranger that may come in his way, whatever may be his rank, wealth, or informa- tion' (1825: 204). It is not only that the black Haitians had changed, but also that Europeans and other whites knew of the new Haitian phi- losophy and were wary of the behaviour of the people. As Mackenzie (1830 1: 24) grudgingly conceded, the people of Haiti were no longer to be referred to as 'subjects' but were now all 'citizens' – the French word *citoyens* had become the term of choice after the French Revolu- tion and also after the Haitian Revolution. It was not a trivial designa- tion of status, for even the otherwise sympathetic Dauxion-Lavaysse did not agree with it: 'Let it not be supposed from what I have said above, that I approve of the opinions of those who, in the revolutionary delirium, liberated the slaves without modification, and raised them to the rank of citizens' ([1820] 1969: 393). The vision of blacks and mulattos as emperor, president and king, dressed fully in regalia adopt- ed from the French, was shocking to the white colonial world and was interpreted as a threat to whiteness and civilisation, two concepts that were synonymous for them. It was even more immediately a threat to the colonial structure and system in which the white Creole had had to act as a kind of broker to maintain the status quo. The white Creole no longer had any such role in Haiti.

Yet, not all European views of Haiti were negative and hostile. For example, the words of the editor of the Frenchman Dauxion-Lavaysse's manuscript presented the situation in Haiti in the years after the Revolution in a positive light:

> These opinions of the author are fully borne out by the astonishing spectacle of a black dynasty in St. Domingo, unquestionably the most extraordinary event to which the French revolution has as yet given rise. When we reflect on the abject state of that fine island in 1789, and view the richest portion of it in 1819, governed by a **legitimate** monarch, who is not ashamed of his origin, will any one deny that the age of revolution has not at length arrived? Leaving this part of the wonder to its own merits, we have only to contemplate the able organization of the new kingdom, and the talents displayed by the members of its administration, and fresh sources of amazement burst upon the mind! Parochial and primary schools, on the Madras system, in every part of King Henry's dominions; a royal college, with annual prizes given to the most distinguished students. Academies for music and painting, a regular national theatre, and royal residence, which, for elegance and chasteness of design is not inferior to many of the palaces of Europe, a numerous clergy, and a long train of nobles, are but a few of the wonders to which our attention is now so irresistibly excited in that interesting quarter of the globe. (Dauxion-Lavaysse [1820] 1969: 371–2)

The positive picture presented here as well as Dauxion-Lavaysse's sentiments that inspired it clearly contradicted the negative images of Haiti that became common in French and European writing generally after the expulsion of the French from Haiti. In Venezuela and Trinidad, where Dauxion-Lavaysse spent several years and about whose inhabitants he was writing, visions of independence and removal of European rule made Haiti into a progressive place and a positive example, because, especially for the Spanish elements in those countries, it turned the tables on the French (and English), who in the minds of Spanish Creoles had always been abusive.

While attitudes about Haiti had been changed by force of circumstances and while notions of equality had been forced upon many, the appearance of equality on the part of owners was really only an immediate consequence of the economic principle of supply and demand, as Candler practically admitted:

> We could not fail here to be struck with the entire equality that seems now to subsist in Hayti between servant and master. Every workman that made his appearance was addressed in the courteous language, 'Mon fils,' and on inquiring the cause, we found it to be that the profits of planting were good, labourers scarce, and that it was necessary to conciliate all by kindness, or no work would be done. (1842: 37)

This was an economic situation in which ethnic/generational differences were seen to play a major part, as Mackenzie explained:

> The very little field labour effected is generally performed by elderly people, principally old Guinea negroes. No measures of the government can induce the young creoles to labour, or depart from their habitual licentiousness and vagrancy. (1830 1: 100)

It is the creole blacks, who had been born into slavery, who were, apparently, less willing to return to the land than the Africans who had been enslaved. So, even though a considerable number of the slaves and former slaves, many of whom were creole, had lost their lives in the fight for Haiti, there is no indication that those who survived had any strong attachment to the land itself (i.e. to agriculture), which they had won. In fact, for former slaves and descendants of slaves, the land, the sun and other features of climate represented a harsh reality, which they had no fantasies about.

Other differences between Africans and their creole offspring were noted by writers at the beginning of the century. Descourtilz ([1809] 1935) identified some of these differences as follows:

> *Infatués de la supériorité de leurs costumes, on voit près d'eux, dans le même sillon, leurs parents guinéens, le corps nu, avec un seul **tanga** qui dérobe leur sexe aux regards ...* (p. 122)

> [Infatuated by the superiority of their clothes, one can see near to them, in the same furrow, their Guinean parents, naked except for a 'tanga' hiding their genitals from view ...]

> *La passion de la danse est tellement impérieuse chez les nègres créolisés, qu'ils s'y livrent à l'excès ...* (p. 125)

> [Passion for dancing is so strong among the creolised negroes that they give themselves to it in the extreme ...]

> *Les Guinéens s'entr'aident dans l'infortune, mais les nègres créoles sont plus égoïstes, et la plupart sans charité.* (p. 141)

> [Guineans help each other in misfortune, but creole negroes are more self-centred and most of them are uncharitable.]

In addition, the creole black was generally said to be more civilised than the African, with 'positive' examples given to illustrate the claim, and, when writers made negative statements about the Creole, as those given above by Descourtilz, it was not usually admitted that the creole blacks were following the examples of the whites in society. Another Creole practice identified by more than one writer is that men openly lived with several women. This was, of course, regarded as consonant with the licentious behaviour of the uncivilised, but was also explained as a consequence of the major loss of men during the Revolution and the resulting unbalanced ratio of men to women.

The consequences of the assertion of blackness and the exclusion of whites were devastating to Haiti as a nation because the documentation of events, as a history of the people, was not in the hands of the protagonists of the Revolution. The press, which was absolutely controlled by whites, unleashed a barrage of virulent, racist comments from all quarters of Europe and the USA, which were based on a philosophy that conceived Africans generally as a sub-human species outside civilisation and the people of Haiti specifically as participants in voodoo, cannibalism and licentiousness and as having no family values. Confronted by the virulent rhetoric and inducted into an educational system that reinforced it, mulattos and those blacks who were exposed to it were propelled even faster on a path of self-hatred, denial and escape, in a search for an aesthetic of whiteness. The killing of mulatto and black by black and mulatto was made to seem bestial, in contrast to the Spanish and American annihilation of the indigenous inhabitants, which in the national literature of the day was glossed over or made to appear a heroic struggle of the civilised over the uncivilised. The treatment of women by the black King Henri of Haiti could not rival that of white Henry VIII of England, but no such comparison could have been admitted.

On the other hand, there were immediate positive consequences of the Revolution. Within Haiti, the lack of access to 'civilised' education among the blacks meant that African cultural features were able to make a more positive and penetrating mark on the evolving culture of the population. In other words, Haiti emerged with its own distinctive, island culture with a greater retention of African characteristics because white European leadership and many of its values had been removed. Its language, as part of the culture, was also quite distinctive, although it was not restricted to that island. Outside Haiti and in spite of the negative European rhetoric, leaders in the islands and elsewhere began to consider with more emotion the possibilities of self-determination, rejection of European control and assertion of creole identity. The words of Dauxion-Lavaysse illustrate these growing national sentiments: '*Somos Americanos y no Gachupines;* "we are Americans, and not Spaniards," the Creoles of Venezuela and other Spanish possessions will frequently exclaim in a tone of ill-humoured haughtiness' ([1820] 1969: 178). Black Haiti exemplified the contrast in identity between the Americas and Europe more sharply than did the white United States, and, in the context of the bigger Caribbean islands, it gave those 'above' black on the scale of 'lineal ascent' more confidence to pursue national goals as well as to assert their own identities as Creoles or Americans.

# Language and identity in Haiti

Haiti was clearly defined by the native language of its people, which was known to all from at least the last quarter of the eighteenth century as *créole*. It was repeatedly referred to in newspapers and in accounts by visitors. Its function as the general language of communication and its possibilities as a literary medium were well known, especially after the initial discussion of its structure, status and function in the works of Moreau de Saint Méry and Girod Chantrans. It had a historic role in the Haitian Revolution in that it was used in an 1801 Proclamation that announced that General LeClerc was coming to Haiti with a naval force. In fact, as Beaubrun Ardouin explained, there were actually two proclamations, each with a translation:

> *La députation avait reçu de Leclerc des exemplaires de la proclamation du Premier Consul et d'une autre qu'il avait rendue lui-même, pour mieux expliquer les intentions de la France.*

> *[L'une et l'autre proclamation avaient des exemplaires imprimés en langage créole. Quelque colon s'était exercé à la traduire ainsi, afin d'assurer un plein succès à l'expédition.]* ([1853] 1958 vol. 5, p. 7)

> [The deputation had received from Leclerc copies of a proclamation by the First Consul and of another which he had made himself to explain the intentions of France better.

> [Both proclamations had copies printed in creole. Some colonist had made the effort to do this in order to ensure that the expedition was completely successful.]]

There was some difference and deception in the translation, however. The original French version was addressed only *Aux Habitans de Saint-Domingue*, and as Ardouin says:

> *Dans le langage colonial, on entend par **habitans** – les propriétaires. Comme c'est à eux seuls que la proclamation s'adressait, **les ennemis** contre lesquels on voulait les protéger étaient **les noirs**, destinés à être replacés dans l'esclavage, a leur profit.* ([1853] 1958 tome 5, p. 6)

> [In colonial language 'habitans' means 'property owners'. Since it was to them alone that the proclamation was addressed, the 'enemies' against whom one wanted to protect them were the blacks, who were to be put back into slavery for their profit.]

The creole version was addressed not only to *toute Zabitans Saint-Domingue* but also *vous tous qui dans Saint-Domingue* [all of you in Saint Domingue]. It is clear that as far as the latter were concerned, the sting of the proclamation was in the tail, that is in the threat that anyone not complying would be punished.

The translation of an official proclamation into a creole language had no parallels outside Haiti. In a sense, the emergence of the official use of *créole* in a written document was preceded historically by two sets of facilitating circumstances. First, there was the *Code Noir*, which was a set of stipulations dedicated exclusively to the control of slaves, without any parallel among the other European colonisers, whose laws for the government of slaves were given piecemeal. It is not that the *Code Noir* was in creole, but that the matter was serious enough for the French that it should be addressed fully and directly with an expression of concern for the slave. The second preceding historical circumstance was that what was now being called *créole* had long become familiar to the French reading public in the early works of French colonisation in the smaller islands, works written by their Catholic priests and De Rochefort. In fact, the way in which the word *Français* (the French) is used in the *langage corrompu* is virtually identical to the way in which it is used in the proclamation – there was the absence of the creole noun marker and the attitude of paternalistic condescension. So, for the *créole* to be put in print to address its speakers was probably not as revolutionary as it seems today.

Generally, newspapers, pamphlets and other literature current in Saint Domingue at the end of the eighteenth century and the beginning of the nineteenth century were in French. The fact that the proclamation was translated into creole meant that there was a clear perception that French would not serve the purpose of urgency and general application because it was not intelligible to the majority of the people, those who were not *habitants*, and that ad lib translations at the time of the proclamations could not be depended on. It is also very likely that the French mind, which had soaked itself from the seventeenth century in the notion of the superiority and magnificence of the French language, automatically excluded French as the medium to deal with a country of blacks and mulattos. Creole was the medium for direct communication in a situation of urgency, as opposed to French, which could only have effected 'trickle-down' communication.

At the time of the event, a written or printed translation of a general proclamation must have had little significance; it probably was excused as a practical necessity by those who noticed. Subsequently, the appearance of the creole proclamation in print was meaningful only to the small percentage of those who were literate in Haiti and, even for these, creole in print must have been no more than what they had seen in Girod-Chantrans, in Moreau de S. Méry and in journals and newspapers of the time. Yet, no matter how non-historic the creole form of the proclamation was at the time, it is noteworthy that in this country, which was the first in which the term *créole* was consistently applied to a language, an official proclamation was actually the start of its official representation.

Attitudes towards the creole language of Haiti continued to be mixed throughout the century on the part of both locals and foreigners. For example, at the very beginning of the century, Michel-Étienne Descourtilz, a Frenchman, who regarded himself as a friend of Toussaint Louverture, inserted the creole fairly freely into his text, without consistent comment about its worth or insinuations about the mentality of its users. On the other hand, Haitian mulattos, imbued with ideas of 'lineal ascent', tended towards French and French culture. Accordingly, Beaubrun Ardouin, who, though a mulatto and though partial towards the black leaders of the early years of independence in Haiti, in his huge history (1853) hardly ever mentioned or cited the creole language. Nearing the end of the century, Sir Spenser St. John, an Englishman whose work is coloured by the typical notions of black inferiority, included a chapter on literature in Haiti in which the creole language featured prominently. In this chapter he claimed that: 'President Geffrard ... used to extol the Creole as the softest and most expressive of languages, and his countrymen are unanimously of his opinion ...' (p. 279). St. John (1884: 303) also acknowledged literary variation in the creole language when he contrasted 'the **cultivated** Creole of the present day with the true ring of popular Creole'. Essentially this was a contrast between the kind of creole to be found in songs that were deliberately composed by authors and the kind that occurred naturally in folklore, that is traditional songs and proverbs.

The general concept of the creole language, as far as its origin and structure are concerned, was the traditional one of *langage corrompu*, that is that it was a kind of 'broken' French. A slight modification of this, which in a sense revealed a higher level of understanding was that it was partly African, as is seen in the following comments by Candler:

> Urchins of boys, as is almost always the case in these expeditions, ran
> before, or behind, and everywhere. 'Bon jour, Monsieur,' 'Bon jour
> Madame,' were the cheerful salutations that met our ear, accompanied
> sometimes by a sentence of unintelligible Creole, half French, half African,
> that amused us from its oddity. (1842: 26–7)

and St. John: 'this uncouth jargon of corrupt French in an African form ... the negroes imported into Hayti learned French words and affixed them to the forms of their own dialects' (1884: 299). However, even where negative comments were made about the creole, there was no consistent implication that it was a hodge-podge. The general absence of comments about lack of uniformity indicates that, to those foreigners who lived in Haiti over a number of years, it appeared to have stability. Comments were made about variation, which related to differences between urban and rural speakers, the latter being said to be more African and more difficult to understand (St. John 1884: 300). Overall, then, the creole, which internally of course had no problems of validity as a language in its own right, when assessed by outsiders also enjoyed a measure of integrity to the extent that it was known to be significantly different from French and assumed to have African elements because the majority of its early speakers had come from Africa.

There was also an awareness by the second half of the century that creole was spoken outside Haiti in Guadeloupe, Martinique, Trinidad, Grenada and in Guiana. J.J. Thomas's (1869) analysis of creole in Trinidad was known in Haiti and in fact drew the following concession from St. John:

> As this Creole language is spoken by about a million and a half people in the different islands of the West Indies, it merits the attention which Mr. Thomas has bestowed upon it; and I would refer those curious on the subject to this elaborate work, in which everything is done to raise the status of a patois ... (1884: 300)

The regionality of the creole, its repeated even if sporadic appearance in written form and discussions about its merits not only improved its status as a language but also heightened the perception of it as a major characteristic of Haitian identity specifically. Yet, it was still very far from supplanting French as the official language of the Black Republic.

Even though the imperial constitution of 1805 meant to exclude white people from the nationhood of Haiti, and this meant primarily the French, there was no stipulation about the rejection of French as the language of written or official communication. There was no stipulation that the language used and understood by the majority of the population was to be the sole official language. French continued to be used by Haitians for literary purposes, historical and creative. Those Haitian leaders who were illiterate, or almost, had secretaries to write letters for them and they were written in French. It is only when Toussaint Louverture was pleading for leniency that he is said to have written the following letter himself:

Premier Consul,

Pere de toute les militre, De fanseur des innosant, juige integre, prononcé
dont sure un homme quie plus mal heure que couppable. Gairice mes plai,
illé tre pro fond, vous seul pourret porter les remede saluter, et l an pé ché
de ne jamai ouver, vous sete medecien, ma po sition, et més service merite
toute votre a tantion, et je conte an tier ment sure votre justice et votre
balance.

*Salut et res pec.* (Ardouin [1853] 1958 tome 5, p. 48)

[First Consul,

Father of the whole military force, defender of the innocent, impartial
judge, give your judgement on a man who is more unfortunate than guilty.
Heal my wounds; they are very deep. You alone can bring them healing
remedies and prevent them from ever opening; you are a doctor. My
position and my service deserve your full attention and I am counting
entirely on your justice and fairness.

Greetings and respect.]

In spite of the problems with orthography and word structure, the letter is essentially in French rather than creole. It shows that this man,
who had been born a slave with creole as his native language, had managed to acquire a reasonable degree of competence in French. It is likely
that most of those who had moved from the bottom socially up the ladder through the Revolution would have had the same kind of competence in French.

Henri Christophe, according to Mackenzie (1830), was fluent in
both French and English, although he preferred to use an interpreter
rather than to speak directly to English visitors. Mackenzie (1830 1:
130–1) also claimed that, although Christophe adopted Roman
Catholicism as the religion of his kingdom, he seriously considered the
establishment of Protestantism and the introduction of the English
language to facilitate this. Greenfield quoted a 'writer in the Fife Herald
as saying: "Christophe, the sovereign of Hayti, wished to change the
language of the black population of that island from French to English
…"' (1830: 70–1). Some years later another English-speaking emissary
to Haiti, John Candler, also claimed (1842: 35) that Christophe, in his
education policy, hoped that by promoting English as a subject in the
schools it would come to supersede French. One of the factors that
could have influenced Christophe, if these claims were not exaggerated,
is that, first, he was born and raised in Grenada, an island that shifted
back and forth between the French and the English, and, second, he
had been given support by the English and the Americans in his fight
with the French. Indeed, the English and Americans were allowed to set

up and expand their Churches without overt hostility from the Haitian rulers. Moreover, some of the schools that were opened were based on English systems of organisation. Consequently, there seemed to be throughout the century at least a small number of persons, especially of the higher classes, who learnt to speak English in this way, and there also seemed to be some measure of diffidence about the status of French as the official language.

On the other hand, Madiou points out that Christophe, by a proclamation of 23 May 1819, decreed that English subjects from neighbouring islands would not be allowed to enter Haiti and that if they did they would be deported back to their country of origin:

> *Christophe tenait à la bienveillance et à l'amitié de l'Angleterre à un si haut degré qu'il interdit l'entrée de sa royaume aux sujets anglais quel qu'ils fussent, qui avaient commis des vols dans les colonies de Sa Majesté Britannique ou qui en sortaient comme fugitifs.* (1848 vi: 38)

> [Christophe valued the benevolence and friendship of England so highly that he prohibited from entering his kingdom all English subjects, whoever they were, who had committed robberies in the colonies of His British Majesty or who were fugitives from these colonies.]

This was certainly a change from the early days of independent Haiti when all blacks entering Haiti were promised freedom: it suggests that Christophe, in his pragmatism, no longer saw his country as a global beacon of blackness and was not going to promote 'English' through immigration.

The major prop of colonisation in the island, the Roman Catholic Church, conducted its services principally in French during the colonial period, but, since it was a French establishment, it also suffered a slap in the face in Haiti as a result of the Revolution. It was not until 1860 that a Concordat was signed with Rome to allow the Catholic Church in Haiti to be regulated from outside. Candler captured the foreigner's view of the pre-1860 situation when he said:

> Since the union of both divisions of the island under the republic, the jurisdiction of the Pope at Rome has been repudiated; the Archbishop has banished himself to a distant country, and the President, following the example of Henry the Eighth, has become head of the church. (1842: 94)

As a result, the quality of priests was said to be very low, the Church was weaker and Protestantism was allowed to gain a foothold. Moreover, more importantly, local control allowed for a continuation and development of syncretism of Catholicism and African religions and a more general use of the creole language in religious practice.

The general dominance of creole among the native population of Haiti, the power of the French language at the official level and competence in English among a small percentage of the society constituted the framework of the overall linguistic situation in Haiti for the early part of the century. Yet, a more heterogeneous and exotic picture of the situation occurs in M'Queen (1825), quoting from an article in the *Edinburgh Magazine* for December 1823:

> and to be found, a people speaking a variety of languages: fugitives from Cuba, who can only speak a kind of Spanish jargon; from Jamaica, whose language is a sort of broken English; emigrants from Curaçoa, talking Dutch; and the original blacks and mulattoes of the island, whose language is a sort of broken French. There are, besides, several white people settled in Hayti, natives of Europe and America ... (p. 202)

This view of the Haitian language situation was possibly true for certain urban areas in the first quarter of the century, but it certainly did not capture the reality of Haiti as a whole because it made the immigrant groups seem much bigger in number than they actually were. By the third quarter of the century, the situation had become much more straightforward, according to St. John: 'There are two languages spoken in Hayti, French and Creole. French is the language of public life and of literature, whilst Creole is the language of home and of the people' (1884: 299). It was a situation in which the black part of the population was associated with the creole language and the mulatto part with French.

In Haiti in the nineteenth century, the term *Créole* consolidated itself as the name of the language of Haitians. On the other hand, it lost its significance as a term of identity for people as a result of the Revolution because the distinction between black African and black Creole declined both socially and economically. Some writers continued to make a distinction in customs and attitudes between the two, but the society itself did not accord any major advantage to Creoles. In the course of the century therefore the word *Créole*, which had never had much relevance to mulattos, had decreasing reference to blacks as the number of Africans naturally declined after the Revolution, and it had little or no use as a word for whites because most of these had migrated. The gradual restriction in the reference of the term to language also constituted a decline in its status because the creole language itself was not officially identified as a major characteristic of the people in the same way that the name of the island and the race of the people were. Thus, like the Black Caribs in St Vincent, Haitians became 'doubly savage' – the name they chose for their country and consequently themselves was indigenous, and the race they identified themselves as was

African. Freedom and independence for them meant becoming natives of their Caribbean island rather than maintaining themselves in a contrast with the Old World and Europe specifically as Creoles.

# The *criollo* and a notion of race in the Spanish islands

Slave society brought into being a system of classification that used 'White' as the point of reference on one side and 'Black' and 'Indian' as other poles. 'White', 'Black' and 'Indian' all had antecedent histories though given different values, which made it easy to link persons to concepts associated with these races, concepts that European expansion had kindled and rekindled with fervour. While it was possible to situate the product of a white and a black halfway between the two traditional concepts, repeated intermixing made any such assessment virtually impossible. In those Caribbean island societies where the numbers of racially mixed people were comparatively large, they came to have a crucial input in the conception of national identity by being a challenge to historicity (i.e. the direct link to white European or to black African exclusively) as the determining feature of race and thus identity.

The Spanish contributed to the 'problem' by making white a flexible category because they had encouraged the practice of according white status to different groups, as Dauxion-Lavaysse noted: 'But the kings of Spain gave diplomas of whites (a kind of scandalous whitewashing) to certain persons who had rendered, or were supposed to have rendered important services to the state' ([1820] 1969: 71–2). Consequently, in the Spanish islands, the concept of white expanded as distinctions of parentage became more and more blurred and as the pressure to be white increased. So, white became a more inclusive category in Spanish America, especially when jobs and positions that were previously reserved for Spaniards were taken over by Creoles.

In the Spanish islands where persons of mixed race were numerically dominant, the word *mulato*, as a racial subcategory of *criollo*, did not and could not become a preferred classification. It was unsatisfactory as a term of identity. The Frenchman Dauxion-Lavaysse reflected the contemporary attitude to mulattoes as well as to the word itself when he said: 'people of mixed blood … in the European languages are stigmatised with the insulting denomination of Mulatto' ([1820] 1969: 395). The analogy between mulatto and mule arose out of the conviction that human beings were naturally divided into distinct races, each with its own purity. So, as Granier de Cassagnac explained: '*les mulâtres sont un produit bâtard, qui n'a pas d'aïeux de son espèce, et qui ne peut pas se reproduire sans s'effacer.'* ['mulattos are a bastard product which has no ancestors and which cannot reproduce without obliterating itself.'] (1842: 102).

In addition, where the distinctions in the social system were set up in a bi-polar contrast (e.g. slave–free; black–white), as the English especially seemed to prefer, the 'proliferation' and preponderance of mixed or intermediate beings posed a major difficulty and a threat to the preservation of the system. The purity of the white race, which constituted a powerful ethereal notion in social stratification in the sugar-plantation societies of the French and English throughout the nineteenth century, remained a sore point for the Spanish. Madden's observation in 1835 (1: 89) that the child of a mustee and a white person was fairer than the Spaniards gave an indication of what other Europeans continued to think of the Spanish. In addition, because their New World societies generally were not dominated by sugar plantations or the plantation social structure, they seemed, in the eyes of the French and English, to encourage consorting with mulattoes and other 'lower' people. Toward the end of the nineteenth century, the distinguished English historian, Froude, referred to the island of Hispaniola, which he actually visited, as being 'now divided into the two black republics of St. Domingo and Hayti' ([1888] 1969: 129). Even though he had read the book of Sir Spenser St. John on Haiti, which was partial towards the Spanish side of the island, he obviously did not think of the people of St Domingo as being racially different from those of Haiti.

In the context of Santo Domingo, the early use of the term *Indo-Hispanos* by Sanchez Valverde in 1785 gave some bibliographical validity to the later use of the term *Indios* to identify non-white Dominicans, and at the same time to contrast them with the black Haitians next door. Ironically, the notion of Indian, as a social category distinct from Negro, came out of Saint Domingue itself. Lespinasse, in his work on the treatment of manumitted people in St Domingue under French rule, pointed out that it was Louis XV in 1767 who, in a policy decision for the French colonies, made a sharp distinction between Indians and Negroes, in which he accorded Indians full privileges – *'ceux qui proviennent d'une race indienne doivent être assimilés aux sujets du Roi, originaires d'Europe'* ['those from an Indian race must be classed the same as subjects of the King from Europe'] (Lespinasse 1882: 235–6). This sowed the seed for non-whites on the eastern side of the island after the Revolution to become Indians.

In the nineteenth century therefore, because of the numerically dominant position of mulattoes in the Spanish islands, a certain flexibility developed that allowed for an extended category of whites, an extended notion of mulatto, a reorientation of mulatto, which evolved into a reconstruction of the concept of Indian and the suppression of the presence of black race or Negro within the Spanish islands. This was at the same time a conscious response to the threatening presence of the black Republic, Haiti, and a way of allowing the citizens of the Spanish islands, of all colours, to identify themselves as different from Haitians.

## The *criollo* and a notion of place

From the beginning, place of birth (i.e. *Indias*) was generally understood to be an integral part of the shaping of the *criollo*, but in the Spanish islands it came to be even more consistently identified through the use of *la tierra*. In what is regarded as the earliest important piece of literature coming out of the islands, the *Espejo de paciencia* ['Mirror of patience'] (1608), a laudatory sonnet preceding the poem itself speaks of *criollo de la tierra* and this theme of *tierra* is repeated in a number of subsequent works as well as presented in others in the nineteenth century in terms of the climate having the most profound effect on the formation of the *criollo*. Arrom also cites an even earlier work, from 1568, in which the author, *el sabio Sahagún* [Sahagún, the sage] identifies *clima* and *tierra* as forces of change: '*los que en ella nacen, muy al propio de los indios, en el aspecto parecen españoles, y en las condiciones no los son ... y esto pienso que lo hacen el clima o constelaciones de esta tierra*' ['those born there, very much like the Indians, in looks appear to be Spaniards and in conditions are not ... and I think that this is brought about by the climate and constellations of this land'] (Arrom 1961: 313). This was the kind of reasoning that was part of a tradition that regarded different races as direct products of climate and one that had been rekindled in the sixteenth century by the appearance together of different races of man in the exploitation of the New World. It reached the academic/scientific level in the nineteenth century. It was in the nineteenth century that the notion of the effect of nature on the evolution of living things had become a very powerful and compelling argument principally because of Darwin's work of 1859, which was based on a study of animals. So, assisted by scientific thought and propelled by the need to get closer to whiteness or make the black African disappear, the *criollo* began to be transformed in the Spanish islands to become the biological product of race and climate.

The most extreme example of this mode of thinking was the conscious recreation of the 'Indian' in the Dominican Republic by intellectuals who saw this type as the result of the influence of the climate on the offspring of white and black. Despradel in an explanation of the development of the myth of the 'Indian' as a racial type in the present-day Dominican Republic writes:

> *Luperón escribió: '... como los aborígenes habían desaparecido, la población de la isla se formó de dos razas distintas tanto por su origen y apariencia como por sus costumbres y preocupaciones. Se trata de las razas europea y africana que, mediante la mezcla, produjeron una mixta, que participa de las otras dos, según la preponderancia de la una o de la otra sangres. Y esta última tiende, por la ley de los climas, a reencontrar la raza primitiva de la isla.'* (1973 i: 16)

[Luperón wrote: '... as the aborigines had disappeared, the population of the island was made up of two races, distinct as much because of their origin as because of their customs and preoccupations. It is a matter of the European race and the African race which, by miscegenation, produced a mixture which, according to the ratio of the blood of the one to the other, exhibited both. And this latter [the mixed blood person], through the law of climate, evolves to become the same as the primitive race of the island.']

Despradel identifies this statement appearing in 1896 as the successful completion of *un trabajo extraordinario de mistificación cultural* [an extraordinary work of cultural mystification], which had started about the middle of the century.

Luperon's writings were actually published in Puerto Rico and it is interesting to note that Puerto Ricans also came to embrace this myth of the 'Indian' as part of their biological make-up. The idea of new races coming into existence in Puerto Rico as a result of mixture of Spaniards with indigenous inhabitants and Spaniards with Africans was mentioned as early as 1797 by Ledru. As was the case with those in the Dominican Republic a little later, Ledru believed that the emergence of the new *races* resulted not only from mixture but also from the effect of the climate – 'Ces *mélanges, joints aux effets du climat, ont produit plusieurs races d'hommes dont chacune a sa couleur, son caractère propre*' ['These mixtures, together with the effect of the climate, produced several races of man, each with its own colour and characteristics'] (Ledru 1810: 161). Some 25 years after Ledru's work first appeared, the *jíbaro* started to become a symbol of Puerto Rican identity and this, in an indirect way, permitted the indigenous inhabitant to come back to life, since the word itself, being indigenous, encouraged such a belief.

In the case of Cuba, although the 'Indian' re-emerged as one of the roots in Cuban identity, *la tierra* as a determining factor was not highlighted. Arrom (1971: 184–214) associates the development of the presumed tripartite racial make-up of the Cuban people with religious mythology and integrationist tendencies. In an explanation of the development of *los tres Juanes* (*Juan Criollo, Juan Indio* and *Juan Esclavo*) [the three Johns – Creole John, Indian John and Slave John] in the *Virgen del Cobre* story in which the three are used in a generic way to represent the three races (White, 'Indian', Black) respectively, what started out as a story with two 'Indians' and a little black boy, *un moreno criollo* (Arrom 1971: 196), in Bernardo Ramirez's account of 1782, gradually changed to one with three men, in which one 'Indian' disappears, *moreno* becomes *Esclavo* and *criollo* becomes *Criollo*, a generic term for whites in Cuba. It was therefore a matter of the 'Indian' surviving rather than being a climatic product.

As a result of the perpetuation of the belief in the genetically transforming effect of the climate on living things, the words of López de Velasco (1898: 37–8), in whose book the word *criollo* first appeared, were made to become almost prophetic 300 years later. On the other hand, it is quite possible that the reappearance and publication of López de Velasco's manuscript in 1898 were not entirely unrelated to the statement of Luperón and more specifically to the re-emergence of the 'Indian'. Whether this was so or not, López de Velasco's link between *criollo* and the effect of the climate was to re-echo through the works and thoughts of those writing in the Spanish Caribbean. Place of birth, as a feature of *criollo* identity, was therefore not simply an accident with political significance, but was regarded as a basic formative feature of that identity and at the same time was a more palatable explanation of the darker-than-European skin of the people.

Beside the beliefs about the effect of *la tierra* on the racial development of the *criollo*, there was a kind of reverse process in which *la tierra* was characterised to reflect the racial features of the *criollo*. This followed from the everyday equating of *tierra* and *criollo*, as is evident in the nineteenth-century definition of *criollo* by Esteban Pichardo – '*en este concepto es lo mismo que decir de la Tierra*' ['in this concept it is the same as saying "of the earth"'] (Pichardo 1976: 191) and conversely in his definition of '*la tierra – Lo mismo es decir Criollo que de la Tierra; pero no se usan indistintamente sino en ciertos casos, por ejemplo, Caña Criolla o Caña de la tierra, Abeja Criolla o Abeja de la Tierra ...*' ['To say "criollo" is the same as saying "de la tierra", but they are not used interchangeably except in certain cases, for example "caña criolla" [= creole cane] is the same as "caña de la tierra", and "abeja criolla" [= creole bee] is the same as "abeja de la tierra" ...'] (1976: 577). As a result of this equational relationship, there developed a further link between *criollo* (race) and *tierra* (soil). The types of soil in Cuba were identified by Pichardo as *Tierra Negra, Tierra-Bermeja* [reddish] and *Tierra Mulata*, the last of which was explained as a mixture – '*como si las dos primeras mezclasen ...*' ['as if the first two mixed ...'] (1976: 577). Pichardo then went on to say: '*En concepto de muchos inteligentes la **Tierra Mulata** es la mejor y predispuesta a todo cultivo*' ['In the view of many intelligent people, "mulatto soil" is the best and the most suitable for any cultivation']. The concept of *la tierra mulata* as the best soil for planting provided an interesting concretisation of the concept of the *mulata* as the perfect fusion of contrasting races and was consistent with Lespinasse's view (1882: 15) of the mulatto as the catalyst for the Haitian revolution.

# The *criollo* and the notion of native language

Language, as the fundamental link between different ethnic groups, was always a powerful force in the colonies, but its evolution to become a peculiarity of *criollo* identity was portrayed in the literature mainly as a matter of addition of local indigenous words to the Spanish language and variations in pronunciations resulting from poor education. Even with these additions and modifications, there was a perception of uniformity of the Spanish language across many different ethnic groups in the colonies and former colonies. There was therefore no special language or variety of Spanish that was promoted in any of the islands as typically *criollo* and referred to by a special name. It is quite paradoxical that the language of the weakest of the three main colonial powers in the Americas was the one that was seen to diverge least in the colonies. Even where Spanish colonial neglect was at its greatest, as in Santo Domingo, there was no mention of a creole language to match that of Haiti. Even though in all three Spanish islands there was a diversity of population matching the continual interest of, and rivalry between, the English, French and Americans, neither English nor French was said to be a widely spoken language in any of these islands. Even though there was in the Spanish islands the promotion of the notion of a new race formed from the mixture of three different races and modified by the climate, generally there was no corresponding promotion of a new language evolving from the mixture of different languages and shaped by local circumstances. Spanish continued to be seen as the dominant language culturally, and the different ethnic groups in the islands apparently identified with it.

There was in each island a vernacular that allowed *criollos* to differentiate themselves from others and gave them a characteristic identity, but it was the common practice of portraying the language of the people in the other islands, especially Haiti, as a sort of degraded or backward (version of a) European language that was without doubt a major obstacle to the positive recognition and presentation of dialectal variation in the Spanish islands themselves. However, as their societies continued to evolve in the nineteenth century, writers in the Spanish islands eventually began to present a *criollo* identity through the use of features of local vernaculars. The vernacular was used with social intent in the early cases, appearing mostly in 'literary' works such as poems, plays and stories.

By the end of the century there was at least one voice proclaiming the beginnings of a new language and intending to trace its evolution. The voice was that of Juan Ignacio de Armas:

> Llamo **lenguaje criollo**, a falta de mejor nombre, al conjunto de vozes i construcciones peculiares, de uso corriente i general en las islas de Cuba, Santo Domingo i Puerto Rico, en las repúblicas de Venezuela i Colombia, i en alguna parte de Centro América.
>
> Empezó a formarse en las Antillas, sobre la ancha base del idioma castellano, desde los primeros días del descubrimiento; se propagó con la conquista al continente, siendo designado en sus principios con el nombre de **lengua de las islas**; se enriqueció a su vez con multitud de vocablos de las nuevas rejiones conquistadas; adquirió homojeneidad i un carácter distintivo, con los primeros criollos; allegó a su formación los más variados componentes; i hoi constituye un cuasi-dialecto castellano, que comprende el litoral del mar Caribe, i que será sin duda, para una época aún remota, la base de un idioma, hijo del que trajeron los descubridores i conquistadores de América. (1882: 5)

> [I am calling 'creole language', for want of a better term, the ensemble of peculiar words and structures, in current and general usage in the islands of Cuba, Santo Domingo and Puerto Rico, in the republics of Venezuela and Colombia and in a part of Central America.
>
> It began to form in the Antilles, on the wide base of the Castilian language, from the first days of the discovery; it spread with the conquest of the continent, at first being called 'the language of the islands'; it was enriched in turn by the multitude of words from the newly conquered regions; it acquired homogeneity and a distinctive character from the first Creoles; in its formation it gathered the most diverse components; and today it constitutes a quasi-dialect of Castilian, which includes the Caribbean seaboard, and which in some distant time no doubt will be the base of a language, the offspring of the one the discoverers and conquistadors brought to America.]

He saw this language developing in the same way that Spanish itself had done:

*El castellano, llamado a la alta dignidad de la lengua madre, habrá dejado en América, aún sin suspender el curso de su gloriosa carrera, cuatro idiomas, por lo menos, con un carácter de semejanza jeneral, análogo al que hoi conservan los idiomas derivados del latin.* (1882: 6)

[Castilian, called to high dignity as mother language, will have left in America, without curtailing the course of its glorious career, four languages, at least, preserving in them a general character of similarity, in the same way that the languages derived from Latin do today.]

Here, as was the case with many other writers in the Spanish colonies, there was a desire to extol the greatness of the Spanish language, as a result of which the emerging languages were put in a subordinate position. Yet the scientific approach to evolution, which Ignacio de Armas claimed to be following, indicated that there was a creole dialect of Spanish in the islands which matched to some extent the creole culture and ethnicity of the people.

On the other hand, during the eighteenth century there had been continued migration into the Spanish islands of Spanish-speaking people, especially from the Canary Islands. This migration, even though it must have consolidated divergent linguistic features that had been brought by the earlier migrants from these islands, at the same time would have neutralised linguistic creolisation to some degree and made it easier for nationalist types in the agricultural belt, like the *jíbaro* and *guajiro*, to appear to be countrified Europeans in their speech.

# The vicissitudes in the evolution of *Dominicanos*

In order to understand some of the background forces at work in the development of identity in the nineteenth century in the eastern part of the island, which had been the first colony to be established in the New World, it is helpful to look first at the simplified outline of its history given in Table 5.1 . This outline gives only a hint of the political instability that plagued the island from the beginning of its colonial history through the period of the revolution of its western third and afterwards. It also has to be considered against the background of the early decimation of the indigenous population by disease and war, the addition of thousands of Africans as enslaved workers and the introduction of thousands of colonists, adventurers and workers of various social levels from Europe and the Canary Islands over the colonial period.

From the start of the nineteenth century the identity of the eastern part of the island of Hispaniola was overwhelmingly shaped by its

**Table 5.1**   Simplified history of Hispaniola

| | |
|---|---|
| 1496 | Spain establishes colonial rule in Hispaniola |
| Sixteenth century | Spanish colonisation in Hispaniola weakens as Spain looks to the mainland to the west |
| 1603 | Spanish devastation of the northwest of Hispaniola to control piracy and encroachment |
| 1640 | France controls Tortuga and begins to encroach on northwest Hispaniola |
| 1697 | Saint Domingue (the western third of Hispaniola) becomes a French colony |
| Eighteenth century | Continuing decline of the Spanish part of the island contrasts with the western part becoming the richest sugar colony in the New World |
| 1791–1804 | Haitian Revolution |

struggles with the western part of the island. At first, these were not really struggles between the masses of the population on the two sides; these were struggles between European powers for the control of the island as well as attempts by a white and mulatto ruling class on the eastern side to distance itself from the black revolution that took place on the other side. In any case, for much of the nineteenth century the people of Santo Domingo were not in control of their own affairs, a fact that affected their vision of themselves.

From 1801 to 1809 the French were effectively in control of them. From 1809 to 1821 the Spanish were again in control of them. In 1821 they declared themselves independent, but in 1822 Haiti took over and ruled the whole island until 1844, with the mulatto leader, Boyer, at the head. Then, after declaring themselves independent of Haiti in 1844 and struggling to remain independent for about 16 years, in 1861 they invited Spain to rule their country again, essentially to protect themselves against Haitian domination. This lasted until 1865 and was a period during which the Spanish, with a slave-holding mentality supported by the continued existence of slavery in Cuba and Puerto Rico, came into conflict with the people of a former colony, most of whom had African ancestry and who had been free for most of the century. In 1869 Santo Domingo sought to be annexed to the USA (again to be protected from Haiti), but was not accepted, in no small measure because the population was said to be '*compuesta de una raza cuya sangre tiene dos tercios de africano nativo*' ['made up of a race whose blood has two thirds African native in it'] (Saviñón 1994 1: 426). At the end of the century, from 1882 to 1899, the country was under the control of a black man, Ulises Heureaux, whose father was Haitian and who was subsequently portrayed as a savage dictator.

Not only, then, was there no consolidation of the political identity of Santo Domingo during the nineteenth century, but also many of the citizens on that side of the island came to see themselves almost as helpless victims who had to be protected by outsiders from their constantly threatening neighbour, one of whose sons had become their president and whose raison d'être in the eighteenth century (i.e. sugar with its black workers) was spreading in their own country at the end of the nineteenth century.

Political instability in Santo Domingo in the nineteenth century was accompanied by a substantial increase in the population. In 1789 Moreau de Saint Méry gave the population as 125,000. At the end of the century it was a little less than 490,000, which was almost a 400 per cent growth. Much of this came from immigration, which took place after the initial emigration, principally of the elite, at the beginning of the century during the Haitian Revolution and during the rule of the whole island by Haiti. However, while the white and mulatto elite were leaving during the first decades, there was immigration of blacks from various places. On the invitation of Boyer, who promised them equality and a better life, free English-speaking blacks came from the United States in 1824 and 1825 to the area of Samaná, where they remained a distinctly different element in the population of Santo Domingo for some time. Other English-speaking blacks also came later in the century to the north of the island, to Puerto Plata. This led Hazard in 1871 to note that 'Here in Puerto Plata there are a large number of negroes from the English islands Nassau, St. Thomas, Jamaica, &c., most of them speaking English quite well; in fact, a large number of the coloured people speak some little of two or three languages' ([1873] 1974: 181). With the growth in sugar cultivation the number of these immigrants increased.

Most significant of the non-Hispanic additions to the population of Santo Domingo, however, were the Haitians who came across to the eastern side for various reasons and were populous in border areas such as Dajabón. The non-Spanish-speaking migrants contributed some of their own linguistic features to their new homeland, but more importantly they constituted a challenge and contrast to the educated Dominican *criollo*, who, in reaction, began to assert his identity more forcefully by highlighting differences of race and identity.

Although its original source and meaning are unclear, it seems as if the term *cocolo* might have developed to label disparagingly the nonwhite, non-Spanish-speaking migrant into the Dominican Republic.

The remark '*Te felicito a ti y a todos los dominicanos por haber sacu-dido el yugo de los mañeses-cocolos ...*' ['I congratulate you and all Dominicans for shaking off the yoke of the "mañeses-cocolos" ...'] is said, by Pedro Mir (1977), to have been made by a Peruvian priest in a letter to a Dominican in 1844 after the Dominican declaration of inde-pendence from Haiti. The tone of the remark makes the negative con-notations of the term very clear and the date and circumstances of it indicate that *cocolo* was being used to refer to Haitians. Even though the term *cocolo* is better known as a designation for the sugar workers from the Lesser Antilles and the Bahamas who came to the Dominican Republic and Puerto Rico at the end of the nineteenth century and after, it seems likely that this latter use was an extension of the earlier one, that is its reference to Haitians. According to Hoetink (1982: 217), the term was also applied to the descendants of the Samaná colonists, who were also non-white and non-Spanish-speaking.

In his attempt to assert his identity more forcefully, the educated Dominican *criollo* was aided in some measure from 1861, when Spain took over control of Santo Domingo, by the advent of *criollos* from Cuba and Puerto Rico. Notwithstanding the abolitionist and nationalist activities of Eugenio María de Hostos, one of those who came, the Cuban and Puerto Rican *criollos*, according to Hazard, constituted a threat to non-white Dominicans: 'hordes of officials came from the two slaveholding islands of Puerto Rico and Cuba, and were placed in authority over the heads of free citizens, many of whom, from their col-our, they professed to look upon as no better than bondsmen' ([1873] 1974: 258).

At the same time *peninsulares* from Spain, principally military peo-ple and administrators, came in thousands to Santo Domingo to re-establish Spanish control of the island. In addition, *isleños*, people from the Canary Islands, came especially to the southern part of Santo Domingo, around the capital. So, it was not only factional fighting and difficulty in communication within Santo Domingo that inhibited the development of cultural uniformity but also the fact that during the course of the nineteenth century, people came from several different countries in numbers large enough to significantly increase and frag-ment the population and identity of Dominicans. The fragmentation was also aided by the fact that the various groups tended to settle in separate communities. Changes in agriculture did nothing to improve the situation and in fact caused a sharp difference in identity to develop between the traditional tobacco- and cattle-rearing *campesinos* on the one hand and, on the other, those newer elements associated with sugar cultivation.

One inspiring factor in the evolution of ethnic identity in Santo Domingo was that its citizens were often reminded that it was Spain's first colony in the New World and that, as a result, it was closer to Spain, the Spanish language and Spanish culture than any other place in the New World. The preservation of such a vision of Santo Domingo seemed to become even more necessary when it was taken over by its black neighbour, Haiti. A difference in race, language and culture had to be promoted for the eastern side of the island to retain a sense of difference from the western side and to prevent itself from being absorbed. However, the highlighting of its early colonial status and history was a backward-looking response, which became increasingly unhelpful to the people of Santo Domingo in the nineteenth century as Spain declined as a major economic and cultural power. Ironically, Cuba, which did not free itself from Spanish domination until the end of the century, had eclipsed Santo Domingo to become the premier Spanish-speaking island and it made Santo Domingo look poor in comparison, according to Hazard ([1873] 1974: 229–30). Thus, Santo Domingo's early history proved to be of little practical value in its turbulent and impoverished state in the nineteenth century.

Ironically, there was no vigorous promotion of the notion of *criollo* in Santo Domingo, especially in the first half of the nineteenth century because *criollo* by definition encompassed all those born in the island and, more specifically, the word *creol* was being used on the western side to refer to the language spoken by the blacks. In such a situation, the name applied to and used freely by Haitians could not emerge as a national term of identity on the eastern side. Furthermore, within Santo Domingo itself there was no general and uniform opposition to non-Creoles: the factional fighting was among Dominicans themselves. For the elite and intellectuals in Santo Domingo in the nineteenth century the most emotive contrast was not generally between *criollo* and European; it was more specific – it was between Spanish-speaking *criollos* and others. Therefore, for them, of the three features of identity – place, race and language – the one that was most critical in the evolution of national identity in Santo Domingo was language.

The topographical differences between the two sides of the island were not enough to inspire illusions of a difference in the emergence of the physical and moral features of the populations on the two sides. However, the division of a single island into two, which was rare in the Caribbean, no doubt created a keen sense of *mi tierra* and *su tierra*, which made unification and division of the island constant thoughts in the minds of those who ruled on the two sides of the island. This was a case where each side had been in charge of the whole island at some time. The island had been used by the Spanish to establish the first colony in the New World, but when, because of the small size of the widely dispersed population, they could not control the whole island, they laid waste the *banda norte*, which led eventually to the establishment of a colony there by the French and French expansion over the western part of the island. The thrust of the Spanish into the continental lands to the west of the islands, the subsequent neglect of the islands by the Spanish and the decline of the Spanish as a colonial power allowed the French, English and Haitians, after the Revolution, to set their eyes on the eastern part of the island and to its eventual take-over by the Haitians. The re-establishment of the division between the eastern part and the western part in the middle of the nineteenth century was probably seen by both sides at the time, and certainly by some of those on the east, as the latest act in a historical saga of dominance.

The racial components of the populations on the two sides were not different even if the ratios of these components were. The circumstances that led to the emergence of the western third of the island as a French colony initially involved a greater complexity of Europeans, but one from which one – the French – emerged dominant. By the eighteeenth century Saint Domingue (the western side of the island) had become the premier sugar-producing, slave colony of the New World. It was the slave colony par excellence, dominated by visions of brutalised and dehumanised black African slaves. It was this vision of the African and the black man that was to persist and to conjure up fear and aversion in the minds of ruling class on the other side of the island, which had not had extensive sugar plantations numerically dominated by black African slaves. It was a vision rendered more horrible by the fact that those blacks had broken free and would work for the white man no more. The identity of those on the eastern side of the island was therefore cultivated and developed as an antithesis of this vision. Even before that, there had been a general belief (see Lemonnier-Delafosse 1846: 198–9) that the rural, mostly cattle-rearing slaves on the Spanish side had life relatively easy and had amicable relations with their masters, with the slave 'lying in peace by the side of his master'. In the contrast with slavery on the French side, Lemonnier-Delafosse, in his account of the first few years of the nineteenth century, presented an idyllic picture of equality and pride on the Spanish side:

> *Blanc, jaune, cuivré ou noir, c'est un Espagnol fier de lui! Malheur á qui lui donne sa dénomination véritable, car alors la colère remplace son flegme, et noir comme l'ébène, il vous répond, en se frappant la poitrine avec un orgueil plus grand encore aux Indes-Occidentales qu'en Europe même:*

*"Yo, yo, soy blanco de la tierra ! ... " parce qu'il sera né créole et non africain.* (1846: 198)

[White, yellow, bronze or black, he is a Spaniard proud of himself! Woe to him who gives him his true designation, for then anger replaces his coolness, and black as ebony, he replies to you, beating his chest with a pride even greater in the West Indies than in Europe itself: 'Me, I am a native white', because he would have been born a Creole and not African.]

This picture suggests that a distinct Spanish creole identity had developed on the eastern side, where a great percentage of the people were native-born, presumably in contrast to the other side where a great percentage of uprooted Africans lived. The astounding claim by the black Dominican to be a white creole shows the extent to which the assimilation process had succeeded. Three years after this claim appeared in Lemonnier-Delafosse's 1846 account, a similar claim appeared in a letter from the US Commissioner in the Dominican Republic to his Secretary of State:

the cruelties of the Haitians toward all who spoke the Spanish language have given such force and universality to the feeling in favor of the whites in the Dominican Republic that it is not uncommon to hear a very black negro, when taunted with his color, reply: 'Soy negro, pero negro blanco'. (Welles [1926] 1966 1: 103–4)

For this kind of claim to make sense, the word *blanco* has to be interpreted not simply as a skin-colour term, but one that signified 'non-slave', 'non-sugar plantation' or *'campesino'*. It is an expansion of the concept of *blanco* that, no doubt, resulted both from antipathy toward Haiti and from the decline of the sugar plantation in previous centuries in Santo Domingo itself. This latter meaning is indirectly implied in the phrase *de la tierra* in the words given by Lemonnier-Delafosse. The use of the phrase *de la tierra* also underlines the integral relationship between *tierra*, *criollo* and ethnic identity in Santo Domingo, as in the other two islands.

The presentation of the contrast between the two sides of the island did not always favour those on the Spanish side, for the latter were not seen as a very hard-working people. The same Lemonnier-Delafosse summed up the difference at the beginning of the century in the following way:

*la parte española sin tener casi haciendas, naturaleza silvestre, ningún cultivo, indolencia del hombre; la parte francesa, bellas y numerosas haciendas; campos bien cultivados, arados, ayudados por el hombre que vive en sociedad y por todos. En la primera, pereza; en la segunda, actividad.* ([1846] 1946: 151)

[the Spanish side with hardly any plantations, natural vegetation, no crops and human indolence; the French side, beautiful and numerous plantations; fields well cultivated, ploughed, aided by people who live socially and for everyone. In the former, laziness; in the latter, activity.]

Of course, Lemonnier-Delafosse's view was coloured not only by the fact that he was French but also by his attitude to what he referred to as *pereza*. For those on the Spanish side of the island it was the very back-breaking, hard work of the sugar plantation on the western side that was degrading. They therefore saw themselves as superior, some slaves included, because they were not involved in it.

Another external view of the identity of Dominicans, which was negative in the extreme and which did no good to their attempts to be annexed to the USA, was one expressed by Representative Wood in a speech to Congress in 1871. It was a racial and cultural categorisation of Dominicans:

> *La población es de un tipo degenerado en grado sumo, estando principalmente compuesta de una raza cuya sangre tiene dos tercios de africano nativo y un tercio de criollo español, a diferencia de cualquier raza de color conocida en ese país o en cualquier parte del mundo. Esta es una mezcla completamente incapaz de asimilar la civilización ...*
>
> *... y así se produjo una mezcla de su sangre* [nativos de Africa] *con la del criollo español, que es todavía más bárbara y salvaje que la del africano.* (Saviñón 1994 1: 426–7)

[The population is of a type that has degenerated to the lowest level, being made up principally of a race whose blood contains two thirds African native and a third Spanish Creole, different from any coloured race known in that country or any part of the world. It is a mixture completely incapable of assimilating civilisation ...

... and so was produced a mixture of their blood with that of the Spanish Creole, which is even more barbarous and savage than that of the African.]

This view not only characterised Dominicans as racially mixed (black + white), but repeated the idea expressed several times before that the result of such a mixture was worse than black. In addition, the apparent insinuation about the degrading effect of Spanish creole blood meant that for Wood (and presumably many like him) the Dominican people as a racial group were beyond redemption. Such a characterisation, which effectively ranked Dominicans lower than Haitians, no doubt infuriated the elite in the Dominican Republic, and the more so since their request for annexation was rejected. It clearly led to more concerted attempts in the literature to promote the vision of a pure Spanish society in the Dominican Republic, and one with little trace of African-ness.

The turbulence and complexity of evolution of identity that took place across the eastern two thirds of the island, especially from the time of the early beginnings of Haitian revolution until a few years before the middle of the nineteenth century when it proclaimed itself the independent Dominican Republic, were reflected in different ways in language and literature. The vicissitudes of Hispaniola and the confusion of languages involved in the events of the first part of this period can be understood from the sentiments expressed in the oft-cited verses credited to Padre Juan Vásquez:[1]

| | |
|---|---|
| *Ayer español nací,* | [Yesterday I was born Spanish, |
| *a la tarde fuí francés,* | By afternoon I was French, |
| *a la noche etiope fuí,* | By night I was Ethiopian, |
| *hoy dicen que soy inglés:* | Today they say I am English; |
| *No sé que será de mi!* | I don't know what will become of me!] |

In addition, the departure from Santo Domingo of many of the white, Spanish-speaking families during this period actually reduced the influence of Spanish colonialism in the emergence of Dominican identity. In fact, the abandonment of the colony by Spain fostered a negative attitude toward Spain and growing feelings of being *dominicano* as opposed to being *español*.

Negroes on the Spanish side of the island benefited from the revolution on the other side and indeed emerged from being an almost invisible force. After the Haitian Revolution, Dominican negroes obviously developed positive sentiments for Haiti and especially its president, Boyer, who gave them freedom (one of his first acts) when he took over the whole island in 1822. At that time, Dominican blacks had their own characteristic linguistic identity, which is seen in the *coplas* [ballads] that they composed (Rodriguez Demorizi [1938] 1979: 52–3) to celebrate their emancipation by Boyer as well as to make fun of their former masters. Caamaño de Fernández (1989: 34) says of the language in these '*coplas ... hacen eco determinados rasgos expresivos del viejo afroespañol, ya en evidente trance de descriollización ...*' ['there is an echo of specific expressive features of the old Afro-Spanish, already clearly in the process of decreolization ...'] What this comment suggests is that during the period of slavery creolised Spanish had developed among the slaves, but that around the middle of the nineteenth century the process of decreolisation was evident. This characterisation of the language of black Dominicans in the middle of the nineteenth century

---

1. Dawn Stinchcomb in *The development of literary blackness in the Dominican Republic* (2004) says that these verses 'were credited to Mónica in the nineteenth century'. Meso Mónica was a freed black man who was a popular but unpublished poet, whom Stinchcomb calls 'the country's first Afro-Dominican poet'.

therefore reinforces the idea that their identity was somewhat different from that of their white masters and as such is evidence of diversity on the Spanish side.

After 22 years of freedom under Haitian rule, the former slaves were fearful of an independent Dominican Republic because, according to Carlos Esteban Deive (1992: 74), they felt that their side of the island would fall into the hands of those who still had strong ties with Spain and that it would revert to being a colony with their subsequent re-enslavement. At that point, liberty was a more powerful force in their minds than any feelings of solidarity with the ruling class on the Spanish side or any concept of being independent Dominicans. In fact, the link with Haiti, which had a majority of people like themselves, and free, may have been a more satisfying reality and a continued prospect of identity, even in spite of a difference in language. However, the diffusing of the rebellious intentions of the ex-slaves by a categorical resolution of freedom meant that in the middle of the century there was a rapprochement between the various classes and races in the newly proclaimed Republic, which in itself pointed towards a uniformity of vision.

Even among the white and mulatto population of the Republic generally, while the Spanish language was the most salient feature of their identity, it could in no sense conform to any notion of purity when used by *criollos*. In fact, it was its Dominican characteristics that gave it its significance. The speech of the traditional *criollo*, who was generally uneducated, was presented in the works of Francisco Javier Angulo Guridi and Juan Antonio Alix. About the latter, Rodriguez Demorizi says: '*Alix es el más criollo de los poetas nacionales, porque es el que define mejor la frontera de lo dominicano y de lo haitiano, de lo propio y de lo exótico*' ['Alix is the most creole of the national poets because it is he who best differentiates what is Dominican from what is Haitian, the native from the foreign'] (1944: 20). In the assertion of this version of Dominican *criollo* identity, the feature that was most valuable to Alix, according to Rodriguez Demorizi (1944: 20), was the '*habla rustica puesta en boga por Guridi*' ['rustic speech made popular by Guridi']. In other words, the characteristic speech of rural, uneducated Dominicans was being featured in the second half of the century as part of the core of Dominican identity.

The *habla rustica* became a central feature in the identity of *dominicanos* not only because it showed the contrast with Haiti and other non-Spanish elements, but also because the feeling of abandonment by Spain led to the highlighting of the dialectal contrast between *criollo* and *peninsular*. This aspect of the emergence of Dominican identity is highlighted by Rodriguez Demorizi in his comments on the work *Cacharros y manigüeros* by Angulo Guridi:

> [*los* manigüeros = *los dominicanos*] *hablan* en criollo, *con sus autóctonos matices:*
>> *Epaña otra ve no güeiva*
>> *a pisai nuetro derecho* ...
>
> *La lengua, pues, siguió siendo distintivo de los dominicanos aun frente a la misma noble madre en cuyo seno la aprendieron nuestros padres.* (1944: 20)

> [[the 'manigüeros' = Dominicans] speak creole, with their own shades of meaning:
>> Would that Spain not again step on our rights ...
>
> The language of Dominicans, then, continued being distinctive vis à vis the very noble mother [the Spanish language] in whose bosom our parents learnt it.]

The *criollo* dialect of the nationalist and revolutionary *manigüero* was seen as the language of the true patriot: it was the dialect that gave Dominicans their unique identity. As is to be expected, it was the variety that was most strenuously avoided and suppressed by the literate elite, especially those who favoured continued political links with Spain or those who believed in a 'pure' cultural tradition in the Republic.

The vernacular of Dominicans also featured earlier in the nineteenth century in early printing and newspapers in Santo Domingo. Rodríguez Demorizi (1975: 28–9) cites a *Dialogo* from an 1821 newspaper as well as a *Diálogo entre dos soldados dominicanos* [conversation between two Dominican soldiers], which appeared in *El Oasis* of 10 February 1855. The latter citation starts in the following way:

| Javier: | *Rumaldo, qué bravo etás!* | [Romualdo, how bold you are! |
|---|---|---|
| Romualdo: | *Pos no he de estailo, Javiei!* | Why shouldn't I be, Javier? |
| Javier: | *Qué te pué acontecei?* | What happened to you? |
| | *A mi no me lo diras?* | Aren't you going to tell me? |
| | *Romualdo ¿Qué ha de sei?* | Romualdo, what can it be? |
| | *Que habiendo epuesto* | after you and I and all |
| | *tú y yo, y toos mis paisanos* | my countrymen risked |
| | *la via contra los haitianos,* | our lives against the Haitians |
| | *que nos peidían ei repeto* | who lost respect for us.] |

(Rodríguez Demorizi [1938] 1979: 76–7)

The soldiers were not identified as rustic and there is no indication that they were meant to be other than typical *dominicanos*. This variety of Dominican Spanish, which was not different from that presented in Guridi and Alix, must be assumed to have been the normal vernacular of Dominicans at the time. This dialect of Spanish, presented by Guridi, by Alix and in the *Diálogos*, was most likely more creolised in areas where sugar was the main crop and blacks constituted a higher percentage of the population. The variety that appeared in the *coplas* of the ex-slaves after they were freed by Boyer might have been more marked, but was not fundamentally different from that of the rest of the native-born population. Overall, then, the use of the vernacular brought out into the open a Dominican whose language had been evolving on its own for over 300 years.

In contrast to this variety, which was popularized in the literature as *habla rustica*, the language of the literate elite in Santo Domingo was cultivated as a standard form of Spanish, especially in extended literary works, and differed little from peninsular Spanish. The most famous example of a Dominican novel with a national theme written in a variety of Spanish that did not really reflect national identity was *Enriquillo*. *Enriquillo* was important for other reasons, however. It resuscitated the myth of the heroic 'Indian' (of Las Casas) and encouraged non-white Dominicans to conceive of themselves as descendants of this 'Indian' hero and 'Indians' generally, because, as Mir points out, '*el indio tiene, aunque no tanto, la piel oscura como el negro y, aunque mucho más, el pelo lacio como el blanco. Ajusta racialmente como símbolo de un pueblo que es, a la vez negro y blanco, como las piezas de dominó*' ['the Indian has, although not very much so, a skin that is dark like the negro's and, although much more so, a skin that is pale like a white person's. It [the Indian] racially fits as a symbol of a people that is at one and the same time black and white, like dominoes'] (1978: 168). This concept of the identity of Dominicans was fostered by the unfortunate (for them) events of the nineteenth century, which caused Dominicans to see themselves as a beleaguered people deprived of their freedom either by their neighbours, by the French or by the Spanish. This was an 'internal' explanation of identity that spiritually linked the sixteenth-century native of Hispaniola to nineteenth-century *criollo* Dominicans by highlighting an indomitable desire to be free or at least not to be enslaved. *Enriquillo*, as a literary character, of course had a long (European) history preceding it – Montaigne's *bon sauvage* in the essay 'Des Cannibales' (circa 1578), Voltaire's *L'Ingénu* of 1767 and J.J. Rousseau's 'noble savage' – which gave it credibility and its romantic appeal. The use of the historical 'Indian' was also a very popular technique in Spanish Caribbean literature in the nineteenth century to get past censorship in order to express contemporary anti-colonial, anti-Spanish and anti-government ideas.

The novel *Enriquillo* itself was in part said to be inspired by its author's direct experience of the abolition of slavery in Puerto Rico in

1873. A bond was thereby established between the victimised but indomitable 'Indian' figure at the beginning of the colony, the black slave at the then current time (in Puerto Rico) and the Dominican at large. Mir, however, is not very kind to Galván, the author of *Enriquillo*, about whom he says: '*Galván era un Enriquillo a la moderna*' (['Galvan was a modern Enriquillo'] (1984: 166), meaning that Galván was a person who betrayed his own people and blacks for money and fame. Yet, no matter how influential Galván was, the fact is that the Dominican people were only too willing to hold on to the Indian as a symbol, as Mir himself admits: '*El pueblo dominicano de todos tiempos acogió la leyenda con singular calor. Las fuentes donde se había nutrido Galván: Oviedo y Las Casas, no eran del acceso popular, de modo que era posible, en ausencia de todo control, encubrir la verdad verdadera y crear, como creó para un futuro infinito, la verdad convencional*' ['Dominican people of all times warmly welcomed the legend. The sources from which the material came – Oviedo and Las Casas – were not generally accessible, as a result of which it was possible, in absence of all control, to conceal the real truth and to create, as it did for an infinite future, a conventional truth'] (1984: 167).

In nineteenth-century Dominican literature the 'Indian' was not an undistinguished ethnic stereotype; there were several historical 'Indian' characters who were revived, which meant that the use of the 'Indian' became more than an artistic technique – Enriquillo, Iguaniona and Anacaona were re-born and refashioned in the literature and presented as Dominican heroes. It is in this context that Santo Domingo was seen as the scene of the first clashes between native and foreigner as well as the scene of the initial and continued victimization of the 'Indian'. In the novel, no peculiar variety of Spanish was used to represent the speech of the 'Indian', Enriquillo. This is because such would have been inconsistent with the vision of a hero, for Enriquillo was a legendary figure of the sixteenth century. In fact, his very name with its pleasant diminutive (-illo) converted him into a sympathetic Antillean Spanish/ Dominican figure.

In addition to named 'Indian' heroes and heroines in the fictional literature, an ethnic indigenous name also had some popular currency in Santo Domingo. According to Hazard ([1873] 1974: 341), *guajiro* was the common local (Spanish) name applied to the cattle rancher and tobacco grower. The *guajiro* was perceived by Hazard to be a native of the place and to be a mulatto; he gave no indication of anything indigenous in the *guajiro*. Likewise, no indigenous features were highlighted by Alix at the end of the century when he contrasted the *guajiro dominicano* with the Haitian *papá bocó*. However, as a generally used, stereotypical name for the *campesino* in the Dominican Republic, *guajiro* did not emerge as prominently as it did in Cuba and as *jíbaro* did in Puerto Rico. The 'Indian' was perpetuated in Santo Domingo more in relation to political struggle and racial type than as a socio-economic and geographical type. No name of a specific ethnic group was converted into the national identity, but the term *indio*, toward the end of the nineteenth century, was used as a general designation for the mulatto and black in Santo Domingo to perpetuate the notion of victim as well as to contrast with the perceived (African) enemy on the other side of the island. *Indios* were therefore perpetuated as a group within the broader Dominican identity, thereby complementing those who saw themselves as pure Spanish.

According to Rodriguez Demorizi ([1938] 1979: 110), Luis José Peguero was the first (literary) person to use the term *dominicano* and this was in 1762/3. Hazard, in contrast (in his comment 'Dominicans, as they now called themselves' [1873] 1974: 248), associated the name directly with the middle of the nineteenth century when independence was proclaimed, when the country was officially identified as *La república dominicana*. Hazard, who was one of the American commissioners who went to Santo Domingo in 1871 to assess its suitability and willingness for annexation, recorded in his account that the people of Cibao (a mainly white area at the time) were even told that they could maintain their national identity, which they were proud of, after annexation:

> if they earnestly and honestly wished to enter the American Union, and were accepted, they would have the satisfaction of knowing that, while they thus became Americans, they would none the less remain Dominicans, a name of which they seemed so proud. (p. 334)

This not only indicated the contradictions in ethnic and national/political identities but also presaged the Puerto Rican situation of the twentieth century.

# Puerto Rican *jíbaros, criollos* and *bozales*

The Haitian Revolution changed Puerto Rico economically and socially, as it did the other Spanish islands, in that from 1815 to 1850 sugar and slaves became increasingly prominent in the image of Puerto Rico – Puerto Rico became one of the major producers of sugar in the Caribbean. An important consequence of the change to sugar was that in itself it involved a dramatic increase in the slave population, it fostered attitudes and relationships that were dominated by the harshness of the plantation system and it established an increasingly strong economic bond between Puerto Rico and the United States, its major export market. Puerto Rico was therefore fundamentally affected by the slave-holding mentality of both its colonial master, Spain, and its economic master, the United States. In addition, in the years immediately after the Cedula of 1815, which encouraged migration to Puerto Rico, thousands of mostly coloured immigrants – in fact, the majority of immigrants – came from the smaller islands of the Caribbean, English, French and Dutch. (As was the case in Santo Domingo, migrants from the smaller islands to Puerto Rico came to be called *cocolos*.) Then, between 1825 and 1835 thousands of slaves were brought into Puerto Rico, many of them by way of the French islands, in spite of British attempts to cut off supplies by termination of the Slave Trade in 1807. The population of Puerto Rico thus increased from 183,000 in 1812 to 443,000 in 1846 and it kept increasing at the same rate during the rest of the century.

In the face of massive immigration and change of economy and life-style, native-born Puerto Ricans, who were pushed off the more productive lands into the centre of the island, became more nationalistic as they tried to preserve their traditional way of life. The name which in the nineteenth century came to be associated with this nationalistic Puerto Rican identity was *jíbaro*, a name that Murillo Velarde in 1752 said *Criollos* in Hispaniola, Puerto Rico and other islands were generally called. It was one of the few names in the hierarchical nomenclature for racial mixture used in the islands that had an 'Indian' component. While Long in 1774 (2: 260–1) said that *givero* was the product of sambo de Indian + sambo de mulatto, in Mexico it was given as a much more complex development:

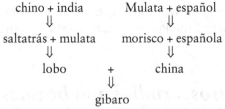

(From Aguirre Beltran [1946] 1989: 176–7 and Léon 1924)

It is not surprising that as a contrast to the thousands of black, non-Spanish-speaking *cocolos* from the smaller islands and elsewhere, the *jíbaro* had already begun by the 1840s to be converted into an 'Indian' figure. The conviction that *jíbaros* in Puerto Rico were part 'Indian' had grown so strong by the middle of the nineteenth century that J.T. O'Neil, a Puerto Rican living in New York, seemed to have conceded it when he said: 'The **Jíbaro**, or native of the rural districts, is supposed to be, generally speaking, of Indian descent; but of the pure aboriginal breed, said to have numbered 600,000 at discovery, we do not believe that one remains' (1855: 148). At the same time, doubt about the truth of this was evident in O'Neil's statement.

It was the name *jíbaro* itself that was largely responsible for this conviction, especially in the minds of those who knew little about Puerto Rico. For instance, before he visited the island, Granier de Cassagnac believed the *jíbaros* to be the only remaining natives of the region:

> *La seule île de l'archipel des Antilles que les anciens habitants n'aient pas abandonnée, c'est Porto-Rico. Il y reste encore environ vingt-deux mille individus nommés Ibaros, et qui descendent de la nation sur laquelle les Espagnols prirent l'île.* (1842: 100)

> [The only island of the Antillean archpelago that the former inhabitants did not abandon is Puerto Rico. There still are around twenty two thousand individuals, named Ibaros, who are the descendants of the nation from whom the Spaniards took the island.]

Because he saw them, during that time of emancipation and imminent emancipation of the slaves, as a foil against the slaves and the emancipated slaves, these 'red skins', who had been universally characterised by European writers as indolent, became in his eyes a wonderful, hard-working people. Then, after he visited the island, the *Ivaros* changed to become a *'population mixte, formée d'indigènes, de Canariens et d'Espagnols'* ['a mixed population made up of indigenous inhabitants, Canarians and Spaniards'] (1844: 190). His view of the Puerto Rican population was that it was made up of *'soixante-mille esclaves, de vingt-deux mille Ivaros, de vingt mille Européens ou Canariens, et de trente mille nègres libres'* ['sixty thousand slaves, twenty two thousand Ivaros, twenty thousand Europeans or Canarians and thirty thousand free negroes'] ' (Granier de Cassagnac 1844: 191). Apparently, then, in spite of the fact that there were so many black people, enslaved and free, and so many *Européens ou Canariens*, there were no longer any of the mulattos that Abbad y Lasierra in 1788 said made up the majority of the population, only *Ivaros*.

Granier de Cassagnac's picture of *jíbaros* contrasted with a more general one given in 1855 by O'Neil, whose characterisation maintained the traditional formative theories. For O'Neil, there was some slight difference between the rural Puerto Rican (the *jíbaro*) and the urban Puerto Rican, a difference that he attributed to degree of *civilisation*. About Puerto Ricans generally he said: 'The Puerto Riqueneans, like the inhabitants of all warm countries, are generally indolent, owing both to the enervating character of the climate, and the bountiful provisions of nature against their wants ...' (p. 152). His fully articulated, stereotypical picture of the rural Puerto Rican or *jíbaro* was as follows:

> they are temperate, honest at heart though thriftless, courteous, hospitable, and devout, especially in the rural districts. The poor (?) tenant of a hut built entirely of the palm tree, and bound together with the strong and pliable **bejuco**, whose only habiliments are a check shirt, osnaburgs pantaloons, straw hat, and an innocent **machete** strapped to his waist; who spends most of the time in his hammock smoking and playing on the **tiple** (a small guitar), doing nothing, or sleeping; divested of care for the future by the possession of a few coffee and plantain trees, a cow, and the indispensable horse, and anticipating the pleasures of the next holiday's cock-fight or dance, will extend the most cordial and polite welcome to the benighted traveler, set before him the best of his plantains, milk and cheese; relinquish to him his rustic bed; unsaddle and feed his horse, which at break of day he will have in readiness, and dismiss his guest with a 'vaya usted con Dios,' refusing with a gesture of pride or offended delicacy all proffer of payment. Such is the **Jíbaro** of the interior. (1855: 152)

O'Neil further went on to say about the *jíbaro* 'but being indolent, impatient of control, and having few wants to supply, they will seldom be induced to work more than three or four days out of the week, and not very hard then' (p. 156). In contrast, the coastal Puerto Rican was tempered by 'civilization: Proximity to the coast, and therefore to **civilization**, renders him interested, salutationless because not answered on first occasions, and deceitful in ratio of the increase of his wants' (p. 152). This picture of Puerto Ricans generally and *jíbaros* specifically perpetuated the same views that were given by European writers a hundred years before. It signalled a discomfort that Puerto Ricans outside of Puerto Rico had with their countrymen at home; it also exemplified the desire, on the part of some colonial writers, to conform to metropolitan views.

The disparity between the two views (Granier de Cassagnac's and O'Neil's) was attributable to difference in viewpoint – the one was tied to factors internal to Puerto Rico and the other was the result of comparisons made between Puerto Rico and metropolitan areas. It is not surprising that it was only sophisticated Puerto Ricans who were taken up with the latter viewpoint, whereas Puerto Ricans generally were more conscious of the former. Within Puerto Rico, therefore, the native-'Indian' view began to assume national significance, and by the end of the nineteenth century the *jíbaro* had acquired a romantic literary image far removed from the reality of everyday life. Having become a symbol of Puerto Rican identity, he was integrally being associated with *la tierra*, as is seen at the start of José de J. Domínguez's explanation of the word: '*El nombre de Gibaro corresponde al indígena de Puerto Rico. Debe de escribirse Hibaro, aspirando la H, y vale tanto como Hijo de la Tierra*' ['The name *gibaro* corresponds to native of Puerto Rico. It should be written *Hibaro*, aspirating the "h", and it is the same as "son of the soil"'] (Font 1903: 138). Here *tierra* had a double significance – native and agricultural. So, the *jíbaro* was essentially a rural peasant, a product of the very land that he was farming, an independent and self-reliant individual.

However, since the prevailing philosophy in the Spanish Antilles among the ruling class was *lineal ascent* to whiteness, sources for the characteristics of the *jíbaro* were sought and found in Europe. This was explicitly done by Domínguez under the guise of a philological explanation in the following:

> *Pero Gíbaros hubo en Europa mucho antes de haberlos en Puerto-Rico. Para hacer esta afirmación, tenemos pruebas filologicas decisivas. Dice el Padre Iñigo - nuestro único historiador - que los indigenas de Puerto-Rico se deformaban la cabeza, dándole la forma de una cuña: no se sabe dónde obtuvo el ilustre benedictino eso dato; pero es lo cierto que la filología lo confirma. En efecto la expresión latina Gibber, equivale á la castellana corcova. Ahora bien: Gibber es apocope de Gibber-o, y Gibbero es flexión de Gibbaro: la duplicacion de la b, sólo tiene por objeto robustecer la pronunciación de la simple que en su origen tenía. Nuestro voz castellana Giba, apócope de Gibara, sirve para llamar á la corcova con la Tierra primordial.* (Font 1903: 139)

[But there were Gibaros in Europe long before those in Puerto Rico. To substantiate this claim, we have decisive philological proof. Father Iñigo, our only historian, says that the indigenous inhabitants of Puerto Rico deformed their heads, giving them the shape of a wedge; it is not known where the famous benedictine got this information, but philology certainly confirms it. In fact, the Latin word *gibber* is equivalent to Castilian *hump* [curvature]. So now, *gibber* is a shortened form of *gibber-o* and *gibbero* is a variant of *gibbaro* – the only reason for the doubling of the 'b' is to strengthen the pronunciation of the single 'b' which it originally had. Our Castilian word *giba*, a short form of *gibara*, serves to name the *hump* [curvature] with primordial Earth.]

Castilian Spanish, *la Tierra primordial* and the *jíbaro* were presented in an association and vision that far transcended the notion of a simple *campesino* who had come from lowly roots and had habits that were generally denigrated.

Ironically, in his determination to establish a link between the *jíbaro* and the white European, Domínguez's explanation drew attention to a link much closer at hand, that between the *jíbaro* and the Caribs (Red and Black) of St Vincent. Domínguez noted that Iñigo Abbad y Lasierra had claimed that the original inhabitants of Puerto Rico deformed their heads: '*la cabeza aplanada por delante y por detrás, porque al nacer se las formaban apretándoselas por el cogote y por la frente, dejándoselas de figura cónica, harto desairada y fea para los ojos que no fuesen de indios ...*' ['the head flattened in front and behind, because at birth it was given this shape by pressing on the nape and the forehead making it into the shape of a cone, very unattractive and ugly to non-Indian eyes ...'] (Abbad y Lasierra 1971: 24). These words were written at a time when the Black Caribs in St Vincent, who were noted for the same deformation of their heads, a practice they had adopted from their original hosts (i.e. the 'Yellow' Caribs), were in conflict with the British. This supposed feature of identity of the natives of Puerto Rico points to the recurring image of the Carib in the development of the nineteenth-century Caribbean island identity as well as to a continuous demographic and historical link between the smaller islands and the larger ones.

Contrary to what Granier de Cassagnac and others believed, however, *jíbaro* was not the name of an indigenous Puerto Rican whose traditions had been preserved. The only indigenous inhabitants who were so called were South Americans of Ecuador. What seems more than coincidental is that Domínguez pointed to Abbad y Lasierra's claim of deformation of the head as a defining feature of the indigenous inhabitants of Puerto Rico and what was true about the Ecuadorean *jíbaros* and what they were infamous for was that they decapitated their victims, removed their skulls and shrank their heads. Yet, it seems unlikely that, when the *jíbaro* became the image of the Puerto Rican native at the beginning of the nineteenth century, head shrinking was part of the consciousness or image of Puerto Ricans.

The real root of the *jíbaro* image was that it was a term of racial classification identifying a Black + 'Indian' + White mixture that seemed to typify the most common phenotype in the Spanish islands. Alvarez Nazario (1977: 68, note 52) believes it to be Brazilian Portuguese in origin. The central factor in his explanation of the spread of the word was the slave trade and thus black slaves, but ironically the central factor in its later glorification by nineteenth-century writers at a time when the number of black slaves was assuming spectral proportions was the notion of native of the Americas and consequently the idea that the people it described were (in part) 'Indian'. Thus the word, as a national image of the Puerto Rican, provided a link with the other Spanish islands, Mexico and South America and at the same time a contrast with the Black Republic of Haiti as well as with the great number of *bozales* who were introduced into Puerto Rico in the first half of the nineteenth century.

The popularisation of the *jíbaro* as a national figure in Puerto Rico was altogether a matter of the romanticisation of the *campesino* and *la tierra*, the apparent appropriation of a cultural and historical figure of Spanish America, the adoption of a name that symbolised admirable characteristics and the assertion of a fundamental claim to the island by being a native of it in the face of the many immigrants who had come into it. The mulatto/*mestizo* of Puerto Rico, who in the European (primarily English and French) scale of racial classification, hierarchy and values was without historicity, that is without a geographical origin as a race, was therefore, through a name, converted into a fiercely independent *hijo de la tierra*, thereby identifying America as the place of origin and presenting a challenge to sharp and old concepts of race. In essence, the challenge involved a shift from apparently inherent and immutable characteristics of race to formative processes, especially those occasioned by Nature.

In Puerto Rico there was another major link generated to establish a continuity with the original natives of the region and more specifically of the island, the original name of which was *Borinquen*. *Jíbaro*, like *criollo*, was a generic term rather than a national one – it was not restricted to Puerto Rico and it did not refer to Puerto Ricans exclusively. Moreover, it was not a comprehensive term in that it suggested a rural essence, one that contrasted with urban sophistication and worldliness. *Puerto Rico* and *puertorriqueños* were the names for the country and the people respectively, but they did not seem to fully capture the romantic imagination. As a result, writers began to resuscitate the old indigenous name for the island (*Borinquen*) to get over the deep attachment they felt to their island and homeland. Among a group of young Puerto Rican-born writers based in Spain one produced a piece in 1846 called *Cancionero de Borinquen*. The element of romanticism and nostalgia in those writing in Spain is seen also in Gautier Benítez's lines:

| | |
|---|---|
| *por un puñado de tierra* | [For a handful of soil |
| *de mi tierra borincana* | of my boricuan [Puerto Rican] land] |
| (From the poem *A Puerto Rico*, cited in Shimose 1995: 90) | |

Ironically, one of the early poems highlighting the designation *borincano* – by Alejandro Tapia y Rivera titled *El último borincano* (reproduced in Rivera de Álvarez and Marcel Álvarez Nazario 1982: 104) – in its title and content was not primarily nationalistic in intent, but presented *borincanos* as the original indigenous inhabitants of the island. In spite of this, the word or a variant of it (*borinqueño*) continued to be used as a name for contemporary Puerto Ricans, thus preserving a 'native' link between the original 'Indians' and the predominantly mulatto popula-

Figure 5.1    A Spanish Planter of Porto Rico, luxuriating
in his hammock: *Dolce far niente.* From Waller (1820).
British Library.

tion of Puerto Rico of the nineteenth century. Even so, *jíbaro, puertorriqueño* and *borincano* seemed sometimes to be interchangeable, and did
not always present different aspects of the island and its people.

From fairly early in the nineteenth century, sources within Puerto
Rico revealed that the absence of comment on language in the accounts
of Abbad y Lasierra and Ledru did not mean absence of a Puerto Rican

variety of Spanish or aspirations towards a specifically Puerto Rican identity. The revelation of linguistic identity was helped forward, as was the case in most places in the Caribbean, with the establishment of a local press and the publication of newspapers. In Puerto Rico this happened in the very first decade of the nineteenth century. Soon after, in 1820, *Coplas del jíbaro* [Jíbaro 'ballads'] appeared in the newspaper, *El Investigador*. The tone of these *Coplas* was the satire and ridicule that had pervaded the literature in the previous century in Europe, but, more importantly, they were written in a dialect that imitated the speech of the *jíbaro*. The presentation of the features of speech of the rural and uneducated *jíbaro* was meant to have a comic effect and to belittle the common man, including the Negro. By maintaining the fundamental concept of *langage corrompu*, one which pervaded the region, this language at the time struck hard at the notion of such speakers being the power within a country.

More importantly, the effect of events in Haiti and notions of revolution and independence were clearly evident in the *Coplas*, which appeared less than 20 years after the declaration of independence in the former St Domingue. The restoration of the name *Haiti* was an act of rejection of the European name, but at the same time it gave the country and the people an indigenous name. The theme in the *Coplas* was the restoration of the Constitution, that is, rejection of absolutist rule by the king and respect for the rule by the people. At the time in Puerto Rico and in the context of surrounding circumstances, this was a topical and politically divisive subject. When the author of the *Coplas* gave the common Puerto Rican the indigenous name *jíbaro*, he was not only mocking the speech of the common man and the idea of such a person having great freedom and power, but was also indirectly mocking the Haitians, who had rejected their European name and given themselves an indigenous name. Very significantly also, the use of the word *suidadanos* corresponded directly to the revolutionary *citoyen* [citizen] of France and Haiti as well as to *ciudadanos* used in the 1821 independence constitution of the Spanish side of Haiti. Furthermore, the *suidadanos* in the *Coplas* were going to listen to Juan Congo beating the drum, a clear symbol of solidarity with the *negro bozal*, blackness and the spirit of revolution. The *Coplas del jíbaro* can be interpreted as a direct response to the Cuban José María Heredia's (1820) *Himno patriotico – Al establecimiento de la constitución* [patriotic hymn/national anthem – on the occasion of the establishment of the constitution], which highlighted in its chorus the words *Ciudadanos* and *Libertad*, as well as *El Dos de Mayo* [the second of May – 1821], which also highlighted the word *ciudadanos* and in its theme recommended fighting for freedom:

> *'Antes que esclavitud, muerte suframos!',*    ['Death before slavery!'
> *clamara sin temor, y del tirano*    he will cry out without fear,
> *hundió en el polvo la soberbia fiera.*    and the tyrant's fierce
>                          arrogance bit the dust.]
> (Heredia 1970)

These sentiments were clearly too dangerous for many in Cuba and Puerto Rico at the time and conjured up pictures of black and coloured hordes overrunning white achievement and supremacy.

The *Coplas* were a significant landmark in the portrayal of Puerto Rican identity because they exposed for the first time the language of 'real' Puerto Ricans. The hostile reaction that the *Coplas* provoked was stated as purely political in nature, but the language in which they were written no doubt made them more painful for those who would not have wanted their identity to be associated with such speakers. More generally, what was starting to emerge was that social criticism was being allied with the vernacular and this meant that language was beginning to become and be used as a serious feature of identity. In fact, the combination of local vernacular and social criticism in the *Coplas* was to become typical of the genre of newspaper dialect that came into widespread use across the many islands of the Caribbean in the late nineteenth and early twentieth centuries.

In spite of the non-standard identity exposed by the *Coplas* and in spite of the striking appearance of a racially mixed people, the apparent absence of a distinctly creole language was at the root of contradictions of national identity in Puerto Rico. For while the peculiarities of character and customs of the people of Puerto Rico were apparent by the end of the eighteenth century and fostered the notional identity of *criollo* as well as *jíbaro*, general facility in the Spanish language allowed for the easier transmission, filtering down and maintenance of European ideas and value systems among most of the population. It also allowed the literate and conservative to foster an identity with Spain. Moreover, there was no obvious general set of linguistic features to tie the people to and constantly remind them of African and other non-European parts of their heritage. It was not therefore difficult for the concept of *Puertorriqueño* to emerge in nineteenth-century Puerto Rican literature as a climatically modified European and for that of *jíbaro* as a rustic white speaking an old dialect of Spanish.

The language of the *jíbaro* was resurrected in the 1840s by Manuel Alonso, one of the earliest writers associated with the delineation of Puerto Rican identity. Alonso, writing from Spain, like O'Neil writing from New York, painted a picture of the *jíbaro* as a rustic, ridiculous figure in obvious contrast with sophisticated Puerto Ricans like himself.

It was toward the middle of the nineteenth century that the model identity of a Puerto Rican was put forward in verse for the first time by Alonso in the sonnet *el puertorriqueño*. As José Luis González argues, Alonso's vision of a Puerto Rican and a Criollo was essentially white:

> The reference at the beginning of the sonnet to 'dark in color' shouldn't be misinterpreted, since what it refers to is a **white** Puerto Rican whose skin has been tanned by the tropical sun (as the 'pale' complexion and a 'well-proportioned nose' confirm). The 'country,' subject of an 'unsurpassable love,' is obviously no longer Spain, but Puerto Rico. The 'soul tremulous with dreams,' the wit that is 'free and proud,' the 'restless thought' and the 'ardent mind,' speak of an inquisitive and challenging attitude toward reality. In fact, the portrait is of a social class as a whole: the new creole bourgeoisie already conscious of its historical destiny. ([1980] 1993: 41)

It is quite interesting and revealing also that in the sonnet Alonso did not identify a special variety of language as a characteristic of the Puerto Rican. Obviously, the characteristics of the *jíbaro* were not consistent with the sophisticated image that Alonso was trying to put forward. According to Modesto Rivera Rivera the sonnet *el puertorriqueño* was basically a self-portrait: '*Podemos considerar como sintesis de esta personalidad un soneto:* **El Puertorriqueño**, *autorretrato fisico y espiritual del autor y definición del criollo ...*' ['We can regard the sonnet 'El Puertorriqueño' as a synthesis of this personality, a physical and spiritual self-portrait of the author and a definition of the Creole ...'] ([1952] 1980: 40). Alonso, often regarded historically as the first figure in Puerto Rican literature, was a first generation *Criollo*, born in 1822 of white Spanish parents – his father from Galicia and his mother born in North Africa. Alonso spent a significant proportion of his adult life in Spain and wrote his most famous pieces – *el puertorriqueño* (1844) and *El gíbaro* (1849) – in Spain as a young man. The Puerto Rican being projected then was an educated, high-society person rather than a middle-class or lower working-class one; it was the image of a transnational Puerto Rican rather than a native Puerto Rican.

It is in his *costumbrista* work (pieces collected and republished in 1849 as *El Gíbaro*) that Alonso presented the characteristic speech of the native *jíbaro*, a character who was clearly not synonymous with *criollo*. The spectators encouraging the cocks in the cockfight with the words:

| | |
|---|---|
| *Pica gayo, engriya jiro,* | [Pick him, cock; get ready to strike, giro (a kind of cock) |
| *mueide al ala renegao,* | Bite the damned wing |
| *juy que puñalon de baca* | Wooie! What a brute of a blow] |
| (Alonso [1849] 1992:92) | |

were obviously Puerto Rican, but of a lower class, whose language could be presented as 'original' and amusing. Even so, the *jíbaro*'s speech was portrayed only as a variant of Spanish and not as a creole with African or other characteristics. In fact, the variation was predominantly phonological. This kind of portrayal of the language of the uneducated by the educated in the same society (i.e. dropping of [s] at the end of words, non-articulation of consonants coming between vowels, change of [r] to [l], change of [l] and [r] to vocalic [y]) intended to signal class distinction above all else.

Considering the geographical context (i.e. the nearness of Haiti), it was not strange that natives of Puerto Rico in the nineteenth century were being presented in the written literature as descendants of the original indigenous inhabitants and as climatically modified white persons of different classes speaking different varieties of Spanish. In Haiti, the mulattos, who had had a major and prominent role in the revolution at the start of the nineteenth century, were in no way presented as representatives of Haitian identity. So, the projection of a Puerto Rican as only a white *Criollo* and a white or indigenous *jíbaro* in the nineteenth century, especially in the first half of the century, was not an innocent reflection of a demographic reality. As José Luis González explains:

> when Alonso wrote his sonnet, one of every two inhabitants of Puerto Rico wasn't white and approximately one out of every ten was a slave. And matters hadn't changed much five years later when Alonso published *El gíbaro*, a book that in no way reflects the true racial composition of Puerto Rican society. (1993: 41)

Alonso's portrayal of a Puerto Rican as white was quite clearly an assertion of whiteness in a geographical and recent historical context of the contrary. It was in clear contrast to the assertion of the black national identity of Haitians.

This portrayal was also a consequence of Spanish colonialism in the islands. While Spanish literature did not contain the level and extent of vitriolic attacks on non-white peoples which that of the French and English did and, while Spain had had a long and involved history with non-white peoples in and out of Spain, the fact is that African slavery began first and was abolished last in the Spanish islands – ending in Puerto Rico specifically in 1873. In other words, African slavery, and consequently Africans, were a part of the reality of Puerto Rico from the first quarter of the sixteenth century till almost the last quarter of the nineteenth, that is over a hundred years more than in the French and English colonies, and most of these years were before the nineteenth century. The psychological consequences of enslavement of another people were therefore more deeply ingrained in the Spanish psyche than in that of any other Europeans. In fact, the absence of vitriolic emancipation arguments in Spanish literature may have reflected a lack of discomfort with the actual practice of slavery. So, the invisibility of the black and mulatto that characterised Alonso's portrayal of a Puerto Rican was a consequence of this deeply entrenched perception of the unimportance of the enslaved.

Africans and their descendants in Puerto Rico were not absent from the literature, but the distinctions in speech given created the impression that those who spoke a deviant form of Spanish were not fully Puerto Rican, that they were *bozales*. While the speech of the white *Criollo* in Puerto Rico was given no special distinguishing characteristics and that of the *jíbaro* was presented as phonologically deviant, that of the *negro bozal* appeared to be significantly different and somewhat unstable. Substantial instances of speech, which obviously was regarded as typical of certain negroes, occurred in the 1852 play by Ramon Caballero *La juega de gallos o El negro bozal* in which José (*negro bozal*) was one of the main characters. One point that is evident throughout the play, however, is that the speech of all blacks was not the same, for the other slave in the play, *Nazaria* (*negrita*), was not made to speak in any way different from the whites in the play. The *habla bozal* in the play had several features that are similar to those identified as features of creole languages elsewhere:

(a) Use of a word (e.g. *ta*) to indicate tense or aspect with an uninflected verb following

| | |
|---|---|
| Yo *ta queré* mucho á ti | [I am loving you very much] |
| siempre *ta regalá* dinero á mi | [always (s)he is giving me money] |
| mi corasó *ta sufrii* mucho | [my heart is suffering much] |
| Toro dia, miamo, yo *ta disé*: Nasaria: po qué tu no *ta queré* á mi? | [Every day, master, I am saying: Nasaria, why you are not loving me?] |
| Ella *ta jasé* bula, miamo | [she is making joke, master] |

(b) One verb form (e.g. *son*) used for all persons (first, second, third) whether singular or plural

| | |
|---|---|
| ese no *son* cagüeteria | [that is not pimping] |
| niña Fererica *son* bueno amo! | [little Federica is a good mistress] |
| y to *son* pa ti Nazaria | [and everything is for you Nazaria] |
| ese Nasaria *son* mugé malo | [that Nasaria is a bad woman] |
| Yo no *son* negro nan casa | [I am not a negro in the house] |

(c) a form of 'go' + verb to indicate future tense (*ba* < Spanish *va* [go])

| | |
|---|---|
| Yo *ba* libetá á tí | [I go free you] |
| Yo te *ba* jasé uno baile | [I was go make a dance] |
| ¿Como *ba* queré señorita que son tan bonita ... | |
| | [How I go love a young lady who is so beautiful ...?] |

(d) a preposition *nan* [in]

| | |
|---|---|
| Lamo ta *nan* gallera | [The master is in the cockpit] |
| Yo no son negro *nan* casa | [I am not a negro in the house] |
| Ahi ta *nan* galería, *nan* covesació con uno músico | |
| | [He is in the gallery in conversation with a musician] |

(e) A tendency to retain subject–verb–object word order when pronouns are the objects, in contrast to Spanish, which puts an object pronoun before the verb (even if it is repeated after the verb)

Yo ta queré mucho _á tí_
Yo ba libetá _á tí_
po qué tu no ta queré _á mi_?

In addition to these features there were many in the pronunciation, e.g. changes in consonant sounds, loss of consonants occurring between vowels, reduction of consonant clusters, loss of unstressed sounds occurring at the beginning and end of words. These were much more concentrated and frequent than the ones given in *jíbaro* speech, and resulted in a variety of speech that seemed significantly different from normal Spanish. The term *bozal*, which was applied to this variety of speech in Puerto Rico, is misleading and cannot be taken to mean the speech of recently arrived Africans or even Africans specifically. The similarities between this variety and others spoken by slaves elsewhere in the Caribbean islands during the nineteenth century suggest that it was the characteristic speech of a social class and that it had taken some time to develop. It would be more appropriate to identify it as the speech of the most African part of the population in Puerto Rico at the time.

As far as real additions to Puerto Rican speech are concerned, there were some from those migrants from the smaller French islands who were permitted to come to Puerto Rico by an 1815 Cedula. A few words, for example those having to do with the *bomba* dance, attest to this influence. Of greater significance was the influence from Curacao. According to Alvarez Nazario 1974: 69–70, there was documented migration from Curacao to Puerto Rico from 1766 up to the nineteenth century, resulting in several concentrations of '*criollos de Curazao*' o 'mulatos holandeses' in San Juan and on the south coast of the island, with smaller groups in Mayagüez and Arecibo and lesser numbers spread out all across the island (Alvarez Nazario 1970: 2). Furthermore, Alvarez Nazario (1970) identifies and comments on one of the earliest texts in Papiamentu, which was produced in Puerto Rico in 1830. Probably the most significant fact about *José*'s variety of speech in *La juega de gallos* is that it is presented in a play, presumably to a white audience, without any comment on or explanation of the speech of the *negro bozal*, which clearly meant that such speech was general and normal enough to be understood by Puerto Rican audiences of all classes. Without even speculating about the influence of Papiamentu speakers from Curaçao in Puerto Rico or in Arecibo specifically, the setting of Caballero's play, it seems reasonable to conclude that this variety of speech was normal in Puerto Rico and, as such, was an integral part of Puerto Rican identity.

In Puerto Rico in the first half of the nineteenth century, although race and language (*jíbaro* and *bozal* speech) were salient characteristics, they were not the kinds of characteristics to be displayed proudly. Puerto Ricans, especially sophisticated ones, were not ready to come to terms with their own peculiar identity; they were at a stage where they were embarrassed by, and making a mockery of, their own characteristics, which were not typical of the elite. At the same time, they were presenting an image of the elite Puerto Rican as a natural modification of a Spaniard. There was, however, in the *Coplas del jíbaro* a glimmer of recognition that the rural peasant had some measure of influence and would emerge as a voice to be heard. As far as the linguistic identity of Puerto Rican types was concerned, the *jíbaro* and the *bozal* could be seen not to be separate and distinct, in spite of efforts to make them seem so, because the characteristic features of pronunciation highlighted in the *Coplas del jíbaro* (1822) also occurred in Alonso's rural characters and were the same as those identified in the speech of *bozal* Africans in Puerto Rico, Cuba and Santo Domingo. Indeed, in the *Coplas del jíbaro*, the fact that the author started out by identifying a person who clearly was African – Juan Congo playing his drum – suggests that the variety of speech being portrayed and the African Juan Congo, as well as the Puerto Rican type, the *jíbaro*, were known to be related.

By the end of the century, the *jíbaro* and the identity of Puerto Ricans had been transformed – the *jíbaro* first becoming 'Indian' before being changed into a climatically altered white person with an indigenous name. There was little trace of the mulatto as a promoted racial type. The spectre of blackness occasioned by the negative vision of

Haiti as well as by the re-kindling of the hierarchical value system of the sugar plantation, under influence from the US South, led to a conscious masking of racial characteristics associated with blackness. This masking was assisted by the addition of Europeans from various countries to the labouring classes in Puerto Rico.

# From Criollos to Cubanos

In Cuba arguments for and against the benefits of the Constitution raged in the newspapers from 1820 when the Constitution was re-stored, and it was these arguments that led to the kind of debate that took place in Puerto Rico and the appearance of the *Coplas del jíbaro* there. However, in Cuba there was no parallel to Puerto Rico's *Coplas del jíbaro* in that no symbolic figure was either used positively or mockingly rejected as a representative of Cubans and the arguments on either side were not presented in language that was understood to be characteristic of a *campesino* or any other quintessential Cuban. The Cuban elite and intellectuals, who were the ones arguing among them-selves, had reason to be much more conscious of and wary about liber-al ideas because they had seen what had happened in Saint Domingue. With a bigger slave population than Puerto Rico and Santo Domingo and with plantation slavery as a powerful determinant in their con-sciousness and intellectual positions, the white elite and intellectuals did not associate independence or any form of their political future with lower social classes. On the contrary, what emerged in nineteenth-century Cuba as a dominant philosophy of white *criollos* was that in order for Cuba to be saved the population had to be whitened, a phi-losophy that was quite indicative of the thinking and identity of this group.

The word *criollo* continued to be used to refer to native-born Cubans, without distinction of race, right through the nineteenth centu-ry. However, it also had a narrower meaning: it was used to refer spe-cifically to whites and blacks whose parents were European and African respectively. The children of *criollos* (in this narrow sense) were called *Criollos rellollos*, according to Pichardo ([1875] 1976: 191). This narrower meaning of *criollo* apparently was needed to preserve an original feature of the word when it was first used, that is it tried to identify a third category of person in between indigenous inhabitant and foreigner – a first-generation native who was by parentage part of the Old World and by place of birth part of the New World. Carpentier ([1946] 1972: 89) therefore summarises De la Torre's (1854: 53) explanation of the successive phases in the acclimatisation of the slave in Cuba in the following way: '*bozales, cuando llegaban del Africa y sólo se expresaban en sus dialectos; ladinos, cuando comenzaban a hablar el castellano; criollos, los hijos de éstos, y reyoyos, los hijos de criollos*' ['*bozales* when they arrived from Africa and could only speak their own dialects; *ladinos* when they began to speak Castilian; *Creoles* the children of these latter, and *reyoyos* the children of Creoles']. While both Africans and Europeans (parents) were identified in this definition of *Criollo*, thereby suggesting that the intermediate person regardless of colour was distinctive, the connection to the Old World allowed the white Creoles, much more so than the blacks, to preserve and highlight the cultural and linguistic traditions of their parents. In fact, one of the early reasons for the definition of the term as *hijos de españoles en Indias* was legal: it meant that officially the children retained all the privileges of their parents and all the entitlements of European-born children. Even when the distinction between *criollos* and (*criollos*) *rellollos* was lost, the European cultural and linguistic heritage of *criollos* remained as a highlight of their identity. This link with ancestry, however, did not diminish the sharp distinction between *criollos* and *peninsulares* in Cuba.

Within the black population, the term *criollo* continued to be needed to distinguish between groups, because after the Haitian revolution Cuba was transformed into the major sugar-producing country of the Caribbean with an attendant massive influx of African slaves, who had to be distinguished from those who were native-born. Lachatañeré ([1939] 1992: 186) points out that, during the period between 1805 and 1826, 151,330 *bozales* came into Cuba, changing the slave population from about 32 per cent of the population in 1804 to about 40 per cent in 1826. This trend continued throughout the first half of the century. As a result of this quick and significant increase, the *bozales* became a very noticeable group in Cuba throughout the nineteenth century. Their numbers fuelled fears of a black Cuba and led to calls for the whitening of Cuba. Saco (1879) recommended that blacks be sent back to Africa. In his mind, and presumably in the minds of many others, blacks in Cuba were not Cuban but merely servants who could be dismissed, sent back to their homes and replaced.

Opposed to this terrified view of blacks was the view that Cubans exhibited far less prejudice against Africans than anyone else. This was consistent with a widely held, traditional view that slavery was milder in the Spanish colonies. Sugar production and the plantation system were thought to have produced the brutal mentality and sugar had not been the main crop in the Spanish islands. When Madden, in the middle of refuting this general traditional view that slavery was milder in Cuba, claimed that Cubans were less prejudiced towards Africans, his reasoning was very different. He was merely perpetuating a belief that his English predecessors had reiterated several times before:

> And the only statement that is really correct in the whole passage, is contained in these words – 'in these colonies the distinction between blacks and whites was less than in all others,' presuming the meaning of the observation to be that, amongst the Spaniards, the prejudice against the stolen people of Africa, on account of their complexion, is less than amongst the colonists of other European States. Such unquestionably is the fact, and there is too much Moorish blood, in the veins of the descendants of the old 'Conquistadors,' for the feeling to be otherwise. (1849: 115–16)

Froude repeated the same belief about 40 years later: 'There is no jealousy, no race animosity, no supercilious contempt of whites for "niggers." The Spaniards have inherited a tinge of colour themselves from their African ancestors, and thus they are all friends together' ([1888] 1906: 303). However, in his desire to malign the Spaniards and their descendants, Madden did not seem to realise that what he was saying was contradictory, because it was sugar that was now dominating Cuba and producing the harshness of slavery. Madden himself (pp. 82–6) pointed out that Cuba was becoming Americanised and that it was the slave-holding whites from the southern States with a colour-segregation mentality who were the ones coming to Cuba to consolidate the position of sugar.

As sugar increasingly began to dominate, a contrast between the image of sugar and those of the other main crops developed in which the others were seen as being more positive. In terms of people, tobacco's image was the *veguero* or rural peasant whereas sugar's image was dominated by the black slave together with the white master and the corrupt product, the mulatto or, more specifically, the mulatta. Throughout the century, caricature of the negro and the mulatta were rife and intense in pictorial illustrations, engravings and lithographic work (see Kutzinski 1993: 43–80). In Cuba negative attitudes to blacks were intensified by emancipation of slaves elsewhere in the Caribbean and in the American south. White Cuban slaveowners were virtually the last set of persons in the New World to come to terms with the idea that blacks were human beings when they were finally forced, after almost 400 years, to give up slavery in 1886. There was therefore little logic in Madden's concession, for it was clear that as a result of sugar a high level of viciousness and negativity had developed in Cuba toward the black population.

In Cuba the nineteenth century accentuated not only racial and social differences in the population but also political affiliations and sentiments. In fact, it is the political situation, starting at the beginning of the century with Napoleon's invasion of Spain in 1808, that is seen to have been responsible for the fundamental change in identity from *criollo* to *cubano*:

> *Así, Sergio Aguirre ha señalado que 1808 significa el momento en que el criollo culmina su transformación en cubano, y que es a partir de esa fecha 'cuando se inician casi simultáneamente las tres corrientes políticas cubanas: reformismo, anexionismo e independentismo'.* (ILLACC 1983: 152)

> [Thus, Sergio Aguirre has pointed out that 1808 is the year that the Creole culminated his transformation into Cuban, and it is from this date 'that the three Cuban political movements – reformist, annexationist and independence – started almost simultaneously'.]

The three-directional pull in the political sphere is further summed up by Max Henríquez Ureña:

> *frente a los que sustentaban la adhesión al régimen español y reclamaban, dentro del mismo, reformas liberales, se encontraban los que defendían el ideal separatista y, en consecuencia, la independencia absoluta, aunque no faltaban quienes, faltos de fe en esta idea, abogaban por la anexión de la isla a los Estados Unidos.* (1963 1: 120)

> [as opposed to those who supported ties with the Spanish regime and, within this, were calling for liberal reforms, were those who defended the separatist ideal and, consequently, complete independence, although there were also those who, lacking faith in this idea, advocated annexation of the island to the United States.]

David Turnbull, after his visit to Cuba in 1838, had a more specific interpretation of two of the factions in Cuba, one which he related directly to the slave trade and one which consequently can be related to the blackening versus the whitening of Cuba. Turnbull's interpretation ([1840] 1969: 171) was that the Spanish rulers in Cuba wanted to keep the island a colony. On the other hand, the white *criollo* element wanted independence. In order to maintain the *salutary terror* and to frustrate the desires of the *criollos*, the Spanish government, according to Turnbull, supported the maintenance of the slave trade and exercised a heavy censorship on dissenting opinion. The desire for white labour on the part of the white *criollo* population was not at any time regarded by Turnbull as part of an intention to whiten the Cuban population but as a better economic strategy – a paid labourer would be cheaper than a slave.

In general, Turnbull's partiality towards the white *criollo* and Cuban independence was consistent with his schema of the hierarchical structure of Cuban society, for in it he saw the *peninsulares* as a repressive top level of the structure. He observed first that 'The distinctions of ranks among the various classes of society is as carefully kept up in Cuba as in the most aristocratical countries of the Old World' ([1840] 1969: 49). He then gave (pp. 49–50) the following social hierarchy for whites in Cuba:

Level 1     The thirty or so resident grandees of Spain – the *Titulos* of Castile (Spaniards)

Level 2     The civil functionaries in the public offices – the *Empleados* (Spaniards)
The officers of the army and navy (Spaniards)

Level 3     The Merchants (Spaniards, *criollos*, foreign)

Level 4     Merchants' clerks (French, English, North American, German, Spaniards)

Level 5     Retail merchants and shopkeepers (From the Canaries, Catalonia, Biscay, North America)

Level 6     Gallegos

Turnbull ([1840] 1969: 50) further pointed out that the only immigrants who were a permanent addition to the population were those from 'Old Spain and the Canaries, but especially the Catalans and Gallegos'. This was a social hierarchy that positioned the majority of white *criollos* and the coloured population below the foreign-born residents of Cuba and clearly implied an opposition between *peninsular* and *criollo*.

Madden, in his attack on slavery in Cuba, was not so concerned with the difference between *peninsular* and *criollo*, but saw an American element as partially responsible for the maintenance of slavery, and for him therefore American presence was very visible: 'some districts on the northern shores of the island, in the vicinity, especially, of Cardenas and Matanzas, have more the character of American than Spanish settlements' (1849: 83). Even if the number of Americans in Cuba was not as great as Madden claimed, there is no question that the appeal of America was attractive to some factions in Cuba.

Both Turnbull's and Madden's primary and passionate role as antislavery activists caused them to polarise the factions in Cuban society almost exclusively in terms of slavery, and did not allow them to highlight the contradictions in Cuban society, for example the fact that many *criollos*, whether they wanted to be independent or attached to

America, wanted to remain Spanish in language and culture. In fact, generally across the society, the concept of *cubano* developed strongly, embracing a notion not only of place but also of race and of language. This was expressed in the words of Saco, quoted by Max Henríquez Ureña:

> *La anexión, el último resultado, no sería anexión sino absorción de Cuba por los Estados Unidos. Verdad es que la isla, geográficamente considerada, no desaparecería del grupo de las Antillas, pero yo quisiera que si Cuba se separase, por cualquier evento, del tronco a que pertenece, siempre quedase para los cubanos, y no para una raza extranjera ... No olvidemos que la raza anglosajona difiere mucho de la nuestra por su origen, por su lengua, su religión, sus usos y costumbres ... (1963 1: 138)*

> [Annexation, the ultimate result, would not be annexation but absorption of Cuba by the United States. The truth is that this island, from a geographical point of view, would not disappear from the Antillean group of islands, but I would like that if Cuba were to separate, as a result of whatever event, from the trunk to which it belongs, it should always remain for Cubans and not for some foreign race ... Let us not forget that the Anglo-Saxon race differs a great deal from us in origin, language, religion, practices and customs ...]

For Saco and others like him, the *cubano* could be nothing else but a descendant of the Spanish, born and brought up in Cuba, speaking the Spanish language and practising the Catholic religion.

The image of the *cubano* and Cuba was promoted by an active intellectual elite, several institutions of learning and a vibrant literature. In Havana alone, according to Turnbull ([1840] 1969: 205), there were a number of 'public establishments (e.g. the University, the Colleges of San Carlos and San Francisco de Sales, the Botanic Garden, the Anatomical Museum and Lecture-rooms, the Academy of Painting and Design, a School of Navigation, and seventy-eight common schools for both sexes)' This view was also given by Madden (1849: 106), in a comparison of Cuba with the French and English colonies in the Caribbean. Cuba, then, in the nineteenth century was like an intellectual beacon in the Caribbean with connections to the centres of scholarship and scholars in Europe, North America and throughout Spanish America.

Expressions of national feeling were evidenced in nineteenth-century Cuban literature at first in a romantic celebration of *la naturaleza* and *la tierra nativa*. Many verses were written extolling the natural beauty of the Cuban countryside, the bounty of the land and the pleasantness of the elements. In time, this focus expanded to include the Cuban people and their customs on the one hand and the Cuban people and their aspirations on the other. The former tended to be light or comic in tone, whereas the latter was serious, political and nationalistic. It was the shift in the literature from universal, European and Spanish themes to Cuban themes that really marked the beginnings of Cuban consciousness.

Focus on the Cuban people and their customs featured both the contemporary and the historical. One of the earliest texts that was explicitly a delineation of Cuba and Cubans by a Cuban for Cubans was written by J.M. de la Torre. There was a full version of the text and a kind of catechetical version for schoolchildren. As part of his characterisation of Cubans de la Torre said: '*Los cubanos son generalmente de buena figura y claro entendimiento, aficionados á la poesía, al baile y á la música. Se les acusa de ser amigos de aparentar, poco aficionados á trabajos mecánicos y mucho á pleitar*' ['Cubans are generally of a good build and keen intelligence, lovers of poetry, dance and music. They are accused of being fond of affectation, not keen on manual work and given to litigation'] (1868: 57). The traditional musical and cultural side of Cubans had now acquired sophistication, the traditional 'indolence' had been refined to a lack of affection for mechanical work, and more sophisticated social and legal behaviours had developed. In truth, this was more a picture of urban Cubans than Cubans generally, one driven by a need to present a positive national image to match the vibrant economy.

The reconstruction of Cuban history and Cuban identity in refined Cuban literature in part involved the use of indigenous figures from the distant past. Some people at the time believed that they could still see indigenous ancestry among the people in those areas of Cuba (Guanabacoa, Caney and Jiguani) where the indigenous inhabitants had survived the longest, according to Turnbull ([1840] 1969: 233), and de la Torre (1854: 54) claimed that a few indigenous inhabitants were still alive in Cuba. Interest in the historical inhabitants of Cuba was obviously not separable from the rise of the *guajiro* (whose name was indigenous) as a contemporary Cuban figure. Paradoxically, there was in the poetic productions a certain precision and accuracy in the identification of the indigenous Cuban as *siboney*.

Henríquez Ureña (1963 1: 174) regards *ciboneyismo* as the bringing together of two traditions (*tendencias*) – *la tendencia indianista* and *la tendencia criollista*. However, though Henríquez Ureña identified the contradiction between what was Cuban in name but Spanish American in reality as well as what was indigenous in name but non-indigenous in substance, he did not highlight the contradiction between the name *siboney*, which was truly Cuban historically but as a recreation seemed false, and the name *guajiro*, which was not the name of Cuban natives but which had come to symbolise Cuban-ness. As Henríquez Ureña points out, evidence of *la tendencia indianista* had appeared earlier in the century in the work of José María Heredia and it is interesting to note that it was as a result of his participation in a plot in 1823 to seize independence for Cuba and establish it as *la República de Cubanacán* that he was exiled from Cuba (Henríquez Ureña 1963 1: 119–20). The intention to restore the longer version of the island's indigenous name was akin to the restoration of the indigenous name to Saint Domingue and was seen as signalling a definitive break with Spain.

*Ciboneyismo* is also analysed by Remos (1958: 119) as a way of dealing with the harsh censorship in Cuba in the nineteenth century. For him, it was not a romanticisation of the past but a way of criticising the contemporary political situation, especially for those who did not have the safety and freedom of being in another country and having their work published outside Cuba. *Ciboneyismo* was not considered to be good literature, but was very popular in the middle of the nineteenth century. Its success is explained as due to a number of factors, but, above all, to the political situation: '*La ingenuidad de su alegorismo, el amor por la naturaleza cubana y sobre todo las evidentes connotaciones políticos de aquellos poemas en aquellos momentos de férrea censura, explican su éxito*' ['The ingenuity of its allegories, the love of the natural beauty of Cuba and above all the evident political connotations of those poems in those times of harsh censorship explain its success'] (ILLACC 1983: 256). The words of the poem *La Bayamesa* by Pedro Figueredo, ostensibly exhorting the Indian *bayameses* to throw off the Spanish yoke and to free their country, Cuba, were made into a song that became the most famous song during the revolutionary period in the nineteenth century (from 1868 on) and the first two stanzas of the same poem are now the national anthem of Cuba. *Ciboneyismo* therefore provided a strong image to link the sixteenth-century *siboney* to the nineteenth-century Cuban in their fight against the Spanish oppressor and has survived into the twenty-first century.

The *siboney* was a deliberate literary recreation but the *guajiro* was the most characteristically Cuban *criollo* to emerge in the nineteenth century. The *guajiro* was essentially rural and tied to the land: '*en la Isla de Cuba, principalmente en la parte occidental es mui comun y distinta su significacion. Aquí **Guajiro** es sinónimo de **Campesino**, esto es, la persona dedicada al campo con absoluta residencia en él ...*' ['In the island of Cuba, principally in the western part, its meaning is very common and distinct. Here "guajiro" is a synonym of "peasant", that is, a person dedicated to the country and living there all the time ...'] (Pichardo [1875] 1976: 296). Alternative names for *guajiro* given by

Pichardo were *montuno* and *jíbaro*. As was the case in Puerto Rico, the 'Indian', who had evolved into a *campesino*, was accepted as the consummate and folkloric native of Cuba. One of the *sainetes* (one-act farces) of Francisco Covarrubias, who started his stage career in 1800, was *El guajiro sofocado*. It was Covarrubias who, according to Remos, was responsible for bringing into being '*un teatro genuinamente cubano, que reflejó con gracia las particularidades criollas, satirizando aquéllas que era preciso ridiculizar*' ['a theatre genuinely Cuban, which tastefully reflected creole peculiarities, satirising those that were correct to ridicule'] (Remos 1958: 93).

Though by the beginning of the nintteenth century the *guajiro* had become a traditional Cuban figure, the name *guajiro* still retained its original reference, for the 'Indians' whose name had been adopted in Cuba were still alive in their own native country. The following was a contemporary description of the *Guahiros* in Venezuela:

> The Spanish writers of this age, as well as the English and French who have copied them, speak of them as of a horde of ferocious robbers, who have resisted all the efforts made for civilizing them. The Spanish geographers rank them among the **Indios bravos**, a name which they give to all the tribes they have not been able to subject. The Spanish historians of the sixteenth century, relate that the Guahiros were, at that period, the friends of the Spanish inhabitants of Truxillo; that the missionaries had converted almost all of them to Christianity; that they shewed more capacity and taste than the other Indians for the arts of civilization, in which they had made a rapid progress in a few years. But the libertinism of the inhabitants of Truxillo caused bloody quarrels between them and the Guahiros. The former did not desist from debauching their wives. One day a gang of Spaniards had the audacity even to go and carry them off by force from one of their villages. The nation or the tribe of Guahiros rose unanimously to revenge this outrage: the warriors entered Truxillo, sword in hand, and made great slaughter among the inhabitants. They declared solemnly that they renounced the religion of men so corrupt, for that nothing was sacred to them. All the efforts made by the Spanish missionaries, since that period, to reconcile them to their nation, have proved fruitless; and they have remained implacable enemies to the Spanish name. Every time in which Spain and Great Britain have been at war, the British government has profited of this antipathy, to excite the Guahiros to commit hostilities against the colonists of the province of Maracayabo, which is the cause of its depopulation. The Guahiros, however, are more civilized than the other Indians, their neighbours; they cultivate their land, weave stuffs of cotton and wool for their clothing: they also rear herds of cattle, which form objects of a very considerable trade between them and the English in Jamaica; they receive in payment spirituous liquors, fire-arms, and gun-powder. All their warriors are mounted. They are the true Caribs, possessing their tall stature, manly,

haughty and independent character. (Dauxion-Lavaysse [1820] 1969: 186–7)

It is quite clear, then, that, when the word was popularly featured in nineteenth-century Cuban literature as the name of the Cuban campesino, Guahiros were prominent in Venezuela and their confrontations with the Spanish well known historically. It was not only that the economic characteristics of the *Guahiros* identified in the last part of the extract admirably fitted the *guajiros* of Cuba, but also that the decline of Spanish colonial power, the factional fighting that ensued, and above all the securing of independence in Venezuela in 1811 encouraged the *campesinos* of Cuba to be confirmed in the mould of the *Guahiros*, in other words, as the independent and true natives of the island. Furthermore, it may not be totally coincidental that Domingo del Monte, the most important literary influence in Cuba in the 1830s and whose *romances cubanos*, according to Henríquez Ureña (1963 1: 159), '*dieron auge en Cuba a la poesia **criolla** de caracter narrativo*' ['brought narrative creole poetry to its zenith in Cuba'], was born in Venezuela. One of the *romances* which del Monte wrote was called *El Guajiro*.

By the middle of the century there were a number of writers commenting on the image, role or character of the *guajiro*. Don José Maria de La Torre, a sophisticated Cuban, gave his view: '*El criado en el campo ó sea **guajiro**, es de constitucion fuerte, y aunque perspicaz es indolente y rutinero*' ['The peasant or 'guajiro' has a strong constitution and although perceptive is indolent and ordinary'] (1854: 55). This was a view that was reinforced in the minds of educated Cubans because he repeated it in his school text book for children, which went through a considerable number of editions. As was the case with Alonso in Puerto Rico, Torre's comment implied that there was a sharp distinction between the educated, urban Cuban and the rural peasant. Moreover, Torre, in repeating the familiar characterisation of the Spanish *criollos* as indolent, assigned this indolence specifically to the *guajiro*.

For Granier de Cassagnac, the *Ibaros* of Puerto Rico were an excellent foil against the tendencies and actions of the black slaves. For Ballou (1854: 142) the same was true of the *guajiros* in Cuba, whom he called *monteros*. Kimball, in contrast, gave a much less tough picture of the *guajiro*, whom he included among the lower classes of Cuban society:

> The Guagiro, with his wild, dark eye, wonderfully expressive gesture, and usually imperturbable self-possession, becomes ridiculously silent and shy in his courting. In a richly-worked shirt of fine linen, worn upon the outside as a sack; a long, and often elegantly embroidered cambric sash – fastening to his side the silver-handled sword, or 'machete,' silver spurs, and low slippers, he will sit for hours opposite his lady-love, only

venturing now and then a word of reproof, to be interrupted in affectionate playfulness, and to which she retorts in the same style; yet, now and then, at a glance, and when unobserved, they do venture to exchange some tender word. But gestures, shrugging of the shoulders, little dashing airs of coquetry in the lady, and bashful approaches on the part of the gallant, fill up the measure of the wooing of the Cuban peasant. (1850: 149–50)

This soft side did not alter the fact that by the middle of the century, as was the case in Puerto Rico, the Cuban peasant called *guajiro* or *montero* was presented as having aligned himself with the interests of whites and as being naturally distinct from blacks in a society that was racially polarised.

The name *guajiro* was from early recognised as a word coming from the indigenous inhabitants of the region, but the geographical origin of the *guajiros* was a source of controversy among nineteenth-century academics in Cuba. Bachiller y Morales, however, was clear in his own mind. Using the work of the Italian Codazzi on Venezuela as one of his sources, he argued: '*Encontramos en el continente una tríbu numerosa de indios llamados **guagiros**, existe una peninsula que lleva el nombre de la **Goayira**. Es, pues, evidente, que de ese punto hubimos el nombre*' ['We find on the continent a populous tribe of Indians called *guagiros*; there is a peninsula that bears the name *Goayira*. It is evident, then, that the name comes from this point'] (1883: 109). Bachiller y Morales then proceeded to account for the application of the name to Cubans who were descendants of Europeans:

¿pero qué tienen de comun con los indios del continente hombres descendientes de Europa? ¿se llamaron así los cubanos alguna vez? – Creemos que si supiésemos el significado de la palabra, fácilmente resolveríamos la cuestion. Pero si acudimos á anologías, desde luego podemos decir que los indios llamaron **guagiros** á nuestros campesinos, por reconocer que eran semejantes á esos seres que sostenian un activo comercio con todas las islas, y que aún en la actualidad se les reputa por uno de los más inteligentes é industriosos naturales ... y los aborígenes que vieron una raza de más poder moral é inteligencia, no pudieron dejar de hacer comparaciones con objetos que les eran conocidos. (1883: 109–110)

[But what do European descendants have in common with Indians from the continent? Did Cubans have this name some time? – We believe that if you knew the meaning of the word, the question would be easily resolved. But if we look to analogies, of course we can say that the Indians named our peasants *guagiros* because they recognised that there were similarities with those beings who carried on an active trade in all the islands and that even in reality they were reputed to be one of the most intelligent and industrious group of natives ... and the indigenous inhabitants who saw a race of greater moral and intellectual power could not but make comparisons with things that they knew.]

It is only by arguing that it was the indigenous inhabitants who gave the *campesinos* an indigenous name that Bachiller y Morales could make his next point.

Contrary to what most of the literature suggested, Bachiller y Morales (1883: 109–10) claimed that the Cuban *campesino* was insulted by being called by an indigenous name and thus being compared to the *indios bravos*. This, no doubt, was an end-of-century interpretation of a development that had taken place at a much earlier time in Cuban history, for there is no indication of this attitude in the definition of *guajiro* in Pichardo's dictionary, the first version of which appeared in the first half of the century. Apparently, feelings of racial superiority had become so deeply entrenched in Cuba among whites in the last half of the century that no white could be admitted to have willingly adopted an indigenous name. Yet, at the same time indigenous inhabitants were believed to be a part of the genetic make-up of the Cuban population. Bachiller y Morales did not even question the idea of 'Indian' genetic survival in contemporary Cubans when he said: '*Parece que estábamos condenados á no llegar al conocimiento de estas cosas por la confusion de las dos razas española é indiana que hoy forman una sola* ...' ['It seems as if we were condemned not to arrive at an understanding of these things because of the mixing of the two races – Spanish and Indian – which today form a single race ...'] (1883: 108). Presumably the sustained preoccupation with the 'Indian', in *siboneyismo* with its political overtones and in the *guajiro* with his admirable qualities, had regenerated the 'Indian' to account for the physical appearance of a sizeable sector of the Cuban population whose hatred of blackness caused them to downplay the genetic link with the large slave and former slave population in their midst.

An incident that took place a few years later again shows the prominence of the concept of 'Indian' in Cuban identity. When the famous actress, Sarah Bernhardt, visited Cuba in 1887 to give some performances, shortly before her departure she is reputed to have insulted the fashionable society of Havana by calling them *indios con levita*. The actress denied that she had used those words, but in her denial it is interesting to note what she is reported by Leal (1982: 203) to have said: ' "*Indios*", *en todo caso, aceptó Sarah, "pero con levita jamás*" ' [' "Indians", of course, accepted Sarah, "but with frock coats, never" ']. Presumably, then, it was not the racial identification that was thought to be offensive to the fashionable society of Havana but the dress.

The *siboney* and the *guajiro* were seen to be essentially Cuban: the *siboney*, for literary and political purposes, was recreated from history as a brave and noble warrior while the contemporary *guajiro*, who was never really promoted as 'Indian' racially, nevertheless acquired the moral strength and intelligence associated with the real *guagiros*. The Cuban *guajiro*, however, did not shed his traditional image. The admirable qualities of the *siboney* and the *guajiro* were transferred, by suggestion principally, to the nineteenth-century *cubano* recognised as white. The relationship between 'white' Cuban and 'Indian' was therefore an undercurrent throughout the nineteenth century. The image of the Negro, on the other hand, was negative and, while those blacks born in Cuba were included within the term *criollo*, they were not featured prominently as *cubanos*. In the case of the *negro bozal*, he was by his very designation identified as non-Cuban and his different identity was projected above all else through his speech.

In 1854 Torre made the statement that '*El idioma usado en toda la Isla es el **castellano**'* ['the language used throughout the whole island is Castilian']. However, during the nineteenth century the linguistic identity of the various social classes and groups in Cuba was less uniform than it had been previously; it was a period of adaptation and assimilation within a social system that was increasingly being dominated by the sugar plantation. It was also a time when language, as a reflection of the identity of different classes in Cuba, began to be highlighted and reflected in popular literature. At the very end of the previous century, Fray José María Peñalver had become better known for what he proposed to do about language in Cuba than what he actually did. In 1795 he had proposed the compilation of a Cuban dictionary in which there would be creole words, Spanish words that had changed in sound and meaning in Cuba, comical and amusing words commonly used, words used by the negroes occurring in the speech of whites, as well as indigenous words that had survived in common speech. It is this list that reveals the linguistic fabric of nineteenth-century Cuba – the original Spanish element had in itself evolved and had also become enmeshed with indigenous survivals as well as with the African element. The combination of all these was being identified as *cubano* and it is this same concept of Cuban identity that was repeated in the next major lexicographical work.

The most important work on language in Cuba in the nineteenth century that was actually completed was the *Diccionario provincial casi razonado de vozes y frases cubanas* [Provincial dictionary almost fully illustrated with examples of Cuban words and phrases] by Esteban Pichardo. It was first published in 1836 and then had three later editions in 1849, 1862 and 1875, each with a slightly different name. In this dictionary differences in speech and differences between people were assessed in relation to good and Spanish. Throughout the dictionary (i.e. after the listing for each letter) the author gave additional words, which he identified as corruptions (*vozes corrompidas*) in the speech of Cubans. Obviously, for him, a substantial portion of what was Cuban Spanish was 'corrupt' and many of these words he said that he could not include in the dictionary, only those which were '*mui generalizados aun entre personas cultas*' ['very generalised even among sophisticated people'] (p. 10). It is only in the definition of words like *tierra* and *guajiro* that Pichardo addressed the question of Cuban identity with a degree of admiration and positiveness. His approach, dominated as it was by the notion of corruption, virtually excluded the reality of linguistic evolution and the acceptance of distinctiveness between dialects. From a broad perspective, Pichardo's intention was to provide a Cuban addendum to the Spanish language, rather than to validate a Cuban dialect of Spanish. His dictionary was not an expression of independence, but a confirmation of deviance, for there was no suggestion that the cubanisation of Spanish was something that Cubans should be proud of. For his time, Pichardo was certainly quite typical in this regard. The positive aspect of this work as a whole was that it showed the regional and ethnic diversity of Cuba, thereby moving beyond the simplistic and uniform identity that dominated earlier European views. So, even if the dictionary could not reflect the political and economic factions of Cuba of the time, it indicated some plurality in the identity of Cubans.

Although he did not say so directly, there is little doubt that the *guajiro* produced most of what he gave as *vozes corrompidas*. The words that were said to be indigenous in origin were not treated as corruptions, but they were regarded as lesser than Castilian words. In the case of the black *criollos*, he said that except for the areas of Matanzas and Havana, where they had some similarities with the Africans, '*Los Negros Criollos hablan como los blancos del pais de su nacimiento o vecindad*' ['Creole negroes speak like the whites of the country of their birth or neighbourhood'] (p. 12). The linguistic hierarchy in the dictionary, then, was largely a mirror of the social hierarchy. In addition, the dictionary contained etymological information about most of the words that reflected the history of the different social, racial and geographical groups in Cuba.

The major group which Pichardo identified as having speech peculiarities was that of the *bozal* slaves, whose numbers had increased dramatically. In describing the linguistic characteristics of *bozales* Pichardo was detailed. He said:

*Otro lenguaje relajado y confuso se oye diariamente en toda la Isla, por donde quiera, entre los Negros Bozales, o naturales de Africa, como sucedia con el Frances CRIOLLO de Santo Domingo este lenguaje es commun e idéntico en los Negros, sean de la Nacion que fuesen, y que se conservan eternamente, a ménos que hayan venido mui niños ... (p. 11)*

[Another corrupt and confused language is heard daily throughout the island, wherever you go, among the bozal negroes or natives of Africa as was the case with the French creole of Saint Domingue; this language is common and identical among the negroes, no matter which nation they are from and they keep it forever, unless they came as very small children ...]

There are several interesting points in these comments that shed light on the nature and development of the language used by these slaves. The fact that, according to Pichardo, the language was heard '*en toda la Isla, por donde quieraI*' ['throughout the island, wherever you go'], that it was '*commun e idéntico en los Negros, sean de la Nacion que fuesen*' ['comon and identical among the negros, no matter which nation they are from'], and that the Negros '*se conservan eternamente*' ['they keep it for ever'] gave this variety a level of stability that belies the idea that idiosyncratic attempts at a second language suggest. In other words, from what Pichardo says, it definitely had become a fixed variety, which Africans acquired when they came to Cuba. More importantly, José María de la Torre said: '*Los negros criollos se distinguen tambien en nacidos en los poblaciones, y nacidos ó criados en los campos [que se dicen **criollos de campo**] pues éstos tienen un lenguaje y maneras peculiares y mas rùsticas*' ['Creole negroes can also be divided into those born in towns and those born or brought up in rural areas [called country Creoles]; these latter have a language and ways that are peculiar and more rustic'] (1854: 54). About this Castellanos y Castellanos says: '*De la Torre confirma, simplemente, un hecho indiscutible: en ciertas zonas rurales, donde la población de color se hallaba alejada del blanco, muchos hijos de africanos utilizaban el **bozal**'* ['De la Torre confirms simply an unquestionable fact: in certain rural areas, where the coloured population was separated from whites, many children of Africans used bozal language'] (1992: 352–3). There seemed to be little doubt, therefore, that both the African and the *criollo* used what was nevertheless called *habla bozal*. It is quite likely that the fear of blackness, the caricature of the black, the desire to send blacks back to Africa and the desire to whiten the population all together led to a necessary association of *habla bozal* exclusively with African.

Pichardo gave the following excerpt as an illustration of the speech of the *bozales*:

*yo mi ñama Frasico Mandinga, nenglito reburujaoro, crabo musuamo ño Mingué, de la Cribanerí, branco como carabon, suña como nan gato, poco poco mirá oté, cribi papele toro ri toro ri, Frasico dale dinele, non gurbia dinele, e laja cabesa, e bebe guariente, e coje la cuelo, guanta qui guanta* ... (pp. 11–12)

Some of the words in this passage, which is meant to illustrate, in Pichardo's view, a kind of confused, distorted kind of Spanish spoken by slaves,are difficult to recover and the meanings are not clear. An attempt at a rough English and Spanish translation is:

*yo me llamo Francisco Mandingo, negrito \*reburujaoro, esclavo de mi amo Señor Dominguez, de la \*Escribanería, blanco como carbón, las unas como un gato, poco poco mira usted, escribe papeles todo el dia todo el dia, Francisco le da dinero, un cubilla dinero, el relaja cabeza, el bebe aguardiente, el coge la cuelga, aguanta que aguanta* ...

[My name is Francisco Mandingo, \*confused/tangled up negro, slave of my master Mr. Dominguez of the Notary's Office, white as coal, fingernails like a cat, he looks at you very little, he writes papers the whole day, the whole day, Francisco gives him a \*money plant, a \*rockcress, \*he relaxes his head, he drinks alcohol, he takes the \*bunch, \*what patience ...]

Pichardo also gave the following interpretation of this bozal speech:

*es un Castellano desfigurado, chapurrado, sin concordancia, número, declinacion ni conjugacion, sin R fuerte, S ni D final, frecuentemente trocadas la Ll por Ñ, la E por la I, la G por la V &; en fin, una jerga más confusa miéntras más reciente la inmigracion; pero que se deja entender de cualquier Español fuera de algunas palabras comunes a todos, que necesitan de traduccion.* (p. 11)

[it is Castilian which is disfigured, broken, without concord, number, declensions or conjugations, without strong 'r', final 's' or 'd', 'll' frequently exchanged for 'ñ', 'e' for 'i', 'g' for 'v' and so on. In short, a jargon which is the more confused the more recently the person has arrived, but which can be understood by any Spaniard except for some words common to all of them which need to be translated.]

Pichardo's comment that the newer the slave, the more 'confused' his speech was must be interpreted to mean that the recently arrived Africans retained more of their native language in their version of this variety. Furthermore, contrary to what Pichardo was suggesting, it would have been very difficult for a Spaniard to understand even the more uniform version, that is the one illustrated above.

The concept of *el negro bozal* as an African whose characteristic speech was *el habla bozal* was really captured and crystallised in the fictitious character, Creto Gangá, created by Bartolomé José Crespo y Borbón and used as his mouthpiece to comment on and caricature social customs in Cuba during the 1840s. Crespo was a Gallician who had come to Cuba at a young age and, according to Cruz (1974: 68), as an underdog himself (being a Gallician) in Spain, sympathised with the *bozal* in Cuba and managed to master the *bozal*'s speech in his writings. However, Crespo was said to be following the example of, or imitating, Covarrubias, who is regarded by many as the first to use contemporary Cuban characters and speech on stage (*'Gracias a él se habló en cubano por primera vez en nuestra escena'* ['Thanks to him Cuban was spoken for the first time on our stage'] (ILLACC 204). Covarrubias was both an actor and a playwright, who dominated the Cuban stage from 1800 to 1850 but none of his works has survived. He is said to have been one of the first actors to paint himself in blackface (in 1815), which was even before Thomas Rice presented his Jim Crow character in 1828 (see Cruz 1974: 279–80). Crespo's character, Creto Gangá, who was presented as an *escritor bozal* [a bozal writer], was therefore said to be Cuba's counterpart to America's minstrel, for the character, Creto Gangá, was just one example of a stereotype that was common in popular nineteenth-century Cuban literature and theatre. It is also said that Crespo, a *peninsular*, was continuing a tradition that had been started by his own compatriots (e.g. Lope de Vega, Góngora and others) in Spanish Golden Age literature and that his character was a local version of the *ladino* negro that occurred in that theatre.

Since Creto Gangá was a comical stereotype of a black Cuban and one presented by a non-native white, his linguistic features were exaggerated in many respects and misrepresented in some cases, for, as Cruz (1974: 57) concedes, *'la deformación fonética y sintáctica ... son apoyos de la comicidad'* ['phonetic and syntactic distortion ... are props of comedy']. The concept of *bozal* went hand in hand with the notion of *langage corrompu* and these two taken together kept the person identified as having these characteristics in the cultural sphere of 'otherness', in the social sphere as a buffoon, and in the context of Cuba separate from the true *cubano*. It was not accidental that the *negro bozal* became popular in Cuba after the suppression of anti-slavery movements. Yet, though he was substantially a white creation, the *negro bozal* no doubt reflected actual cultural and linguistic features of blacks in Cuba. The cultural features noted in other islands – the penchant for derision, the imitation/mocking of whites and what Long in Jamaica saw as 'catching at any hard word [which] they alter and misapply in a strange manner' – were all part of the *negro bozal* stereotype and the *habla bozal*.

No matter what the intention of the author was in his social comments, the black in Cuba was, by this created variety of language, converted into or, for some, confirmed as an inferior being. The *habla bozal* therefore fuelled the view that Cuba should be whitened in order for this corruption of the Spanish language and this stain on the image of Cuba to be removed. In fact, throughout the century the notion of pure Castilian Spanish was conceived, and promoted, as the ideal. Pichardo, for example, in his dictionary used a hierarchy of language in which much of the usage in Cuba was contrasted with (Castilian) Spanish. Preceding the actual definitions of the words, there were in the introduction comments on the different varieties of speech to be found in Cuba at the time. In that section Pichardo was more explicit about what he considered to be good Spanish and corrupt Spanish as he made reference to the characteristics of speech used among various groups. In addition, for those in the literary sphere, Domingo del Monte, the major force in 1830s Cuba, suggested that: '[*el poeta castellano*] *hará que la lengua castellana* **resplandeciente como el oro puro y sonoro como la plata** *y en toda su pulcritud, pero también en toda su libertad, sirva de magnifico engaste a sus concepciones ...*'['[the Castilian poet] will make the Castilian language, resplendent like pure gold and sonorous like silver and in all its beauty but also in full freedom, serve as a magnificent setting for his thoughts.'] (Quoted in ILLACC 1983: 194–5 nd, Rivas 1990: 94–5). The contrast between the *habla bozal* and Castilian Spanish therefore was intended to parallel the contrast between the black African and the white *criollo*.

At the end of the century, in discussing versions of Spanish produced by blacks in Cuba, Bachiller y Morales was still making a clear distinction between black *criollos* and Africans. He said: '*El negro bozal hablaba el castellano de un modo tan distinto al que sus hijos usaban, que no hay oido cubano que pudiesen confundirlos*' ['The bozal negro spoke Castilian in a way that was so different from his children that no Cuban could confuse them'] (1883a: 99). For Bachiller y Morales, the African's speech was characterised by the use of [o] for [u] at the beginning of words, as in *oté* (< *usted*), and the confusion of pronouns and the gender of the pronouns. On the other hand, the characteristic feature of the black *criollo* was the changing of [l] at the end of words to [i], as in (*el* >) *ei*. This same feature of the black *criollo* had earlier been identified by Pichardo but at the same time he indirectly linked its source to Spain. In giving qualifying information about the *negros criollos*, he said: '*aunque en la Habana y Matanzas algunos de los que se titulan* **Curros***, usan la i por la r y la l v.g. poique ei niño puee considerai que es mejoi dinero que papei ...*' ['although in Havana and

Matanzas some of those called 'Curros' use 'i' for 'r' and 'l', e.g. ...'] (p. 12). The identification of some *negros criollos* as *Curros* therefore led to the belief that the features identified, which were clearly the same as those used among the *bozales*, were really Spanish in origin, because, as Pichardo said '***Andaluz** y **Curro** han venido a convertirse en sinónimos*' ['*Andaluz* and *Curro* have become synonymous']. Such a belief helped to separate *bozales* not only from black *criollos* but also, more importantly for white Cubans, from white *criollos*, who had the same features in their speech. No attention was paid to the fact that Andalucia had, from even before Columbus's voyages, a substantial number of *ladino* Africans and, as Bachiller y Morales himself pointed out, the features identified in Spanish and Portuguese literary works as typical of Africans in the peninsula were the same as those in Cuba.

To complicate matters further, one of the same distinctive pronunciation features credited to Andalucian and African sources was proposed as Canarian by Alvarez Nazario (1972), dealing with Puerto Rico – change of /r/ and /l/ at the end of syllables to /i/, e.g. *poique* (< *porque*); *cueipo* (< *cuerpo*). In addition, the (subject + verb) word order in questions such as *Que tu quieres?* and *De donde ustedes vienen?* is also said to be Canarian in origin by Alvarez Nazario (1992: 461). However, this very word order in questions is identical to that in creole languages throughout the Caribbean. Another example given by Alvarez Nazario as Canarian – '*voy a ir buscándolo*' ['I am going to go looking for it'] – is matched by Jamaican '*mi a go go look fi it*'. Equally interesting is the fact that the word *ñoca*[without fingers], which Alvarez Nazario gives as Canarian, occurs in the non-standard speech of Barbadians with a slightly different pronunciation of the nasal sound at the beginning of the word but with the identical meaning. For Barbados, the source of the word has been posited as West African (< Gã *noko*; Ewe *núkè*).

Another term used in Cuba in the nineteenth century that successfully masked black and African elements of Cuban identity was *mambí*. This was a term said to have originated in Santo Domingo but became better known in its reference to Cubans. It was said to have been initially used by the Spanish in a derogatory manner to refer to those Dominicans who refused to submit to Spanish rule in the period 1861 to 1865 – those Dominicans who had most to fear from the Spanish and their slave-holding practices still vibrant in neighbouring Puerto Rico and Cuba were the ex-slaves and their children who had been free from the beginning of the century. The Spanish soldiers who left Santo Domingo and went to Cuba applied the same term *mambí* to the Cuban insurgents, who had much in common with the Dominicans.

However, for the Cuban revolutionaries who were fighting against Spanish rule *mambí* soon became a positive term.

There is no exact certainty about the origin of the word as a term of identity and there seemed to be a desire among established Cuban and Spanish writers to find an indigenous origin for it. For instance, Bachiller and Morales, among other things, said the following about the origin of the word:

> D. Antonio González escribió un folleto **Los Mambises** (Madrid, 1874) y dice que son varias las etimologías de la palabra: para unos es el nombre con que se llamaban los indios rebelados contra los caciques que se ocultaban en los bosques ... (1883: 318)

> [D. Antonio Gonzalez wrote a pamphlet *The Mambis* (Madrid, 1874) and says that there are many etymologies of the word: for some people it was the name for the Indians who rebelled against the caciques and hid in the woods ...]

On the other hand, Ortiz more recently identifies it as an African word: 'Por *nuestra parte hemos hallado la palabra* **mambí** *con varios significados en los lenguajes de África ... Los esclavos congos llamaron* **mambí a los rebeldes, en su lengua** ...' ['For our part, we have found the word mambi with various meanings in African languages ... Congo slaves called rebels *mambi* in their language'] (Ortiz 1974: 337). Ortiz's explanation gets strong corroboration from the first-hand experience of James O'Kelly, an adventurous Irishman who, as a correspondent for a New York newspaper, visited 'Mambi-Land' in 1872 and said of the people there:

> Indeed, considering that very many of these people had been slaves, – all of them except De la Torre were coloured, – their conduct contrasted very favorably with what I have since observed among the white Catalans and Castilians, who contemptuously look upon them as barbarians and *negros sueltos*, or to translate the idea, 'runaway niggers' ... This little nameless hamlet belongs to a numerous class of settlements scattered over Cuba Libre. (O'Kelly 1874: 182–3)

Viewed in the light of O'Kelly's description, which associated the people in Mambi-Land with runaway slaves, the relationship between *mambí, manigua* and the older word *maniel* (a place of refuge for runaway slaves) may not therefore be only an accident of phonology and may be used to show the long and crucial role that the black population played in the ethnic identity of Cuba.

In the second half of the nineteenth century the basic concept of the *mambí* was principally one of a guerilla warrior, operating from the hills and inaccessible places, referred to as *la manigua*, from which derived the word *maniguero*, the word used in Santo Domingo as the equivalent of the *mambí*.

The association of the *mambí* with the struggle for national independence brought an element of exoticism and romanticism to the word, a fact which was evident in the account of the same O'Kelly:

> The land of the Mambi is to the world a shadow-land, full of doubts and unrealities. It is a legend, and yet a fact. It is called by many names, yet few know where begins or ends its frontier. Spaniards call it the Manigua, or Los Montes, Americans talk of it as Free Cuba, and those who dwell within its confines, Cuba Libre, or the Mambi-Land ... Its limits may be vaguely marked by the shores of Cuba; for even in the Spanish strongholds the dominion of the Mambi is spread over Cuban hearts. (1874: 12)

While O'Kelly may have been seeking to add an air of mystery and adventure to his account, the connotations of 'shadow-land' and its relationship to the idea that 'the dominion of the Mambi is spread over Cuban hearts' may truly have reflected the essence of the identity of many non-white Cubans in the nineteenth century.

*Criollo* with both its general and narrower meanings, *tierra* in its basic and figurative senses, and *mulato/mulata* with its varied denotations and strong emotional connotations were all contributory elements in the concept of *cubano*. These together with the use of local varieties of the Spanish language gave the term *cubano* a fundamental but at the same time a fairly broad scope of reference. In addition to these intrinsic characteristics, however, confrontation with two other major identities within Cuba at the time sharpened the concept of *cubano* and the consciousness of it. These two identities were *peninsular* on the one side and *bozal* on the other. In addition to these two, there were people of various other nationalities in Cuba and there were various interests outside of Cuba that had their eyes on Cuba. As a result, different economic, social and political attachments caused people within Cuba and outside Cuba to visualise Cuba as either Cuban (independent), American or Spanish, but this indeterminate vision of the political future of Cuba in the nineteenth century did not in itself diminish or determine the concept of *cubano*, for the concept of *cubano* started with and was squarely based on notions of race, place and language.

## *Criollismo* in las tres Antillas

In the nineteenth century, *criollismo*, as an ideology, symbolised independence in the Spanish Caribbean islands. It created itself out of several elements – first, the native element, which attracted to itself the notion of 'Indian' and contrasted itself to foreigner (including both *peninsular* and *bozal*); second, the mulatto element; third, the natural element; and, fourth, the linguistic element. The promotion of *criollismo* was an attempt to assert a Spanish Caribbean identity by those who for political or economic reasons no longer wanted to be Spain's dependents. Yet, while *criollo* in the Spanish islands was repeatedly said to refer to all races, *criollismo* was more of a white intellectual movement than a spontaneous expression of national or regional political identity among the people at large.

The re-creation of the 'Indian' (in *ciboneyismo*) was the doing of intellectuals, but was very nineteenth century in that this new 'Indian' represented the victimised Caribbean native who was then re-emerging, in the person of the *criollo*, as a fiercely free and independent character. At the popular level, the victimised ethnic group that emerged out of oppression in the nineteenth century was actually the African slave, but at the intellectual level, in the Spanish islands, this emergence of the African was transposed to the historical 'Indian', almost producing an intellectual equivalence between the real (African) *mambí* guerrilla and the mythical ('Indian') *siboney* warrior. Before Ortiz at the end of the nineteenth century, the very word *mambí* had come to be regarded as a word of indigenous origin and thus the concept of *mambí* came to be associated with the 'Indian' warrior. All this was possible because in the Spanish islands two worlds had persisted – the world dominated by sugar and the African slave on the one hand and, on the other, the non-sugar world, which at the beginning of colonisation had practically enslaved the indigenous inhabitants and in which indigenous concepts and themes had endured. The two worlds had persisted both in terms of the actual economic/agricultural reality and because of the need on the part of a white-oriented world to make a separation from Haiti, African, black and slave. Yet, as was inevitable, the two worlds had become intertwined.

While at the intellectual level (i.e. as a result of research) the 'Indian' had been re-created with 'accurate' names (e.g. *siboney*), at the popular level the indigenous names that had become popular (*guajiro*, *jíbaro*) were difficult to explain and justify as native to Cuba and Puerto Rico, so much so that there were attempts to give them a European origin. Paradoxically then, the *criollo* in Cuba and Puerto Rico took on a contradictory ethnic aspect through both intellectual and popular adoption and use of specific indigenous names. The *criollo*, as if by absorbing characteristics of the indigenous inhabitant, had changed ethnically without becoming 'Indian'. In the Dominican Republic at the intellectual and political levels the 'Indian' was systematically imposed on the people as a racial category and so *indio* began to become a general racial type replacing black and its derivatives. In this context *indio* was understood to be a euphemism, but at the same time it was felt to be part of the non-sugar, non-slave world in the Dominican Republic.

In the relationship between the two identities (*criollo* and 'Indian'), therefore, there was a reflection of the contradictory realities and attitudes of the people. On the one hand, a term like *siboney* was accurate in its reference to the original natives of the islands, but, on the other, was mythical (untrue) in its reference to the then current natives (*criollos*) of the islands. The word *criollo* itself exemplified in the nineteenth century a proud recognition of reality while the indigenous names demonstrated both a need for the exotic and a need to preserve an integral part of the culture of the islands. However, what was very significant in the popular use of the names *guajiro* and *jíbaro* was that there was no consciousness of them being indigenous in origin or connotation. These were names that had genuinely evolved with the people, the *criollos* of the islands.

The words *guajiro* and *jíbaro* had also survived as words for domestic animals like horses, pigs and cats that had strayed away and had become tough and independent. It is this very usage that sheds some light on the development of the terms to refer to characteristic types of people in the Spanish islands. This usage recalls the application of the term *cimarron* to indigenous inhabitants who had escaped from enforced labour in the early Spanish settlements and who in European eyes became wild (again). So, those outside the discipline and control of the European colony were likened to the indigenous inhabitants. In other words, the original concept of *cimarron* had been shifted to and preserved in the indigenous terms *guajiro* and *jíbaro*. It is not simply that the tough, independent life of the *campesino* attracted the label of *guajiro* and *jíbaro*, but that the concept of *cimarron* had from the earliest days (on the *banda norte*) been associated with wild cattle and a rough unsophisticated life. It was therefore those persons who were involved in cattle rearing, above all, who attracted the labels *guajiro* and *jíbaro*.

The 'Indian' element of the *criollo* was basically, primarily and inescapably linked to the notion of native of the Americas. It came to be associated with cattle rearing and tobacco growing in spite of the fact that the indigenous inhabitants themselves had had little to do with cattle rearing and tobacco growing. It was as if they became a part of this life because they, as enforced labour, contrasted with the African, who was inseparable from sugar. In other words, since the world of sugar contrasted with the world of tobacco growing and cattle rearing, and the 'Indian' contrasted with the African, the 'Indian' came to be associated with the non-sugar world. The *criollo*, as native of the Americas and bolstered by the element of 'Indian', inevitably also came increasingly into conflict with the European, whose culture had continued to diverge over a period of centuries from that of the *criollo*. As a result there was quite a deliberate formulation of a *criollo–peninsular* contrast, with a resentment of the *peninsular* and of Spain, which in itself was to become a major spur in the development of *criollismo*.

The *criollo,* as native, came into conflict with *bozal* and by extension with the negro generally. The evolution of *criollo* as white native started from the concept of *criollo* as new native – a contrast with old native (indigenous inhabitant) on the one hand and immigrant/foreigner (European) on the other. The concept of *criollo* as first-generation American linked to two worlds (the Old and the New) persisted not only because first-generation *criollos* no doubt continued to be a sizeable part of the population but also because ties with Europe (e.g. language) were important. At first, by and large, *indio* ('Indian') was seen as different, as constituting a different race and as having no immediate cultural connection with the Old World. However, with the increase in the number of mulattos, with the extension of *criollo* beyond first-generation American, with the increase in the importance attributed to *tierra*, and with the fluidity of the concept of 'white', *criollo* became nearer and nearer to mulatto.

In the Spanish islands it was the 'mulatto', who, above all others, by force of circumstances had to be the first to regard the New World as patria/native land, for the white *criollo* of the upper class looked to Europe as the source of enlightenment and the African continued to believe in the soul returning to its homeland (Africa) after the body died. Even some of the creole slaves held on to the beliefs of their African parents and grandparents and as such were not as tied spiritually to their New World place of birth as they are often assumed to have been. It is not accidental therefore that Hazard, for example, equated 'mulatto' and 'native' ('Both the man and the woman were mulattoes or natives ...' (1873: 341). Through consistent interbreeding, the 'mulatto' moved away from being a direct product of black and white to become what was identified as a distinguishable racial type. The presence of this racial type was dominant in the Spanish islands because it constituted the overwhelming majority of the population, and it was this numerical dominance of mulattoes that initially propelled the move toward an 'internal' explanation of *criollo* as opposed to an 'external' explanation that used historicity of race and linked people either to Europe or to Africa. Eventually, this explanation reverted to historicity, through the use of the concept of *Indio*, to give further credibility to the mulatto.

The evolution of the concept of *criollo* as a New World phenomenon was accompanied from the earliest times by the notion of natural evolution and had *tierra* as one of its critical determiners. When the mulatto emerged as the dominant racial type in the Spanish islands, the notion of *tierra* was central to the explanation of uniformity of outlook and behaviour across all *criollos*. While the Darwinian approach provided a general background for the centrality of *tierra*, it is the preponderance of mulattos in the Spanish islands that heightened its importance as a determiner – this new racial type required an 'internal' or *criollo* validation. The 'internal' explanation thus used *la tierra* or the influence of climate as the critical ingredient in the make-up of what was perceived as a distinctive racial type. It was essentially an intellectual and literate explanation, which through repetition came to be accepted by the people themselves. It can be interpreted as an attempt to de-emphasise the historicity of race and to promote nature or, more specifically, place (*la tierra*) as one of the most important features in the New World construction of identity. It can also be interpreted as a particularly Spanish phenomenon – a need by people whose vision had always been regional rather than central to establish their own *pueblos* wherever they went.

The preference for an 'internal' explanation and the highlighting of *tierra* was also a result of the fact that the appeal to historicity did not give Spanish descendants at that time any comfort in the notion of purity of race. From at least the last quarter of the eighteeenth century (in Sanchez Valverde 1785) to the middle of the nineteenth century (in Lemonnier-Delafosse 1846), the notion that the Spanish were not of pure race was expressly stated and seemed to be a very strong belief among the other European nations, who used it both to account for the decline of the Spanish and to explain their barbaric annihilation of the indigenous inhabitants. In fact, Lemonnier-Delafosse (1846: 223) specifically said that they had inherited their bloodthirstiness from their Bedouin African ancestors (i.e. the Moors).

Yet, there seemed to be other factors that led to the preoccupation with *tierra* as an explanatory force. From the beginning the Spanish were fascinated by the flora and fauna of the New World: these were the products of, or directly dependent on, the earth. The indigenous inhabitants were also seen to be close to the earth. The effect of the climate and the land was regarded as overpowering by Europeans, as, for example, the heat of the sun, the power of hurricanes, the bright colours of the flora, the fierceness of the insects. These things and others suggested that the European had to be modified to survive. Even the indigenous inhabitants were seen as too weak to survive the rigours of the climate if they had to work hard. White *criollos* adjusted to the climate because they were bred up in it and mulattos were ideal for it, presumably because they would have inherited from the African the physical toughness needed for survival. Such practical reasoning may have been supplemented by more metaphysical beliefs. In commenting on the myths and beliefs of the Tainos, José Arrom says: '*Yúcahu Bagua Maórocoti, el Señor de los Tres Nombres, el Icono de las Tres Puntas, en sí resume los tres factores primordiales que felizmente se armonizan en las Antillas: tierra, mar y hombre*' ['Yúcahu Bagua Maórocoti, lord of the three numbers, icon of the three points, in him come together the three primordial factors which happily harmonise in the Antilles – land, sea and man'] (1989: 43). The role of *tierra* in this trinity, which had become a part of the cosmos of the Tainos, probably emerged as self-evident also to *criollos*. This belief in *tierra* as a spiritual and formative force is quite distinct from the strong territoriality instinct in Europeans, which equated the identity of people with land (e.g. English – England; French – France).

The language presented as the soul of *criollo* identity in a sense contradicted the 'internal' explanation, in that Castilian Spanish was not the native language of the *criollo*. However, there did not seem to be a significant enough difference between the language of the mother country and the language of the white *criollos* of the colonies for them to proclaim independence and distinctiveness for *criollo* language. *Criollo* linguistic identity therefore continued to depend on historicity to confer validity and prestige, and this meant that deviance from the perceived European norm had to be socially linked to inferiority. In other words, the elite *criollo* linked himself to the European norm and considered peculiarities (consistently regarded as peculiarities of the lower social classes) as deviations from the norm or necessary for local communication. The peculiarities of Puerto Rican, Dominican and Cuban speech were linked to varieties elsewhere in the Americas, and then they were all linked to regional varieties in Spain. Where differences were undeniable and non-Spanish, they were treated as foreign, either borrowings for indigenous items or the production of unassimilated Africans or *bozales*. In fact, the easy acceptance of the notion of 'broken' language (similar to the *langage corrompu* in the French islands) as the medium of communication among *bozales* precluded serious consideration of any other source but the European language for local varieties of language, especially those spoken by white *criollos*. The fact is that most local peculiarities could not but be seen as 'corruptions' of Spanish, and corruption was seen as the consequence of black presence and general lack of education and sophistication. Identity of *criollo* (person) and actual *criollo* (language) was therefore not tenable in the Spanish islands since *criollo* (language) suggested black slaves and uneducated *campesinos*. In any case, there was no identity of name of language and name of people, because there was neither the one nor the other. So, though *criollo* was promoted as a conception of the new independent native of the Spanish islands, in direct contrast to Haiti it was not extended to the actual language spoken by Cubans, Puerto Ricans and Dominicans generally.

Castellanos and Castellanos (1988: 263–4) put forward the argument that the Cuban intellectual elite in the nineteenth century, being aware of the fundamental unifying function of language in the development of national identity, regarded slavery and the presence of thousands of Africans in Cuba as a serious threat to Cuban identity. According to this argument, the anti-slavery sentiments of the well-known essayists of the day were really the outward expression of a deep-seated desire to remove Africans from Cuba and to get rid of the native languages of the Africans and their *habla bozal*, which were an obstruction to an emerging national identity. The intention was not just to whiten Cuba but to have a pure Spanish language spoken by all. It can be said that attempts to whiten the other two islands had the same objective. In the case of Santo Domingo, this nexus between whiteness and Spanish was almost obsessive among the elite and led to the elimination of *negros* from the country and their replacement by *Indios*. It would seem as if this were a perpetuation of the mentality of the Spaniard who had expelled the Moor from his country and wanted to behave as if the Moor had never existed, had never blackened his language, and that his language had resisted the African and would do so again. There was therefore little acceptance of Ignacio de Armas' idea of a *lenguaje criollo*.

While there were leading nineteenth-century writers whose work can be judged as linguistically prejudiced and racially insidious, there were others, like Esteban Pichardo, whose Cuban dictionary was a major step in the illustration of a more representative *criollo* identity. Moreover, it demonstrated a recognition of the fact that specifically Cuban linguistic and cultural features had developed across the country in significant numbers to constitute a peculiar linguistic identity. In addition, the fact that a *criollo*, who had been born in Santo Domingo and was living in Cuba, was able to complete this work meant that *criollo* letters had reached a reasonable level of sophistication, a fact that was evident from the recognition of the dictionary by the Spanish Academy in 1849. While Pichardo viewed what he called corruption of the Spanish language as typical of the other races present in Cuba (i.e. the Africans and the Chinese) and deplored it among people he regarded as cultured, his identification of indigenous words in the normal vocabulary of Cubans suggests that the notion of recovering pure Castilian was not any deep-seated desire on his part.

Writers used the term 'creole' to link all America, and *criollo* was used to link the countries of Spanish America. Cuba, Santo Domingo and Puerto Rico were three different entities, but they were also perceived as having a special link. Maldonado-Denis, commenting on Eugenio María de Hostos, one of the most noted nineteenth-century promoters of *criollismo*, said: '*Hombre de profundo sentido patriótico cree, al igual que Martí, que no hay mar entre Cuba, Santo Domingo y Puerto Rico*' ['A man of profound patriotic sentiments who believes, like Martí, that there is no sea between Cuba, Santo Domingo and Puerto Rico'] (1992: 46). Such a belief was based primarily on the existence of the Spanish language in all three of these places, the nearness of them and the fact that they had had a similar history and position in Spanish colonisation, all three having been tied in their early colonial history to the *audiencia* of Santo Domingo. When Santo Domingo asked to be annexed to Spain in 1861, not surprisingly it was the Spanish press that highlighted the importance of linking the three islands: 'And with this island [Santo Domingo] stretching its right hand to Puerto Rico and its left hand to Cuba, we commence a new system, giving us the control of the Gulf' (Quoted in Hazard [1873] 1974: 255 from the *Cronica* of 1861). No doubt, the geo-political importance of the three islands as a block led Hostos, who took up residence in the Dominican Republic toward the end of the century, to see the three islands as a cultural bloc. The perception of them as a unit could not have been based on any similarity of economy at the time or the state of politics or common aspirations for reform or independence. In effect, it was the perception of a common link in language and in geography that led intellectuals to propose *las tres Antillas* as a single unit. *Antillas*, as a name for the islands, was being resuscitated, but at the same time it was being limited by language to three of them.

*Las tres Antillas* were seen not only as a unit in themselves but also as the centre of Spanish America (or *Nuestra América*) and the birthplace, as it were, of the *raza latina en el Nuevo Continente* (Hostos: Diario tomo 1, p. 253; Maldonado-Denis 1992: 48). This *raza latina* was a concept based on a perceived commonness of racial mixture, climate and language (i.e. *mulato, naturaleza, español*). *Las tres Antillas* and the *raza latina* were projected as the place of the coming together of and the final melting-pot of the different races, as is clear in Maldonado-Denis' explanation of María de Hostos' thought: '*medio geográfico natural entre una*

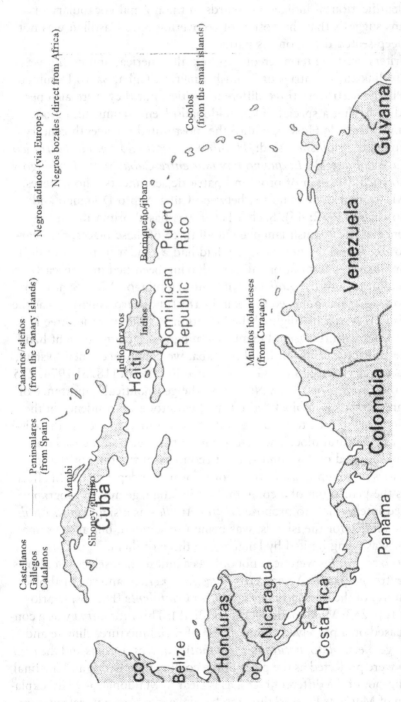

**Figure 5.2** A simplified picture of some of the ethnic influences in *las tres Antillas* (Cuba, Dominican Republic, Puerto Rico)

*y otra fusión transcendental de razas, las Antillas son políticamente el fiel de la balanza. El verdadero lazo federal de la gigantesca federación del porvenir; social, humanamente, el crisol definitivo de las razas.'* ['natural geographical medium between two transcendental fusions of races, the Antilles are politically the needle of the balance, the true federal bond of the gigantic federation of the future; socially and genetically the definitive melting pot of the races.'] (1992: 46). In this conception, *raza latina (en el Nuevo Continente)* could not simply mean 'white Iberians' or Europeans. The word *raza* was actually intended to refer to a new racial type and in the context it had to be *criollo* (European + 'Indian' + African). The word *latina* obviously meant Spanish and was intended to add sophistication as well as confer historicity, but, as a melting-pot feature, it had to be nearer to *ladino* (foreign Spanish) rather than to Castilian or European Spanish, even though the notion of pure Spanish was strong among Antillean intellectuals at the time. The word *naturaleza* preserved the notion of *tierra* and its role in the formation of the *criollo*. In fact, it was given an even grander role in this nineteenth-century formulation of identity – '*En las Antillas, la nacionalidad es un principio de organización en la naturaleza ...*' ['In the Antilles, nationality is a principle of organisation in Nature ...'] (Maldonado-Denis 1992: 46). So, the intellectual elite of the Spanish islands managed to construct an identity in which they occupied the central spot in a newly envisioned world. It was an identity in which a specific language was the unifying force and the tropical climate was the modifying force; it was for them a perfect union of the internal (language), the human being (three races in one) and the external (Nature).

One difference that separated Puerto Rico from the other two islands was the lesser intensity of the spectre of the sugar plantation with the black African slave there. It is not simply that the Puerto Rican population had been more successfully whitened than that of its neighbours and that it was the most recent layers of white, most of whom had no raw feelings generated by the sugar plantation, that came to dominate the country. It is also the case that, unlike Cuba, Puerto Rico did not see a return to a dominance of the sugar plantation and its attendant mentality, and, unlike the Dominican Republic, the spectre of the black African was not an ever-present threat in the imagination of the people. Ironically, it was the spectre of the sugar plantation and the sentiments that it wrought in all who were affected by it that impelled people more strongly to assert independence. It is not accidental therefore that Puerto Rico never forcefully asserted its political independence as the Dominican Republic had done and Cuba was to do.

# West Indians and Créoles in the smaller islands

## Diversity, uniformity and 'improvement' in the English islands

> If ever the naval exploits of this country [England] are done into an epic poem – and since the Iliad there has been no subject better fitted for such treatment or better deserving it – the West Indies will be the scene of the most brilliant cantos.

This statement, written with a feeling of sadness and nostalgia at the end of the nineteenth century by Froude ([1888] 1969: 10), reflected the feelings of decline that Englishmen were experiencing and their desire for a glorification of past exploits, especially their greatest triumphs. France was even worse off because it lost its North American colonies, it lost some of its island colonies and, the greatest blow of all, it lost Saint Domingue. Aube expressed similar feelings of nostalgia and sadness:

> *La Martinique et la Guadeloupe sont, avec quelques îlots insignifiants: les Saintes, la Désirade, Saint-Barthélemy, Saint-Martin, **tout ce qui reste à la France de cet empire colonial** qu'elle possédait autrefois en Amérique; empire, le mot n'est que juste. Le Canada, la Louisiane, Saint-Domingue, la reine des Antilles sous le ministère du duc de Choiseul, en attestent la vérité. Les souvenirs glorieux que ces noms évoquent sont à la fois la consolation et la condamnation de notre orgueil national ...* (1882: 1)

> [Martinique and Guadeloupe are, with a few little insignificant islands – the Saints, Desirade, St. Barthelemy, St. Martin – all that remains to France of that colonial empire that it once had in America; empire, that's the correct word. Canada, Louisiana, St. Domingue, the queen of the Antilles under the ministry of the Duke of Choiseul, attest to the truth of this. The glorious memories which these names evoke are simultaneously the consolation and condemnation of our national pride ...]

For Aube, the islands had become an annoying reminder of the former greatness of the empire of France. As a result, throughout the nineteenth century, they assumed greater symbolic significance as a reminder of what could have been, but could not make up for the great losses suffered. Yet, while there was Froude's grandiose vision of the West Indies, it is also the case that what Dauxion-Lavaysse said of St Vincent at the beginning of the nineteenth century was true of all the Caribbean islands of Britain at that time: 'Though it has belonged to Great Britain for a long time, it is only since the American war it has acquired its actual colonial importance' ([1820] 1969: 439).

The concept of national identity, which had become a political reality in the United States in the last quarter of the eighteenth century and in Haiti in 1804, gradually began to spread across all the islands in the nineteenth century. As a result, *West Indian*, as a political term, came into its own more and more to identify Creoles in the British islands as opposed to those in the French and other European colonies. At the same time, influence from the United States led Day to claim that West Indians were becoming more American and Froude and Trollope to despair in the same belief. The influence from Haiti led Hearn, at the end of the century, to voice his resignation at the prospect of the take-over of the islands by the negro race. Individual island identities were becoming more distinct, but neither regional nor island identity superseded the kind of racial identification that had developed as a fundamental part of the conception and description of the slave colonies. Race continued to be the most dominant characteristic of identity. Emphasis was not on uniform characterisations of all the natives of specific places: descriptions in books continued to list racial and ethnic groups in each island and to describe them in turn.

In the nineteenth century the English and French islands were seen as being made up of Jamaica to the north west, Barbados to the south east and in between a string of islands referred to as the *Antilles*. The Bahamas and other small islands to the north of the big islands did not figure prominently in the literature and for the most part seemed to be non-existent for Europeans. While Jamaica and Martinique were generally known to belong to the English and the French respectively, the number, smallness and lack of distinctiveness of most of the other islands preserved general ignorance about their existence, their names, their owners, their location and their sense of identity. The inability of foreigners to distinguish between the many islands led to the practice of stereotyping people across the chain of islands. This was epitomised in Froude's statement at the end of the century – 'The characteristics of the people are the same in all the Antilles ...' ([1888] 1906: 51).

Lack of discrimination, generalisation and stereotyping materialised in the repetition of material taken from preceding works as books about the English islands increased in number. It was normal for authors of the time to repeat incidents, names and examples of language that had occurred in contemporary works written in another language – English authors quoted from the French and vice versa. In many cases the repeated material was not only unsubstantiated for its accuracy but also was presented by the author as if it were new and specific to the context. The clearest case of reproduction of previous material is in Wentworth (1834). Wentworth in his travels went from island to island and for each place produced local speech as if it were speech he had heard and noted down. Not only did he get material from lesser known predecessors but he also rearranged it to make it fit his own account. Note the similarity in the following relating to Antigua. Johnson in 1830 [the pages are not numbered] wrote:

> On such occasions they will say 'Massa, him no fu' kill, him b'long me granny' or 'him too ñoung' or 'him too ole fu' kill' or 'me keep him fu' luck massa, him odd 'un'

Wentworth (1834 2: 204) had:

> 'Him too ñoung fu kill, or him too ole fu kill, aw keep him fu breed' or, 'him no 'blong me, him 'blong me granny' or ... 'aw keep him fu luck, him odd un'.

Clearly Wentworth looked at the earlier description of Antigua and repeated what was said with only slight variation. The most extensive case of 'borrowing' by Wentworth is in relation to St Kitts – in this case it was Mathews (1822) that was the source. In addition to 'borrowing' from these writers, Wentworth repeated material with slight variations for different territories. This tendency to replicate incidents throws doubt not only on the authenticity of them but also on the accuracy of the citations. Arising out of this kind of copying and repetition, there developed certain attitudes, styles and practices, which succeeded in representing not only the customs and culture of the slave population but also some characteristics of the white Creoles in a generalised and familiar way.

Paradoxically, the island that was most closely associated with England was the one that had the greatest measure of autonomy, the one whose natives were the most identifiable and the one whose natives consistently identified their island as their native land. Less than a hundred years after the settlement of Barbados by the English there was a big enough community of native-born whites for the term *Barbadian* to establish itself, as a reference by John Oldmixon attests. What was even more extraordinary was that Oldmixon also associated the enslaved population with the term: '40000 of them [Negroes] are Natives of the Island, as much Barbadians as the Descendants of the first Planters, and do not need such a strict Hand to be held over them as their Ancestors did ...' (Oldmixon 1708 2: 14). *Barbadian* was therefore one of the earliest new national identities to be mentioned by European writers writing about the New World.

The white natives of Barbados, at this early stage, saw themselves and their island as special and distinctive in a whole 'new world', where the major distinctions of identity were still between the indigenous inhabitants, the people who came from 'old world', and the children of these latter who were born in the 'new world'. By the end of the eighteenth century, this sense of distinctiveness was being strongly expressed across the Barbadian population, even by the slaves, according to Pinckard:

> This sense of distinction is strongly manifested in the sentiment conveyed by the vulgar expression so common in the island – 'neither Charib, nor Creole, but true Barbadian,' and which is participated even by the slaves, who proudly arrogate a superiority above the negroes of the other islands! Ask one of them if he was imported, or is a Creole, and he immediately replies – 'Me neder Chrab, nor Creole, Massa! – me troo Barbadian born.' (1806 2: 76)

Pinckard observed that the poor whites also expressed strong sentiments about their nationality: 'They do not even admit themselves to be **Creoles**, but they are "Barbadians" – a something distinct and superior – a something different from, and unlike the inhabitants of the other West India islands!' (1806 2: 133). The identity of the people became so well known that they were referred to by familiar names – *Badian* (Dickson 1789: 108, Pinckard 1806 2: 133), *Bimms* (Day 1852).

As a result of their fame or notoriety, general characterisations of Barbadians became common in nineteenth-century accounts. A typical though lengthy one was that by Frederick Bayley:

> I will take this opportunity of giving the character of a Barbadian, in which hospitality forms no inconsiderable trait.
>
> A Barbadian resembles in no point a Creole of the other islands; his manners, his feelings, his ideas, and even the tone of his voice is different. He possesses much good nature, an open heart, strong feelings, and, generally speaking, is greatly attached to his family. He is also somewhat patriotic; and it would require much argument to convince him that any other island in the West Indies can be of consequence equal to the land of his birth, which he proudly remembers is the first and most ancient of all colonies of the mother country. Barbados was always in the possession of England; was never attacked by the armies of a foreign power; and only once by the forces of the long parliament. There are, therefore, not a few Barbadians who can boast a long pedigree, and trace their line of descent as far back as the times of the first and earliest settlers in the island. They

also justly pride themselves on their loyalty to their sovereign; and on having afforded, in ancient times, a refuge to the royalists. A Barbadian has much firmness of character, and what the world would call a nice sense of honour. Indeed, I know no class of people more impatient under an injury, or more quick to avenge an insult. Attachment to the Church of England, and detestation of the Methodists, are marked features in his character, over which prejudice has some influence. I never knew a Barbadian easy out of his own country; in other places trifles annoy him, and his general remark is, 'How different from Barbados!' nevertheless, he is industrious, and strangers may live in his island without troubling himself with their affairs, I wish I could say as much for some others. His countrymen make use of some odd phrases, and have a tone of voice entirely peculiar to themselves; so that after having visited the island you might easily know a Barbadian, if you met one in society, in any other quarter of the world; in such a case, address him frankly; tell him where you have been; talk to him of his own dear country, – praise it (for you may do so without flattery), and you will make a friend of him at once. (1833: 48–49)

This kind of positive characterisation of the natives of Britain's favourite colony by a British writer no doubt was partly the result of a mirror-image perception and a perpetuation of the eighteenth-century concept of Barbados as 'Great Britain itself in miniature' (Keimer (1741: iv).

In contrast, the people of Jamaica were not even called *Jamaicans* normally. Admittedly, there are in the first half of the eighteenth century two references to *Jamaicans* in discussions of maritime activities – 'the Jamaicans fit out vessels to fish upon her' (Atkins 1735: 225); 'which the Jamaicans use in boats to kill guanas' (Atkins 1735: 252) – but in both cases the term seems to mean no more than people from Jamaica. In Long's (1774) three-volume work on Jamaica the designation 'Jamaican' was not used; neither was it used more than 50 years later in Madden's two volume work. Creoles were being divided and described separately within colour divisions, as if they belonged to different castes, and non-natives were identified as belonging to different nationalities. At the beginning of the nineteenth century Renny did use the term: 'The manners of Jamaicans, both Creols and British settlers, differ very sensibly from those of their fellow-subjects in Europe' (1807: 209). For Renny, however, *Jamaicans* was a British colonial term rather than one that indicated uniformity across the population.

Around the time of emancipation, Madden presented Jamaica as a plural society with 'English, Irish and the Scotch residents ... the native Creoles, the French emigrants from St. Domingo, the Spanish settlers from the Main, the Jews, the free coloured population, and the negro slaves/the apprenticed labourers' ([1835] 1970 1: 80). He summed it up by saying: 'There is therefore a multiplicity of complexions, a multiplicity of creeds, a multiplicity of tongues, and, most unfortunately, a multiplicity of interests' ([1835] 1970 1: 80). Madden remarked that the divisions in the society were too deep to allow for general cooperation: 'the ignorance of the negro, the arrogance of the brown man, and the pride of the white, will continue for some time to baffle the endeavour to amalgamate their interests ... The blacks dislike the browns, the browns look down upon the blacks, and the whites have no love for either' ([1835] 1970 1: 144–5). No strong white identity was seen to be there; mulattos were seeking to consolidate their social position; there were unconquered Maroon elements within the island; and the black (former slave) population had been fashioned into plantation and village communities or had acquired a certain measure of independence for themselves.

It is not that there had never been strongly expressed local sentiments in Jamaica, for, as Patterson points out:

> By 1700 local patriotism was so pronounced among the early settlers, and their children who became known as the 'creole party', that on several occasions they attempted to exclude English-born persons from filling posts in the island and even went as far as declaring that they 'will not allow themselves to be called Englishmen'. (1967: 34)

These sentiments, however, had come from the old West Indians 'transplanted' from Barbados and other eastern Caribbean islands, whose dominance declined as they died out or left. Then, in the eighteenth century in contrast, return to England with wealth was not only the desire but the reality for many whites in Jamaica.

Accordingly, this was an island in which whites held on longest to the notion of England as home. Trollope observed that:

> Nothing is more peculiar than the way in which the word 'home' is used in Jamaica, and indeed all through the West Indies. With the white people, it always signifies England, even though the person using the word has never been there. I could never trace the use of the word in Jamaica as applied by white men or white women to the home in which they live, not even though that home had been the dwelling of their fathers as well as of themselves. The word 'home' with them is sacred, and means something holier than a habitation in the tropics. It refers to the old country. (1860: 100)

Although Trollope was repeating a point that had been made by previous writers, his direct application of it to Jamaica was based on his own experience and reflected reality among whites in Jamaica, as attested in the degree of absenteeism among Jamaican plantation owners. Bear in mind, however, that James (1963: 56–7), in reference to French proprietors in pre-revolution Haiti, had related degree of absenteeism to degree of economic success.

Ironically, Trollope felt more comfortable in Jamaica than anywhere else, regarding it as even more English than Barbados:

> In Jamaica too there is scope for a country gentleman. They have their counties and their parishes; in Barbados they have nothing but their sugar estates. They [Jamaica] have county society, local balls, and local race-meetings. They have local politics, local quarrels, and strong old-fashioned local friendships. In all these things one feels oneself to be much nearer to England in Jamaica than in any other of the West Indian Islands. (1860: 96)

This was indeed a contradiction of the generally held views about Barbados and Jamaica and their degree of acculturation towards England. Nevertheless, it pointed to a kind of social polarisation that indeed characterised Jamaican life and identity. For instance, after emancipation the term *Creole* came to be associated with just the non-white part of the population in Jamaica, according to Underhill: 'It is greatly to be regretted that it has become the habit of the coloured Creoles to speak of the white population as foreigners, although a large proportion of the whites are as much natives, being born in the island, as the Creoles themselves ...' ([1862] 1970: 225). Trollope's comments, then, about the English-ness of Jamaica could be taken as applying to the lifestyle of Jamaican whites.

Though writers did not convey a characteristic feeling of Jamaican-ness or Creole-ness across the whole population, yet throughout the nineteenth century the familiar belief persisted that climate moulded all native-born inhabitants and made them similar. At the beginning of the century, Dallas suggested the following as an explanation for the behaviour of (white) Creoles in Jamaica: 'Many causes of national character are so mixed as to be almost inscrutable. It may, perhaps, be partly ascribed to the sensibility that a warm climate excites, that Creoles are said to be impatient of subordination, and addicted to juridical controversy' (1803: CX1V). This comment showed the extent to which belief in the effects of physical elements on man had come to dominate contemporary thinking. Indeed, Dallas went on to identify Creoles as a distinct race with similar social characteristics.

An earlier eighteenth-century view of the character of creole slaves had highlighted the acculturation process by putting emphasis on the influence of the master in the shaping of the creole slave: 'Like master,

like man' (Long [1774] 1970 2: 404). Long's intent was really to validate a distinction he made between African slaves and creole slaves, by
claiming that creole slaves, in their upbringing, developed an attachment to their master: 'On every well-governed plantation they eye and
respect their master as a father, and are extremely vain in reflecting on
the connexion between them' ([1774] 1970 2: 410). He went on further
to argue that creole slaves even had a measure of patriotism: 'They
seem also to feel a patriotic affection for the island which has given
them birth; they rejoice at its prosperity, lament its losses, and interest
themselves in the affairs and politics that are the talk of the day'
([1774] 1970 2: 411). Another eighteenth-century view, given in
M'Neill (1788/1789), presented the creole Negro in Jamaica as influenced by whites but also as an evolved composite of different Africans:
'The character of the Creole may be considered as the general character
of the Negro. He possesses the cruelty of the Coromantee, softened by
his intercourse with the Whites; and the subtility and craftiness of the
Eboe are in him enhanced by a cunning which propensity and constant
practice render superlative ...' (p. 26).

Association with whites as well as the perceived salutary effect of
the climate was also seen by *Marly* to produce a creole negro, superior
and also perceptibly different in physical appearance:

> The Maroons and the long-creolized negroes, have greatly improved, both
> in features, in size, and in strength from the African race; and in the course
> of years they will become the Georgians of the negro race. The thick lips,
> and the broad nose, are giving place to what Europeans consider beauty;
> and probably, after two or three generations have elapsed, they will have a
> resemblance to their masters in every respect, except colour. (1828: 87)

Wentworth repeated the same idea a few years later, though not specifically about Jamaica: 'The difference between a creole and an African
negro, as well in respect to their disposition and habits, as to their
physiognomy and general character, is palpable in a striking degree ...'
(1834 2: 219). This idea of 'improvement' in the Creole of course was
in keeping with the overall philosophy of *lineal ascent* to whiteness, but
it was as if the *ascent* was being brought about additionally by climate
and upbringing. So, in Jamaica on the one hand there was identification
of diversity according to race and language and on the other uniformity
according to ecological and social experience.

In the presentation of identity in Trinidad in the nineteenth century,
there was also a contradiction between uniformity and diversity.
Gamble used the term *Trinidadians* and characterised them thus:

> The Trinidadians are a tall, well-made people, rather showy in their
> persons, fond of dress, music, and amusements generally, especially

dancing and theatricals. They have a frank manner about them, but are somewhat fickle, and lack stability of character and firmness of purpose. The liberated classes and their descendants cannot be said to have attained to any high standard of moral character. If the truth must be told, they are litigious, and somewhat lax in the principles of honesty... It is not from want of example, that the Trinidadians are not more frugal and thrifty; for most of the immigrants, the Coolies in particular, are saving and frugal. But notwithstanding example set them, and in spite of the consequences, the Trinidadians remain as I have described them. (1866: 30)

This characterisation of the people of Trinidad in the nineteenth century was remarkable, not in terms of its truthfulness then or subsequently, but because it was made in a context of unusual ethnic diversity and newness of population.

What nineteenth-century English writers indicated above all about Trinidad, the most southerly of the islands, was the diversity of the people. Collens (1888: 9), for example, reproduced the following from the *Trinidad Chronicle* of 28 January 1876:

The people of all nations, countries, races,
French, English, Spanish, Scotch, or Portuguese,
From Afric's or fair India's hotter shores;
Creoles, Coolies, Chinese, their language,
Manners, customs, everything so different.

In addition to the *nations* identified, there were a few families of aboriginal people who still survived. They were not only mentioned by De Verteuil but were also characterised as 'all speaking the Spanish language, and having preserved Spanish habits – fond of smoking, dancing, and all kinds of amusements, but above all, of the **dolce far niente**' (1884: 158). The overlap between what De Verteuil called 'Spanish habits' and Gamble's 'fond of dress, music, and amusements generally, especially dancing and theatricals' is clear, and indicates a continuity of ideation in the characterisation of the people of Trinidad in spite of diversity and constant change.

When the British took over the island in 1797, there were said to be about 18,000 inhabitants there, and in about a hundred years the population had swelled by 200,000. Trinidad became throughout the nineteenth century a country of immigrants. The non-native population in Trinidad thus was an influential proportion of the total population and by the end of the century it was actually more than the native population, many of whom were no more than first-generation Trinidadians. This remarkable diversity of ethnic groups, with considerable variation within each major group (European, African, Creole, East Indian [= from India] and Chinese respectively), living in a relatively small island, was unique in the Americas in the nineteenth century. Even more extraordinary is that even though the island passed from Spanish hands to British in 1797, it became even more French in culture and remained so for much of the nineteenth century, in spite of the fact that English was adopted for all public purposes in 1823. There was no other case in the Caribbean of a country being dominated by a culture of a specific European country without having been officially ruled by that country at some time. Even so, Trinidad still had three European cultures on show at the same time, which also made it unique.

An illustration of the differences between the Spanish, French and English habits in Trinidad, where all three could be easily compared in the first part of the nineteenth century, was given by Joseph:

> During the nights here, the English inhabitants seem to enjoy the delicious climate gravely, silently, and inactively; the French promenade and serenade each other; the few Spaniards we have here, often in the country spend half the night in playing the guitar, while their dames enjoy the luxury of a nocturnal bath, like the fair creoles of the neighbouring continent.
>
> And here it may not be out of place to remark, that of all the European nations which have colonized in this part of the world, the Spaniards seem to have adopted habits the best suited to the climate, while the customs of the English are the least wise of any.
>
> The food of the Spaniard is light; he is generally temperate in his drink, and dresses elegantly, yet his clothes are made of the slightest materials; he never sets the heat of the climate at defiance for the sake of fashion.
>
> The Englishman in this part of the world generally grumbles that he cannot get Leadenhall beef; he endeavours to have on his board the repast he has been accustomed to at home; drinks what is called London particular Madeira, i.e. Madeira genuine, save that one third of it is brandy; and were he to approach a dinner-table without a long coat made of broad cloth, he would, in this Island, be thought a barbarian by his countrymen. (Joseph [1838] 1970: 32–3)

The overall picture of language that was presented matched the diversity of the population, but French Creole was the dominant native language in Trinidad and the one that immigrants confronted and had to learn for practical purposes. English, however, was a growing force, for Gamble noted that while French creole was the most widely spoken language, nearly everyone could speak English. It was the very diversity of language that led to the decline of French creole in Trinidad and the beginning of the emergence of a variety of English as a general language of communication for all groups. In the case of those African, Chinese and East Indian languages where there was not a sufficient community of speakers in Trinidad, they had little chance of survival themselves, but they limited the use of French creole principally to inter-group communication. French creole itself, in spite of its obvious validity and pronouncements to this effect, was still regarded, by even its strongest supporter, J.J. Thomas, as principally a *spoken idiom*, a corruption and derivative of French, and the result of the adoption of a European tongue by a barbarous nation (Thomas [1869] 1969: 1). Even so, it was not this attitude alone that was responsible for its decline, bearing in mind its much longer survival in St Lucia and Dominica, where English came to be the official language. An additional factor was that since East Indians did not mix freely with (negro) Creoles, it meant that there was no general acquisition of French Creole among East Indians. According to Gamble (1866: 33), English became a lingua franca for East Indians who did not understand each other's languages.[1]

In the face of the cultural diversity, Froude toward the end of the century could not contain his distaste and that of his countrymen for Trinidad:

> We could not colonise it, could not cultivate it, could not draw revenue
> from it. If it prospered commercially the prosperity would be of French
> and Spaniards, mulattoes and blacks, but scarcely, if at all, of my own
> countrymen. ([1888] 1969: 66)
>
> The English come as birds of passage, and depart when they have made
> their fortunes. The French and Spaniards may hold on to Trinidad as
> home. Our people do not make homes there, and must be looked on as a
> transient element. ([1888] 1969: 74)

In repeating the familiar notion that home for the English was England, Froude may in fact have been conceding defeat in the matter of cultural ascendancy in Trinidad, but, on the other hand, it may have been a matter of maintaining a view of the uniqueness of the English. Note, for example, that in accounting for the beauty that had bloomed among French creole women, the Englishman Coleridge had earlier attributed it to elements of nationality and race other than English:

> The colored women here [in Martinique], as in St. Lucia and Trinidad, are
> a much finer race than their fellows in the old English islands ... I think for
> gait, gesture, shape and air, the finest women in the world may be seen on
> a Sunday in Port of Spain. The rich and gay costume of these nations sets

---

1.　Mohan, however, underscores the role of Trinidad Bhojpuri as a general language of
　　communication among East Indians in Trinidad speaking different languages: 'The
　　sudden confrontation between all these diverse varieties in Trinidad and the absence of
　　any accepted standard variety of Bhojpuri, the majority language, to unify the relo-
　　cated community led, not surprisingly, to linguistic adjustments ultimately aimed at
　　producing a single, somewhat homogeneous code from out of the diglossic zoo' (1978:
　　10).

off the dark countenances of their mulattos infinitely better than the plain dress of the English. ([1826] 1970: 141–2)

Coleridge made his point about race mixture and nationality even clearer when he said: 'The French and Spanish blood seems to unite more kindly and perfectly with the negro than does our British stuff' ([1826] 1970: 141). This was a view that was repeated later in the century 'so that it would really seem, as observed by Coleridge, that "the French and Spanish," and I would add the Danish "blood, seems to unite more kindly and perfectly with the negro than does our British stuff"' (Baird 1850 1: 29). Of course, this view of blood compatibility had an underlying suggestion that there was something unique about English blood that did not allow it to mix with others.

There was an intention then, partly out of pique and partly out of feelings of racial uniqueness, to dissociate the English from the diversity of Trinidad. This reflected depth in the cultural divisions and diversity in the island, but Gamble's general characterisation of *Trinidadians* pointed to a perceptible uniformity in the people that could only be explained as the result of local adaptation.

In some of the other older British islands, in contrast, writers were not too sure what the names of the people were – *Antigonians* or *Antiguans*? *Kittiphonians* or *Kittitians*? This uncertainty meant that only then, after two hundred years of colonial existence, was island identity starting to emerge in each island and people to be regarded as single national groups.

In the case of Montserrat, national identity was retarded because the people were constantly associated with the Irish in their speech. In commenting on the island of Montserrat, Coleridge made the following statement about the speech of the slaves: 'The negros here have an Irish accent, which grafted on negro English forms the most diverting jargon I ever heard in my life' (1826: 170). What was striking to Coleridge and others, like Wentworth, who called Montserrat 'Little Ireland' (1834 2: 269), was the association of a British accent with negro speakers. Evidently, there was no parallel of this elsewhere in the islands. Montserrat's special peculiarity was explained by its population composition in which the Irish had had early dominance. In the middle of the seventeenth century, Gardyner (1651: 75) pointed this out. When Sloane passed Montserrat in 1687 on his way to Jamaica, he also did the same (Sloane 1707: 41). As a result of this early and continued dominance, there was, according to nineteenth-century reports, a relatively strong linguistic influence of the Irish on the slave population.

Twenty-five years after Coleridge, Baird (1850 1: 43) made the same comment about Irish linguistic influence in Montserrat. He then went on to comment on Coleridge's statement and to give an anecdote in support of the Irish-ness of the Montserrat accent:

> Mr. Coleridge says of the accent, that it forms the most diverting jargon he ever heard in his life; but the following anecdote, well known to those who have visited the island, will best illustrate both its nature and its extent. Viewing, as the inhabitants of the Leeward Islands generally do, Antigua as the capital and head-quarters of their number, the negro who has 'emigrated' from Antigua to Montserrat talks of the length of time he has been 'out,' just as the Canadian or Australian emigrant does of the length of time that may have elapsed since last he saw the bold mountains of his native Scotland. And it is said that many years ago, when an emigrant from the Emerald isle was about to settle in Montserrat, he was surprised to find that the negro who was rowing him from ship to shore spoke with as pure a Milesian brogue as he did himself. Taking the negro for an Irishman, though a blackened one, and desirous of ascertaining the length of time that it took so thoroughly to tan the 'human face divine,' the Patlander addressed his supposed countryman with the question, 'I say, Pat, how long have you been out?' 'Three months,' was the astounding answer. 'Three months!' ejaculated the astonished and alarmed son of Erin – 'three months! And as black as my hat already. Row me back to the ship. I wouldn't have my face **that black** for all the rum and sugar in the West Indies.' (Baird 1850 1: 43)

Of course, the anecdote commented not only on Montserrat speech but also on Englishmen's beliefs about the Irish. Beside that, the Montserrat accent must have been regarded as very infectious for it to have been acquired in three months by the boatman. However, such exaggerated claims about the Irish-ness of Montserrat attested to a contemporary belief which had persisted from earlier times and shaped the self-image of Montserratians in the nineteenth century.

Montserrat was the only one of the northern islands that had speakers who are reported to have sounded strikingly distinct. No other island accent in the northern islands was singled out by writers of the day, but Day (1852) linked a Cockney characteristic of pronunciation to the whole group of islands:

> St. Kitts is the first island, coming north, where the v's and w's become perverted: 'vel,' 'vy,' 'vot,' 'wessels,' 'wictuals,' &c., roll out here in true cockney style. How this originated it is impossible now to tell. The inhabitants of St. Barts also are conspicuous for this unenviable peculiarity, and certainly with females setting up for ladies, it is rather remarkable.(2: 213–4)

No doubt the long colonial link of the Leeward Islands to Britain led to the association of British speech characteristics with these islands, as was the case also with Barbados. For this reason, whether the claims were erroneous or not, they kept the identity of the islands within the ambit of English colonial culture.

Nothing in particular was said about the speech of Antiguans, but Antigua was favourably compared in its uniformity of colonisation and in its progress to Barbados:

> Not only has this island been less subjected to invasion, and all the consequent casualties of warfare, than any other colony in these seas, Barbadoes only excepted, but it appears to have enjoyed greater freedom from domestic dissension, and a more general diffusion of intelligence among the leading members of its community, promoting a greater degree of social order, and a stronger bond of union, than have generally existed among the inhabitants of the neighbouring British islands. (Wentworth 1834 2: 178–9)

In the period after emancipation the positive image of Antiguans continued:

> The natural advantages of that lovely island; its climate, situation, and scenery; the intelligence and hospitality of the higher orders, and the simplicity and sobriety of the poor; the prevalence of education, morality, and religion; its solemn Sabbaths and thronged sanctuaries; and above all, its rising institutions of liberty – flourishing so vigorously, – conspire to make Antigua one of the fairest portions of the earth. (Thome and Kimball 1838: 52)

The projections of the anti-slavery movement apparently had been fulfilled in this exemplary colony, which had, unlike the others, moved directly from slavery to full freedom without going through a stage of apprenticeship and which had come under the presumed salutary influence of the Methodists.

In the case of St Lucia and Dominica, their identity was indeterminate: they were no longer French and they were not English and there was no uniform and strong feeling of being St Lucian or Dominican. In the case of St Lucia specifically, the game of musical chairs between the French and the British continued over a longer period than it did in Grenada, but with the French having the greater influence on the culture and identity of the island. It was in 1815 that St Lucia changed hands for the last time, finally being taken over by the British. After 1815, the British had an almost insurmountable task in trying to fashion St Lucia in their own image, because French law, religion and language were already deeply embedded into the culture of the people.

The English thought that the best way to overcome French influence in the former French islands was through the school system. They therefore concentrated their attention on children and disregarded adults, whom they regarded as incorrigible. This plan met with varying degrees of success, depending on the amount and extensiveness of schooling. It was hampered by the fact that schooling at the elementary level was usually purveyed through the Church, and in Grenada, Trinidad, St Lucia and Dominica, the people, and consequently their children, were officially Roman Catholic and did not willingly attend the Protestant schools. It was the view of English-speaking missionaries like Underhill that they were encouraged in this by Catholic priests: 'They speak a French patois, in which the priests encourage them, to preserve them from the influence of the English clergy and missionaries' ([1862] 1970: 96). By the end of the nineteenth century, however, the English policy was succeeding in Grenada and Trinidad, both of which were noticeably changing from French creole to English creole. The same was not the case in Dominica, because of a greater degree of general neglect on the part of the British there.

For the British, Dominica was not in any sense a major colonial acquisition; it had been more a matter of definitively ousting the Caribs from the best lands in the island. It is not surprising therefore that at the end of the first quarter of the nineteenth century Coleridge regarded it as 'first divided by language, then by religion, and the inconsiderable residue, which is supposed to represent the whole, is so torn to pieces by squabbles as bitter as contemptible, that the mere routine of govern-ment was at a dead stand' ([1826] 1970: 146). Toward the end of the century Froude's assessment of the island was 'Dominica is English only in name.' ([1888] 1969: 145). In the interim between the time of Coleridge's comment and Froude's the island did not appear to have become nationalistic or assertive in its identity.

Right around the time of emancipation in the British islands, four books by British writers appeared – Lewis (1834), Carmichael (1833), Wentworth (1834) and Madden (1835). Lewis (1834) was a posthu-mous publication by an absentee plantation owner who recorded his experiences of his visits to his plantation in Jamaica in 1815 and 1817 when slavery was still in full swing. Carmichael (1833) was a two-vol-ume narrative by a plantation owner's wife who lived first in St Vincent and then in Trinidad. Wentworth (1834) was a travelogue that was sup-posed to be a relation of actual experiences, but Wentworth's con-sciousness of and dependence on other writers seemed to have been disturbed by the appearance of the first two works when he had already written his. Madden (1835) was an account in the form of a series of letters by a stipendiary magistrate who was actually a doctor and who was sent to Jamaica to monitor what was seen as the historic changeover from slavery to apprenticeship and to adjudicate in matters thereafter.

Wentworth thought of his own work as a new beginning, a move away from the unpleasantness of slavery. His main concern was 'taste', which for him meant a concentration on society and culture rather than on politics and economics. The focus of Carmichael (1833), according to its title, was 'Domestic manners' and this title also wanted to give a picture of contented slaves in the West Indies as seen in their everyday behaviour and speech. However, while Wentworth saw himself as producing an account that was 'tasteful' for the reader, Day, writing about two decades later and identified as the 'author of a book on etiquette', was already by that time dispensing harsh criticism of contemporary, post-emancipation West Indian behaviour. Interestingly enough, Thomas Russell, a Jamaican writing in 1868, in the *Preface* to his book, which was on the vernacular language of his own people, apologised for 'sacrificing "Delicacy" at the shrine of "Truth"'.

The intention (under the guise of 'taste' or 'manners') on the part of Wentworth and Carmichael to move the identity of the West Indies away from one that was the result of colonial exploitation to one that was a product of its own innocence and inadequacies was probably not principally to justify slavery, but to show that the African had benefited greatly in the New World, that the 'savage' had been changed by the civilising process. The intention was also to show how those who had come into contact with the African had changed, and how the products of the two had evolved in their own society.

Several improvements in the negro were said to have taken place as a result of conversion to Protestantism. Gurney, in the context of Methodism in Antigua, said: 'The moral improvement of the negro population is amply evinced by two facts — the increase of marriage, and the decrease of crime' (1840: 61). Day conceded 'improvement' in St Vincent when he said: 'Strange to say, the negroes, a little way out of the town, are much improved in morals, not stealing so much as formerly, and getting married instead of living in a state of concubinage' (1852 2: 107).

Nevertheless, because changes in social status and political power became more visible and real in the way they affected visitors to the West Indies, foreign writers increasingly expressed their negative reactions to them. These reactions really came out of the conviction, especially on the part of the English, that, although Europeans had converted Africans into Creoles and made them better human beings, Creoles had become ungrateful. Day's assessment of New World negroes is one example of disgruntled English reactions and convictions:

> It is a psychological fact, a peg on which philosophers may hang a theory,
> that the American negro, as well as the French, is much more clever and

intelligent than the British. The Yankee negro speaks with a greater fluency and a far better pronunciation; he expounds, demonstrates, and lays down the law in a very ingenious way. The French negroes (Martinique and Guadeloupe) are also far more intelligent than the English, but, at the same time, much more depraved and demoralized; whilst, of the British negroes, the Barbadians are the most impudent, and the Jamaica negro is esteemed the most intelligent. It seems to be a disagreeable fact that, in a barbarous state, superior intelligence is invariably linked with a higher degree of depravity. Whenever, in the West India islands, we find a negro 'character' talkative, noisy or abusive, we may be sure he is either a Barbadian, an American, or a French negro. (1852 2: 108–9)

Here, what were essentially positive characteristics of identity were regarded as depraved because it was not in the best interests of whites for negroes to be articulate and show intelligence: they were to be seen and not heard, to be docile and obedient as they had been before.

*Congo saw* and talkativeness had always been seen as depravity rather than as viable methods of survival for the powerless. Now, even the negroes' penchant for ridicule was set in a framework of dishonesty in order to discredit it: 'The negro population of Barbuda does not differ essentially from negroes elsewhere. They are very civil but steal wherever they can, and, when detected, make satirical songs on the manager for stopping their pay' (Day 1852 2: 296). These reactions formed a consensus opinion among whites and led to the desire, on the part of the whites, to modify the 'natural depravity' of the negroes in the 'right' direction, that is through the media of religion and education. Notions of natural forces and theories of the effect of the climate on people received impetus from the academic world, especially from the theories of Darwin. Ideas of 'civilisation' as a conflict between the natural and the rational were the basis of the belief that improvement in the identity of West Indians meant controlling the natural (i.e. African-ness) and increasing the level of the rational (implicitly and explicitly linked to European). The conviction that the former 'savage' could be brought under control through education and religion was strengthened by the example of Antigua, which was moulded into a model colony.

# Division, *fusion* and 'improvement' in the French islands

Social changes and changes in the identity in the French colonies in the nineteenth century were substantially attributed to ideas and events in

France rather than to social and economic forces within the islands themselves, in spite of the fact that the same bitter hatred between whites and *gens de couleur*, which had precipitated the fall of the colony Saint Domingue, undermined the social hierarchy in the smaller French Caribbean colonies. Changes in government in France resulted in expulsion or migration of people to the colonies and consequently in transfer of contemporary ideas there. Identity in feeling between the colonies and the people of France was noted by Boyer-Peyreleau: '*A la Martinique, à la Guadeloupe, à la Sainte-Lucie, à Tabago, la révolution fut accueillie avec le même enthousiasme et souleva, dans toutes les classes, un intéret aveugle et des passions violentes.*' ['In Martinique, Guadeloupe, St. Lucia and Tobago, the revolution was welcomed with the same enthusiasm and excited in all classes a blind interest and violent passions'] (1823 2: 356).

In the colonies political changes brought about a direct confrontation between French revolutionary ideals and the institution of slavery with its attendant colour hierarchy. As a result, according to Boyer-Peyreleau, Guadeloupe was said to have been converted into a place apart by the 'despotic' rule of Victor Hugues, the leader of the French Revolution in the West Indies: '*La Guadeloupe n'offrit plus l'aspect d'une colonie française; elle devint une sorte de puissance, isolée au milieu des mers, ne conservant le nom français que pour le faire redouter*' ['Guadeloupe no longer looked like a French colony; it became a sort of power, isolated in the middle of the seas, preserving the French name only to make it an object of fear'] (1823 3: 39). The elimination of social distinctions (including those of slavery) and the introduction of ideals of the French Revolution into Guadeloupe by Hugues were interpreted as 'despotic' by the planter class, who no doubt had the image of Haiti on their minds. Martinique came under the control of the British from 1794 to 1802 and so escaped the direct introduction of republican ideals into the island. Even when it reverted to French rule, the planters there began to promote a concept of autonomy within the French world as a kind of rebellion against the revolutionary leadership in France. In fact, from the start of the French Revolution and throughout much of the nineteenth century, developments in Martinique did not follow the same path as those in Guadeloupe.

As to the social and cultural manners of *créoles* in the two islands, various explanations were given to account for them and for differences between the people. Granier de Cassagnac, whose comments about whites in both islands were generally positive, noted social differences between the people but tried not to favour one island over another:

*Les diverses colonies françaises des Antilles se sont toujours distinguées entre elles par le caractère de leurs habitants, et cette différence existe encore. La Martinique est plus cérémonieuse, la Guadeloupe plus cordiale. On disait autrefois: nos seigneurs de Saint-Domingue, messieurs de la Martinique, les bonnes gens de la Guadeloupe.* (1842: 102)

[The diverse French West Indian colonies have always been differentiated among themselves by the character of their inhabitants, and this difference still exists. Martinique is more formal, Guadeloupe more cordial. Once people used to say: 'our noblemen of St. Domingue', 'the gentlemen of Martinique' and 'the good people of Guadeloupe'.]

In his hierarchy of formality, Saint Domingue was on top and Guadeloupe at the bottom. From this perspective the hierarchy represented the relative social and economic importance of the islands:

*La population de la Martinique se distingue par des qualités qui la font aisément reconnaître, même en France.*

*Les habitants y ont ce qu'on peut appeler une distinction personnelle plus marquée que dans les autres îles. L'habitant de la Guadeloupe est plus spécialement bon et affable; celui de la Martinique est plus spécialement homme du monde. Cette différence se retrouve partout, dans le maintien, dans le langage, dans l'accent, dans la toilette, dans l'intérieur du ménage. A la Guadeloupe, vous sentez que vous êtes chez vous; à la Martinique, vous sentez que vous êtes chez vos hôtes.* (1842: 344)

[The people of Martinique are noted for qualities which make them easily recognisable even in France. The people there have what one would say is a personal distinctiveness which is more noticeable than in the other islands. The native of Guadeloupe is more specifically a good and affable person; the native of Martinique is more specifically a man of the world. This difference is to be found everywhere – in their bearing, language, accent, clothes, and in the inside of their houses. In Guadeloupe, you feel at home; in Martinique, you feel like a guest.]

Since Granier de Cassagnac's intention was to be complimentary to both islands,[2] the hierarchy, stood on its head, was one of affableness, in which case the people of Guadeloupe were on top. So, both islands, from this characterisation, could claim superiority – the people of Guadeloupe as more down-to-earth and hospitable, the people of Martinique as more sophisticated and better known.

Victor Schoelcher, in his comparison of the two islands in 1847, came to a different conclusion. Speaking of Guadeloupe, he said:

---

2. Dessalles (1996:156) noted the duplicity of Granier de Cassagnac: 'I already expressed my opinion of M. de Cassagnac, who is a mere confidence man, although his fiery pen can usefully serve our interests. It is really quite humiliating to see a man of talent sell his writing. M. de Cassagnac would write equally well against us if he found people who were willing to pay him for it.'

*Le luxe des habitants ... se pouvait surtout juger à la comédie, où dans une salle bien décorée, abondamment éclairée, on voyait chaque soir deux rangées de loges remplies de femmes vêtues avec une telle recherche qu'on aurait pu se croire en Europe, n'eût été la variété de couleur des spectateurs.*

*Il faut reconnaître qu'il y a un caractère de civilisation bien plus avancé à la Guadeloupe qu'à la Martinique.* (Schoelcher 1847: 368)

[The luxury of the people ... could be judged above all at the theatre, where every evening in a well decorated, amply lit hall, you could see two rows of boxes filled with women so elaborately dressed that you would believe you were in France, were it not for the variety of colours of the spectators.

It should be noted that there is a character of civilisation that is much further advanced in Guadeloupe than in Martinique.]

In this comparison, which regarded Guadeloupe as more advanced than Martinique, Schoelcher used women as his point of comparison and Europe as the yardstick for civilisation. As an abolitionist, it was not accidental that Schoelcher mentioned that there were ladies of various colours in the theatre in Guadeloupe. Schoelcher conceded, however, that Guadeloupe had been set back in its development by the devastating 1843 earthquake, which caused extensive damage and loss of life.

Another comparison of the people of the two islands favouring Guadeloupe occurred at the end of the century in the observations of Garaud, who made a sharp distinction between the whites of the two islands, in which he castigated the whites in Martinique for their isolation: '*Il est cependant regrettable que les anciens maîtres, blancs Créoles, appelés **békés** dans le patois local, se soient prématurément résolus à vivre à l'écart. Cette classe s'est enfermée chez elle et laisse dédaigneusement les mulâtres se débattre à leur guise*' ['It is regrettable, however, that the former masters, the white Creoles, called *békés* in the local patois, have prematurely decided to live apart. This class has kept to itself and disdainfully allowed the mulattos to do as they please'] (1895: 216). In contrast, he was complimentary to the whites in Guadeloupe, saying: '*Les blancs de La Guadeloupe sont restés dans la vie publique et malgré leur petit nombre ils exercent sur la marche des affaires une influence heureuse. L'attitude des blancs de La Martinique mérite d'être sévèrement jugée*' ['The whites of Guadeloupe have remained in public life and in spite of their small number they have a positive influence on the outcome of matters. The attitude of the whites in Martinique deserves to be severely censured'] (1895: 217). In Garaud's assessment, therefore, the *békés* of Martinique were haughty and acted out of an over-riding sense of class and colour rather than out of any sense of national responsibility.

Because it was normal for both islands to be mentioned together, they were treated as if they were identical in language. Breen (1844: 359) referred to them as 'the purely French communities of Martinique and Guadaloupe' and contrasted them with the 'two purely British communities of Barbados and Antigua', and, though Granier de Cassagnac asserted that the two islands had differences in language, he gave no evidence of these differences. William Paton, an American on a trip through the islands, perceived similarities between Martinique and Louisiana: 'The whites speak French with an accent that very closely resembles the speech of the creoles of Louisiana' (1896: 110–1). Yet, there is nothing to validate this remark and it may simply have been that for an American the connection between Martinique and Louisiana suggested itself rather than that there was any close identity between the language and people of Martinique and Louisiana.

In the French islands identity was constrained within the parameters of racial identity. There were the beginnings of a sense of national identity when the terms *martiniquais* and *guadeloupéen/guadiloupier* appeared, but the old system of identity based on race persisted with distinctions across the colonies between *Créoles* (whites), *gens de couleur* and *noirs*. Cassagnac's characterisation of identity in Martinique and Guadeloupe before emancipation, for example, concerned itself almost exclusively with whites. At the other end of the social spectrum the *noirs* were presented as the descendants of people who were the cultural opposite of *békés*. At the time of emancipation and after, there was a perceived necessity for them to show themselves worthy of their freedom by behaving in a civilised manner. The following is a *béké*'s account of an incident where a prominent politician, a *gens de couleur*, was advising negroes about how they should behave:

> The hungry negroes had fallen upon the meat that had been distributed to them. Bissette was so displeased that he cried out to them: 'what are you doing, my friends? You behave like cannibals, like savages! The more I try to raise you up, the more you lower yourselves. You make me ashamed. Am I not a negro like you? Then do as I do, imitate the whites! They alone will civilize you. Do not imitate the mulattoes. What use is the drum? Don't you see what the whites use for their dances? Like them, use the violin.' (Dessalles [1808–1856] 1996: 260)

The likening of the negroes to cannibals and savages clearly established an unbroken link between the negroes and the original Island-Caribs, who were so described. More interesting is the fact that this was a *gens de couleur* claiming to be negro and advising negroes not to behave like mulattoes.

The division between *créoles* (whites), *gens de couleur* (coloureds) and *nègres/noirs* (negroes) was sharp, so much so that special attempts were made to break down the barriers immediately after emancipation through a policy of *fusion*. It was illustrated by Dessalles ([1808–1856] 1996: 207–8) in the following: 'M. Husson threw himself into the arms of M. Pory Papy, the mulatto lawyer, and then into those of the negro Alexandre Sauvignon; this reunion of the three colors cemented the harmony among the inhabitants. Truly, a great sight!' The symbolism here extended beyond this simple incident in 1848. The *tricolore*, the French flag, had developed in 1789 as a symbol of the unity of the monarchy (white) and the people (blue and red).

In 1850, Rufz presented a new type – the homogeneous *créole*. Rufz's focus was on the transformation that had taken place in the preceding two hundred years – between the time of DuTertre (1654), whose early description was of newcomers (Europeans and Africans) to the islands, and 1850, when the islands had assumed an identity of their own. For Rufz, this had been a period that brought into being the *créole*, who appeared in different skin colours but inside was the same, markedly different from the original Europeans and Africans respectively:

> *Sous le ciel des tropiques, la race africaine, comme l'européenne, se modifia sensiblement par la reproduction. Les uns et les autres donnèrent naissance à un être nouveau. Il y eut le créole africain, comme le créole européen plus particuliers au sol. De même que des Européens venus des divers points de la France, il est sorti un rejeton tellement identique, qu'il n'est pas possible d'en démêler la provenance originaire, de même de l'Africain long et élancé du Sénégal, du Congo lourd et trapu, du Mandingue plus vif et plus délié, est sorti le nègre créole, tellement fondu, homogène et approprié au sol, qu'il n'est pas possible de retrouver sur son front, ses pères et mères, sa souche naturelle, sa filiation. Cependant sur le terrain même de l'Afrique, comme nous l'avons dit, les races sont aussi diverses que peuvent être en Europe le Lapon et l'Espagnol; mais aux îles, sous l'influence d'une temperature égale, tout est devenu créole.* (Rufz 1850: 148)

[Under the tropical sky, the African race, like the European, was noticeably modified through reproduction. Both of them gave birth to a new being – the African Creole and the European Creole, who were more characteristic of the soil. Just as from Europeans coming from diverse points of France, there came shoots so similar to each other that it is not possible to discern their original sources, so too from the tall and slim African from Senegal, the heavy and thick-set African from the Congo, the more lively and nimble Mandingo, there came a creole negro so blended, so homogeneous and so adapted to the soil that it is impossible to look in his face and see who his mother or father is, or his natural stock or his line of consanguinity. Note that in the lands of Africa, as we said, the races are as diverse as they can be in Europe as between the Laplander and the Spaniard. However, in the islands, under the influence of an even temperature, everything became Creole.]

*Sous le ciel des tropiques le produit de la race africaine, le créole africain, subit les mêmes modifications que le produit de la race européenne, le créole européen. Sous le masque de l'epiderme, ils présentent des caractères communs, fraternels, c'est un air de famille qui les rapproche, comme les plantes d'un même sol ... C'est la même constitution physique, le même tempérament, nerveux-lymphatique, les mêmes qualités, les mêmes défauts. Ce ne sont plus l'Européen ni l'Africain, c'est l'enfant du territoire. Cette action du sol qui s'approprie les races et qui les frappe pour ainsi dire à son effigie, amène pour les nations, comme pour le genre humain, cet admirable résultat, la variété dans l'unité.*

*Le nègre créole est élancé, il a des proportions belles, les membres dégagés, le col long, les traits de la face plus délicats, le nez moins aplati, les lèvres moins grosses que l'Africain; il a pris du Caraïbe l'oeil grand et mélancolique; son regard s'est attendri, se prête mieux aux émotions de la vie civilisée. On y retrouve rarement la sombre fureur africaine, l'air ténèbreux et farouche; il est brave, communicatif, fanfaron. Sa peau n'a plus la teinte aussi noire que celle de son père, elle est plus satinée; ses cheveux sont encore laineux, mais d'une laine plus souple, sa sclérotique est encore bistrée, ses formes plus arrondies; on voit que le tissu cellulaire prédomine comme dans les plantes cultivées, la fibre ligneuse et sauvage se transforme.* (Rufz 1850: 150)

[Under the tropical sky, the African Creole, product of the African race, underwent the same modifications as the European Creole, the product of the European race. Under the mask of their skin, they present common, fraternal characteristics; it is feeling of family that brings them together, like plants from the same soil ... It is the same physical constitution, the same highly-strung lymphatic temperament, the same qualities, the same faults. It is no longer a matter of European or African, it is a matter of a child of the soil. That action of the soil that appropriate races and fashions them, so to speak, in its own image, brings about for nations, as for human beings, this wonderful result, variety in unity.

The creole negro is slim, beautifully proportioned, with supple limbs, a long neck, finer facial features, a nose less flat and lips less thick than those of the African; he has acquired big and melancholy eyes from the Carib; his look has softened and lends itself better to the emotions of civilised life. Seldom does one find the sombre fury of the African or his dark and fierce look. He is hearty, talkative and boastful. His skin no longer is as black as that of his parent and it is smoother. His hair is still wooly, but a softer wool. His eyes are browner and rounder in shape. His cellular tissue predominates, as in cultivated plants, as the ligneous and savage fibre is transformed.]

It is very clear from this the extent to which Rufz regarded climate as the main force in the process of genetic modification, as well as his view of the similarity in evolution between human beings and plants. For him, the *créole* was *l'enfant du territoire*, an almost perfectly adapted new being.

Revolutionary ideals and metropolitan education put men to the fore in the development of national identity in Martinique and Guadeloupe, but what differentiated the French creole world from the Spanish and the English especially was the prominence of women in the vision of ethnic identity. In the Spanish Caribbean islands in the nineteenth century, the religious concept of trinity was extended to symbolise the union between the three races – white, indigenous inhabitant and negro – a union graphically illustrated in William Blake's 1796 egalitarian engraving (*Europe supported by Africa and America*) in Stedman's Suriname, which had the female as the representative of each continent/race. It was not accidental that the indigenous inhabitant in the Spanish islands was replaced by the mulatto in the French islands and that in the French islands the female mulatto came to represent their image of identity. French creole women acquired a glamorous and exotic reputation of their own, and it was Empress Josephine who was largely responsible for this reputation. Martinique was her birthplace, as a result of which the island fortuitously came to be associated with her beauty, personality and actions. This charming Creole, who had won the heart of Napoleon and was crowned Empress in 1804, became the pride of Martinique, helped her native island as much as she could, after which an imposing statue of her was erected in the capital of the island. She it was who contributed substantially to the identity of Martinique not only in France and the French world but also in Martinique itself. She became a symbol of exotic beauty and of achievement for the women of Martinique. At the end of the century Josephine was no doubt in Lafcadio Hearn's mind as he described the women of Martinique, and generally in accounts of visitors in the last part of the century, the identity of Martinique was dominated by descriptions of women's physical features, their costumes and their adornments.

By the second half of the century, then, focus had shifted to the *métis*. While Rufz, following Schoelcher, was captivated by ecological determinism, Cornilliac, doubtful that the African could be an element in any mixture resulting in beauty, resurrected the Carib to account for this beauty:

> *Lorsque parmi les populations des Antilles on remarque de ces métis au teint olivâtre, à la taille svelte et élancée, dont le profil droit et les traits*

*réguliers rappellent les habitants de Madras ou de Ponichéry, avec lesquels on les confond souvent, on cherche pensif en contemplant leurs longs yeux pleins d'une douce et étrange mélancholie, chez les femmes surtout, leur chevelure noire et abondante, aux reflets soyeux, tombant avec profusion sur les tempes et sur le col, à quelle race peut appartenir cette variété, dans laquelle domine un caractère qui semble indélébile et parait plus marqué à mesure qu'on s'éloigne de l'élément africain: c'est le sang caraïbe, qui s'est mêlé au sang européen et à celui du noir et qui, malgré tous les croisements ultérieurs, quoiqu'il n'ait pas été renouvelé depuis plus de deux cents ans, conserve encore, comme aux premiers temps de mélange, ce cachet particulier qui le décèle chez tous ceux dans les reins desquels il coule.* (1867 1: 130)

[When among the peoples of the Antilles you notice those olive-skinned metis, svelte and slim, whose profile and regular features remind you of the people of Madras or Pondicherry, with whom they are often confused, you wonder, while looking at their long eyes filled with a strange sweet melancholy, especially among the women, their long black hair with a silky look, falling profusely on their temples and their neck, to which race belongs this variety in which there is a dominant, apparently indelible character, one which seems more marked as it moves away from the African element. It is the Carib blood, which is mixed with European and negro blood and which, in spite of all the subsequent mixtures, although it has not been renewed for more than two hundred years, still preserves, as in the first years of the mixture, that particular stamp which reveals it among those in whose loins it courses.]

Cornilliac's argument was a more direct insertion of the Carib/indigenous inhabitant into the mix, in comparison with previous arguments which, using ecological determinism, had merely drawn parallels between the evolution of the indigenous inhabitant and the evolution of people of mixed race in the Caribbean.

There is no evidence that Cornilliac's argument gained any support in the years following, and in his accounting for the evolution of the *fille-de-couleur*, Hearn implicitly rejected it:

Considering only the French peasant colonist and the West African slave as the original factors of that physical evolution visible in the modern *fille-de-couleur*, it would seem incredible; – for the intercrossing alone could not adequately explain all the physical results. To understand them fully, it will be necessary to bear in mind that both of the original races became modified in their lineage to a surprising degree by conditions of climate and environment. (1890: 317–18)

In other words, Hearn, at the end of the century, wholeheartedly accepted Rufz' argument for genetic improvement through climatic forces.

In accounting for the love of adornment, which was another aspect of the appearance of French creole women highlighted, climate was also the reason put forward by Bird: 'Perhaps it ought not to be a matter of surprise that in a climate almost invariably bright, love of ornament should amount to a weakness, on which immense sums are continually expended ...' (1869: 324). Negro women were included in this love of adornment by Paton, who was obviously struck by their appearance when he said: 'With a Franco-Africaine, love of personal adornment is a passion in the gratification of which she displays a reckless extravagance ...' (1896: 109). Significantly, Bird did not spontaneously link the love of adornment to a French cultural tradition. However, he did not think to explain why climate had this effect on French creole women more than others.

Climate was also seen, by Hearn, to effect intellectual improvement in the person of mixed race. An implicit contrast in this case was that, whereas beauty was related to women, intellectual improvement was related to men:

> the mulatto began to give evidence of those qualities of physical and mental power which were afterwards to render him dangerous to the integrity of the colony itself. In a temperate climate such a change would have been so gradual as to escape observation for a long period; – in the tropics it was effected with a quickness that astounds by its revelation of the natural forces. (319–20)

This purported intellectual improvement in mulattos was actually a variation of a claim made earlier by Edwards, who spoke of climate contributing to 'the early display of the mental powers in young children' (1794 2: 12). As far as the mulatto was concerned, this improvement was seen as a threat to the hierarchical racial structure of colonial society, as evidenced in Haiti.

Ecological and other arguments used to account for presumed genetic improvement in the racially mixed person ran counter to older, purely genetic arguments that the mulatto had acquired the worst of both parents. It was a contradiction that could be explained by the fact that 'deterioration' arguments were generally directed towards men or mulattos as a social group, whereas 'amelioration' arguments were an attempt to explain the physical beauty of women and the hold that it had over white men, without making white men seem lesser than they were presented to be. At the same time, however, the *métis* acquired the same kind of contradictory reputation in the French islands that the *mulata* did in the Spanish islands in that her beauty was seen as both phenomenal and destructive. She, as a mixture and an enticement to mixture, was seen as a threat to slavery, which required sharp stratification, and to family values, which were threatened by the *ménage à trois.*

Continuing the focus on the genetically mixed person, Hearn created the impression that the population of Martinique was dominated by them – 'the general dominant tint is yellow, like that of the town itself — yellow in the interblending of all the hues characterizing *mulâtresse, capresse, griffe, quarteronne, métisse, chabine,* – a general effect of rich brownish yellow. You are among a people of half-breeds, – the finest mixed race of the West Indies' (1890: 38). This impression managed to sustain itself even though in 1848 the negro population in Martinique was more than ten times the white population. The *gens de couleur* had come to the forefront as the racial type that represented French Caribbean identity in the nineteenth century, so much so that Meignan underscored the relationship between *gens de couleur* and the Caribbean as *patrie*: '*Les hommes de couleur seuls ont continué à regarder cette terre comme leur patrie, l'ont aimée, ont peuplé les villes et sont devenues les véritables maîtres de la colonie*' ['People of colour alone have continued to regard this land as their native land; they have loved it, have peopled its towns and have become the true masters of the colony'] (1878: 49).

As was the case in the Spanish islands, the mulatto/métis, as a New World product, came to be seen as the pre-eminent native of the islands: '*Le véritable Martiniquais, ce n'est même plus le créole; nouvelle transformation, c'est l'homme de couleur ...*' ['The true Martinican is indeed no longer the Creole; it is the man of colour, a new transformation ...'] (Meignan 1878: 152). The indigenous inhabitants were excluded as a component of this identity '*du Caraïbe il n'en est plus question. Il n'en reste rien que son ancien territoire, son ancien soleil, ses anciennes forêts*' ['There is no longer any idea of the Carib. The only thing that remains of him is his former territory, his former sun and his former forests'] (Meignan 1878: 152). Remarkably, as Meignan was announcing the pre-eminence of the *gens de couleur*, he was also at the same time predicting their disappearance: '*Or, cette race, crée par les hommes, ne peut durer qu'un temps comme tout ce qui vient des hommes. Livrée à elle-même, elle est fatalement destinée à disparaître et à redevenir noire!*' ['Now, this race, created by man, can last only as long as what comes from man. Left to itself, it is fatally destined to disappear and to become black again'] (1878: 49–50). This prediction about the disappearance of the *gens de couleur* was of course based on the unstated but quite evident fact that the majority of the population was really negro, a fact that was considered to be of little value and prestige in nineteenth-century presentations of French Caribbean identity.

Consciousness of a regional French creole identity across the smaller islands was sustained by the use of a common language by the people in Guadeloupe, Dominica, Martinique, St Lucia, Grenada and Trinidad. The language of official communication (or at least one of them) in all these islands in the early part of the nineteenth century was identified as French, though little was said about its actual form in the Antilles. In contrast, the language of informal communication was labelled with negative stereotypes because it was known to be associated in its origin with the negro population. In reality, the use of the creole language had moved beyond its humble beginnings, as Breen noted: 'it [the Negro language] has now almost entirely superseded the use of the beautiful French language, even in some of the highest circles of colonial society' (1844: 185). As to the nature of the language, Breen's comments are revealing, even if essentially personal:

> As a **patois** it is even more unintelligible than that spoken by the Negroes in the English Colonies. (1844: 185)

> I can say for myself that, although possessing an extensive knowledge of the French language, acquired during a sojourn of five years in France, I have failed in obtaining anything like an adequate notion of this gibberish, during a residence of nearly fifteen years in St. Lucia and Martinique. (1844: 185–6)

If Breen was not simply exaggerating his own abhorrence for the creole language, then his comments indicate that there was a deep structural gap between French and French creole in the smaller islands in the first half of the nineteenth century as there was in Haiti. This contrasted with the English islands, in which there was a continuum-type language situation stretching from a standard form of English spoken by the elite to a very divergent variety spoken by those at the bottom of the social ladder.

As the century progressed, the linguistic uniformity across the six French-creole-speaking islands started to break down as a result of changes that took place in those of them that were actually under English control. In reference to Grenada near the end of the century, Hearn said: 'In the mixed English and creole speech of the black population one can discern evidence of a linguistic transition. The original French *patois* is being rapidly forgotten or transformed irrecognizably.' (1890: 91). This fairly objective observation about Grenada was followed by a much wider one, extending presumably to St Lucia, Dominica and St Vincent:

So often have some of the Antilles changed owners, moreover, that in them the negro has never been able to form a true *patois*. He had scarcely acquired some idea of the language of his first masters, when other rulers and another tongue were thrust upon him, – and this may have occurred three or four times! (Hearn 1890: 92)

Hearn's assessment of the consequences of these linguistic experiences is summed up in the same term used by the first European writers on the region: 'The result is a totally incoherent agglomeration of speech-forms — a *baragouin* fantastic and unintelligible beyond the power of any one to imagine who has not heard it' (Hearn 1890: 92). This evaluation cannot be summarily dismissed as a typical European reaction because Hearn himself demonstrated his familiarity with the French creole of Martinique. Yet, there can be no doubt that by his use of the word *baragouin* Hearn intended to put the language of these islands at the end of the nineteenth century in the same category as that of the indigenous inhabitants of the region in the sixteenth century.

Consciousness of a regional French creole identity across the smaller islands was not consolidated by the use of a single regional name (similar to 'West Indian' for the English islands).

In spite of the lack of a regional name, there was a uniformity of culture in those islands, which emerged more prominently than anything from the English islands. The vision of beautiful women of mixed race, the practice of Carnival and its links to the Catholic Church, social activity with balls, dancing, lively conversation, repartee and colourful dress, all these features were connected together by the French creole language. This was not a political identity, but, as a white French creole phenomenon, it looked backwards as it celebrated grandeur and *joie de vivre* and, as a 'coloured' phenomenon, it got new energy from the mass of the population, whose status changed during the nineteenth century and who sought to celebrate those changes. The same love of witticisms and masquerade that were seen as French and European had also been noted among the slaves. It was therefore a strong cultural identity that developed. The movement from island to island that characterised the French world in the first half of the nineteenth century spread and fed the French creole culture and identity. At the end of the century, Paton's (1896: 109) term *Franco-Africaine* also suggested a vision of comprehensiveness, but it also showed that race continued to be the most dominant classification and that, more specifically, women had emerged as the highlight in the French islands. It would seem that the major factor that determined the absence of a French creole regional identity was that the French world in the Caribbean contracted significantly in the first quarter of the century, leaving Martinique and

Guadeloupe as the only French islands of note. With the loss of Haiti and the concession of some of the small islands to the British, there was politically no real vision of an expansive French Caribbean in the nineteenth century and so no common regional designation emerged.

# Refugees in the British islands

## Passparterres

In the period between 1834, when the British abolished slavery, and 1848, when the French did, a number of slaves escaped to nearby British islands from Martinique and Guadeloupe. The fact that they acquired a label, *passparterres*, meant that they became noticeable in the islands they went to, especially St Lucia, as is evident in the following comment: 'The refugees from the foreign island of Martinique, commonly designated by the derisive appellation **Passparterres**, constitute a characteristic ingredient in the population of St. Lucia' (Breen 1844: 169). Capadose was a little more explicit about these residents of St Lucia when he said:

> South of Tapion is an abandoned Fort called Ciceron, and in its vicinity, rich lands without any apparent owners: but inhabited by a number of labouring people who have built cottages and cultivated some acres of ground for their own support, and are supposed to be runaway slaves from the French islands of Guadeloupe and Martinique and style themselves, or are styled, *Passe par terre* people. (1845 2: 154)

Nearly 25 years later, without any reference to St Lucia or any other island, Thomas gave as an example of words in Trinidad that were not easy for strangers to understand '*yon passe-pâ-tèr,* "a pass by land," i.e., one who has come from out the Bocas' ([1869] 1969: 83). Apparently Thomas meant that the person was coming from the South American mainland, which, if it were true, extended the application of the word. He seemed to be unaware of Schoelcher's (1842: 114–19) restriction of the term to slaves escaping from the French islands still under slavery at the time.

It is unlikely that the term spontaneously came into existence in two different places. In 1844 Breen, who had spent 13 years in Martinique and St Lucia before he wrote his book, did not know the true origin of the term but, like Schoelcher (1842: 118), surmised that it was used by the refugees from Martinique to deceive St Lucians (for Schoelcher it was Dominicans, i.e. from Dominica) about the way they had come to St Lucia. In addition, it did not seem to have been a very recent word

when Breen was in St Lucia and explained the precautions taken to prevent escape and the consequences:

> And truly, when we consider the difficulties of escape – the vigilance of the
> authorities of Martinique – the system of espionage employed by the
> planters to check desertion amongst their slaves – the strict surveillance of
> the 'guarda-costas,' and the distance and dangers of the passage between
> the two islands, it is matter of surprise that so many of them should have
> succeeded in accomplishing their object ... Still, it is no unusual
> occurrence to see twelve or fifteen men and women land on the coast of St.
> Lucia, from a canoe in which five persons could not sit at their ease. We
> know that numbers perish in the attempt, either from the roughness of the
> weather or the wretched condition of their boats; and that many, upon
> being closely pursued by the guarda-costas, plunge into the deep, never to
> rise again, preferring death and a watery grave to the life and labour of
> bondage. (1844: 169–70)

Breen's words were similar to those of Schoelcher (1842: 114) and before him to those of Turnbull, who was impressed by the courage of the
escapees: 'How much more justly were those poor negroes entitled to
the reputation of undaunted courage than the mere mariners of old!'
([1840] 1969: 563). Turnbull himself did not cite the name *passparterres*, which suggests that it had not yet become a well-known designation or at least not well known outside the islands where the escapees
settled.

Another possibility for the source of *passparterre* may have been
European ignorance of Caribbean geography. It was Wentworth who
made the following comment:

> A few of our West Indian colonies are governed by orders in council from
> England, and the palpable misconception and errors, and the total
> inapplicability of many of their provisions, have been clearly
> demonstrated during the last few years by their repeated reversal. We may
> here mention an instance of ignorance which occurred some years ago on
> the part of an individual, whether in Parliament or not, we forget, who
> interfered in the regulation of the Post office packets [boats] ... This is,
> however, surpassed by the transmission of a government despatch to
> Honduras, *via* Jamaica, with the instruction on the envelope, 'if the packet
> should have sailed to be forwarded *over-land*!' (1834 2: 83–4)

Following from this, one can well imagine an explanation given by or
given to a French official, which stated that the slaves came *par terre*.
In other words, it seems as if the *over-land* or *par terre* notion was a
common joke associated with ignorant Europeans.

Another consideration in accounting for the term *passparterre* is one
suggested by an erroneous etymological explanation of the origin of the

Caribs: 'Labat was a student of languages before philology had become a science. He discovered from the language of the Caribs that they were North American Indians. They called themselves *Banari*, which meant "come from over sea"' (Froude [1888] 1969: 131). What Labat said was: '*Le nom de Caraïbe et celui de Banaré qu'ils donnent aux Européens et qui veut dire homme qui est venu par mer, est chez eux un titre honorable ...*' ['The name *Carib*, as well as *Banaré*, which they gave Europeans and which means "people who came by sea", is an honourable title ... "] ([1722] 1979: 240). In this consideration the important factor is the notion *homme qui est venu par mer* [come from over sea]. The equivalent of *banare* in the seventeenth-century literature was given as the French word *compere* in a context where Europeans were being welcomed and accepted into the local community as brothers. It means that the *passparterres* were, in the middle of the nineteenth century, the latest instances of *banare*, even though there is the question whether it [*passparterre*] was a designation given to them or one they gave themselves.

The reality captured in the term *passparterres* was most keenly felt by slaves in Martinique and Guadeloupe between the years 1833 and 1848, especially since it was in the French colonies at the beginning of the century that there was a brief experience of freedom and especially since Martinique and Guadeloupe were the immediate neighbours of St Lucia and Dominica in the chain of islands. Such a situation clearly did not endear the enslaved class in the French islands to their masters and it (re-)kindled in them the idea that their freedom and practical salvation was through migration overseas. Schoelcher (1842: 114) gave an estimate of 5,000 slaves who fled from Martinique and Guadeloupe to the English islands – between 700 and 800 were said to be in Dominica, 600 in St Lucia and 600 in Antigua. In the case of those who escaped to Antigua, according to Schoelcher, most of them re-migrated to Dominica, St Kitts and Trinidad, where they felt more comfortable because of similar language and religion. Schoelcher therefore, contrary to Thomas, identified the origin of the *passparterres* in Trinidad as the French islands.

As to the social connotations of the term, Breen referred to *passparterres* as a 'derisive appellation', but derision may have been directed at the person who believed the explanation rather than at the person being identified by the term. Thomas made no explicit comment on the attitude of persons toward the refugees and, even though Schoelcher acknowledged that it was a mild term of disparagement, he gave evidence to support the idea that the refugees fitted in well in their new communities and that they made a significant contribution in St Lucia specifically. From the point of view of the *passparterres* themselves, the term reflected a spirit of defiance and deception.

## Wilberforce negroes and King's negroes

Quite distinct from the creole population (white, coloured and black) were the Africans who were introduced into the West Indian colonies during the nineteenth century, not as slaves but as 'liberated' persons. Those Africans who were intercepted by the British and brought to the colonies after the abolition of the slave trade in 1807 were regarded as apprentices who had to serve up to seven years before they were given total freedom. They continued to be introduced into the British colonies for much of the century, but they did not come to constitute a great percentage of the population in any colony. They were important because they contrasted with all other groups, especially in the first half of the nineteenth century. As a group, they were noticed by several writers, who referred to them by designations that highlighted their role and status, but did not remove them from the negative stereotyping that Africans had suffered from the beginning of colonisation.

Coleridge's comments ([1826] 1970: 254–5), in reference to Antigua, were quite informative about contemporary attitudes and beliefs about the Africans in the first quarter of the century, when they were seen as people who simply did not fit into a well-articulated system. The Africans were made to seem unreasonable, in spite of their well-founded suspicions, and were portrayed as a disruptive element. This characterisation was based on a deep-seated notion, oft repeated by European writers, that Africans were intellectually and morally no good, but, more importantly, it constituted an attempt to preserve the colonial hierarchy of identity, which situated the African at the bottom of the society. These Africans were a threat, from the English and Creole point of view, because they were re-introducing features into the British islands that the creolisation process had been 'successfully' eliminating in the majority of native-born slaves.

It was almost inevitable that 'liberated' Africans in the British colonies in the decade before emancipation would attract attention from abolitionists and anti-abolitionists. For Coleridge ([1826] 1970: 256) the situation of the 'liberated' Africans presented a preview of the total emancipation of the slaves. Coleridge's view made little allowance for the fact that at the beginning of the nineteenth century the majority of the slaves in the British islands were native-born and somewhat removed from the experiences and views of the 'liberated' Africans:

their prospective reactions to emancipation were regarded in the same way as those of these new Africans, who were in a state of culture shock. Presumably, the reports from Haiti and the news of widespread refusal of the liberated to continue to do what they had been doing before was the crucial feature in this preview of liberation.

The 'liberated' Africans were given a new status with two different designations: 'Those attached to military departments are called by the natives, "King's niggers;" and those who are free, by the termination of their apprenticeship, bear the appellation of "Willyforce nigger"' (Bayley 1833: 455–6). Though the designations *King's nigger* and *Willyforce nigger* were said to have been given by 'the natives', the Africans were identified partly according to their perceived owner or benefactor in England and partly according to the creole valuation implied in the word *nigger*.

While Day noted that the term *nigger* was 'merely West Indian nomenclature' (the West Indian version of *negro*), he interpreted it as having negative connotations in Barbados:

> The Barbadian negroes have a great dislike to be called by the whites 'niggers,' although towards each other they use the word freely as a term of disparagement. 'She is de most worthless nigger in all de town, de most worthless,' said a young lady, in my hearing, *talking to herself*. You will see a fellow, himself as black as a saucepan, shaking his fist at 'dose lazy niggers.' (Day 1852 1: 33)

There is little doubt that in combination with *Willyforce* and *King* in reference to the 'liberated' Africans *nigger* had the negative connotations that Day indicated. This is even more evident from observations noted down by other writers:

> These Africans are very much disliked by the Creole slaves. It is common to hear two of them quarrel bitterly with each other, when all the curses of England and Africa are mutually bought and sold; but your right Creole generally reserves his heaviest shot for the end. After pausing a moment and retiring a few steps, he saith ... 'You! You!' with the emphasis of a cannon ball; 'who are *you*, you – Willyforce nigger?"' Whereat Congo or Guinea foameth at the mouth, Creole eyades rejoicing in the last blow. (Coleridge [1826] 1970: 256 [note])

> I once heard a free African call a young slave a 'wicked little picaninny,' as it appeared to me in joke, and I was astonished at her answer. 'You curse me eh! You curse me! – you dam Guinea nigger! – You Willyforce-Congo! I make you sabe how for curse me.' (Bayley 1833: 582)

For the creole slave, then, *Congo*, *Guinea* and *nigger* all had the same low valuation.

For his part, the 'liberated' African felt himself to be superior to a slave, according to Joseph: 'Formerly the 'King's man,' as the black soldier loved to call himself, looked (not without reason) contemptuously on the planter's slave ... ([1838] 1970: 263). From the English point of view, as represented by Joseph in 1838, the African had superior status before emancipation, for, even though Joseph scoffed at the African's contempt for the slave, saying 'he himself was after all but a slave of the State' (p. 263), 'he conceded that a soldier in a black regiment was better off than a slave' (p. 262). However, Joseph disagreed with those who still saw the African soldier as superior after emancipation, saying: 'a free African [i.e. a former slave] in the West Indies now is infinitely in a better situation than a soldier, not only in a pecuniary point of view, but in almost every other respect' ([1838] 1970: 262). Yet, the view of the African soldier as superior to other negroes seemed to be the more abiding view, for in the middle of the century Day expressed that view very clearly:

> Strange to say, the black troops (being native Africans) have no sympathies in common with the colonial negroes. They are, on the contrary, extremely proud of wearing a red coat, and of being 'Queen Victoria's soldiers," therefore they look down upon the colonial blacks; and having no sort of delicacy of feeling or of compunction, they would as soon thrust their bayonets into a 'common niggar,' as look at him. (1852 1: 284)

The fundamental conflict in status was between soldier and African. On the one hand, bearing arms was seen throughout the slave period as a privilege accorded according to status; on the other hand, the African was culturally on the lowest rung of the social ladder because of his perceived total backwardness. Yet, converting the African into the *King/Queen's soldier* and viewing him as superior were not contradictory in the colonial society because, for the Englishman's well-being and peace of mind, it guaranteed a continued conflict, which militated against the development of homogeneity in identity among all Africans and African descendants. It is not surprising therefore that English writers highlighted the idea of conflict between the 'liberated' Africans and the slaves: it was a portrayal of animosity and conflict between different groups within the negro population in the colonies that had been noted and repeated from the earliest days by European writers.

Besides this, the 'liberated' African was portrayed as ungrateful towards the English:

> The last batch of native Africans, liberated from a slaver, and brought to St. Vincent, when told what they were to do, (i.e. to labour in the fields as

apprentices for one year, the price of their liberation), drew their hands across their throats, to signify, as an old Eboe woman interpreted it, that 'Dem cut dere troat fus;' yet each of these fellows had cost the government £40. (Day 1852 2: 163)

What this meant was that not all 'liberated' Africans gained comfort or satisfaction from being dubbed *King's men* in the West Indian islands. Gamble (1866: 29) observed that between 1834 and 1866 about 4,000 liberated Africans had been added to the Trinidad population. A little more than a decade earlier, Day (1852 1: 274) had provided a detailed description of the arrival of Africans in Trinidad in which he maintained the Creole-over-African hierarchy. This was done, not by reference to culture, but by reference to mentality and resultant behaviour. His contrast between the newly arrived African, characterised by 'the absence of all power of thought' and presumably therefore sub-human, and 'Our highly civilized colonial negroes' reinforced the negative stereotype of the mental level of Africans. Nevertheless, by the time the 'liberated' Africans had been converted into one of 'Queen Victoria's soldiers ready to thrust their bayonets into a "common niggar"', Day had acquired a measure of respect for them. The view of the 'liberated' African that prevailed eventually was that the experience in the colony brought about improvements in him, and that the donning of the King/Queen's military uniform transformed him into a proud being, even if he retained his brutality.

In spite of the fact that knowledge of the various African ethnic groups was vague and often inaccurate, their influence on language was keenly perceived. Russell in Jamaica:

> With regard to accentuation, it must be remarked that people who live on sugar estates make a peculiar stress on nouns following the indefinite article *one*; (especially on such estates which employ Africans) thus: One *man* was ya all *a* de time da look pon we (looking upon us all of the time).
>
>  This method of euphonizing by accentuation is not to be found amongst the settlers in the more mountainous parts of the island. This difference can only be accounted for in this way, I think: The estates usually employ, besides Creoles, Africans, and these latter, even after they can manage to speak 'creole,' still retain the deep and harsh accentuation of their own language: the Creoles imitating them become, after a time, in some measure "infected." This is the more evident when we consider that people in the more remote mountainous parts, most of whom have never seen an African young man or woman, never accent words in this manner. (Russell 1868: 1–2)

According to this, therefore, there was a difference in speech between the creole negroes in the mountainous interior, who generally did not come into contact with Africans, and those on the coastal estates who did. This was a very interesting explanation of regional variation, which contrasted not only with the usual general association of conservatism with the rural and the inaccessible but also, and more specifically for the Caribbean, with the accepted view that it was the Africans who imitated the Creoles. What one may conclude from Russell's explanation, then, is that in the post-emancipation period African speech, consciously or unconsciously, became a model for creole negroes. It can be said to have constituted a resurgent model of African-ness in the nineteenth century.

The *passparterres* (escaping slaves) and the 'liberated' Africans were unofficial and official refugees respectively who were scattered across the English islands, except Barbados. The *passparterres* were seen to be operating within a sphere of French creole language and culture, while the Africans represented a resurgence of African influence as well as the African unchained, both of them within islands that were moving towards English manners. The *passparterres* were presented as courageous to seek freedom while the 'liberated' Africans were regarded as ungrateful for not appreciating their liberation. Both groups were refugees from slavery who added to the diversity of the islands they went to, but, more importantly, they were symbolic of the conflicts between European nations in the matter of slavery in the nineteenth century and as such were evidence of the different stages relative to each other in English and French policy and thinking.

## Creole language – validation, corruption and diffusion

Language emerged as an indicator of island identity and even as a unifying force in the face of hostility. Following on from Mathews (1793) and Moreau de Saint Méry (1796), both of whom wrote in praise of the 'genius' of the creole language but within larger works, three creole writers within a short space of time in the second half of the nineteenth century dedicated whole books to the analysis of the creole language. Russell published his small work on the language of Jamaica (*The etymology of Jamaica grammar*) in 1868 and that of Thomas (*The theory and practice of Creole grammar*) on French Creole in Trinidad followed in 1869. In 1874 Jean Turiault published his study of French Creole in Martinique (*Étude sur le langage créole de la Martinique*). Though these writers praised the 'genius' of the creole language, they maintained the notion that it was limited and a corruption of the European language, English and French respectively. Furthermore, it was the very divergence from the European language (interpreted as *unintelligible gibberish* and *baragouin*) that drove these writers to produce extensive explanation and justifications of the creole language. Explanations in written form having thus been provided, the written form in turn confirmed the creole language as an entity in itself, different from others. It consequently gave the users a negative (broken/corrupt) identity of their own, more so than Creoles in places such as Barbados, Antigua and other islands whose vernaculars did not seem to need any extensive explanation.

The negative view of the vernaculars in the islands was captured in such terms as *vile patois* and *creole drawl* – the former identifying French Creoles and the latter English Creoles. The *creole drawl* in the anglophone Caribbean was more than an unsophisticated comment by a single writer: it was a repeated assertion for about a hundred years. Initially it must have been the result of a sudden confrontation with pronunciations that had generalised across communities – confrontation with new 'accents' in which specific vowel articulations and intonational patterns had come together in a systematic way. Words such as *want, water, talk, walk, lost, draw, boss, sir* were pronounced with a long vowel, and, since such words were very frequently occurring, they created an overall impression of drawling. British perceptions of drawling were of course based on frequency of occurrence of pronunciations that were different from their own and seemed longer in West-Indian speech. However, some of the 'drawn out' pronunciations of West-Indian nineteenth-century English were inherited from Irish and Scottish English, and their generalisation in West-Indian English reflected the historical influence of these ethnic groups as well as the conservatism of West-Indian society. English writers commenting on West-Indian English were, knowingly or unknowingly, repeating the same negative remarks that had been made by the educated of London society about regionalisms.

Beyond pronunciation there were also differences noted between British and creole speech. Wentworth, in the following observation, which did not refer to a specific island, conveyed the experience of sharp difference, though not complete intelligibility, in his exposure to the speech of the negro population: 'Frequent intercourse with the negroes is required before you clearly understand their Anglo-patois, particularly that of the Africans, owing chiefly to the number of odd expletives they employ ...' (1834 2: 17 [footnote]). So too was the following remark made by Day, which highlighted the different discourse pattern of the negro population: 'They are, at the same time, exceedingly emphatic; so as to be, from their noisiness, most disagreeable neighbours. Every second word is emphasized and jerked out with a vehemence almost distressing.' (1852 2: 113–14). Such comments were as much a reflection of the inability on the part of British writers to accept the reality of language differences as they were an attestation of prominent general features in the speech of the black populations across the English islands.

Connections between the speech of white and black Creoles in the English islands had been commented on in the previous century in reference to white women's speech. Day pointed out the following about whites, without distinction: 'It is a curious custom of the whites to speak broken English to the negroes, by way I suppose of making themselves better understood ...' (1852 1: 91). Although it is possible that in Day's time white men were speaking to the coloured people in their dialect and that this was not generally so in the eighteenth century, it is more likely that from the beginning both men and women spoke alike (or at least had the competence to). Some eighteenth-century writers (in the *Barbados Gazette* 1732, Poole 1753, Thompson 1770 and Long 1774) had explained the speech of white women as the result of interaction with the slaves and lack of education. This was also the view of Resident referring to the 1790s: 'From her necessary intercourse with the female slaves, she also acquires a portion of West Indian manner ...' (Resident 1828: 223). On the other hand, Wentworth (1834 2: 183) claimed that it was for the sake of their own children, whose comprehension had been affected by their upbringing by negro or mulatto wet-nurses, and Day (1852 1: 91) said that the use of the slaves' dialect by whites was for the sake of the slaves. In short, these later writers saw it as, or changed it to, a deliberate act for practical purposes. It is unlikely that women's language in the English islands differed markedly from men's at any stage. The probable difference was in what men expected of women and what women actually did.

The vernacular language of the French Creole Caribbean, as presented by various European writers, had a negative evaluation when considered as a means of communication in the public formal arena. In 1835 in the context of St Vincent, Madden (1970: 45–6) commented: 'The vile French **patois**, peculiar to the West Indies, is still spoken a great deal by the negroes, – indeed some of those I conversed with talked nothing else.' As was the case in Haiti, the creole language had to be understood by all residents of the French islands because it was the native language of the majority of the people. Acquisition of competence in creole language became part of the creolisation process for whites going there to reside. However, the creole language was socially subordinate to French. French as opposed to creole therefore had great social significance when used by *gens de couleur*, as Dessalles indicated: 'His bastard girl speaks only French and has learned at Mme Bissette's certain mincing ways that will elicit mockery and hatred from her class' (1996: 291). Consistent use of French expressed the desire to be recognised as a person of higher class.

As was the case in Haiti where the creole was used by the French in a proclamation at the beginning of the Revolution, the creole received some measure of validation as a language when it was used in 1848 in an emancipation proclamation for the French colonies: *Written in French and Creole, this piece informs the negroes of the government's intentions concerning their liberation* (Dessalles 1996: 208). It was a quarter of a century later that the creole language of Martinique received its first 'scholarly' validation from Turiault, one of its own speakers. Turiault's assessment was similar to that of Moreau de S. Méry, in his assessment of Haitian creole about 80 years before – acceptance of the European view of the low intellectual level of the language but a respect for its 'genius' culturally: '*C'est donc un patois, un jargon; mais si ce patois est capricieux, désordonné, enfantin, s'il est drôle, amusant pour les Européens, il a aussi un caractère d'originalité qu'on ne saurait méconnaître ...*' ['It is therefore a patois or jargon; but if this patois is capricious, confused and childish, if it is amusing to Europeans, it also has a character of originality which cannot be mistaken ...'] (Turiault 1874: 6). The same kind of sentiment was expressed in 1863 by Budan about his own creole in Guadeloupe: '*je plains sincèrement ceux qui, ne connaisant pas notre langue, n'en pourront qu'imparfaitement apprécier les finesses*' ['I am sorry for those people who, not knowing our language, can only imperfectly appreciate its finer points'] (1863: 25).

At the end of the century, Garaud's view of French creole language was that it was at its best when it was spoken. For him there was a perfect union between the negro creole woman in full rage and the creole language: the language was ideally suited to passionate, demonstrative behaviour:

> *La langue créole est bien faite pour servir de telles colères. Elle est rapide, bizarre, imagée. Mais pour être vivante, il faut qu'elle soit parlée; quand elle est écrite, elle est inerte et morte. Il lui faut l'accent, les intonations, le geste, les poses, les éclats de rire, les interjections, dont les Créoles émaillent et animent leurs conversations. Mais aussi quelle agitation dans ces phrases qui se heurtent et se pressent! Quel sang coule dans cette ardente pensée!*

*Les Européens, malgré leurs efforts et leur persévérance, n'arrive que difficilement à se servir de cet idiome. Ils le comprennent bien, mais le parlent mal. Cela se conçoit: c'est une langue mimée plutôt qu'une langue parlée.* (1895: 23–9)

[The creole language is tailor-made to serve such outbursts of anger. It is rapid, bizarre, colourful. However, to be alive, it must be spoken. When it is written, it is inert and dead. It needs the accent, intonation, gestures, poses, outbursts of laughter, interjections, with which Creoles decorate and enliven their conversations. But what excitement in sentences which collide and crowd each other! What blood runs through that ardent thought!

Europeans, in spite of all of their efforts and perseverance, only with difficulty manage to master this language. They understand it well enough, but they speak it badly. That is understandable, for it is a performed language more so than just a spoken one.]

Its presumed scholarly limitations were identified by Hearn:

Metaphysical and theological terms cannot be rendered in the patois; and the authors of creole catechisms have always been obliged to borrow and explain French religious phrases in order to make their texts comprehensible. (1890: 222)

Yet, because of the total number of works that were dedicated to the study of one creole language or another in the nineteenth century, creole languages began to attract the attention of the academic world.

In addition to the analyses of creole languages by Creoles themselves (i.e. Russell, Thomas, Turiault), there was one which had appeared earlier in the century (1842) done by the Abbé Goux. Later, with the information provided him by previous studies of creole languages, Addison Van Name was able to do a comparative study of a few creole languages in the Caribbean in 1870. Two years thereafter, another study of a French creole language appeared, undertaken by Auguste Saint-Quentin.

These works focused on creole languages of the Caribbean, but for Europeans the word *creole* had come to acquire a wider, colonial meaning. The Spanish *Diccionario de autoridades* (Real Academie Española) defined *criollo* as '*hijo de padres europeos nacido en cualquiera otra parte del mundo*' ['a child of European parents born anywhere else in the world']. García Icazbalceta, writing in the 1890s as a native of the New World (Mexico), disagreed with this wide definition and gave his own restrictions: '*La definición no es exacta, porque la palabra está confinada á la América española ó francesa, y á las Islas Filipinas*' ['The definition is not exact because it [creole] is confined to Spanish or French America and the Philippine Islands'] ([1899] 1975: 127). His definition showed that in the Spanish world itself the colonial view-point had also extended the word *criollo* beyond the Americas to the Phillippines, which had been under Spanish control from the sixteenth century up to 1898. For the French, *créole* included colonies in the Indian Ocean as a result of French perceptions of similarities across all the various French colonies. In the preface to his book, Turiault (1874: 5) said that *créole* meant '*né en Amérique, dans la colonie*' ['born in America, in a colony'].

As a result of this extension and with the appearance of analyses of apparently similar languages in other parts of the world, in the early 1880s the German linguist, Schuchardt, who had become interested in colonial languages, outlined relationships between vernaculars in the Indian Ocean, some in the Caribbean and others elsewhere. Schuchardt used the term *creole language* in his writings to cover all these colonial languages, including the 1880 study of Mauritian patois by Charles Baissac. What may have led Schuchardt in this direction was the article on *la langue créole de la Guinée portugaise* by Bertrand-Bocandé in 1849. In this article Bertrand-Bocandé not only applied the term *créole* to varieties of language in West Africa but also gave a definition of creole language, using the West African varieties without direct reference to the Caribbean.

# The identity of Creoles – emerging precision and diffusion in their identity

As a result of the legal and moral changes that ended the institution of slavery and as a result of the fact that in the English and French islands negroes were the vast majority of the population, it was inevitable that the dominant racial identity at the national level would shift from white to non-white. In the case of the English islands, consciousness of the possibility of negro dominance, as distinct from a spectre solely intended to frighten, had already taken hold 50 years after emancipation. In the French islands, the shift from white to negro in the vision of writers had reached only halfway by the end of the century. Thus, the typical colour image of the Creole in most of the English islands was becoming black and that in the French islands brown. In the French islands there was no sustained portrayal of multiple ethnicities and diversity in the composition of the population and there was even a tacit rejection of the Carib as a major component in what was portrayed as the typical racial type. Racial mixture was therefore viewed much more positively in the French islands than in the English ones. Separation of racial types was much more the practice among the English, which meant that preservation of a racial mentality and the racial hierarchy of slavery was still a priority for those who were the first to stop slavery officially. The mentality of sharp racial division, which was characteristic of the English, points to an interpretation that the dismantling of slavery was principally the dismantling of an economic system rather than the pursuance of racial equality and brotherhood. Improvement, as a societal concept, was in the English perspective exclusively moral. In the French islands, in spite of the separateness of the *békés*, improvement was accepted as a racial reality, which meant that for them there was a stronger belief in genetic combination than in genetic separation.

The people in the smaller islands, like those in the bigger ones, fell under the general designation 'Creole', which according to its origin was a non-specific, geographical designation. Within this general designation were sub-divisions according to European 'mother' country, but while the people in the English islands were called 'West Indians', there was no comparable name for those in the French islands, even though the islands were quite normally referred to as *les Antilles*. Among 'West Indians' it was only in Barbados, contrary to expectation and to the English image of the island, that a national, geographical name (*Barbadian*) was commonly used to identify the people. For English writers Jamaica had no particular national identity, Montserrat was linked to the Irish, St Lucia and Dominica were rejected as non-English, and so too was Trinidad. Antigua was seen as a model colony in a kind of sanitised way, that is without any special local characteristics. The people of Trinidad were called *Trinidadians* by one writer (Gamble), even though it was difficult to know whom he was referring to in the midst of the great diversity and newness of population in that island. In the French islands, as a result of colonial competition between the two main ones, Martinique and Guadeloupe, a geographical name for the people in each island began to emerge, but by the end of the nineteenth century it had not yet become fixed and uniform in either case.

In both the English and French islands there was a strong belief in the influence of place on people. Rufz and Hearn saw physical 'improvement' in whites, negroes and above all *gens de couleur*. Marly and Wentworth preferred to focus on physical and cultural 'improvement' in negroes alone, probably assuming that whites were perfect. Dauxion-Lavaysse was even more specific, making a claim for improvement in 'inferior races of negroes: The inferior races of negroes improve in the colonies, in respect to intellect, either by their mixture with the superior ones, or by a better climate than that of Guinea' (Dauxion-Lavaysse [1820] 1969: 370). Yet, at the same time, the idleness mentality of Creoles was reaffirmed as the psychological result of the influence of the land. This was thought to be so in two ways. First, as had been the belief from the earliest years, whites in their origin and fibre were believed to be temperate climate people and so to be affected adversely by tropical climate: *I well know that the heat of the climate inclines them* [white Creoles] *to indolence* ... (Dauxion-Lavaysse [1820] 1969: 181). Second, the bounty of the land, the absence of extreme cold and of other unfavourable climatic conditions meant that human beings could exist comfortably without having to work hard. As a result, the idleness culture developed. This latter argument was in nature the same as that which said that whites were spoiled by the privileges of slavery (e.g. Raynal 1773, Bossu 1777). De Molinari's own addition to this well-known argument at the end of the nineteenth century was that the privileges of Nature guaranteed the inferiority of those who were reared with them and prevented the crossover from savage to civilised:

> *L'état d'inferiorité des races qui se sont formées dans les régions dites privilégiées de la nature provient pour une bonne part de ce privilège. Elles ont manquée du stimulant nécessaire pour franchir la première et laborieuse étape qui sépare l'état sauvage de la civilisation.* (De Molinari 1887: 223)

> [The state of inferiority of the races which formed in regions regarded as 'privileged by Nature' comes in good measure from the very privilege. They lacked the necessary stimulant to surmount the first and arduous stage which separates the savage state from civilisation.]

The fact that the strong belief in the influence of the ecology on the formation of the identity of the people was not parallelled by the greater use of precise, geographical designations for the people indicates that neither genetic and cultural uniformity nor belief in ecological determinism was powerful enough by itself to bring about the use of national names. It suggests that such use was more the result of political perceptions and assertiveness, as was the case in Barbados. At the general level, the extension of the meaning of the word *creole* by the French especially and the Spanish constituted a major contradiction because it reduced to insignificance the original meaning and purpose of the term *criollo*, which was to distinguish people and things of the New World from those coming from the Old World. In the context of the late nineteenth-century Caribbean therefore, what it meant was that as island identity with Creoles at the centre started to become more precise, the general term *criollo/creole* in the writings of Europeans became more diffuse.

# Conclusion

## Dramatic contrasts in external characterisations of the Caribbean

When Columbus and other Europeans set out on their voyages at the end of the fifteenth century, they were modern-day versions of the classical Ulysses and they became even more so when, after a period of time, they re-appeared in Europe telling stories of islands and people far away. Because they had had sense enough to take back some of the things and people of the New World with them when they went back to Europe, their stories were more believable than those of Ulysses. The Homeric nymphs and sirens and places in the islands did not find embodiment in specific persons and places in the Antillean islands, but they found confirmation in a general sense in the almost nude women there whose presumed innocence and availability were no doubt titillating and enchanting to Europeans at home. The association of islands with odyssey and enchantment thus grew. It is not accidental, although quite intriguing, that at least two well-known Homeric words have counterparts in the islands, without one being able to establish beyond a doubt a direct historical link between them. Charybdis and Calypso are paralleled by Carib and calypso, the latter of which is somewhat removed from the Greek image but still bearing some relationship to it. Such correspondences therefore conjure up a link between Ulysses and Columbus, between an old mythological world and a new golden world, and between the creative imagination and reality.

From the time of the earliest accounts of the region the general image of the people of the islands was an exotic one because it was presented from the point of view of Europeans. It was an image dominated by the sea, the land, the nature, exuberance and colours of the plants and the brightness of the sun, all of which in some way were transferred to the people, either to their appearance, their spirit or their behaviour. In other words, the identity of the people was seen to correspond to their tropical habitat. In this colourful correspondence between people and habitat, instead of subtlety in variation it was a dramatic contrast of opposites that was noted. The dramatic contrast was consolidated as factual in the literature through a practice of repetition, or 'borrowing'. Writers on the New World reproduced stories, events and ideas that appeared in earlier works, often modifying them to make them seem as if they were their own experience, thereby establishing a continuity of 'evidence', which converted opinion into reality and simplicity into greatness.

A classic example of this practice of 'borrowing' in the early works on the New World is in the repeated account of flying fish. The flying fish, in itself, together with the flying fish–*dorado* [dolphin]–seagull relationship, was a phenomenon that attracted the attention of Europeans as they neared the islands after their long voyage across the Atlantic and became a welcoming sign of the islands. The first account is in Oviedo (1526 fol. xlviii–xlix) in which the size and shape of the fish are given, including the fins/wings; the 'flight' of the fish is explained, with some reference to the wings/fins; the habitat is identified; the context of activity is described, in which the fish are pursued as prey by bigger fish (*doradas*) in the sea and by birds when they leap into the air; an analogy is established between the flying fish and man in which the vulnerability of both is highlighted; the analogy is given a Christian message. A similar account is given by Thevet (1575), then Acosta (1590) and it was reproduced in the seventeenth century by Ligon (1657) and a number of French writers, including Jean Barbot. The wonder of a fish 'flying' was still being noted in the nineteenth century by Pinckard (1806 1: 214 et seq; explanation of the mechanics), by Monk Lewis (1834; in a three-page poem) and by Bridgens (1837; pictorially).

The similarity of detail across these accounts eliminates any possibility of them being independent experiences. The most important point to understand from this 'borrowing' is that writers read previous works very carefully and used them freely in producing what were supposed to be personal accounts. What therefore started as a simple visual experience of an early writer was converted into a general and familiar Christian image and analogy, moving from the literature of Spain into that of France and others.

Another example of this kind of 'borrowing' or repetition of images in the early works is in the description of the preparation of yucca/cassava,[1] a favourite subject in several histories whose authors seemed to be fascinated by the fact that the plant that was the staple food was also very poisonous and that the poison was converted into food almost by alchemy:

---

1. There were several names for this plant in the various languages and there were also different varieties of the plant, which at first the Europeans could not have known about.

> A certaine marvelous industry of Nature lieth hidd in the use of the roote
> jucca. Beeing put into a sacke, it is pressed with great waights layd thereon
> after the manner of a wine presse, to wringe out the juice thereof. If that
> juice bee druncke rawe, it is more poysonous the aconitum, & prēsetly
> killeth, but being boyled it is harmles, & more savery the the whay of
> milke. (Anghiera 1628: 288)
>    ... and more excellent breade made of the roote jucca bruised smale,
> and dryed, which beeing brought into cakes which they call cazzabi, may
> safely bee kept two yeere uncorrupted. (Anghiera 1628: 288)

Such accounts identify the method of removing the poisonous juice
from the root; the baking of the cakes involving the use of a stone; and
especially incidents, involving animals, illustrating the potency of the
poison. Here again the details which filter from Martyr (1516) to the
eighteeenth century are the result of reading and not of actual experience, and the repetition is accompanied by only minimal advance in
scientific accuracy over the years.

The paradox of a staple food and a deadly poison all in one was no
doubt symbolic of the New World for early writers and it is no accident
that as the New World developed, European concepts of it were dominated by polar extremes and excesses. The flying fish–*dorado* [dol-
phin]–seagull relationship was a visual experience of marine life that
was converted into a philosophical reflection on the plight of the weak
in the face of the strong, on the plight of the indigenous inhabitant in
the face of the European, and on the plight of human beings in general.
A contrast of opposites was also seen in the indigenous inhabitants,
who initially (1493) were said by Columbus to be fearful and then later
generous, who were later divided by the Spanish into two groups, one
friendly and the other hostile, and later were presented by the French as
on the one hand noble and unspoilt, belonging to the golden world of
Greek mythology, and, on the other, savages and cannibals from an
uncivilised world. Later, the most fundamental contrast came to be the
one symbolised by the sugar cane, which presented a sharp contrast
between sweetness and wealth on the one hand and slavery and
degradation on the other.

When the debate about the morality of slavery raged at the end of
the eighteenth century and the beginning of the nineteenth century,
engravings illustrating the viciousness of slavery provided the most vivid and abiding images of the Caribbean. Blake's *A Negro hung alive by
the ribs to a gallows* (1792/3), which was done for and appeared in
Stedman (1796), was a harshly critical image, one which must have
inspired one of the poems (Sonnet 6) on slavery by Robert Southey in
1794:

**Figure 7.1** Blake's *A Negro hung alive by the ribs to a gallows*. From Stedman ([1794] 1806). British Library.

High in the air exposed, the Slave is hung;
To all the birds of heaven, their living food!
He groans not, though, awaked by that fierce sun,
New torturers live to drink their parent blood;
He groans not, though the gorging vulture ear

The quivering fibre. Hither look, O ye
Who tore this man from peace and liberty!
Look hither, ye who weigh with politic care
The gain against the guilt! Beyond the grave
There is another world: bear ye in mind,
Ere your decree proclaims to all mankind
The gain is worth the guilt, that there the Slave,
Before the Eternal, "thunder-tongued shall plead
Against the deep damnation of your deed."
(Robert Southey, Bristol, 1794

In contrast, there was the coconut tree image of the Caribbean, which
evolved from the eighteenth-century depiction of the sweet life of the
West Indian planter. This latter in fact preceded the *dolce far niente* image, which in nineteenth-century Europe represented one view of *time-
less southern* [European] *indolence* and *a Golden Age* (Ormond and
Blackett-Ord 1987: 9). The expression *dolce far niente* was taken from
Italian and, probably because of this, the image was inextricably associated with a Latin way of life or preference. It was a positive image in
nineteenth-century Europe, illustrated in a painting by Frans Winterhalter in 1836, which Ormond and Blackett-Ord describe as follows:
'Ostensibly a scene of the siesta hour during the grape picking season, it
is, in reality, a peasant version of the *fête champêtre*, in which the

**Figure 7.2** *A West India Sportsman* (The coconut tree image & *Dolce far
niente* Caribbean style). Published by William Holland. National Library
of Jamaica.

figures abandon themselves to the pleasures of the senses' (1987: 9). In the case of the Caribbean, however, *dolce far niente* was never a positive image when the focus was on Creoles or natives of the region. From the earliest descriptions of the indigenous inhabitants, it is clear that, because Europeans saw other races as workers and servants, the idea of leisure among them was intolerable. This is evident in Charlevoix's version of eighteenth-century *criollos* (see Chapter 4).

The *dolce far niente* way of life in the Caribbean among Creoles was attributed to three sources: indigenous, Spanish and environmental. In the case of the Spanish colonies, it was considered to be a feature, as it were, of the Latin 'diaspora' as well as a way of life of the indigenous forebears and, because the English and the French regarded the indigenous inhabitants as uncivilised and because they despised what they interpreted as an extension of the Spaniards' way of life to negroes, *dolce far niente* was neither an upper-class image nor a positive one. Furthermore, racial difference was maintained in the application of the concept to the negroes of the Caribbean by Day when he remarked that 'The *dolce far niente* of Naples is prompt activity compared with the indolence of the negroes' (1852 1: 249). In addition, when Creoles no longer tolerated European domination and differed in lifestyle, *dolce far niente* was explained as ecological. It was thought to be sustained by natural bounty – the Caribbean was portrayed as a place where food provided itself, thus fostering idleness.

However, there was a more fundamental reason for the French and English dislike of the Spanish colonial way of life, one which may be seen as superseding the colonial context. As Diamond argues, humans have always had preferences for different means of food gathering:

> men hunters tend to guide themselves by considerations of prestige: for example, they might rather go giraffe hunting every day, bag a giraffe once a month, and thereby gain the status of great hunter, than bring home twice a giraffe's weight of food in a month by humbling themselves and reliably gathering nuts every day. (1997: 108)

As a result of these preferences, human societies come to have negative opinions of means of food gathering that differ from theirs, as Diamond goes on to explain:

> throughout human history farmers have tended to despise hunter-gatherers as primitive, hunter-gatherers have despised farmers as ignorant, and herders have despised both. All these elements come into play in people's separate decisions about how to obtain their food. (1997: 108)

When, therefore, Spanish Creoles were reported by Charlevoix (1731 2: 479) to have said that in their societies *ay Hombres*, it can be seen to be consistent with Diamond's argument. The same sentiment was implicit in the Dominican negro's *Soy negro, pero negro blanco* (Welles [1926] 1966 1: 104). These sentiments were coming from a country that produced the original buccaneer, who hunted and skinned wild cows for a living. It was the element of excitement and danger in hunting and even gathering which made the life of a farmer mundane in contrast. The farmer, however, never understood more than security and material gain at the end of the day and eschewed any other option.

Even more negative images of Caribbean people came out of their practical relationships or responses to Europeans, especially the French and English. In these images European politics, economics and social class created a subordinate and even non-existent identity for the people of the Caribbean, which the latter constantly had to react against or try to overcome. With the advent of the Europeans in the region at the end of the fifteenth century, the indigenous inhabitants, not even where they so desired, were not allowed to exist on their own in peace. Bouton's (1640: 135–6) reflection on their preference (*'Ils disent que c'est nous qui avons besoin d'eux, puis que nous venons en leurs terres, qu'ils se sont bien passez de nous, & s'en passeront bien encore'* ['They say that it is we who have need of them, since we came into their territory, that they did well without us and will continue to do so']) really underscored the fact that they had little understanding of the juggernaut that was about to crush them; they could not see that an isolated, independent or separate identity was no longer possible. No European writer was going to say, like the writer of Numbers 23:9: 'Lo, a people dwelling alone, and not reckoning itself among the nations!', especially when there was a vision of gold in the offing.

In the hierarchy of civilisation constructed by French writers, the indigenous inhabitants were dubbed 'savages' and placed on the bottom rung, in polar opposition to 'Christians'. This was not simply a religious or cultural assessment, but essentially a political act: it was not fundamentally altered even when Europeans later came to find out about the sophisticated Aztecs and Mayas. For French writers, especially those who were priests, 'savages' were not only uncivilised but they also corrupted Europeans and European things with which they came into direct and intimate contact. From the earliest years, therefore, the Caribbean came to be depicted as a place where the European race, European morals and culture, and European languages were corrupted. A darkened skin, a dissolute life style and vulgar language were characteristics associated with whites in the islands, and they were at the very top of the social and cultural hierarchy. The climate of the islands, which was characterised as hot and unhealthy, was said to contribute to the dissolute lifestyle and deterioration.

Descendants of the Spanish colonists in the Spanish islands were despised by the French and the English not only because they had been racially corrupted through miscegenation (in the eyes of the French and English) but also because they seemed to them to have little interest in Europe, being quite proud of their own *criollo* identity (see Charlevoix 1731 2: 479). Americanisation and independence among *criollos* were obviously conceived of as backwardness, and racial mixture was thought to cause deterioration, the idea being that the bad traits from each of the contributing races came together to overpower and super-sede the good ones. The connection, in the English and French vision, between *criollos* in America and *cologlies/koulouglis/*colollos*, in North Africa (*Argel/Barbary*) seems incontestable. Blofeld, speaking of Algeria, commented that:

> People from almost every part of the world, may be met here. All the primitive types of the African race, all those that the migrations or the successive conquests have transplanted to this ancient soil, some, modified by admixture, others remaining unaltered by living in tribes, and leading a nomade life. (1844: 113)

Similar comments by French and English writers, tainted by negative attitudes to mixture of people and languages and the peoples' *joie de vivre*, characterised accounts of the Caribbean from the earliest days up to the nineteenth century.

The *criollo* lack of interest in Spain, viewed by French and English writers as a descent into backwardness, put into focus the attachment to the mother country exhibited by French and English colonists. The picture of the people of the French and English islands endeavouring to send their children 'home' to be educated and to follow the fashion of Europe, while it suggested that these colonists were sensible, confirmed the superiority of Europe as well as the inferiority and insecurity of colonials. The picture of indolent Spanish colonists satisfied with their life in the Caribbean and proud of the Caribbean as their home was interpreted as the result of their adoption of the mores of the indige-nous inhabitants.

European images of corruption and mixture in the Caribbean (and North Africa) contrasted with and increased Renaissance images of whiteness and purity in Europe. This contrast was needed, maintained and strengthened as the cultural and philosophical underpinning of European colonisation and exploitation – it established Europe at the pinnacle in the hierarchical order of things. All those who came under the influence of European writers, including Europeans themselves, imbibed this philosophy. It was European writers who, using the earliest (i.e. Greek and Roman) written literature of Europe, had earlier begun to construct a cultural view of Europe as one descending from a 'classical' period, a pure culture and pure people, eliminating from this picture the vast cultural and linguistic mixture that Europe really was. European confrontation with the New World and the necessity to assert superiority in endeavours there led to increasing sanitising of European self concepts and a polarisation with those of other cultures. Culture in the colonies therefore had to be maintained as inferior. Thus, the seventeenth-century picture of buccaneers in Hispaniola chasing cows, skinning them and roasting their flesh in the open air was soon matched by that of pirates chasing Spanish ships, robbing them of their gold and silver and returning to Port-Royal in Jamaica to celebrate their conquests in the capital city of debauchery. These pictures were also matched by that of the West-Indian planter returning to Europe flanked by black slaves and flaunting his filthy lucre.

Another contrast with Europe that was used to highlight the poverty of colonial society was the familial structure of society versus the absence of it. Because the Caribbean was peopled and run in a way to ensure maximum economic gain, the family (or at least the European concept of it) was not the base unit in its social structure. In the plantation system of slavery specifically, the preservation of the nuclear family unit among Africans was unimportant and it was repeatedly fractured in the buying and selling of slaves. Consequently, Bird (1869) argued that nationhood was virtually impossible in Haiti since it was not composed predominantly of nuclear families. Additionally, he argued that, because the role of the woman was central in the home and thus the family, any country in which a great number of women were engaged in commercial activities that prevented them from performing their domestic duties could not become a successful nation. So, since women were identified as performing both legitimate and illegitimate (i.e. prostitution) commercial activities in the Caribbean, it was impossible for such societies to become cohesive nations. Furthermore, white men keeping women disturbed the accepted family structure and undermined the nation. In short, because they differed in social structure from European societies, Caribbean societies were regarded from this European perspective as generally baseless.

Yet, the first and probably most sharply contrastive characterisation of Creoles was said to have been made not by Europeans but by Africans, who regarded their creole off-spring as off-shoots and inferior because of their place of birth (a new world without culture). Like the Europeans, who were interested in a closely knit ethnic and family structure, the absence of a genuine, traditional, cultural matrix (i.e. an ancestral home) for their off-spring must have been therefore a great concern for Africans. This absence, of course, contradicts Du Tertre's (1654: 476) cavalier view of Africans (*'ils font de toute terre leur patrie'* ['they make every land their fatherland']) and it may explain the persistent hiatus between the African and the Creole. Initially, the African view of the African–Creole hierarchy was the reverse of what the dominant European view came to establish. The African in the islands who had previous experience of being a free person in Africa is said to have looked down on those who were born into slavery, as the creole slaves were. The African had ritual marks to show proof of status, while the creole slave had none. For Europeans, however, marking was disfiguration and was symbolic of punishment or of chattel, and so, for them, the African bore his abject status visibly so that all could see it constantly. Thus, whether the viewpoint was African or European, there was a distinction between marked and unmarked people, and Creoles were in the latter category.

As to language, a polar contrast between European and non-European was maintained from the outset. There were reports of a great number of languages in use among the ingenous inhabitants and then among Africans, as well as lack of comprehension among their users. In contrast, a notion of uniformity in European languages established itself as a result of the fact that little was said about lack of intelligibility among Europeans. As far as the Africans were concerned, the belief in the lack of unity and lack of a common identity issued from the repeated statements about language diversity, and it persisted throughout the colonial period.

The notion of *bastarda lengua* was applied to language in the Caribbean before there was any clear distinction between the Romance dialects in the Iberian peninsula, dialects that had been under Arabic influence for the preceding seven centuries. The term *langage corrompu* was applied to language in the French Caribbean before there was any official standard variety of language in France, a country with distinctive regional dialects. Makeshift languages for basic communication between speakers of different languages had long been in use among the Europeans and North Africans in the Mediterranean. Language 'corruption' in the Caribbean was therefore not being contrasted with any reality of language 'purity' in Europe; the contrast was a necessary part of the politics of colonisation, and indeed in a convoluted way helped to create the perception of a pure standard in the metropolis.

It was natural for contrasts to be highlighted in the early stages of culture contact when people were seeing each other for the first time.

What was critical in the islands, where new societies were being con-structed, was that sharp, social and racial stratification was put in place to maintain the contrasts permanently. As a result, perceptions of con-trast were prevented from giving way easily to perceptions of uniform-ity. So, it is not that dramatic contrasts were more characteristic of the sixteenth century than any other, for the encounter between new 'migrant' (African, European or other) and New World person contin-ued throughout the period of slavery across different European colonis-ers. In fact, the persistence of the very term 'creole', which was coined to identify contrast, testified to the strength of the perception of con-trast over the centuries. Ironically, the portrayal of contrasts in the lit-erature was of less import in the sixteenth century when there was general illiteracy in the new colonies than in the ninteenth century when the rhetoric of emancipation was very strident and filtered down to all strata of the society. Contrast and uniformity were simultaneous and went hand in hand.

# New and conflicting internal visions of Creoles: ordained divisions versus natural blends

The *criollo* shifting of focus away from Europe was obviously caused by the decline of Spain and the introduction of new elements (indigenous and African) into their own genetic make-up and consequently into their cultural make-up. Thus, exclusive control of the creole mind by Europe began to be 'eroded' by competition from elsewhere. It is not surprising, then, that the European view of racial contamination, which had been popular from the beginning of the eighteenth century, but which con-flicted with the beliefs of *criollos* themselves, was explicitly contradicted by a late nine-teenthcentury view among French and Spanish Creoles, which claimed that mixture conditioned by the Caribbean environment caused spectacular improvement. Even before that, Charlevoix's (1731: 479) negative attitude to a genetically tripartite *criollo*, made up of *Espagnol*, *Afriquain* and *Ameriquain* and occurring in varying combina-tions, was in the late eighteenth century to some extent converted into a positive image in Blake's engraving 'Europe supported by Africa & America' (in Stedman, 1794). The earlier view of *criollos*, which saw the product of mixture as deterioration, thus was no longer uniform even among Europeans at the end of the eighteenth century. In the second half of the nineteenth century mixture was seen as improvement in the French Caribbean (Rufz 1850; Hearn 1890), and as '*fusión transcen-*

**Figure 7.3** *Europe supported by America and Africa* By William Blake. From Stedman ([1794] 1806). British Library.

*dental de razas ... el crisol definitivo de las razas'* ['transcendental fusion of races ... a definitive crucible of the races'] (Maldonado-Denis' interpretation of María de Hostos 1869–1903) also in the second half of the ninteenth century in the Spanish Caribbean.[2] Furthermore, in the post-revolutionary French Caribbean the policy of *fusion* (Dessalles [1847] 1996) was seen as a desirable social and political activity, as if to match *fusión* in the Spanish Caribbean, which was seen as a positive, genetic, historical process.

---

2. This was repeated and extended in Vasconcelos' twentieth-century view of a cosmic race in Latin America made up of all races.

The change of attitude towards mixture in Spanish and French colonies was essentially a graduated change of point of view from outsider to involved outsider to native. In the English colony of St Vincent, however, there was no equivalent change of viewpoint in the case of the other significant mixed group that emerged, that is, the Black Carib. From the start, among the Caribs the racial mixture of Carib and Black was negatively regarded and rendered by the word *chibárali*. It is probably because Black Caribs were a two-part mixture, which did not ostensibly include the European, that there was no improvement in their image. It is also because St Vincent, the place of origin of the Black Carib, became an English colony, and one in which the Black Caribs were standing in the way of English 'progress', that there was no change of attitude to the Black Carib. Moreover, no Black Carib perspective was represented in the literature to counterbalance the European view and provide for dialectical discourse. Yet, there was at least one possible, positive, indirect legacy – the word *jíbaro*, which became a nationalist symbol in the Spanish islands.

The new genetic type in the French and Spanish Caribbean was seen to be the result of a natural process of creation through procreation. This was different from the European view of genetic types, which were seen as divine, ordained creations and divisions. For the European, ordained divisions were not to be meddled with, and so Darwinian philosophy of evolution became heresy. Entrenched notions of 'classical' purity from a by-gone age therefore came into conflict in the ninteeenth century with the concept of a continuously evolving natural world in which there was thought to be improvement in beings through natural selection. Improvement in genetic type in the Spanish and French Caribbean through natural selection and favourable environment was 'scientifically' supported by Darwinian philosophy. On the other hand, divinely ordained separation of races was necessary for European colonisation as well as US dominance of the Spanish islands, and was confirmed by church doctrine. This fundamental conflict in interpretation and vision can be compared with and related to two contrasting views of colours – one that distinguishes between 'primary' colours and other colours, these latter seen as deriving from the former, and the other in which all variations of colour are seen as equal in nature and not really as naturally constituting separate colours or hierarchies of colours.

In the Spanish and French Caribbean the new genetic type identified in the nineteenth century constituted an evolution in vision. Racial identity was no longer totally constrained by the plantation/pyramid model ranging from white on top to black at the bottom with sharp divisions in between; new political orders were emerging in the different groups of islands and with some of them were associated different genetic types or stereotypes. There was also a shift in vision in Haiti after the revolution, but this did not involve the promotion of a new genetic type – it was an attempt to stand the pyramid on its head. The focal racial types that emerged in the nineteenth century in the different islands can be identified as follows:

| Spanish islands | French islands | Haiti | English islands |
|---|---|---|---|
| Mestizo | métis | black | white/black |

It is not that the valuation system that contrasted 'white' through many gradations with 'black' had disappeared in the Spanish and French islands; it is that demographically predominant types there were repainting their own image in an expression of ethnic identity. The 'Indian' (native of the islands) began to be used as a key element to establish identity racially and territorially. In the English islands, except possibly Jamaica, no politically dominant mulatto type had emerged or had ever been accepted as being distinct and therefore this category, as a representative of a new English Caribbean identity, was not prominent. In the English Caribbean the sharpness of the contrast between black and white was eventually dulled only in those islands that had substantial Asian immigration after emancipation. Asians, wherever they went, came to constitute a genetic intermediate zone between black and white, but with a separate cultural identity; in a sense, they were the opposite to the 'Indians' (natives of the islands), who were used to represent a blending of black and white.

There was also a beginning of change in vision in the area of language. The use of Caribbean types of language in written literature began to disturb the exclusive use of the 'ordained' types (French, Spanish, English, etc.). In spite of a continued belief in the intellectual superiority of the languages of Europe, Caribbean varieties were seen as being able to express the essence and spirit of Caribbean people. These constituted counter-balancing entities in a language situation in which stated, formal, societal preferences contrasted with implicit values in informal usage. In the different groups the contrasts were as follows:

| Spanish islands | French islands | Haiti | English islands |
|---|---|---|---|
| Spanish/jíbaro, habla bozal | French/Creole | French/Creole | English/bad English |

Yet, not even in Haiti, where the local variety of language came closest to being recognised as a language on its own, was there any stated political intention to resolve this apparent contradiction, that is to remove the old standard and to make the national language correspond exclusively to the image of the Black Republic. Valuations attached to the 'ordained' languages seemed more immutable than those attached to race and other aspects of culture.

There was also an evolution in the concept of home as well as places to which one was naturally allied. Driven by deteriorating economic circumstances and aided in some cases by the relaxed grip of the mother country, Creoles saw themselves crossing boundaries and forming alliances beyond the traditional ones. In the Spanish islands many saw the United States as a favourable suitor and actively sought to become attached formally to this bigger political entity for security and economic reasons. Other groups in the Spanish islands maintained a more traditional and conservative view, arguing that even if Spain was no longer their beacon, they preferred to form an alliance with only those who spoke their own language. In this respect, the vision of *nuestra América* was essentially a vision of a Spanish America. In contrast, the vision of alliances within the new political order in Haiti was not one determined by language but by race. Haiti saw itself as the capital of the black world in America. On the other hand, those who migrated from Haiti and other French islands to Louisiana in the wake of the French and Haitian revolutions consolidated cultural and linguistic links between the islands and the North American mainland. In the English colonies, pursuit of negro solidarity was not a reason for migration and so it was only poor Jamaicans who, probably because of geographical proximity, were attracted to Haiti. For those at a higher economic level in the English islands, trade made the United States an attractive ally. Yet, even though generally the English colonies remained tied to Britain, overall in the islands there was a more American vision of things, for, as each island (or part of an island) emerged with an individual identity, there was a consciousness of belonging to the wider region of America, of being ethnically New World.

# Creolisation – ethnicity as an environmental process

The change in vision in the Caribbean was not a sudden, all-embracing event; it was a gradual and inevitable consequence of migration and culture contact that started with the very advent of Columbus. When the first Spaniards in the New World accepted the 'name' *guatiaos* and the bond of brotherhood it symbolised, it was the first step in their Americanisation. Besides the exchange of names that initiated it, the bond of brotherhood proved to be an important relationship, and not one merely accepted out of politeness, seeing that it survived to become the *matelot* relationship of young French 'colonists' in St Kitts and the *boucaniers* in Tortuga and northwest Hispaniola more than a hundred years later. It also became the 'shipmate' relationship of the slaves who came across together in the middle passage. This was not a social relationship that the first Europeans brought with them, but one which they adopted from their 'hosts' because it was found to be beneficial. A strong peer-group relationship was an environmental necessity for new colonists and other arrivals who wanted to survive and hopefully prosper in conditions foreign to them and among people it was better to befriend.

Even in a colony like Barbados where there were no indigenous inhabitants to initiate a *guatiao* model of social organisation and where English preferences were dominant from the start, the environmental necessity for interdependence was evident. The success of this colony was the result of its internal structure, which was interpreted by Keimer to parallel that of a family:

> The truth is, the whole community is like a single family. Each individual is known to the rest, and as all are supposed to come at first with an intention, and the hopes of making or mending their fortunes, so by a constant intercourse of business and pleasures, there is an opportunity of conversing almost at once, with men of every condition and circumstance of life. In that island (which has been for near a century more fully inhabited and better cultivated than any part of England except London) the true characters of persons are soon discovered, and their talents or foibles seldom long concealed, which is partly owing to the hospitality of those already settled, who it is acknowledged receive and entertain strangers in the kindest and most generous manner, provided they bring any recommendations, or can give a tolerable account of themselves; always laying out how they may forward the interests and encourage the undertakings of such as propose to reside among them. (1741: iv)

The bonding and building process that knitted together people and business was envisaged as the model to follow. It was a notion of family and a notion of looking out for others, arising out of the difficult circumstances and realities of new societies.

Consciousness of a new or Caribbean identity can be said to have started with the use of the name *criollo*, though this was a name and identity created by others and though initially the term may have been one of disparagement. *Criollos* were not foreign to those who coined the term, but they were 'new' in the sense that their secondary characteristics were different and furthermore it was believed that, even if initially their physical characteristics were the same as those of their European and African parents respectively, their place of birth would gradually change some of these 'primary' characteristics. In other words, *criollo* embodied a strong conviction, first expressed by López de Velasco (1571–4) but later becoming general, about the effects of environmental conditioning on race and colour. So, *criollo* was not a place-name but a geographical term of classification that highlighted ecological factors of determination. The association of place, soil, plant and shoot that related *criollo* to Oviedo's *cogollo* in 1526 reflected an analogy between reproduction of plants and reproduction of people that dominated the thinking of the Spanish for centuries.

Use of terms that compared growth of people to growth of plants was also a mode of thinking among the English when they established their colonies in the seventeenth century. Note the use of *transplantation* and *mother-soil* in following by Hickeringill: 'The major part of the Inhabitants being *old West-Indians*, who, now *Naturalised* to the Country, grow better by their *Transplantation*, and flourish in Health equivalently comparable to that of their *Mother-Soil*' ([1661] 1705: 41–2). Here, by 'old West-Indians' Hickeringill meant the English colonists who had originally been in Barbados or St Kitts but had migrated to Jamaica after the English took it over in 1655. So, even if these colonists had not been born in the islands, at least the land had modified them to the extent that they had become 'old West Indians'. The notion of transplantation (from Europe to the colonies) also occurs in Saintard (1754: 135), where it is identified with loss of love of the *patrie*. The analogy between human beings and plants was therefore a dominant concept in early visions of what was, although it was not explicitly named such, a creolisation process – a process by which Old World stock was converted into Creoles.

The analogy with plants and the influence of the soil on them continued as the focus shifted from transplantation to local modification. This is nowhere more strongly expressed in the nineteenth century than in Rufz's concept of the Creole as *'l'enfant du territoire'* ['the child of the territory'], the result of *'cette action du sol qui s'approprie les races et qui les frappe ... à son effigie'* ['that action of the sun, which takes the races and knocks them ... into its own image']. For Rufz, the process had taken two centuries. Similarly, for Moreau de Saint Méry, creolisation at the community level was gradual:

> *Dans les établissemens par Colonie, au contraire, il n'y a aucun trait, aucun esprit général; c'est un composé informe qui subit des impressions différentes, & qu'il n'est point aisé de façonner. Cela se fait remarquer, surtout lorsque ces Colons vont habiter un climat qui leur est absolument*

*étranger. Chacun conservant alors l'habitude de quelques usages du lieu qu'il abandonne, mais modifiés & appropiés au pays où il est transplanté, la Colonie entiere ne peut offrir qu'un ensemble bizarre. C'est donc au tems seul qu'il appartient d'influer sur des êtres qui se trouvent ainsi mêlés comme par hasard.* (1788: 2)

[In the establishment of colonies, on the contrary, there is no general trait or spirit; it is an informal mixture which undergoes different influences and which is not easy to fashion. This can be seen especially when these colonists go to live in a climate which is totally foreign to them. Each one maintaining the practice of some customs of the place which they left behind, but modified and tailored to the country into which they have been transplanted, the whole colony could not but be a bizarre ensemble. It is time alone which has influence on beings who find themselves mixed together in this way as if haphazardly.]

Moving beyond acclimatisation (a base level at which human beings can be compared to plants) toward homogenisation and community spirit (an exclusively human level) required time for differences to diminish and commonalities to emerge. However, there seemed to be a conviction that the environment would act equally on all and would cause to flourish on the one hand or die on the other those characteristics that were or were not respectively compatible with it.

*Criollo* was a geographical term of classification that linked the children of Old World people to the Americas and, more specifically, to the Caribbean. Place of birth, however, did not in itself absolutely and inevitably create a consciousness of the Caribbean as home among all those born there because the concept of home was as much inculcated as it was a lived experience. For the same reason, place of birth did not absolutely and inevitably exclude all those not born there from a consciousness of the Caribbean as home. While there were might have been several political and social reasons that did not encourage creole whites to identify with the Caribbean islands as home, the key factor was the colonial and therefore subordinate status of the islands. For some creole negroes, there were psychological and religious reasons why Africa represented home, causing them not to completely identify with their birthplace.

As to those whites who were not born there, whether the Caribbean became home was dependent on various personal, social and political factors. Interestingly enough, creolisation or a growing preference for life in the islands affected not only colonists but also military people: 'I have met with many military men who have passed the greater part of their lives in the West Indies, and are now so completely creolized, that I doubt much if they would wish to return home' (Alexander 1833:

250). In this way, soldiers and other professionals, such as priests, for personal reasons became fixtures in the Caribbean islands and consolidated their own roles in the shaping of the colony as well as the influence of the mother country.

As to the colonists themselves, Mintz, in distinguishing between the responses of different Europeans to the colonies, underlines the importance of political factors:

> Spanish administrative control over the colonies was more rigid than that of the French and English. But one may hazard the guess that rigid colonial administration by the metropolis resulted in the swifter growth of a local or 'creole' identity. Whereas the Spanish settlers in Cuba and Puerto Rico soon came to view themselves as Cubans and Puerto Ricans, the French and British colonists apparently tended more to see themselves as Europeans in temporary exile. (1971: 487)

In reference to slaves in the same territories, Mintz continues from this view of whites and says:

> At any rate, I would argue that the more a Cuban slave were to identify with his master, the more Cuban he became; whereas the more a Jamaican slave were to identify with his master, the less Jamaican he would become. (1971: 488)

Mintz thus directly relates the shaping of slave identity, and thereby that of the whole colony, to the political shaping of the identity of masters. In this analysis, political forces are seen to push people in one direction or another by causing psychological reactions. In the case of the Spanish colonies, rigidity in the umbilical cord is seen to have caused the colonists to want to be free of restriction. This psychological reaction to dominance in itself illustrates the theory of contrast or 'otherness' in the construction of identity.

While the same sort of reaction may have been partly responsible on the North American mainland for the Boston Tea Party and US independence, in the British Caribbean islands, as Mintz says, the reaction was virtually the opposite. However, in this comparison and contrast between the Spanish Caribbean island colonies and those of the British and French the element of time should not be forgotten. Moreau de Saint Méry's stress on the element of time in the formation of identity ('C'est donc au tems seul qu'il appartient d'influer sur des êtres qui se trouvent ainsi mêlés comme par hasard' ['It is time alone which has influence on beings who find themselves mixed together in this way as if haphazardly'] (1788: 2)) can be used to make a distinction between the Spanish and the English colonies – the Spanish colonies were more than a hundred years older than the English and the French ones.

In spite of the many statements by English writers and the conviction that the African slaves were happier and better off in their new homes than in Africa, evidence shows that many Africans were spiritually attached to their homeland and that ideas of returning to Africa persisted up to the twentieth century in spite of systematic vilification of Africa. Back-to-Africa ideas were embodied in the widespread belief among the slaves in the transmigration of souls and, with each new boatload of slaves, this belief was kept alive. After the abolition of the Slave Trade by the British at the beginning of the nineteenth century, repatriation of Africans became a reality even if it was not inspired by Africans. At the same time, in Cuba repatriation of blacks was mooted as a solution to the Cuban 'problem', although again this was not inspired by Africans. However, in 1837 in Trinidad the Daaga incident, which featured an African and his companions trying to make their way back to Africa, suggests that among 'liberated' Africans, slaves, ex-slaves and their descendants *Guinea* or *Congo* still continued to represent 'home'. Consciousness that the presence of Africans in the islands was the result of a diaspora meant for many – Africans themselves and others – that Africa was their rightful place of abode.

Creolisation or, more specifically, recognition of self as Creole (parallel to recognition of self as *criollo* or *créole* in the Spanish and French islands respectively) was slow and heterogeneous in the English islands. The process of bonding and 'family' building was fostered only where there was a desire to survive in the foreign environment of the colony and to make that foreign environment home. However, attachment to the islands as home was delayed in the English islands (excluding Barbados) because neither the English nor the Africans willingly and overwhelmingly relinquished their ancestral attachments to other places. The words of the Jamaican slave shortly before emancipation ('You brown man hab no country ... only de neger and buckra hab country' – Marly 1828: 94–5) echo the beliefs of most of his predecessors as well as those of his masters. It was as if *country*, for them, implied attachment to a racial stock in its ancestral environment or was inalienably tied to what Schoelcher (1842: 153) referred to as the '*deux souches primordiales*' ['two primordial roots']. The words of the Jamaican slave thus forthrightly contradict Du Tertre's opinion that the African slaves had no abiding sense of homeland ('*Ils font de toute terre leur patrie*' ['They make every land their fatherland'] – 1654: 476).

# Homeland, geopolitics and nationalism

*Breathes there the man, with soul so dead,*
*Who never to himself hath said,*
*This is my own, my native land!*
(Sir Walter Scott, 1803 *The Lay of the Last Minstrel*, Canto Sixth, verse 1)

There is a suggestion in these Scottish verses that national feeling is spontaneous and that the individual who does not have it, at his death, will be 'Unwept, unhonour'd, and unsung'. By and large, however, in the first 200 years of the history of the 'new' natives of the New World there was little evidence of this kind of sentiment and there was little concept of nation state. It is with the growth of nationalism in Europe that the concept of nation began to crystallise in different parts of the New World and 'new' nations began eventually to emerge. Even so, when the United States declared themselves independent, it may have appeared to many to have been the result of economic conflict more so than the result of a conscious desire to assert nationhood – the US itself was a union of states and not a single 'nation'. However, when Haiti became independent and asserted a new identity through its first constitution, the concept of national identity became crystal clear and the prospect of other places in the Americas becoming politically independent nations began to take hold among Creoles generally as well as Europeans, who had up to then regarded the territories in the New World as projections of themselves. Ironically, visions of nation states then began to control the thoughts of politically minded people and to sharpen divisions among the people, as a result of which the depiction of ethnic identities across islands was disturbed. Thus, in the Caribbean the original concept of *criollo*, which had long preceded notions of national identity, declined in the face of growing nationalism and national identities.

Using the Spanish islands as a model, the most important factors in the development of *criollo* identity and the attachment to the islands as home can be said to have been the emergence of a new genetic type, political separation from the mother country and a longer period of homogenisation. However, while these factors determined ethnic identity, they did not amount to an assertion of national identity. In Cuba in 1608 the first inklings of one of the islands as native land appeared when a heroic figure was dubbed *criollo de la tierra*. Even more so than *criollo* itself, the use of the term *la tierra* meant that the extended meaning of *la tierra* had become a reality for them – they had come to regard their birthplace as 'native land' or 'home'. This was when Cuba found itself in the midst of the conflict between Spain and other European nations in the late sixteenth and seventeenth centuries because it had the most convenient port in the Spanish islands. Cubans were embroiled in conflict and constantly fending off attacks, not from any formal, colonial enemy but from pirates. It was this absence of a formal European enemy that made the conflict more Cuban in nature. Thus, national sentiment or assertion of identity was fostered by and closely associated with conflict in which people banded themselves together for survival and, apparently, to defend *la tierra*. It took more than a hundred years longer in the case of Santo Domingo, but it was also military conflict that provoked the use of the national label *dominicano*, which occurred for the first time in Luis José Peguero's '*Romance en que se dise, que los valientes Dominicanos an sabido defender su isla Española*' ['Romance in which it is said that the valiant Dominicans knew how to defend their island Hispaniola'] (1762 1: 267–71).

If military conflict, according to these examples, was the spur for assertion of nationalism, then national identity must be regarded as negative and divisive, as opposed to ethnic identity, which issued out of a perception of similarities rather than differences. Supporting this view is Saintard's comment that attachment to the land in Saint Domingue was not an inspiration to nationalism but the opposite: '*Ceux qui y naissent recevant de la nature une Patrie fertile, s'annoncent par des moeurs franches & naturelles plus convenables aux lieux; avec moins d'intrigues ont plus d'attachement au sol; sont plus Colons, & par-là plus sujets*' ['Those born there, receiving from Nature a fertile native land, exhibit open and natural customs more suitable to the places; with less intrigue they have a greater attachment to the soil; they are more colonist and consequently more subject'] (1754: 136). This was a variation of the theory behind the coconut-tree image and the notion of indolence in the Spanish islands – fertile land leads to abundance and abundance leads to contentment. So, a person in a colonial situation for whom the land was bountiful accepted domination and had little inclination to be hostile or revolutionary or to want to assert national identity.

Of course, Saintard was speaking of white Creoles, not slaves. For slaves the land was a harsh reality and scarcely a source of bounty. It was commonplace for European writers to note that after emancipation the former slaves wanted to separate themselves from the land, and this was so even in Haiti after Haitians won their independence. Thus, where the land was the source of the daily harshness of life, it evoked less of a sentimental bond with the notion of 'home'. For black Creoles, identity of land and the concept of 'native land' or 'home' was possible where there was no harsh agricultural system like that of sugar cultivation, as in some cases in the Spanish islands. For black Creoles generally, defence of the land could hardly have been a spontaneous reaction and could hardly have provoked national sentiments.

There was, however, one case in the Caribbean islands where 'national' identity was proclaimed from early even by the black population in a way that seemed to supersede ethnic identity and in a situation where there was no actual military conflict – Barbados from the beginning of the eighteenth century. In this case where the 'national' identity embraced both the white and the black population, it can be argued that this came about because many of the early whites there were political exiles or banished convicts[3] who felt a keener sense of separation from England, who had to abandon the notion of returning home to England and who had to come to terms with Barbados as home. In fact, Pinckard, referring to poor whites there, underlined the notion of separation over time:

> throughout many generations, their predecessors have lived constantly, in the island. Some have not been able to trace back their pedigree to the period when their ancestors first arrived, and therefore have no immediate thought or regard, concerning their mother country; but abstractedly consider themselves only in the detached sense of Barbadians, fondly believing that in the scale of creation there can be no other country, kingdom, or empire equal to their transcendant island – to their own Barbadoes ... (1806 2: 132–3)

It was this 'national' sentiment, which grew among the white creole population, that was transmitted to the slave population especially by the poor whites who worked side by side with them. Barbadian nationalism and attachment to Barbados as homeland, among blacks, can be said then to have arisen indirectly from military conflict, but was confirmed by pattern of socialisation and passage of time.

In addition to the element of time, Pinckard explained that national sentiment in Barbados also developed because of the colonial importance of the island:

> From situation [i.e. where it was geographically situated], and from its fine bay for shipping, even independence of its produce, it must ever be valuable to us, and indeed may be considered as the key of the West Indies. Some of the Creoles of the island, not barely sensible of this, commit the excess of attaching to it a degree of importance beyond even England itself. – 'What would poor old England do,' say they, 'were Barbadoes to forsake her?' (1806 2: 78)

---

3. The *Oxford English Dictionary* gives 1655 as the date of the first use of the verb 'to barbadoes' – W. Gouge in Thurloe (1742): 'The prisoners of the Tower shall, 'tis sayd, be Barbadozz'd.'

The same could be said for Cuba in the context of the Spanish Indies. It was therefore this feeling of geo-political importance that fostered and sustained national feeling – it was not the land in itself but where the land was situated.

Another possible though slightly different case of 'early' reference to 'national' identity was Trinidad in the ninteenth century. The fact is that this island had a very diverse population but paradoxically a population referred to as *Trinidadians* by Gamble (1866). This designation, it can be argued, reflected an absence of a strong umbilical link to a European mother country. This was the result of the fact that the hold exercised by the Spanish on the island had been tenuous and then, when it became French creole and French dominated, there was no legal or formal tie between the colony and France, and at the same time there was no spiritual bond between the mass of the population and the British, the legal rulers. In effect, the absence of any strong umbilical tie to any one European nation resulted in a picture of an independent national identity. This, however, was not a case of assertion of national identity by the people themselves.

In reacting to the conflicts that developed in Haiti after the revolution, Bird pointed to what he considered an important requirement for the success of a nation when he said that '... it was the intelligent portion, and not the ignorant masses of the nation, which originally chose the Republican form of Government ...' (1869: 132). He was implying that republicanism (or probably any form of self-government) required intelligent leadership, and where that was lacking (as in the case of Haiti) there would be no real nation. Bird's intention was to link intelligence to nationhood as an economic success, but there is no question that the element of great leadership (military and ethnic) also played a major role in the emergence and recognition of nations, as was the case in Haiti where the Haitians had their own Napoleons. It goes without saying that revolutionary, nationalist leadership was inimical to colonial interests and was consistently vilified by European writers. Saintard's comment – '*Une colonie n'existe point pour elle: elle existe pour la nation qui l'a fondée*' ['A colony does not exist for itself: it exists for the nation which founded it'] (1756: 232) – was a threat, meaning that ideas of independence and revolutionary leadership should be crushed. Leadership was facilitated, however, not by invocation to defence of the homeland but by recognition of similar plight caused by racial factors.

Over and above Saintard's intended meaning in his words cited above, attachment to the land locked the colonist into an (export) economy set up by and for the mother country, and as such into a subject status. In comparison, geo-political importance was paradoxical for, though it promoted nationalism in the colony, it led to intransigence on the part of the mother country, which wanted to maintain the prized possession, or continued hostility on the part of those who wanted to get it. This clearly militated against national independence. Ironically, in the case of Saint Domingue, where the colony was economically important to the mother country, but where there was little national sentiment, on the part neither of those who owned the land nor of those who were forced to work it, national independence was achieved because those who by race could claim no other homeland (i.e. *gens de couleur*) fostered a revolution. It seems therefore that the sentiment in Sir Walter Scott's famous words is more spontaneous and powerful when the element of race is inalienably linked to the notion of homeland, and when the homeland is under threat in some way.

# Psychological traits of Caribbean people

Genetically, there could have been no emergence of new traits in the evolution of Caribbean people; it was a matter of accentuation of some of the traits of the different people that came together in the Caribbean and a variable meshing together of them. The first illustrated image of the indigenous people of the Caribbean (Columbus's 1493 illustration) was one contrasting the fearful with the generous, and it was these two characteristics that persisted long after the decline of the indigenous inhabitants. All components of the population, including the indigenous inhabitants, were in a state of culture shock at the start of European colonisation of the islands, and it was the most common response to culture shock that provided a fundamental trait of Caribbean people – suspicion and hostility towards other elements in the society. Immediately, the whites saw it in their best interest (i.e. for their own protection) to do nothing to reduce hostility between ethnic groups. Yet, within every ethnic group there was an element of weakness, which meant that the hostility was in most cases covert or at least not fully expressed: it was hostility obscured by a mask. Where Africans came to be the majority of the population, the mask was one of gaiety and, even where they were not, the gaiety that they seemed to exude was so infectious that it provided a counterbalance to the hostility brought out by culture conflict.

The violent nature of plantation slave society made for erratic and extreme vacillation between hostility and gaiety – between the intention to remove the enemy physically and the intention to remove the enemy from the mind, however temporarily. The psychological need to create an easier world for the mind was the daily experience of everyone and it is this that led to escapism or self-deception in many guises. Escapism in its most rebellious form was evidenced in the maroon, and in its most compromised form in imitation of upper-class whites, or, in the case of colonial whites, in a prolonged cultivation of the idea of Europe as home. Erratic and extreme vacillation between hostility and gaiety and the tendency towards escapism were fundamental traits inherited by Creoles across the islands and, since the ethnic conflicts did not disappear, these traits persisted over the centuries. If there was any validity to what was called 'that volatile spirit so peculiar to the Creole' (Thompson 1770: 112), it must have been the result of this vacillation.

The survival of the societies with all but the indigenous component still standing meant that gaiety was more dominant than hostility and that compromise rather than conflict was more often the choice, especially among the dominated. These were certainly not societies about which one could say 'the frontier bred individualism and with it, democracy or "the bonds of custom are broken and unrestraint is triumphant"' (Bailyn 1992: 35). In these societies there was among all surviving components a high level of scarring and with it fortitude and resilience, but the experience of being constantly embattled promoted gregariousness and co-existence as a solution.

A characterisation of the early (1630s) planter in Barbados created the picture of a composite European, one who was practical and hospitable:

> the merry planter or freeman to give him a Carecter, I cann call him noe otherwise than a German for his drinking, and a Welshman for his welcome, hee is never idle; if it raines, he keapes securely under his roofe, if faire hee plants and workes, in the feild. He takes it ill, if you pass by his doore, and not tast of Liquor. (BDIB)

According to the writer, what was most important to the planter was hospitality, for he went on to give a case of a planter upbraiding a 'Gentleman' for his 'uncivility' in refusing the repeated offer of a drink.

In the first half of the eighteenth century, Charlevoix sought to explain the characteristic hospitality of the people of Saint Domingue as a heritage from their foreparents: 'L'héritage, qu'ils ont conservé le plus entier de leurs Peres, c'est l'hospitalité; il semble qu'on respire cette belle vertu avec l'air de S. Domingue' ['The heritage which they have preserved most completely from their parents is hospitality; it seems as if it seems as if you breathe this beautiful virtue in the St. Domingue air'] (1731: 483). In presenting the Puerto Rican jíbaro, what was foremost in O'Neil's mind was unselfish hospitality:

The poor (?) tenant of a hut ... will extend the most cordial and polite welcome to the benighted traveler, set before him the best of his plantains, milk, and cheese; relinquish to him his rustic bed; unsaddle and feed his horse, which at break of day he will have in readiness, and dismiss his guest with a 'vaya con Dios,' refusing with a gesture of pride or offended delicacy all proffer of payment. (1855: 152)

Although O'Neil did not give a source for the characteristic hospitality, his presentation of the *jíbaro* as a chivalresque figure put his hospitality in a Spanish tradition. Two slaves in Barbados saw it in a different light, as recorded in a conversation between them and the newly arrived author, Pinckard:

Are the people, here, kind to strangers? 'Oh, yes! It is always our custom: everybody should be more kind to strangers than to their own people.' Why so – should we not be kind to every body? 'Yes! We should be kind to every body, but we should be more kind to strangers, because they come far from their own home, and their friends; and because we may some time travel ourselves, and want kindness from others.' (Pinckard 1806 1: 252)

The very same explanation for hospitality to strangers was given by an old lady in Grenada, born at the end of the nineteenth century: *'u ba kon eti u kay tombe nã etwãnje, u ba kone ki mun kaj leve u. ba pu fe move ba etwãnje, se pu fe i bõ'* ['You don't know where you might fall when abroad, you don't know who will lift you up. You should not do ill to strangers, you are to do them good'] (Recorded 1970; Roberts 1971: 236). The perceived reasons for hospitality may have varied across the Spanish, French and English Caribbean across the spectrum of races and classes, but the constant mention of it from the earliest writings up to the end of the nineteenth century indicates that it was distinctive to foreigners.

These Caribbean societies were, from their very start, places where hospitality was needed for survival by those who came from elsewhere (principally Europeans), and as such it could be accounted for as a trait related to environment. More fundamentally, hospitality was seen to be an ally of gaiety and characteristic of the social way of life said to be typical of people in warm countries, as opposed to that of temperate countries where people tended to be more selfish or individualistic. The contrast between the two was highlighted in Keimer: 'So frank and friendly a converse and treatment in America, makes some of its natives not a little distasted at the very different reception they find here [England] ...' (1741: iv). Thus, when Columbus reached the Bahamas for the first time, he was greeted by Tainos, who extended a hand of welcome to him, thereby facilitating his survival and European prosperity thereafter. This welcome was even more meaningful because it was a sharp contrast to the hostility of those indigenous inhabitants of the region who wanted no part of the intruders. It is not surprising therefore that European writers continually noted hospitality and hostility in Caribbean people.

If Caribbean people exhibited volatility in their psychological make-up as a result of extreme vacillation between hostility and gaiety, they also exhibited a form of degeneracy as a result of the ways in which interrelationships between the two sexes emerged. Early colonial societies in the islands were abnormal in the sense that they were distant from the home bases of those who established them, they were instituted as economic enterprises and they were started off as social entities by men almost exclusively. At the start of these societies there was no natural numerical balance between the two sexes and consequently there was no psychological equilibrium in mores between them. The relative proportion of women who came to be in these colonies varied across the islands and so did their roles. European men, who controlled the societies, initially resorted to whatever women were available (i.e. indigenous and African) for their sexual and emotional needs and continued to do so even when European women became more available. Other men were much more constrained in their choices.

Early colonial societies were abnormal also in the sense that they brought together component cultures that had differing traditions of treatment of women. Generally, there was a cultural–ecological difference between women's roles in temperate European climates and those in tropical climates, where life was more outdoors. In the indigenous and African cultures in question women were treated as workers, both indoors and outdoors, domestic and agricultural; in the European cultures women were treated more as indoor, domestic workers. Beyond this, there were culture-specific traditions. Among the indigenous inhabitants in one scenario young women were given to men as presents by their parents, and in another women were captured in raids and made to serve as one of several wives and servants. In the case of Europeans, many women came into Caribbean society having been banished from Europe as immoral persons or wrongdoers (which in some cases meant no more than being destitute in a city), and were subsequently treated as unfit for marriage and more suited to prostitution. In the case of Africans, many women who came to the Caribbean were merely changing their locus of servitude. In this milieu the role of woman as mother who was central to the family unit (see Bird 1869: 349) was certainly not a conscious goal for many females.

Yet, because the Christian Church became politically dominant and established in these societies and because the woman as mother and domestic was central to its philosophy, especially to Roman Catholics, a wide and distressing gulf developed between the guiding principles of the Church and everyday reality, a gulf to which individuals had to reconcile themselves. Even where women were unmarried and unencumbered, sexual relationships and emotional ties did not necessarily lead to stable or openly recognised unions. Racial and social stratification militated against this. Furthermore, a slave as a *pieza* or chattel was in direct conflict with the family unit as the constituent structure of the slave population, as long as imported *piezas* could be bought in the market more cheaply than they could be produced and reared locally. Then, when in the later slavery period local reproduction seemed to be a viable option for increasing slave numbers and indiscriminate (i.e. not bound by any notion of monogamy) breeding was encouraged among the slaves, with the onus on the women to produce abundantly, mothers may have been central, but a plurality of fathers ensured conflict and instability.

Outside of Church ideals, one of the early goals set out for sexual intercourse and reproduction in Caribbean societies was the improvement of the breed, that is ascent to whiteness. The problem with this was that it was inherently contradictory, that is except where white men were concerned. White men were allowed to have children with coloured women without having to marry them and there was no real penalty attached to what white men did. In all other cases, one of the two parties (man or woman) would have been 'descending' and would thus have been frowned on by their relatives and associates. The distinctions (between *ascent, stationary* and *retrograde*) in Edward Long's (1774) table of mixtures showed quite clearly the primacy of the white man, for it was only where the white man was involved that the issue was treated as *ascent*. A general, socially approved goal, which recommended that all women have sexual intercourse with and be impregnated by one type of man (a white man), was not only psychologically damaging to all men and women in the society but was also unparalleled in human social organisation.

In the islands, then, what developed in the psychological make-up of the people was a lurid and sick 'understanding' of sexuality characterised by various intertwined elements: the white man seen as having an inordinate craving for black and coloured women; rape and coercion of women facilitated by ownership of them or power over them; prostitution as a popular option for women and also as a normal expectation of female slaves among urban slave owners; exploitation of sex by females to gain freedom for themselves or their relatives as well as to gain 'political' influence; double relationships and families involving open unions and clandestine ones; long and intense love relationships obscured and troubled by inequality in status or colour. The mulatto and people of mixed race, as products of these elements, were constant reminders of them, and the *fille de couleur* and the *mulata* came to symbolise poignantly these elements in their respective societies.

If there was *creole degeneracy* in the Caribbean islands, this is where it was. The people were unable to respect the creole essence of the societies and culture they had formed not simply because they had not issued from genetic purity but because of the degeneracy surrounding sexual relationships. Originally *criollo* had meant American version of an Old World race, but by the end of the nineteenth century it had acquired the notion of mixed, as symbolised in the *mulata* and *métis* in the Spanish and French colonies respectively. Paradoxically, the very mixture that was a symbol of degeneracy, and thought to have inherited the worst traits from both parents, was at the same time admired for its unparalleled beauty. The admiration of the woman as a love object was typical of the time, but, as can be sensed in Turiault's verses, the idea of pleasing was a central element in the perceived traits of the *fille de couleur*:

| | |
|---|---|
| *L'amour prit soin de la former* | [Love took pains to shape her |
| *Tendre, naïve, et caressante,* | Tender, guileless, affectionate, |
| *Faite pour plaire, encore plus* | Made for pleasure and even more |
| *pour aimer,* | for love, |
| *Portant tous les traits précieux* | Having all the precious traits |
| *Du caractère d'une amante,* | Of a lover's character, |
| *Le plaisir sur la bouche et* | Pleasure on her lips and love in |
| *lamour dans ses yeux.* | her eyes.] |
| (Turiault 1874) | |

This was of course consistent with the element, hospitality, which was identified as a creole trait generally, hospitality toward the white man.

Admiration for the mulatto woman's physical beauty did not mean that the official view of the ruling classes in the Caribbean moved away from condemnation of white men's preference for slave women, the caricaturing of it and the designation of it as a threat to the social order:

| | |
|---|---|
| *Mulata! Sera tu nombre*[4] | [Mulatta! Will your name be |
| *injuria, oprobio o refrán?* | insult, shame or refrain? |
| *No sé! Sólo sé que al hombre* | I don't know! All I know is that to a man |
| *tu nombre es un talisman* | your name is a talisman |

---

4. The verses as well as the lithograph are cited in Kutzinski (1993: 23). About the lithograph she says: '*El palomo y la gabilana* [Male dove and female hawk] ... is most explicit in its representation of white men as victims of the mulata's ambitions ...'

> *Tu nombre es tu vanagloria*    Your name is your vainglory
> *en vez de ser tu baldón;*    instead of being your disgrace;
> *que ser mulata es tu gloria,*    that being mulatta is your delight
> *ser mulata es tu blasón.*    to be mulatta is your honour.]
> (Muñoz del Monte 1845)

The Spanish, French and English, however, exhibited differences in degree of mixture with non-white groups and in their attitudes towards them. In the Spanish islands, especially Cuba, the notion that the mulata was destructive and dangerous persisted. With the resurgence of sugar there, all the racial hostility was re-kindled and incited by an element from the American South, using Haiti as a spectre and emancipation elsewhere in the islands as a threat to the good life. So while upper-class Cubans preened themselves for 'la finura y suavidad de sus costumbres, su franquesa y celebrada generosidad' ['the refined and sauve nature of their customs, their candidness and their renowned generosity'] (de la Torre 1868:57), they were at the same time corroded by an unbridled hostility towards their black population. It was as if the sophisticated needed an 'enemy' or an opposite to contrast themselves with. Additionally, it was as if the 'stigma' of the Moors was indelible and an everlasting burr, especially since the French and English kept mentioning it.

In the case of the English, Protestantism and the public articulation (literate and oral) of morality using polar concepts of right and wrong as well as 'ordained' divisions exerted a powerful influence in small

**Figure 7.4**  The *fille de couleur. From* Hearn (1890).

**Figure 7.5**  *El palomo y la gabilana.* From Kutzinski (1993). CENDA.

island communities, but it was probably the higher proportion of European women in these colonies that lessened miscegenation. The French articulated general political positions about status of ethnic groups in the colonies (e.g. the *Code Noir*; conferral of full civil status on indigenous inhabitants and refusal to *métis*; abolishing slavery and re-imposing it), but these were essentially pragmatic decisions and not necessarily in keeping with philosophical and moral ideals in France itself. At the same time, the absence of a publicly proclaimed, strong, religious morality in the French colonies allowed for intercourse between ethnic groups and the emergence of the *métis* as an intermediate group. This intermediate group, the *gens de couleur*, as a result of political demarcation on both sides, vacillated, more than in the Spanish colonies, between one pole and the other, for pragmatic reasons, and engendered a general French Creole malaise and ambivalence.

# Psychological characteristics in Caribbean language usage

When Oldendorp in 1777 explained to his readers that 'every European language which is spoken in a corrupted manner in the West Indies is called Creole' ([1977] 1987: 251), his definition established an element of sameness as well as an element of difference in the relationship between European language and Caribbean variant. It also identified, through the concept of 'corruption', an element of subordination in the relationship between the two. For the individual in the early colonial context in the Caribbean, the notion of language corruption was a fact of life because that was how everyone conceived language acquired in adverse learning conditions in a stratified society. In every European colony, there was, or had been at some previous time, a European (variety of) language spoken by whites, which was believed to be the language that those at the bottom of the society were or had been trying to speak. It was inevitable that as long as the power balance remained with white on top and non-white underneath the language produced by those at the bottom would be viewed as a corrupt or bastard version of the European language. It was inevitable also that the individual who was not at the top, believing what he/she spoke among peers normally not to be the best language or not to be a language at all, would in some situations try to produce a better version. While the intention to produce a better version of language in some situations may be a human response characteristic of all societies, when the distance between normal competence and target is considerable and when the belief in the need to demonstrate higher competence assumes an element of compulsion for social and ethnic reasons, a measure of linguistic distress sets in and becomes chronic. It was a distress caused by the inability on the part of adults to accomplish some social requirement; it was a distress that incidents and comments in everyday life tended to aggravate in the colonial situation, as was evident in the comments made by creole slaves about newly arrived Africans, by the cruel fun poked at speakers of *habla bozal* and *jíbaro* Spanish. A significant characteristic that this distress produced among many 'linguistic aspirants' was an almost unabated need to try to equal those above. On the other hand, to recoil into one's own sphere (i.e. to use one's own language among peers) provided the individual with an inestimable measure of comfort. In Caribbean colonial situations, then, linguistic vacillation and linguistic distress went hand in hand for a great number of people.

Historically, the linguistic vacillation between use of native language and use of a language to communicate with foreigners started with the indigenous inhabitants, who, in their own countries, found themselves having to try to speak the various languages of European visitors, who decreasingly showed any desire to learn theirs. Then, it became the lot of the speakers of various African languages introduced in great numbers into the situation to continue this practice of trying to acquire language to communicate with Europeans. These Africans were in a more distressful situation, however, because they did not necessarily have countrymen with whom they could communicate in their own native language. Even when creole languages emerged, thereby making general communication more natural, linguistic vacillation continued both because of the hierarchical levels of Caribbean societies, each with its characteristic form of speech, and because of the belief in the European language as the ultimate goal to master. The only place where the compulsion abated for the majority of the population was Haiti after the revolution: it was in Haiti only that the vision of the European language as a target ceased being part of the consciousness of the majority of the population.

As to the better off and the more standard-speaking members of these societies, a state of denial afflicted them – they either denied that creole varieties of language existed or denied that they themselves used them. The more 'enlightened' among them regarded creole language features in a historical sense as survivals from Europe either legitimately or through corruption, but the majority conceived of them in a more ad hoc way either as the speech of those who were outside the true society or of those who had not yet been integrated into the society (e.g. *bozales*), or, on the other hand, as the result of careless and slovenly speech and as such having nothing to do with anything historical. In the minds of such people, Africa and African linguistic survivals were entirely absent or totally denied. Such denial or ostrich-like responses were means of eliminating the embarrassment of creole varieties and of linguistic vacillation. They portrayed people who were ashamed of their own language competence and deluded themselves that it was otherwise.

Distress was only one of the consequences of the diglossic situation; another was, in a sense, the opposite – the deliberate exploitation by the subordinate of a situation in which they were regarded as unintelligent and powerless. If distress was caused by feelings of incompetence or lack of success in communication, the opposite feeling was caused by being able to get the better of superiors through clever use of language – the use of double meaning, the use of inference rather than explicitness, imitation of the speech of superiors for the purpose of mockery, and masking seriousness of purpose with apparent gaiety. This kind of competence developed and increased as the populations of the Caribbean islands became more and more creole-born. In addition to these speech habits, on which the lower classes had to depend to maintain their self-esteem when interacting with superiors, they also learnt to act out conflict among themselves in exchanges of words and in the competitive use of language, both of which moved the language of these Creoles to higher levels of expressive and artistic function. Eventually, all such uses of language became characteristic of the whole of creole society since even those at the top had to contend with their superiors from the mother country.

## Creoles in the shadow of a crusading European scribal tradition

Caribbean languages and varieties of language did not emerge as instruments of victorious societies. Caribbean colonial societies, especially the smaller islands, were not warrior societies that produced historic and influential exhortations to epic encounters or poetic narratives of great battles or celebrations of great heroes. There was no epic, literary tradition that proclaimed the exploits of those who by their deeds defined the national identity. The only work, besides those featuring indigenous heroes/heroines, that was of this type was the very early work (1589) by Juan de Castellaños, *Elegías de varones ilustres de Indias*, and, in glimpses, in Balboa's *Espejo de paciencia*. After that, it was only Boukman in Haiti who was flirtingly given this kind of image (Fraser 1891:

47–8). Enriquillo, a sixteenth-century *cacique* in Santo Domingo, was recreated in a romanticised version in the nineteenth century, but this, like several others, was part of an 'indianist' tradition to provide a vicarious experience for a people who were generally portrayed as the opposite of warrior. Toussaint Louverture was portrayed, not as an epic figure speaking his own native language, but in decline as an incompetent speaker of French begging for leniency (Ardouin [1853] 1958 tome 5, p. 48). The absence from the history of Caribbean people of epic language in the context of military exploits meant that effectively (that is, in a militaristic world) Caribbean people had no respected language history of their own to refer back to with pride. Their languages continued to be off-shoots and bastards, since they were not attached to any nationalist identity, and Caribbean people did not speak them with the confidence that comes with military victories.

In trying to determine the most important factor in the formation of cultural identity, Diop says:

> Montesquieu would very probably lean toward the linguistic factor, he who wrote that 'as long as a conquered people has not lost its language, it can have hope,' hereby stressing that language is the unique common denominator, the characteristic of cultural identity par excellence. ([1981] 1991: 214)

As brand new colonial societies started at the beginning of the sixteenth century, Caribbean societies had no *unique common denominator* comparable to the Romance languages that linked all Latin people. By the nineteenth century, however, all Caribbean colonies had allegiances to one European language or another. Each group had developed an allegiance to a language comparatively few of them mastered, a language whose cultural matrix was not the Caribbean. Indeed, the fact that many of them were part of an African diaspora meant that officially they could not be regarded as speakers of European languages. They could not appropriate the European language to themselves and dominate it, and so, by and large, remained pseudo-Europeans in language.

For a long time there were no historians who were creole-born and, even when they emerged, they adopted no nationalist or regional posture as was the norm for historians. The historical accounts of the Caribbean islands fit into the epic European tradition in the sense that European historians were tied to their own national identities and presented past events as the exploits of their countrymen in conflict with their enemies. Right from the start, the Caribbean was defined within this framework by Martyr, Herrera, De Rochefort, DuTertre and others. *Nos ancêtres, les gaulois* is what creole children in Martinique and

Guadeloupe were familiar with up to the twentieth century, the same 'mother country' viewpoint ingrained into every colonial child. It is not only that *Antillia*, a figment of the imagination of Europeans, became the Antilles, but also that, as Joseph said, 'when any of the early Spanish historians made an error, all the succeeding writers generally transcribed that error' ([1838] 1970: 87). The real Caribbean was in books written by Europeans and even Caribbean people had to learn it, true or false, from those books. The scribal tradition consolidated itself in Europe – 'scribes', often with very limited or no direct knowledge of the reality of the Caribbean, reproduced and updated histories of previous writers from different languages with editorial changes.

It was indeed rare to find a writer with first-hand experience who even purported to write for people in the Caribbean, as Bird (1869) claimed to do:

> the hope of rendering service to Hayti herself, constitutes one of the leading motives of the work now before us, and may ultimately lead to its translation into the French language. But the fact of seven or eight millions of the descendants of Africa in the new world, speaking the English language, seems to render it desirable that it should first appear as an English work, the more so, as one of the leading objects is, the general interests of the 'Black Man.'

Yet, in spite of the fact that Bird had spent nearly 30 years in Haiti, he was not a nationalist, creole historian. It is not that his missionary intent invalidated his work, because this would have invalidated most of the work on the French Caribbean. It is that Bird fell into a category of person who looked at Haiti as a positive experiment in blackness. He was therefore a white man who was attracted to a new, black phenomenon, became involved with it and championed it as a cause. This kind of writer was outside the literate mainstream that shaped and consolidated views of the Caribbean.

European control of the scribal tradition led to a constant need on the part of Caribbean people for external validation on the one hand and, on the other, to anger against misrepresentation. Caribbean writers, in reacting to European descriptions of Caribbean people, became suspicious and resentful if they did not coincide with theirs but thrilled if they did. Caribbean people did not find much comfort in their own collective views but insisted on European redress in the case of perceived misrepresentation, in the belief that it was only in Europe that Caribbean reality could be given validity. A crucial factor in this belief was the knowledge that Caribbean 'corrections' of misrepresentations had limited circulation. In short, Caribbean people, because of the realities of this literalist epistemology, had and were forced to live an inferiority complex that undermined their self-confidence.

Contrary to what Goody (1977) suggests, in the case of the Caribbean islands accumulation of knowledge did not guarantee the elimination of misconceptions; logical conclusions arising out of close examination of written material did not ensure truth; reflection and criticism facilitated by distance did not have any advantage over immediacy because distance was often supplemented by creativity. In short, the identities created through literacy were not overwhelmingly satisfying to Caribbean people, but were more so to those who created them.

Yet, the absence of a local and a nationalist scribal tradition was not totally debilitating: it aided and abetted a powerful ferment of words. It allowed for the development of a rich fabric of meanings and connotations, as well as an unmonitored spread across geographical space. This happened in these predominantly oral societies where there was no historical documentation of the evolution of 'new' words, and popular etymology made them appealing by attaching them to familiar words and contexts through mostly localised explanations. Because there were many 'new' words of unclear origin in the language contact situation in the Caribbean, the licence of popular etymology facilitated the development of a tendency not to be limited by the kinds of grammatical stringencies that confront scribes but, by stretching their meanings in all directions and by playing with their forms, users made words serve a communicative as well as an artistic purpose. This usage (like jazz) was moulded to suit its speakers more so than speakers being forced into subservient conformance within a set language (like a classical European musical score). This usage is attested in the plethora of names and terms that came into being in all New World societies to identify different groups and types of people.

If the literate in the Caribbean, therefore, were forced to live out an inferiority complex in the shadow of European control of the scribal tradition, the majority, the illiterate, forged their own linguistic and cultural identity almost impervious to its restrictive influence. It is the popular interpretation that gave the islands their unmistakable, kaleidoscopic identity, but the scribal tradition to a varying but increasing degree masked, circumscribed, delimited and suppressed it. For Gaspar de Villagrá, absence of a historian and of a written record constituted a totally unfortunate situation; for others, writing distinguished between savage and civilised society. In an emerging society, however, absence of a written canon, of a Greek or Roman model allowed people from different sources to work out their own identity. The scribe and his peculiar, secluded manner of fashioning national consciousness was superseded by the illiterate, partially literate and the literate in daily, dynamic interaction with all their vagaries and occasional sparks of brilliance.

## Periphery and core – contrasts, reversals and counterbalances in the colonial/national paradigm

The case of the Caribbean islands fits Bailyn's (1992) characterisation of the New World colony as the periphery, with Europe as the core. In this concept of periphery and core, there is both a symbiotic and a hierarchical relationship between the two. In nature, however, periphery and core are part and parcel of the same organism: they are in an integral relationship, the one to the other, not a symbiotic relationship, nor a dependent one, nor a parasitic one. It is only from the point of view of a third party that the one may be said to be more valuable or greater than the other, or it is only when periphery and core are extended, through figurative use, to different organisms, that the nature of the relationship can be viewed as hierarchical, mutual or parasitic. The periphery–core concept is not applied in all cases whenever different organisms relate to each other. Take, for example, the case of a plant that first grows in one terrain and then moves to and is transplanted in a new terrain. It is joined by other plants uprooted from elsewhere and they all now form a new grove or forest. Cross-fertilisation takes place and new plants emerge to mix with the old ones. There is nothing unusual in this scenario either in terms of movement of plants or cross-fertilisation, except that it is usually seeds rather than grown plants that move (borne by animals, the wind or water). In fact, one may say that this is the way of Nature. There is no periphery and core relationship between the old groves/forests and the new ones. It is when human beings contemplate this scenario and when human beings are the 'plants' in question, that the periphery–core relationship surfaces.

The periphery–core relationship starts with a perception of difference. For human beings, apparent differences in physical characteristics are more important than phylogenetic sameness. Neither the awareness that human beings differ like plants/flowers in their colouration and shape (Lopes de Gómara 1552) nor the knowledge that colour does not establish species difference is as powerful as the tendency to assemble according to highly visible sameness and to reject association when there is difference. Added to this, perception of difference is accompanied by a tendency towards dominance, a concept that in plant genetics is considered to be the result of random selection but as a human characteristic is considered conscious and motivated. This combination of an impulse to reject or circumscribe difference and a tendency towards dominance leads to the periphery–core characterisation when two communities relate to each other.

There is another way in which the periphery–core relationship emerges, however – it results from the way in which the 'periphery' comes into being. In this case, colonies are seen, in a purely economic vision of them, as enterprises set up elsewhere for the benefit of an entrepreneur and later taken over by the state. This rules out the notion of an almost unconscious emergence of colonies and explains more simply the matter of dominance or dominion as a factor of ownership. This vision was clearly embraced by Saintard (1756: 232) when he said that a colony does not exist of its own accord, but exists for the country that established it. Ownership of some of the people (slaves) and control of others (servants) was a natural part of this vision.

These two accounts (the psychological and the economic) of the emergence of the periphery–core relationship were intertwined in the history of the Caribbean islands. These islands were part of the New World and, unlike the Old World with its transmitting cultures, developed as recipient cultures. Within these recipient cultures the islands were small entities bounded by sea, thus ideal for a colonial divide-and-rule policy and a subject status. There the tropical habitat did not restrict outdoor life and the ruling classes, whose ideas dominated the society, took some time to come to terms with the difference between temperate and tropical. The dominated classes vacillated between gaiety, rebellion and movement as solutions to their exploitation. There was therefore little impulse and concerted action among them (except in the case of Haiti) to dominate their societies or those of others: there was no military impulse to conquest of territory and consequently no epic national history and identity to celebrate in song and language. The theatres of war and the battlefields were for the most part external to the everyday life and thoughts of the majority of people in the islands, except when the islands changed hands as spoils of victory and the people had to get accustomed to new rulers.

However, the periphery–core relationship was not totally fixed, but changeable in nature based on perceptions of wealth and power. At the end of the eighteenth century when the North American colonies refused to accept political marginalisation, it was the dawning of the realisation that the colonies were not as peripheral and powerless as they were being made to feel. The colonies knew that they provided the raw materials for, and as such were essential to, the core and substantially made it what it had become. At the beginning of the nineteenth century, Pinckard regarded it as the 'veneration of the illiterate' (1806 2: 78) when some Barbadian Creoles were heard to say 'What would poor old England do ... were Barbadoes to forsake her?' (1806 2: 78), but they did so because they had a sense of power. When the economy of Cuba blossomed in the mid-nineteenth century and Cuba seemed to have to aid an ailing Spanish economy (Madden 1849: 53), Cubans no doubt were tempted to express the same kind of sentiment as Barbadians. Indeed, such sentiments had been in evidence in the Spanish island colonies from earlier, according to Walton:

> The Creoles are particularly attached to their own country, which they think the best of any in the world, from its having been in every war, a point of attack to England; the great object of French intrigue, the subject of envy and enterprise to their free neighbours on the north, and in short a bone of contention for them all. When they contrast it with European Spain, they see nothing but poor adventurers, who come amongst them with a view to get riches, by filling the most menial offices; and as ease and affluence are their chief good, they judge of all by the species that come amongst them. They feel pride and consequence from being born in a new hemisphere, and conceive that to Creolism is attached a degree of dignity and honour. (1810 2: 76–7)

In short, the colonial relationship between periphery and core was a symbiotic and hierarchical one in which positions sometimes came to be reversed, especially in the minds of the colonists.

Reversals in perception at the macro level were brought about cumulatively by reversible realities at the component level, as exemplified in yucca/cassava and the sugar cane, which embodied contrasts of opposites and therefore allowed for opposite visions of the same thing. Another example of reversal and contradiction at the component level can be seen in the use of the word 'maroon'. In 1796 Stedman's *Narrative of a Five Years Expedition against the Revolted Negroes of Surinam* appeared and in 1803 Dallas's *The History of the Maroons* (referring to Jamaica). Both of these works gave a detailed picture of the harsh and perilous life of the maroon. In contrast, in Barbados in 1796, Pinckard and his hosts went on what was called a 'marooning party' (Pinckard 1806 1: 327) and then after a good night's sleep were 'prepared for another marooning day' (1: 356). About 40 years later, Mrs Lanigan, describing social life in Antigua, related 'maroon party' to the *fête champêtre* and thereby evoked a comparison with Winterhalter's *Dolce far niente*:

> the Antiguans have other methods for getting rid of the time that hangs too heavy upon their hands. Now and then a *maroon party*, or West Indian *fête champetre*, is given; when groups of beautiful girls and gallant

youths, stayed matrons, and gentlemen of riper years, assemble together, with full purpose to enjoy the passing hours. Some sweet spot, generally near the sea-side, is chosen for the day's resort; or else some

> '– green and silent spot amid the hills,
> A small and silent dell.'

And beneath the shade of some far-spreading trees, whose boughs form natural arcades, the rural banquet is spread. Various pastoral sports are here enjoyed; and although no 'Weippert's band' is in attendance, the sound of the lively violin, or soft-breathing flute, often floats across the blue waters, and mingles with the murmur of the playful wavelets.
(Lanigan 1844 2: 213)

During his stay in Trinidad, Day (1852 2: 86) mentioned that he 'went on a maroon jaunt to Point Hiacos'. Crowley (1955: 118) noted, in reference to St Lucia, that 'Picnics are called "mawon" in patois meaning "wild"' and identified 'the English "maroon" from the Spanish word for runaway slave' as the origin of the term. Redhead (1970) and Smith (1971) exchanged hostile words over the meaning of *maroon* in Carriacou, with Redhead claiming that it was a 'Spring Feast' and Smith insisting that it was a 'co-operative effort to get work done on community projects'. Elder's interpretation (1973: 20) agrees with Smith, equating *maroon* in Grenada with *gayap* in Trinidad, *len' han'* in Tobago and *combite* in the French West Indies.

The shift in vision is dramatic – from the European's account of the rebellious negro to the Creole's version of *dolce far niente* or some other positive activity. There was no great separation in time of the two images. There could have been separation in geographical space if the negative image is seen as belonging to the bigger islands/land masses and the positive to the smaller islands. Identity in freedom from restraint, in coming together and in nearness to Nature was what the Creole saw in the two images of *maroon*, rather than the European's preoccupation with dominance. Contrast in social class, race, economic conditions and big and small geographical entities is what was most apparent in the two images. It is as if, like the slave, the master was attracted to the idea of breaking free and communing with Nature and with like souls.

Attraction to the other side was even better illustrated in cultural activities and Carnival especially, in which social superiors liked to dress 'down' and social inferiors to dress 'up' – in Carnival gentlemen liked to play the role and wear the costume of the *neg jadin* (the field labourer), whereas in their Christmas and other cultural activities female slaves liked to imitate or wear the dresses of their mistresses and mimic their speech. This in itself highlighted the symbiotic nature of the relationship between classes within the colony and especially reflected the up and down shifts in sexual relationships that were a normal part of daily interaction.

Reversals can also be seen in what may be regarded as contradictory identification. As a result of Columbus's initial error, a place that was not the Indies became the Indies and the people immediately became Indian although they were not Indian. Soon, West Indians (who were quite different from Indians, i.e. indigenous inhabitants) came into being, and initially they were not Indians (from India) but later some of them were, so that some West Indians were East Indian. No born East Indians were West Indian, however. Almost as if independent of all this, the Dominican Republic officially converted *negros* into *indios*, with the result that in the Dominican Republic there are Indians who are not East Indian, indigenous inhabitant or West Indian.

Negativity towards the classification *negro* was so strong in the Spanish world that a black slave could claim to be white, as was seen in Lemonnier-Delafosse's 'Yo, yo, soy blanco de la tierra!' ['he, lama native white'] (1846: 198), as well as in Welles's 'Soy negro, pero negro blanco' ["I am a negro, but a white negro'] (1966 1: 103–4). Interestingly enough, the same kind of reversibility was noted by La Courbe in his 1685 comments about mulattos and some black people in Portuguese Senegal ('quoiqu'ils soient noirs, ils asseurent neantmoins qu'ils sont blancs' ['though they are black they are sure that they are white'] – 1913: 193). In the last example *blanc* was interpreted to mean 'Christian'; in the first example *blanco* was taken to mean 'Creole', as opposed to African; in the second example *blanco* presumably meant 'non-plantation/non-Haitian'. In all of these there was a polar contrast between the visual and the reformulated conceptual image.

In the case of the ethnic group called the 'Black Caribs', there were several faces and obverses to their identity. Generally, they did not fit into known ethnic classifications because there was, in such conceptions, no alliance between their genetic/race characteristics (which belonged to the Old World) and their cultural characteristics (which belonged to the New). For the French, they were *Sauvages africains* or 'doubly savage'. For the English, they were allies of the French and spoke some kind of French. They were quintessential Creoles because they combined the Old World with the New, but they were not referred to as such, presumably because they were called 'Caribs'. As far as they themselves were concerned, they were certainly not Africans, whom they associated with slavery; they regarded themselves as Caribs. They became the first forcibly exiled people within the New World, as a result of which they disappeared from the notice of those who painted pictures of the islands. They were converted by chroniclers into cowardly people, in contrast to those (the 'Yellow' Caribs) whom they and the Europeans had superseded and whom the Europeans converted into valiant and noble people.

In the field of language study, also, there was a reverse application of the word 'creole' to the Old World, thereby creating an inherent contradiction but at the same time removing disparities between the Old World and the New. The word *criollo* developed originally to identify children of Old World people born in the New World. However, from the second half of the nineteenth century, as a linguistic term, it began to be used to refer to varieties of language in the Old World. The pre-Columbus history of the form (or parts) of this creole word most likely goes back to Indo-European with a later development through Latin into Spanish and then into the variety of colonial Spanish used in the Canary Islands, to the west of North Africa. It became associated with sugar cane and therefore with Africans through its journey to the Caribbean islands, and it is from Cuba or Hispaniola, most likely, that it was transmitted across the American mainland. Its life as a creole word, that is an Americanism with a specialised meaning, emerged out of an agricultural context, propagated through both oral and written channels. It was later applied by analogy to colonies in the Old World, including Asia, belonging to the two Latin European nations that had colonies in the New World. What became central to the meaning of the word, then, was colonisation. Conceptually, then, this analogy represented an Americanisation of the Old World.

Reversal in interpretation of 'creole language' has continued up to present – the traditional view of *langage corrompu* has been converted into a view of creole language as the human bioprogram of language. One modern view of language, Universal Grammar, distinguishes between core and periphery, identifying human linguistic competence (i.e. the capability for language that every human being has) as core, in contrast to the specifics (which are acquired through exposure to them) of any particular language, which are identified as periphery. In the actual expression of language, of course, core and periphery are not separable. Using this formulation of the nature of language, the Language Bioprogram Hypothesis (put forward by Derek Bickerton) argues that creole languages, which hitherto had been regarded as bastard languages and peripheral to major language study, reflect the core of human language, because they were formed in situations where children had to depend inordinately on their genetic capacity for language (i.e. in those situations there were no mother tongues or models of language for them to acquire wholesale). This interpretation of creole languages therefore demonstrates a reversal in the interpretation of periphery and core.

Two faces of a kaleidoscopic group identity emerged prominently in the Caribbean islands in the colonial period – one transparent and affording a vision of the dangers and effects of social and racial repression and the other reflective and exhibiting perceptions of geo-political self-importance. The one was typically symbolised by Haiti and the other by some Spanish colonies and Barbados. The latter face of identity was self-indulgent and non-violent and was therefore generally disregarded by Europeans, whereas the former was racist, conflictive and had the potential to corrode and spread, and consequently was circumscribed and manipulated by those in control. These two faces of identity existed and persisted in every colony and continually fed each other throughout the colonial period.

Though difference in language between colony and mother country became a feature of identity in all the colonies, local varieties of language, having been portrayed from the start as corruptions, did not assume any level of importance as official marks of identity and none was promoted as such in any colony. Only in Haiti was there the suggestion of an attempt to change the official language, but even this was only to exchange one European language (French) for another (English). Socially, bonding, 'family', hospitality, gaiety, religion with morality and education, and physical beauty were portrayed as characteristics of Caribbean colonial identity, but this was in the face of a historical reality that testified to widespread experience and proclamation of the opposite. Nature and the environment were by far the dominant forces in Caribbean identity in the writings on the Caribbean islands in the colonial period. The notion that Nature was responsible for changes in species for functional reasons was the most powerful and palatable way of explaining old and new ethnicities and justifying the human abuses emanating therefrom. A belief in post-life liberation was the most widespread philosophy adopted by those who had constantly to withstand these abuses, a philosophy that was at the same time the most powerful ally of the status quo. Caribbean people shaped themselves into new societies, not as conquerors of territory, but as resilient survivors where many had succumbed, as beneficiaries and victims of racial and linguistic differences, as survivors with a love for life itself because of their nearness to the very Nature that energised their lives.

The Caribbean grove or forest was not a replica of Europe or Africa or anywhere else. It was not just an unconscious product of its ecology either. The supreme factor in its development was the human mind, the human instinct for survival as well as the process of genetic recombination in Nature[5] and its capacity for reconstruction and reversibility. The Caribbean grove began to reconstruct the world with its own characteristics and identity. With its miscegenation, it began to affect (previously thought of as 'infect') the Old World. The ordained divisions ('purity') of the Old World felt themselves under threat and losing control and, in order to maintain the old order, they fulminated against (sexual) miscegenation and (language) corruption. The threat to the old order, which had provoked Edward Long's 1774 exhortation to ascend to whiteness, increased over the centuries. Then, Haiti, the first nation in the islands, with its new (recombined) language and its black and (recombined) mulatto population, drove the first decisive dagger into the heart of the old order. After that, Haiti's mulatto population led those of neighbouring Spanish islands to begin to proclaim a new cosmic race (*el crisol definitivo de las razas*). So, by the end of the nineteenth century and the beginning of the twentieth there was a reactionary rise in white supremacist political organisation and eugenecist[6] rhetoric, the latter not only from Adolf Hitler but also from the 'venerable' Winston Churchill.[7] Eventually, in the twentieth century there was a racist outburst in white Europe, with the country that had had the least 'infection' from the New World (Germany) reacting the most viciously.

# Place, race, language and the shaping of ethnic and national identities

Of the three elements of identity looked at in this study race appears to be the most powerful because its being a construct is obscured by the fact that its components (skin colour, hair type, etc.) are immutable in the individual. Race immediately comes sharply into focus in considerations

---

5. Jones (1993: 101) characterises genetic recombination in the following way: 'Sex means that new mixtures of genes arise all the time as the chromosomes from each parent recombine ... Sex reshuffles life's cards ... Sex is a convenient way of bringing together the best (some of which may even be better than what went before) and purging the worst. It separates the fate of genes from that of those who carry them. Sex is a kind of redemption, which, each generation, reverses biological decay. In some ways, sex is the key to immortality. It is the fountain of eternal youth – not for the individuals who indulge in it, but for the genes they carry. Sex speeds up evolution because each generation consists of new and unique mixtures of genes, rather than thousands of copies of the same one.'

6. Francis Galton founded eugenics, the main aim of which, according to Jones (1993: 3–6), was to 'check the birth rate of the Unfit and improve the race by furthering the productivity of the fit by early marriages of the best stock'.

7. Jones (1993: 9) quotes Churchill as saying in 1910: 'The unnatural and increasingly rapid growth of the feeble-minded and insane classes, coupled as it is with steady restriction among all the thrifty, energetic and superior stocks constitutes a national and race danger which it is impossible to exaggerate. I feel that the source from which the stream of madness is fed should be cut off and sealed off before another year has passed.'

of national identity as visibly different people confront each other. Part of the reason for this is that in the evolution of human beings there is the appearance that each race evolved separately in a different part of the world and consequently there is a very strong sense of racial separateness and thus purity.[8] Race mixture therefore seems contrary to what was ordained and normal preferences. In the case of the Caribbean, where it was part and parcel of subjugation and slavery, racial mixture was regarded as even more unnatural, and strenuous efforts were made to prevent it by the powerful and the literate. As a consequence, the concept of a national identity was virtually impossible as long as 'unrefined' mixture of races was highly evident. Specific race as a component element of national identity was in the interim superseded by notions of a lurid and sick sexuality on the one hand and visions of a 'purifying' sexuality on the other. This began to change only when a new racial type emerged as dominant or when one of the component races established itself as dominant.

Language is partly invariable (in the sense that it is a characteristic of all human beings) and partly variable (in the sense that languages vary across human groups) and consequently is a less powerful defining characteristic of identity. Language, contrary to its very nature as a human characteristic and its history as the reflection of the experience of individuals and groups brought together, is imbued with a core that is called 'real' or 'genuine' by its native speakers, when it is associated with any one ethnic group. Language is a de facto separator because, when they speak, non-natives and foreigners are immediately identifiable: they are unable to produce the 'real' or 'genuine' language. Paradoxically, however, language without mixture (which is not equivalent to the 'real' or 'genuine' language) is a natural ally of nationalism. As with race, 'unrefined' mixture resulting from inter-group contact is seen as corruption; it is rejected as a component of national identity and such speakers are regarded as not having an identity or a clear identity. In the Caribbean, where language mixture was a reality par excellence, the retention of a 'refined' and 'pure', non-native language as the official language was largely unquestioned, even though it was not the 'real' or 'genuine' language of the people. For nationalism, therefore, language purity (i.e. absence of perceptible contamination by other nations) is a more powerful prerequisite than genuineness, but, on the other hand, genuineness is the central element in ethnic identity.

Place is apparently the weakest of the constituents of identity in that the details of its effect are not immediately recognisable or exclu-

---

8. Note that while many innocently associate this with ecology, Charles Darwin in *Sexual selection and the descent of man* explained it as the result of mating preference.

sive. There is no absolute and unequivocal element of ethnicity conferred simply by being born in a specific place; such elements are not inherent but acquired over a period of time. Ecology determines ethnicity and being a native in a general sense, but the notion of a critical, formative period of time seems more important, even though the beginning, end and length of it are not self-evident. As such, there is no absolute necessity for a person to have been actually born in a place to be a native of that place; the person needs only to have been bred there. Yet, in the best known and earliest term of ethnic identity coming out of the New World (i.e. *criollo*) the critical defining element was place of birth. The major reason for this was that the formative period in the new environment also witnessed obvious generational changes – the older generation (i.e. the migrants themselves) saw their off-spring (i.e. the Creoles) as different and labelled them accordingly. The off-spring of course had no prior knowledge that would have led them to see themselves as different from their parents. In any case, as the number of Creoles increased, as the distance between successive generations grew and as the conflict between the mother country and colony heightened, physical environment, variety of language and the effects of slavery combined to separate Creole from European in their ethnicity.

In spite of its apparent weakness as a criterion, place is also the starting point for national identity. Even though, in contrast to race and language, there is no sense of 'purity' perceived in the alliance of place with the event of birth, yet place of birth is primary because it is not a construct and is immutable. In this case, the primacy of 'place' as the initial element of national identity does not start with the effect that the land has had on the individual, which is really a consequence, but a priori with the animal/human instinct to mark out and protect territory, linked to the instinct for self-preservation. Presumably, in evolutionary terms, the transitory kind of territoriality characteristic of the nomadic turned into a more permanent type characteristic of sedentary peoples as a result of a combination of satisfying elements and restrictive elements. In other words, the urge to move on decreased because supplies of food remained adequate and also because natural geographical barriers afforded protection as well as made moving difficult. Threats from outsiders and even from the elements themselves caused people to be even more protective towards their offspring and their possessions. In short, territoriality was at the heart of the social distinction between 'us' and 'them', which in turn led to a spontaneous search, in a definition of 'other' and of self, for exclusionary criteria (e.g. language; physical characteristics combined into race) and 'inclusionary' criteria.

Analysis of the development of identity in the Caribbean is instructive in that it shows how identity configured itself in various places after mass migration into them, which is, with modification, a recurring human situation. While more than one group was involved in the development of identity in the case of the Caribbean, one group had much greater control than the others in what took place. At first there was no difference between a European in Europe and one who came to the Caribbean. This remained substantially so as long as there was a notion and a practice among Europeans of going back home. For African slaves, it was not the same because the African was not only traumatised by the Middle Passage but also was decisively cut off from home and furthermore was unable to communicate normally in the new society. For Asians, who came later, their umbilical ties to their homeland were not severed in the same way as those of Africans, thereby facilitating preservation of ethnic separateness in their new 'homes'. However, the crystallisation of national identity is not simplistic; it really depends on the configuration and interaction of various criteria, for even the apparently immutable characteristics of race may be subordinated to others, as is evident in the Haitian saying *'Neg riche se milat; milat pov se neg'* ['A rich negro is a mulatto and a poor mulatto is a negro'].

Political and economic rifts between Europeans in Europe and those in the Caribbean eventually led to the perception of 'them' and 'us', principally among whites. At the same time there was an acute perception of 'them' and 'us' within the Caribbean settlements themselves, caused by a more immediate threat to life and limb, which splintered Caribbean societies into groups. For self-protection against the other groups within the society, exclusionary criteria such as purity in race and language were spontaneously used by whites, but this solution quickly turned into a gradient system in order to accommodate the characteristics of mixed offspring and to control the society as a whole. In fact, genetic mixture and linguistic mixture starting from the earliest encounters in the New World resulted in new beings and entities that were central (i.e. in the middle) in the development of a common creole ethnicity. In time, as the different groups in the Caribbean societies became more intertwined with each other, commonness in ethnicity spread across the society and the division between native (i.e. Creole) and foreigner became more pronounced. However, this did not lessen, among whites, the instinct for self-preservation, that is, the need to preserve power and possessions, which required the maintenance of the exclusionary criteria, thereby militating against a common national sentiment across the society. A mentality of barriers thus dominated Caribbean colonial societies. What this meant then was that coincidence of ethnic identity and national identity was not possible, or was so only where the force of the exclusionary criteria and barriers was masked by the perception of a common external enemy and the need to ward off this enemy (e.g. Dominicans' view of Haiti). In any case, all colonies were riven by perceptions of racial enemies, perceptions kept in focus by a predominantly foreign press even after the formal abolition of slavery everywhere.

Assertion of nationalism has as its strongest ally or greatest enemy literacy, because literacy can create out of any past varying identities, even ones that are contrary to what the eye has seen and the ear has heard. In the case of the colonial Caribbean, it was able, by widespread diffusion and repetition, to create identities where no records existed, to nullify records created through an oral tradition, and to foster divisions, in accordance with the points of view and interests of those who controlled it. Nationalism requires an epic record (i.e. one with heroes) and, in the case of most of the colonial Caribbean, literacy was a great enemy of nationalism because it created a historical record without Caribbean heroes, and indeed there was a vilification of those who would have been heroes. It was in the Spanish colonies that nationalism first began to flourish in the press and this was because the Spanish metropolitan press had become weak and had almost been supplanted by the press in Mexico, Lima and other cities in the New World, introduced there even before the middle of the sixteenth century. In the Spanish Caribbean colonies nationalism flourished in the press in the nineteenth century because there was a coincidence of a political enemy and racial enemies. The growth of anti-Spanish and anti-European sentiment on the one hand was accompanied by an anti-negro, anti-African sentiment amongst a people who, by default some would say, chose to be Caribbean and used the indigenous inhabitants as a central core to promote their identity. Black nationalism in Haiti, which came through military struggles and preceded Spanish Caribbean nationalism, suffered because it had little access to the press and less control of it. In addition, unlike in the United States, there was no concerted promotion of a national variety of language to bolster nationalism.

There was, however, from early in Caribbean societies a co-existence of races and mixture of races. Gradually there developed a will to change the gradient system from vertical ('ordained') to horizontal ('natural'), with the vilified 'corrupt' language at the heart and soul of it. In time, wider literacy gave Creoles and nationalists greater control over the shaping of their own identities.

# Thematic glossary of ethnic and national terminology indicating ways of classifying and identifying

## Place names and concepts

**America**  perhaps the only major place name in the New World that derived from the name of an actual person (< Amerigo Vespucci). Compare the much narrower reference of 'Colombia' (< Columbus). As a name for the New World, it was most used outside of Spain partly because of the popularity of the work *Mundus Novus* (credited to Vespucci), which was probably the first printed work to identify the New World as a separate and distinct continent.

**Antillia**  an island or set of islands believed in the fifteenth century to be west of the Iberian Peninsula; from the beginning an element of exoticism attended the belief in the existence of these islands.

*Antisles de l'Amerique*  a name virtually concocted by French writers in the seventeenth century in an attempt to give an understandable etymology to *Antilles.*

*banda del norte*  a northwestern area of Hispaniola (Osorio) that was devastated in 1603 by the Spanish Crown to control contraband but which effectively resulted in the beginnings of Haiti. It is conceptually associated with *maron* (cattle and fugitive slaves), *flibustier – boucanier – habitan*, and *matelotage.*

**Borinquen**  an indigenous name for the island of Puerto Rico, which occasionally occurred in the early literature but which was used in the nineteenth century with the same kind of nationalistic intent that characterised the use of 'Haiti' in the neighbouring island.

**'Carib'**  the name of an island in the general area of his first landing mentioned by Columbus in his letter of 1493.

**Caribana**  a seventeenth-century cartographic (map) designation of the vast expanse of land stretching across the northern part of South America, which included the Gulf of Urabá (in modern Colombia) where the Caribs are supposed to have originated.

**Caribby Islands**  a term of the seventeenth century that referred to the smaller islands in the island chain from the Virgin Islands southwards, said to have been inhabited by the 'fierce Carib Indians'.

*la conquista*   a Spanish term meaning the exploration, appropriation and colonisation of the New World by the Spanish principally, and more generally by all the European nations; this westward 'conquest' was driven by the same kind of fervour and conviction that subsequently characterised 'manifest destiny' in the United States in their acquisitive movement from the Atlantic to the Pacific.

*dolce far niente*   (Italian: sweet + doing + nothing) a European concept originally applied to a form of behaviour in the warmer (Latin) areas of Europe but later transferred to the tropical colonies of the New World; a life style attributed to natives (indigenous and colonial) of the Spanish islands especially, deriving from the belief that the tropics were naturally bountiful and that since, as a result of this, natives of these climates did not have to work hard to get food or clothing, they spent their time in idleness.

**golden worlde**   a classical vision of the world in its earliest epoch as an almost perfect place; a vision applied to the pre-Columbus world of the indigenous inhabitants; a vision of the world contrasting with one of the contemporary world, which was seen to be riddled with corruption and decadence.

**Haiti**   an indigenous name restored to name the island to replace 'Saint Domingue' after the revolution that removed the French from power.

**home**   a word used by colonists and other whites from the earliest days in the English colonies to refer to England or Britain; also used by creole descendants of whites in the colonies to refer to England.

*las Indias*   the popular name used by early Spanish writers (e.g. Enciso 1519, Oviedo 1526) to refer to the New World; a name that had arisen out of the mistaken belief that Columbus had reached an already known place by a reverse route. The persistence of the name through the sixteenth and seventeenth centuries, even after the mistake became evident, is testimony to the power of the written word. The name gradually declined, as a designation for the whole of the New World, in the face of the competing term *America*.

**the Indies**   basically a straightforward English translation of the Spanish term *las Indias*.

**manigua**   the bush, as a place of refuge for rebels or runaway slaves; conceived of as important in the history of independence in the Dominican Republic; consequently (> *maniguero* > 'Dominican').

*palenque*   a palisade, as a place of refuge for runaway slaves (> *palenquero* 'maroon', 'bush negro').

*tierra*   a term used in the Spanish islands with the meaning that the earth/land was imbued with ecological power, conveying the idea that the human being was shaped ethnically and even racially by being born and raised in a certain place. Alternative but less popular terms used to convey this concept were *clima* and *naturaleza*.

*de la tierra*   a (deictic) term usually used with nationalistic intent among natives in each respective Spanish island, it being understood that the speaker was referring to his own native land.

*las tres Antillas*   a term used in the nineteenth century in an attempt to establish Cuba, the Dominican Republic and Puerto Rico as a cultural unit.

**West India**   all of 'the new found lands in the west Ocean'; a term used by English writers subsequent to *the Indies* partially to correct the original mistake; the singular form of the term in English preceded the plural form as a popular term, even if it did not last long.

**West Indies**   even though this term was used originally by English writers to refer to the whole of the New World, it quickly narrowed in its reference to mean the chain of islands in the Caribbean; it increasingly became narrower in meaning to signify the English islands principally and a preceding adjective had to be added (e.g. French, Dutch) in order to specify the islands owned by other European powers.

**west ocean**   (< Latin *Oceanus Occidental*) a vague fifteenth- and sixteenth-century European cartographic conception of the sea west of Europe; the ocean in which the New World was situated.

# Persons

## According to group/interrelationship

*banare/banari* (*compere*)   an indigenous word that seemed to have been used principally in the smaller islands to refer to each person in a pact in a practice that was probably normal across several ethnic groups before the advent of Columbus; in the early encounters between the French and the indigenous inhabitants in the smaller islands the French word *compere* was substituted.

*carabela – batiment –* shipmate (*sipi*) (*mati*)   etymologically unrelated words selected respectively by the slaves themselves from three of the major languages of colonisation in the Caribbean (Spanish, French and English) to identify a fellow slave who made the Middle Passage in the same boat; *sipi* and *mati* seem to have been familiar derivatives of *shipmate*.

*cogollo* (*criollo*)   a Spanish word used simply to mean the shoot of a plant (e.g. a banana plant), but used in the early Spanish colonies by

Acosta to draw attention to the New World offspring of Old World plants and, by extension, people. Also used later in Cuba to mean 'cane top' (= the top of the sugar cane). The word *criollo* solidified this concept, carrying with it the implication that the ecology would have influenced the offspring to turn them into beings different from the parents.

*cristianos*   a term used by fifteenth-century Spanish writers to refer to the people in Spanish exploration parties, as part of the Spanish *conquista*, in an intentional religious contrast that implicitly and explicitly regarded other peoples as heathens or savages.

*flibustier – boucanier – habitan*   Originally these terms summarised a seventeenth-century economic and survival relationship in and around Tortuga and the former *banda norte* of Hispaniola in which one band of adventurers (*boucaniers*) dedicated themselves to hunting; a second band (*flibustiers*) to transportation of the carcasses and skins to serve as food and clothing; the third band (*habitans*) devoted their time to working the land. In time, this relationship disappeared and the first two terms came to overlap in their reference and implicitly contrasted with the last in terms of morality (dishonest vs honest) and location (sea vs land).

*guatiao* (*taino*) (*moitié*)   identified by the Spanish in their earliest encounters with the indigenous inhabitants in the bigger islands as a kind of solidarity relationship 'formalised' by an exchange of names, symbolising the idea that the one person became the other; there was an argument that the word *taino* etymologically had the same significance as *guatiao*. The sixteenth-century French writer Montaigne seems to have interpreted the word *guatiao* as *moitié* [half], as in contemporary English 'my better half'.

*matelot*   a term used to refer to a participant in an economic or working pact or bond between young male colonists in seventeenth-century Saint Domingue and other French colonies.

*tamon*   an indigenous word used by the 'Caribs' in St Vincent and Dominica to refer to Africans used as slaves among them. Out of the *tamons* emerged the 'Black Caribs', regarded as a group with a unique fusion of race and ethnicity.

## According to race

### African word

*bakra/béké*   a term used principally by the slaves in the English and French islands respectively to refer to the white master. In Barbados, the latter term produced *ecky becky*, a disparaging name used to refer to a poor white.

Dominican Republic usage
Euphemistic ('improvement') extension in the use of a word for nationalistic purposes: *indio* (to contrast with *'haitiano'/ 'negre'/ 'noir'*)

European words
Use of basic words: *mestizo* (= 'mixed'); *mulatto* (thought to mean
   'mule-like')
Use of colour adjectives: *moro/negro/noir/*black; white; brown; yellow; olive
Use of concocted European language structures to differentiate racial
   types: *tresalvos* (*tres* + *albo* = "three + white"); *cuatralvos*; *tercero-*
   *na*; *quarteronne*; quadroon; *criollo rellollo/reyoyo*; *gens de couleur*
Use of European feminine forms to identify certain types of
   woman: *capresse*; *métisse*; *mulata*; *mulâtresse*; *negress*; *chabine*;
   *fille-de-couleur*
Use of the known to identify the new: *chino* ("Chinese") > eighteenth-century *morisco* (i.e. mulato + white)+ white; nineteenth-century issue of a mulatto and negro

Indigenous word
*chibárali* (*jibaro*)  an indigenous word of the Lesser Antilles used to
   identify the issue of a Black Carib and a Yellow Carib; most likely
   the linguistic source of the word *jibaro*.

Puerto Rican usage
Whitening of a term for nationalistic purposes:
*jibaro*  a rural peasant type seen in the nineteenth century as essentially native; a result of the post-Haitian revolution whitening of the
   original meaning of *jibaro*.

## According to place (location)

*campesino*  (< *campo* [country, countryside]) a term used in the
   Spanish islands to refer to a rural peasant, often seen as the true
   native in the Spanish islands.
*criollo/Créole/*Creole  a person born and raised in the New World
   whose parents came from the Old World; a non-indigenous native of
   the Americas; later, a non-indigenous native of the Caribbean.
*hijo de la tierra*  a term used in the Spanish islands to refer to a native
   of a country, having the (ecological) implication that the land influenced the formation of the native; the equivalent of this term used in
   the French islands was *enfant du territoire*.

**ladino** a person who had to some degree become acculturated to Latin culture through spending time in Spain; a person who became proficient in Spanish language and culture; a slave who had passed through the 'seasoning' period and become familiar with the local culture in the Spanish islands.

*montañés* (< *montaña* [mountain]) a term used in Cuba and the other Spanish islands to convey the idea of a rugged, tough, rural native from the countryside.

**peninsular** a native of the Iberian peninsula (a Spaniard); used in the Spanish islands as a stark contrast with a native born person.

*veguero* (< vega [fertile lowlands[) a term used in Cuba to refer to a tobacco farmer.

## According to place (country) name

**Antigonian** an early (1740s) name for a resident or native of Antigua; by 1844 'Antiguan' had become the normal name for a native of Antigua.

**Barbadian** a name for a native of Barbados used as early as 1708. A shortened, familiar version, *Badian* (> Bajan), was mentioned in 1789 and another familiar name, *Bimm*, in 1852.

*borincano/borinqueño* nationalistic, unofficial name from the mid nineteenth century for a native of Puerto Rico.

*Camballi* the Italian version of the name, according to Marco Polo, of the inhabitants of *Canbalu* or ancient Peking, from which the Spanish word *canibales* is said by some to derive.

*Caraibe*/Carib French and English versions respectively of the Spanish *Caribe*.

*Caribe* a name used by some early Spanish writers to refer to a native of Caribana or descendant; a native of the smaller islands between Puerto Rico and Trinidad.

**Charib** a form partially influenced (in terms of its location and nature) by the latter of the pair 'Scylla and Charybdis' of European mythology.

**ciboney/siboney** the name of one of the indigenous groups in Cuba at the time of Columbus.

*cubano* a name used for a native of Cuba that was already quite normal in the first quarter of the nineteenth century.

*dominicano* a name apparently first used in 1762; the name that came to be used popularly in a nationalistic way for a native of the Dominican Republic (i.e. after independence) in the middle of the nineteenth century.

**Galibi**   a name thought to be phonologically related to Carib and consequently the people identified by the two names thought to be the same or related.

*Guadeloupéen/Guadiloupier*   competing names (in French) used in the second half of the nineteenth century to refer to a native of Guadeloupe; the latter eventually disappeared.

*Guahiro/guagiro*   an indigenous inhabitant of Goayira in Venezuela.

*Haitien*   the name (in French) that was used from 1805 for a native of Haiti.

**indio**   a term that has survived from the time of Columbus to refer to a native of *las indias*, an indigenous inhabitant.

*isleño*   a name used to refer to a native of the Canary Islands as an alternative to *Canario*. Migrants from the Canary Islands formed a significant part of the population of the Spanish islands, especially Puerto Rico.

**Jamaican**   a name used in 1735 to mean people (presumably white) from Jamaica; a name used in 1807 to mean the (white) residents of Jamaica; not used as a regular name for a native of Jamaica until the second half of the nineteenth century.

**Kittiphonian/Kittefonian/Kittyfonian**   a name, with variant spellings, for a native of St Christopher (familiarly known as St Kitts), which combined the official name and the familiar name; it was mentioned from 1830 through to the end of the nineteenth century.

*Martiniquais*   the name (in French) which became normal in the second half of the nineteenth century to refer to a native of Martinique.

*puertorriqueño*   a name that came to be regularly used from the middle of the nineteenth century for a native (probably mostly white originally) of Puerto Rico. An English equivalent used in 1855 was *Puerto Riqueneans*.

**Trinidadian**   a name used in 1866 to refer to a native of Trinidad probably meaning at that time a descendant of a white or African or a mixture of these two; used to contrast with the majority and rest of the population which was made up of different immigrant groups.

**West Indian**   a native of the West Indies. When used in a 1771 English play, the person identified was a white creole from Jamaica.

## According to race + place

**cocolo**   a person of black African descent from the Leeward or Windward Islands living or working in Puerto Rico or the Dominican Republic.

**Moro/Moor/More** (< Greek *mauros* [black, very dark]) Spanish, English and French versions of the word that was used to refer to a native inhabitant of Africa generally.

*mulato holandes* a term used to refer to people of mixed race from Curacao living in Puerto Rico in the nineteenth century.

*raza latina* a political and ethnic term used in the nineteenth century to identify Spanish-speaking people shaped by the ecology and by racial mixture to become superior and culturally uniform; the Dominican Republic, Cuba and Puerto Rico were seen as the source of this 'race'.

## According to experience/practice

*baquiano* a Spanish word used to mean a Spaniard (or other European) who had become familiar with life in America.

*boucanier* a word deriving from an indigenous word used by the French in the seventeenth century to mean a person who hunted, skinned and barbecued cows in and around northern Hispaniola.

buccaneer a person who gained a livelihood by dispossessing others at sea.

*camballi* according to Vespucci, very savage people who ate human flesh.

*canibale* a word used as an alternative to *caribe* by sixteenth-century European writers to refer to the indigenous inhabitants in the smaller islands of the island chain, who, they claimed, ate human flesh.

*Carabi* according to Vespucci, an indigenous word used to refer to the Europeans meaning 'men of great wisdom'.

*Caraiba* according to De Laet, a word which indigenous inhabitants in Brazil used to refer to the Portuguese because they performed miracles or did things beyond the understanding of the indigenous inhabitants.

*Caribe* a name deriving from an indigenous word thought to mean 'a tough, fierce, warmongering person'; a word thought to mean 'stronger than the rest'.

*chapeton* a Spanish word used to mean a Spaniard (or other European) newly arrived in America and unfamiliar with life there.

*ciboney/siboney* a name given to an independent-minded (revolutionary) native of Cuba in the nineteenth century either as a ruse to bypass censorship or intending to portray Cubans as indigenous.

*cimarron*   a term used by the Spanish to refer to a labourer (enslaved or indentured) who escaped from a European colonial plantation or settlement.

*guajiro*   a synonym of *campesino*; an independent minded, rural native or patriot in (western) Cuba comparable in character and behaviour to the *guagiro* in Venezuela.

*habitant*   a word used in the French colonies to mean a colonist or the resident owner of an estate (*habitation*), thus 'a sugar cane plantation owner'.

*indios bravos*   a name the Spanish gave to indigenous groups they were not able to subject.

*jibaro//ibaro/ivaro/gibaro*   a nineteenth-century concept that saw the real native of Puerto Rico as an independent-minded, white rural peasant; a gauche rural peasant in Puerto Rico associated in one way or another with the notion of independence.

**King's nigger/King's man**   a 'liberated African'; an African seized from the slave ships of other nations by the British after 1807 and set to work for a number of years as an 'apprentice' in a British military department in one of the colonies.

*mambi*   a Dominican who refused to submit to Spanish rule in the mid-nineteenth century; a Cuban rebel against Spanish rule in the nineteenth century; a non-white rebel or guerrilla warrior symbolising a 'Free Cuba' in the second half of the nineteenth century.

*maron*   the French version of Spanish *cimarron* meaning an escaped slave existing independently

**maroons**   used in Jamaica to refer to escaped slaves who managed to establish themselves in their own communities in mountainous and other areas inaccessible to European soldiers and search parties

**maroon**   a term used by British in the colonies to refer to a rural picnic.

*montero*   a rural peasant especially in Cuba who customarily used a horse in his daily work; a figure who came to symbolise independence in Cuba.

*negro bozal*   a slave coming directly from Africa or Cape Verde who was in the colony less than a year; the equivalent in English was 'saltwater negro'.

*negro ladino*   an African who came to the New World via Europe; an African who was in the colony for more than a year.

*passparterre*   a slave from the French islands who arrived surreptitiously or by a clandestine route in one of the English islands: it was understood that the person came by 'sea' rather than by 'land' (*terre*).

*salvaje/cimarron*   alternative words used in Spanish to refer to indigenous inhabitants in the early colonial period. Other European colonisers showed differential preferences for the one or the other word
when identifying people who had escaped from or were outside
European settlements or control.

*Sauvage*   a word that in the colonial period became a name and was
used generally by the French to mean 'indigenous inhabitant'.

*Taino*   an indigenous word or expression initially understood by Europeans to mean 'a good and noble person', which became the name
of an ethnic group; an indigenous group set up in a contrastive relationship with *Caribe*.

**Willyforce nigger**   a name given to a 'liberated' African who had completed the period of 'apprenticeship'.

# Language names and designations

*baragouin*   a term used by the French in the seventeenth century to
refer to the language spoken by the indigenous inhabitants in the
colonies when communicating with Europeans. It was later also
used to refer to the language used by the African slaves to serve the
same purpose.

*bastarda lengua*   the concept of 'bastard language' started in reference
to the kind of language used in the sixteenth century for international communication in the Mediterranean area. It was extended to the
Americas to characterise the language spoken by the indigenous
inhabitants to communicate with Europeans. However, it was the
French who more energetically propagated the concept in the term
*langage corrompu*.

*castellano*   the name used in Spanish historical writing to identify the
language of Spaniards generally (= 'Spanish'): it was understood to
be a variety that started in Castile and spread across and dominated
Spain at the expense of other varieties. Like other European languages, it was held up in the colonies as the model to aspire to.
Sometimes the more explicit form *lengua castellana* was used.

*creol*   the name used for the language in Saint Domingue/Haiti that
had become the general language of communication among the local
population; in some cases in reference to Haiti as well as other
French colonies the longer term *la langue créole* was used to refer to
the local language that had developed among the slaves.

**creole drawl**   originally an observation made in the eighteenth century
about the way in which white creoles in the English islands drew out

(the vowels in) their syllables; by the nineteenth century it had become a repeatedly noted characteristic of creole speakers of English in the Caribbean and beyond.

**Guiney dialect**   This term was used by Edward Long to refer to African languages (more specifically words) in a rare concession by a European writer that the slaves' speech in Jamaica contained African features.

*habla bozal*   a term used to refer to the speech of African slaves in Cuba and other Spanish colonies. Ironically, whereas *bozal* was used to refer to a slave who had been in the colony for less than one or two years, *habla bozal* was used to identify the speech of slaves who had been there for several years.

*hablar franco*   a name given by the Moors and Turks to a language made up in the sixteenth century by Christians out of Italian, Spanish and Portuguese; a language used in business and trade in Algeria and throughout the Mediterranean with variants according to the native language of the respective speakers.

*habla rustica*   a term used to refer to the speech of rural dwellers in Spanish colonies without necessarily specifying their race and as such it was usually interpreted to refer to whites. Some of the features of *habla rustica* were identical to those of *habla bozal*.

*lenguaje criollo*   a term used by Juan Ignacio de Armas to describe the ensemble of peculiar words and structures, in current and general usage in Cuba, Santo Domingo, Puerto Rico, Venezuela, Colombia and part of Central America. This was a singular and basically unpopular attempt to set up the language varieties used by Creoles in the New World in a child–parent relationship with *castellano* with the etymological model of the Romance languages in mind.

*patois*   a term used by nineteenth-century French and English writers to refer to local, non-standard varieties of language in the islands, as in the 'vile French patois'; *Anglo patois*; an alternative term used in the English islands was the 'negro language'.

# Bibliography

A *brief, but most true relation of the late barbarous and bloody plot of the Negro's in the island of Barbados on Friday the 21. of October, 1692. In a letter to a friend.* 1693. London: George Croom.

Abbad y Lasierra, Fray Iñigo. [1788] 1966. *Historia geográfica, civil y natural de la isla de San Juan Bautista de Puerto Rico.* Rio Piedras, PR: Editorial Universitaria, Univ. de Puerto Rico.

[1788] 1971. *Historia geográfica, civil y natural de la isla de San Juan Bautista de Puerto Rico.* San Juan de Puerto Rico: Porta Coeli.

Abrahams, R. and Szwed, John. (eds). 1983. *After Africa.* New Haven and London: Yale Univ. Press.

Acosta, José de. 1590. *Historia natural y moral de las Indias.* Seville: J. de León.

1591. *Historia natural y moral de las Indias.* Barcelona: J. Cendrat.

Adorno, Rolena. 1988. El sujeto colonial y la construcción cultural de la alteridad. *Revista de critica literaria latinoamericana, no.* 28. pp. 55–68.

Advielle, Victor. 1901. *L'Odyssée d'un Normand à St. Domingue au dix-huitième siècle.* Paris: Librairie Challamel.

Aguado, Simon. 1602. Entremés de los negros. In *Nueva Biblioteca de Autores Españoles (Cotarelo y Mori, Emilio. 1911),* vol. xvii, pp. 231–5.

Aguirre, Mirta, Salvador Arias, David Chericián, Denia García Ronda, Virgilio López Lemus, Alberto Rocasolano. (eds). 1980. *Poesía social cubana.* La Habana: Editorial Letras Cubanas.

Aguirre Beltrán, Gonzalo. [1946] 1989. *La población negra de México: Estudio etnohistórico.* Mexico: Univ. Veracruzana, Instituto Nacional Indigenista, Gobierno del Estado de Vera Cruz, Fondo de Cultura Económica.

Albornoz, Aurora de and Julio Rodriguez Luis. 1980. *Sensemayá: La poesía negra en el mundo hispanohablante.* (Antología) Colección Asomant de Antología. Madrid: Editorial Origines.

ALEC. 1986. *Glosario lexicográfico del atlas linguistico-etnográfico de Colombia.* Bogotá: Imprenta Patriótica del Instituto Caro y Cuervo.

Alegria, Ricardo E. 1985. Notas sobre la procedencia cultural de los esclavos negros de Puerto Rico durante la segunda mitad del siglo XVI. *La Revista del Centro de Estudios avanzados de Puerto Rico y el Caribe.* July–December 1985.

1986. *Las primeras representaciones gráficas del indio americano 1493–1523.* Second edition. Centro de estudios avanzados de Puerto Rico y el Caribe.

Alexander, Captain J.E. 1833. *Transatlantic Sketches, comprising visits to the most interesting scenes in North and South America, and the West Indies. With notes on Negro slavery and Canadian emigration.* London: R. Bentley.

Alexandre, Pierre. 1972. *Languages and language in Black Africa.* Translated from French *Langues et langage en Afrique noire (1967)* by F.A. Leary. Evanston, IL: Northwestern Univ. Press.

Alleyne, M.C. 1961. Language and society in St. Lucia. *Caribbean Studies* vol. 1, pp. 1–11.

1988. *Roots of Jamaican Culture.* London: Pluto Press.

2002. *The construction and representation of race and ethnicity in the Caribbean and the world.* Mona, Jamaica: Univ. of the West Indies Press.

**Alonso, Manuel A.** [1849] 1992. *El Jíbaro.* Estudio de Luis O. Zayas Micheli. Rio Piedras: Editorial Edil.

**Alvar, Manuel.** 1987. Las castas coloniales. In López Morales and Vaquero (eds). (1987), pp. 17–32.

1987. *Léxico del mestizaje en Hispanoamérica.* Madrid: Ediciones cultura hispanica, Instituto de cooperación iberoamericana.

**Alvarez Nazario, Manuel.** 1961. *El elemento afronegroide en el español de Puerto Rico.* Second edition 1974. San Juan de Puerto Rico: Instituto de Cultura Puertorriqueña.

1970. Un texto literario del Papiamento documentado en Puerto Rico en 1830. In *Revista de Instituto de Cultura Puertorriqueña,* Vol.13, No.47, pp. 1–4.

1972. *La herencia lingüística de Canarias en Puerto Rico: Estudio historico-dialectal.* San Juan de Puerto Rico: Instituto de Cultura Puertorriqueña.

1973. Nuevos datos sobre las procedencias de los antiguos esclavos de Puerto Rico. In *La Torre,* nos 81–2, pp. 23–37.

1973a. El Papiamento: Ojeada a su pasado histórico y visión de su problemática del presente. [*Atenea,* Mayaguez, P. R., IX, nos 1–2, pp. 9–20.]

1977. *El influjo indígena en el español de Puerto Rico.* Río Piedras, PR: Editorial Universitaria.

1992. *El habla campesina del país: Orígenes y desarrollo del español en Puerto Rico.* Rio Piedras, PR: Editorial de la Univ. de Puerto Rico.

**Amphlett, John.** 1873. *Under a Tropical Sky: A journal of first impressions of the West Indies.* London: Sampson Low, Marston, Low & Searle.

[**Anderson, A.** 1797] Howard, Richard A. and Elizabeth S. Howard (eds) 1983. *Alexander Anderson's Geography and History of St. Vincent, West Indies.* Cambridge, MA: Harvard College and The Linnean Society of London.

**Anderson, Benedict.** 1991. *Imagined Communities: Reflections on the origin and spread of nationalism.* London and New York: Verso.

**Anghiera, Pietro Martire d'.** 1516. *De orbe nouo decades.* Alcalá de Henares.

1530. *De Orbe Novo.* Alcalá de Henares: M. de Eguía.

[1533] 1555. *The Decades of the newe worlde of west India.* Written in the Latine tounge by Peter Martyr of Angleria, and translated into Englyshe by Rycharde Eden. London: William Powell.

1628. *De Orbo Novo. The famous historie of the Indies: Declaring the adventures of the Spaniards, which have conquered these countries, with the varietie of relations of the religions, lawes, governments, manners, ceremonies, customes, rites, warres, and funerals of that people. Comprised into sundry Decades. Set forth first by Mr. Hackluyt, and now published by L.M. Gent.* Second edition. London: L.M. Gent.

[**Anon.**] 1830. *The Negro's Friend Notes on Slavery, made during a recent visit to Barbadoes.* London: Harvey and Darton; Houlston and Son; Bagster and Thomas, Printers.

**Ans, André-Marcel d'.** 1987. *Haiti: Paysage et société.* Paris: Éditions Karthala.

*Antilia and America. A description of the 1424 nautical chart and the Waldseemüller globe map of 1507 in the James Ford Bell collection at the Univ. of Minnesota.* 1955. Minneapolis.

Arbell, Mordechai. 1981. *La Nacion*. In *"La Nacion": The Spanish and Portuguese Jews in the Caribbean*. Tel Aviv: Beth Hatefutsoth, the Nahum Goldmann Museum of the Jewish Diaspora.

Arber, Edward. 1885. *The first three English books on America.[? 1511] – 1555 A.D. Being chiefly Translations, Compilations, &, by Richard Eden*. Birmingham: Turnbull & Spears.

Ardouin, Alexis Beaubrun. [1853] 1958. *Etudes sur l'histoire d'Haïti suivies de la vie du Général J.-M. Borgella*. Deuxième édition conforme au texte original annotée et précédée d'une notice biographique sur B. Ardouin par le Docteur François Dalencour. Port-au-Prince, Haiti: Chez l'Editeur, François Dalencour.

Armas y Céspedes, Juan Ignacio de. 1882. *Oríjenes del lenguaje criollo*. Segunda edición, correjida i aumentada. Habana: Imprenta de la viuda de Soler

Arrom, José Juan. 1941. La poesía afrocubana. *Revista Iberoamericana*, no. 7, November 1941, pp. 379–411.

1951. Criollo: definición y matices de un concepto. *Hispania*, vol. 34, no. 2, pp. 172–3.

1961. Notas sobre la primera generación criolla en Hispanoamérica (1564–1594). *Revista Iberoamericana*, Vol. XXV1, nos 51 and 52, pp. 313–21.

1971a. Criollo: Definición y matices de un concepto. In *Certidumbre de América*, pp. 11–26, Madrid: Gredos.

1971b. Presencia del negro en la poesía folklórica americana. In *Certidumbre de América*, pp. 122–53. Madrid: Gredos.

1971c. La Virgen del Cobre: Historia, leyenda y símbolo sincrético. In *Certidumbre de América*, pp. 184–214, Madrid: Gredos.

1983. Cimarrón: apuntes sobre sus primeras documentaciones y su probable origen. *Revista española de antropología americana*, vol. xiii, pp. 47–57.

1989. *La mitología y artes prehispanicas de las Antillas*. Mexico, D.F.: Siglo Veintiuno Editores.

2000. *Estudios de lexicología antillana*. San Juan, PR: Editorial de la Univ. de Puerto Rico.

Atkins, John. 1735. *A voyage to Guinea, Brasil, and the West-Indies; in his Majesty's ships, the Swallow and Weymouth*. London: Ward and Chandler. Reprinted in 1972 by Metro Books, Northbrook, IL.

Attwood, Thomas. 1791. *The history of the island of Dominica*. London: J. Johnson.

Aube, Le Contre-Amiral. 1882. *Martinique: Son présent et son avenir*. Paris and Nancy: Berger-Levrault et Cie.

Auberteuil, Hilliard d'. 1776-7. *Considerations sur l'état présent de la colonie française de S. Domingue. Ouvrage politique et législatif*. Tome second. Paris.

Bachiller y Morales, Don Antonio. 1883. *Cuba primitiva. Origen, lenguas, tradiciones e historia de los indios de las Antillas Mayores y Las Lucayas*. Second edition. Habana: Libreria de Miguel de Villa.

1883a. Desfiguracion a que está expuesto el idioma castellano al contacto y mezcla de las razas. *Revista de Cuba*, Tomo X1V, pp. 97–104.

Bailyn, Bernard. 1984. New England and a Wider World: Notes on Some Central Themes of Modern Historiography. In D. Hall and D. Allen (eds) *Seventeenth-Century New England*, pp. 323–8.

1992. *The Boundaries of History: The Old World and the New*. Providence, RI: The John Carter Brown Library.

Baird, Robert. 1850. *Impressions and experiences of the West Indies and North America in 1849*. Philadelphia, PA: Lea & Blanchard.

Baissac, Charles. 1880. *Étude sur le patois créole mauricien*. Nancy: Berger-Levrault et Cie.

Balaguer, Joaquin. 1973. *Historia de la literatura dominicana*. Santo Domingo.

1984. *La isla al revés: Haití y el destino dominicano*. Santo Domingo, RD: Librería Dominicana.

Balboa y Troya de Quesada, Silvestre de. [1608] 1941. *Espejo de paciencia*. Estudio crítico de Felipe Pichardo Moya. La Habana: Imprenta Escuela del Instituto Cívico Militar.

Ballou, Maturin Murray. 1854. *History of Cuba: or, Notes of a traveller in the Tropics; being a political, historical and statistical account of the island, from its first discovery to the present time*. Boston: Phillips, Sampson and Company.

Barbot, Jean. 1752. In Churchill, John. *A collection of voyages and travels, consisting of authentic writers in our own tongue, which have not before been collected in English, or have only been abridged in other collections*. Volume V. 1752. London: Thomas Osborne.

Barclay, Alexander. 1826. *A practical view of the present state of slavery in the West Indies; or, an examination of Mr. Stephen's "Slavery of the British West India colonies:" containing more particularly an account of the actual condition of the Negroes in Jamaica*. London: Smith, Elder & Co.

Barrell, Theodore. 1843. Sketches of Demerary Incidents. Unpublished typescript.

Bastide, Roger. 1967. *Les Amériques noires: Les civilisations africaines dans le Nouveau Monde*. Paris: Payot.

Bates, William C. 1896. Creole folk-lore from Jamaica. *Journal of American Folk-Lore* 9–10, pp. 38–42; 121–28.

Baxter, Thomas. 1740. *A Letter from a gentleman at Barbados to his Friend now in London concerning the Administration of the late Governor B…g*. London: J. Roberts.

Bayley, F.W.N. 1829. *The Island Bagatelle, containing poetical enigmas on the estates in each parish in the island of Grenada. In six parts. Interspersed with tales and other miscellaneous poems*. Grenada: W.E. Baker.

1830. *Four Years' Residence in the West Indies During the years 1826, 7, 8 and 9*. London: William Kidd.

1833. *Four Years' Residence in the West Indies During the years 1826, 7, 8 and 9*. Third edition enlarged. London: William Kidd.

Bazin, H. 1906. *Dictionnaire bambara-français, précédé d'un abrégé de grammaire bambara*. Paris: Imprimerie nationale. 1965 Gregg Press: Hampshire, England.

[BDIB] Manuscript, Trinity College Library, Dublin, containing *A Breife Discription of the Ilande of Barbados*, circa 1650. (See also J. Hutson (ed.) *The English Civil War in Barbados* (2001).)

Beaudoux-Kovats, Edith and Jean Benoist. 1972. Les Blancs créoles de la Martinique. In Benoist, J. (ed.) *L'Archipel inachevé: Culture et société aux Antilles françaises*. Montreal: Presses de l'Univ. de Montréal.

Bébel-Gisler, Dany and Hurbon, Laënnec. 1975. *Cultures et Pouvoir dans la Caraïbe: langue créole, vaudou, sectes religieuses en Guadeloupe et en Haiti*. Second edition. Paris: Éditions l'Harmattan.

Beckford, William. 1788. *Remarks upon the situation of Negroes in Jamaica, impartially made from a local experience of nearly thirteen years in the island*. London: T. and J. Egerton, Military Library.

1790. *A descriptive Account of the island of Jamaica*. London.

Belgrove, William. 1755. *A treatise upon husbandry or planting*. Boston.

Benítez Rojo. Antonio. 1989. *La isla que se repite: El Caribe y la perspectiva postmoderna*. Hanover, NH: Ediciones del Norte.

1992. *The Repeating Island: The Caribbean and the postmodern perspective*. Translation of *La isla que se repite* by James E. Maraniss. Durham and London: Duke Univ. Press.

Bethell, Leslie. 1984. *Colonial Latin America, vol.* ii. *The Cambridge History of Latin America*. Cambridge: Cambridge Univ. Press.

Bertrand-Bocandé, M. 1849. Notes sur la Guinée portugaise ou Sénégambie méridionale – De la langue créole de la Guinée portugaise. *Bulletin de la Société de Géographie*, vol. XII, Third Series, pp. 73–7.

Bickell, The Rev. Richard. 1825. *The West Indies as they are; OR A real picture of Slavery: But more particularly as it exists in the Island of Jamaica. In three parts with notes*. London.

Bickerton, Derek. 1984. The language bioprogram hypothesis. *The Behavioral and Brain Sciences* vol. 7, no. 2, pp. 173–88.

Biet, Antoine. 1664. *Voyage de la France Equinoxiale en l'isle de Cayenne, entrepris par les François en l'année M.DC.LII. Devise en trois Livres*. Paris.

Bird, M.B. 1869. *The black man; or, Haytian independence. Deduced from historical notes, and dedicated to the government of Hayti*. New York: Published by the Author.

1876. *L'homme noir: ou, Notes historiques sur l'indépendance haitienne*. Translated from English 1876. Edinburgh: Murray and Gibb.

Blanchard, R. 1910. Encore sur les tableaux de métissage du Musée de Mexico. *Journal de la Société des Américanistes de Paris*, III, pp. 37–60.

Blofeld, J.H. 1844. *Algeria, past and present. Containing a description of the country of the Moors, Kabyles, Arabs, Turks, Jews, Negroes, Cologlies, and other inhabitants; their habits, manners, customs, &c.; together with notices of the animal and vegetable productions, minerals, climate, &c.* London: T.C. Newby.

Blome, Richard. 1672. *A description of the island of Jamaica with the other isles and territories in America, to which the English are related, viz, Barbadoes, St. Christophers, Niebis, or Mevis, Antego, Barbuda, St. Vincent, Bermudes, Dominica, Carolina, Montserrat, Virginia, Anguilla, New England, New Foundland. Taken from the notes of Sr. Thomas Knight, Governor of Jamaica; and other experienced persons in the said places. Illustrated with maps*. London: L. Mibbourn.

Boddam-Whetham, J.W. 1879. *Roraima and British Guiana with a glance at Bermuda, the West Indies, and the Spanish Main*. London: Hurst and Blackett.

Borde, Pierre Gustave Louis. [1876; Pt. 2 1882] 1883. *Histoire de l'ile de la Trinidad sous le gouvernement espagnol*. Paris: Maisonneve et Cie.

Botero, John. 1635. *The Cause of the Greatnesse of Cities. Three Bookes. With Certaine Observations concerning the Sea. Written by John Botero: And translated into English by Sir T. H(awkins)*. London.

Bossu, M. 1768. *Nouveaux voyages aux Indes Occidentales*. Paris.

1777. *Nouveaux voyages dans l'Amerique septentrionale contenant une collection de lettres écrites sur les lieux, par l'Auteur, à son ami, M. Douin, Chevalier, Capitaine dans les troupes du Roi, ci-devant son camarade dans le nouveau monde*. Amsterdam: Changuion.

Boucher, Philip P. 1989. *Les Nouvelles Frances: France in America, 1500–1815. An imperial perspective.* Providence, RI: The John Carter Brown Library.

Bourdieu, Pierre. 1991. *Language and Symbolic Power.* Edited and introduced by John B. Thompson. Translated by Gino Raymond and Matthew Adamson. Cambridge, Mass.: Harvard Univ. Press.

Bouton, Le P. Jacques. 1640. *Relation de l'etablissement des Francois depuis l'an 1635. En l'isle de la Martinique, l'une des antilles de l'Amerique. Des moeurs des Sauvages, de la situation, & des autres singularitez de l'isle.* Paris: Sebastian Cramoisy.

Bowser, Frederick P. 1972. The African in Colonial Spanish America: Reflections on research achievements and priorities. *Latin American Research Review* vol. 7, pp. 77–94.

Boyd-Bowman, P. 1956. The regional origins of the earliest Spanish colonists of America. *Publications of the Modern Languages Association of America, 1956,* LXX1, No.5, pp. 1151–72.

——— 1964. *Indice geobiográfico de cuarenta mil pobladores españoles de America en el siglo xvi.* Bogotá: Instituto Caro y Cuervo.

——— 1976. Spanish emigrants to the Indies, 1595–98: A profile. In Chiapelli, F. 1976 (ed.), Vol.2, pp. 732–3.

Boyer, Paul. 1654. *Veritable relation de tout ce qui s'est fait et passé au voyage que Monsieur de Bretigny fit à l'Amerique occidentale. Avec une description des moeurs, & des provinces de tous les sauvages de cette grande partie du Cap de Nord: un dictionnaire de la langue, & un advis tres necessaire à tous ceux qui veulent habiter ou faire habiter ce païs-là, ou qui desirent d'y establir des colonies. Le tout fait sur les lieux, par Paul Boyer, escuyer, sieur de Petit-Puy.* Paris: P. Rocolet.

Boyer-Peyreleau, Colonel Eugène-Édouard. 1823. *Les Antilles françaises particulièrement La Guadeloupe, Depuis leur découverte jusqu'au 1er Janvier 1823.* Paris: La Librairie de Brissot-Thivars.

Brathwaite, Edward. 1971. *The Folk Culture of the Slaves in Jamaica.* London: New Beacon Books.

Breen, Henry H. [1844] 1970. *St. Lucia: Historical, Statistical and Descriptive.* London: Longman, Brown, Green, and Longmans. 1970 reprint by Frank Cass, London.

Brereton, Bridget. 1981. *A history of modern Trinidad, 1783–1962.* Kingston, Jamaica ; Exeter, NH: Heinemann.

Brerewood, Edward. 1614. *Enquiries touching the diversity of languages, and religions through the cheife parts of the world.* London.

Breton, Raymond. 1665. *Dictionaire Caraibe Francois. Meslé de quantité de Remarques historiques pour l'esclaircissement de la Langue. Composé par le R. P. Raymond Breton, Religieux de l'ordre des Freres Prescheurs, & l'un des premiers Missionaires Apostoliques en l'isle de la Gardeloupe & autres circonvoisines de l'Amerique.* Auxerre: Gilles Bouquet.

——— 1666. *Dictionaire Francois–-Caraibe.*

Bridgens, R. 1837. *West India Scenery, with illustrations of negro Character, the process of making sugar, from Sketches taken during a voyage to, and residence of seven years in the island of Trinidad.* London: Robert Jennings.

Brinton, Daniel G. 1871. The Arawack language of Guiana in its linguistic and ethnological relations. *Transactions of the American Philosophical Society,* vol. 14, pp. 427–44.

457

Brown, Larissa V. 1988. *Africans in the New World 1493–1834*. Providence, RI: The John Carter Brown Library.

Browne, Patrick. 1756. *The civil and natural history of Jamaica*. London.

Brutus, Edner. 1948. *Instruction publique en Haiti 1492–1945*. Port-au-Prince, Haiti: Imprimerie de l'Etat.

Bryan, Patrick E. 1980. The question of labor in the sugar industry of the Dominican Republic in the late nineteenth and early twentieth centuries. *Social and Economic Studies*, July/Sept. 1980, pp. 275–91.

Budan, Armand. 1863. *La Guadeloupe pittoresque*. 1972 reproduction. Basse Terre: Société d'Histoire de la Guadeloupe.

Burnard, Trevor. 2004. *Mastery, tyranny, and desire: Thomas Thistlewood and his slaves in the Anglo-Jamaican world*. Mona, Jamaica: Univ. of the West Indies Press.

Burr-Reynaud, F. and Dominique Hippolyte. 1941. *Anacaona: poème dramatique, en vers, en trois actes et un tableau*. Port-au-Prince.

Burton, Richard. 1997. *Afro-Creole: Power, Opposition and Play in the Caribbean*. Ithaca and London: Cornell Univ. Press.

Butel-Dumont, Georges Marie. 1758. *Histoire et commerce des Antilles angloises*. Paris.

Caamaño de Fernández, Vincenta. 1989. *El negro en la poesía dominicana*. Centro de Estudios Avanzados de Puerto Rico y el Caribe. Santo Domingo: Corripio.

Caballero, Ramon C.F. 1852. *La juega de gallos o El negro bozal: Comedia en dos actos y en prosa*. In *Recuerdos de Puerto Rico*, pp. 43–81.

Caldecott, A. [1898] 1970. *The Church in the West Indies*. 1970 reprint by Frank Cass & Co, London.

Campbell, John. 1763. *Candid and impartial considerations on the nature of the sugar trade; the comparative importance of the British and French Islands in the West Indies*. London: Printed for R. Baldwin.

Campbell, P. F. 1982. *The Church in Barbados in the seventeenth century*. Barbados: Barbados Museum and Historical Society.

Candler, John. 1842. *Brief Notices of Hayti: with its condition, resources, and prospects*. London: Thomas Ward & Co.

Canny, Nicholas and Anthony Pagden (eds). 1987. *Colonial Identity in the Atlantic World, 1500–1800*. Princeton: Princeton Univ. Press.

Capadose, Henry. 1845. *Sixteen years in the West Indies*. 2 vols. London: J.C. Newby.

Carmichael, Mrs. 1833. *Domestic Manners and Social Condition of the white, coloured, and Negro population of the West Indies. In two volumes*. London: Whittaker, Treacher, and Co. 1969 reprint by Negro Univ. Press, New York.

Caro, Pedro. 1996. Entre 'guineos' y 'mandingas': En pos de una identificación nacional que nos defina. In *Listin Diario*, 14 January, 1996.

Carpentier, Alejo. [1946] 1972. *La música en Cuba*. Seconde edition. Mexico: Fondo de Cultura económica.

Cassidy, Frederic & LePage, Robert. 1980. *Dictionary of Jamaican English*. Second Edition. Cambridge: Cambridge Univ. Press.

Castellanos, Jorge and Castellanos, Isabel. 1988. *Cultura Afrocubana (El negro en Cuba, 1492–1844)*. Miami: Ediciones Universal.

   1992. *Cultura Afrocubana 3. Las religiones y las lenguas*. Miami: Ediciones Universal.

Castellanos, Juan de. 1589. *Primera parte, de las elegias de varones illustres de Indias. Compuestas por Juan de Castellanos Clerigo, Beneficiado de la Ciudad de Tunja en el nuevo Reyno de Granada.* Madrid

Castillo, José del. 1979. Las Emigraciones y su Aporte a la Cultura Dominicana (Finales del Siglo X1X y Principios del XX. *eme*, vol. viii, no. 45, pp. 3–43.

Cervantes Saavedra, Miguel de. 1949. *Don Quixote de la Mancha.* A new translation from the Spanish, with a Critical Text based upon the First Editions of 1605 and 1615, and with variant readings, variorum notes, and an Introduction by Samuel Putnam. New York: The Modern Library.

    1916. *Obras de Miguel de Cervantes Saavedra.* Biblioteca de Autores Españoles. Tomo Primero – Sexta edición. Madrid: Imprenta de Los Sucesores de Hernando.

    Novelas ejemplares. Tomo II. Ramón Sopena, Editor, Biblioteca Sopena, Barcelona. (includes 'El celoso extremeño')

    1922. *Obras completas de Miguel de Cervantes Saavedra* (including *Viage del Parnaso*). Edición publicada por Rodolfo Schevill y Adolfo Bonilla. Madrid.

Chanca, Dr. [1494]. A Letter addressed to the Chapter of Seville by Dr. Chanca, native of that city, and physician to the fleet of Columbus, in his second voyage to the West Indies, describing the principal events which occurred during that voyage. In Major, R.H. (ed.) 1847, pp. 18–68.

Chanvalon, Thibault de. 1763. *Voyage a la Martinique, Contenant Divers Observations sur la Physique, l'Histoire naturelle, l'Agriculture, les Moeurs, & les usages de cette Isle, faits en 1751 & dans les années suivantes.* Paris.

Chardon, Daniel. 1779. *Essai sur la colonie de Sainte Lucie. Par un ancien Intendant de cette Isle.* Neuchatel.

Charlevoix, Le P. Pierre-François-Xavier de. 1730. *Histoire de l'Isle Espagnole ou de S. Domingue. Ecrite particulierement sur des mémoires manuscrits du P. Jean-Baptiste le Pers, Jesuite, Missionnaire à Saint Domingue, & sur les pieces originales, qui se conservent au Dépôt de la marine. Tome premier.* Paris: Hippolyte–Louis Guerin.

    1731. *Histoire de l'Isle Espagnole ou de S. Domingue. Ecrite particulierement sur des mémoires manuscrits du P. Jean-Baptiste le Pers, Jesuite, Missionnaire à Saint Domingue, & sur les pieces originales, qui se conservent au Dépôt de la marine. Tome second.* Paris: Hippolyte–Louis Guerin.

Chiapelli, Fredi (ed.). 1976. *First Images of America*, vol. ii. Berkeley and Los Angeles: Univ. of California Press.

Christaller, J. G. 1933. *A dictionary of the Asante and Fante language called Tshi - Chwee, Twi – with a grammatical introduction and appendices on the geography of the Gold Coast and other subjects.* Basel: Basel Evangelical Society.

Cieza de León, Pedro de. [1553] 1945. *La Crónica del Perú.* Buenos Aires and Mexico: Espasa-Calpe Argentina, S.A.

Clarke, Sir Simon. 1823. *Some considerations on the present distressed state of the British West India colonies, their claims on the government for relief, and the advantage to the nation in supporting them, particularly against the competition of East India sugar.* London: C. & J. Rivington.

Clifford, James. 1988. *The Predicament of Culture: Twentieth-Century ethnography, literature, and art.* Cambridge, Mass. and London, England: Harvard Univ. Press.

Coke, Thomas. 1788? *The case of the Caribbs in St. Vincent's.* Dublin?

    1808. *A history of the West Indies.* Liverpool.

1810. *A history of the West Indies, containing the natural, civil, and ecclesiastical history of each island: With an account of the missions instituted in those islands, from the commencement of their civilization; but more especially of the missions which have been established in that archipelago by the Society late in connexion with the Rev. John Wesley, vol.* 2. London. 1971 Reprint by Frank Cass.

Cole, Michael and Scribner, Sylvia. 1974. *Culture and thought: a psychological introduction.* New York: John Wiley & Sons.

*Colección de documentos inéditos relativos al descubrimiento, conquista y colonización de las posesiones españolas en América y Oceanía, sacados en su mayor parte del Archivo de Indias* . . . 1864–1884. 42 vols. Madrid.

Coleridge, Henry. 1826. *Six Months in the West Indies in 1825.* 1970 Reprint by New York Universities Press, New York.

Coll y Toste, Cayetano. 1914a. Memoria y Description de la isla de Puerto Rico mandada hacer por S.M. el rey D. Felipe II. El año 1582. Archivo de Indias – Patronato. *Boletín Histórico de Puerto Rico,* no. 1, pp. 75–91.

1914b. Origen del campesino de Puerto Rico. In *Boletín Histórico de Puerto Rico,* vol. VII.

Collens, J.H. 1888. *A guide to Trinidad. A handbook for the use of tourists and visitors.* Second edition, revised and illustrated. London: Elliot Stock.

Colley, Linda. 1992. Britishness and otherness: An Argument. *Journal of British Studies,* vol. 31, no. 4, pp. 309–29.

Colman, George. 1788. *Inkle and Yarico. An Opera in three Acts as performed at the Theatre-Royal in the Hay-Market on Saturday, August 11th, 1787.* Dublin.

Columbus, Christopher. 1992. *Cristóbal Colón: Textos y documentos completos. Edición de Consuelo Varela. Nuevas cartas: Edición de Juan Gil.* Second edition. Madrid: Alianza Editorial.

*Select letters of Christopher Columbus, with other original documents, relating to his four voyages to the New World.* Translated and edited by R. H. Major, Esq. of the British Museum, 1847. London: Hakluyt Society.

The Spanish letter of Columbus to Luis de Sant' Angel Escribano de Racion of the Kingdom of Aragon. Dated 15 Feb 1493. Printed by Johann Rosenbach at Barcelona early in April 1493. Reprinted in reduced facsimile, and translated from the unique copy of the original edition. Reproduced by George Young (ed.) *The Columbus memorial,* pp. 1–64.

Cooper, Thomas. 1824. *Correspondence between George Hibbert, Esq., and the Rev. T. Cooper, relative to the condition of the Negro slaves in Jamaica, extracted from the Morning Chronicle; Also, a Libel on the character of Mr. and Mrs. Cooper, published, in 1823, in several of the Jamaica journals; With notes and remarks.* London: J. Hatchard and Son.

Cooper, Vincent. 1984. The St. Kitts Angolares: A closer look at current theories on language development in the Caribbean. Paper presented at the 5th biennial conference of the Society for Caribbean Linguistics, Mona, Jamaica.

Cornilliac, J.J.J. 1867. *Recherches chronologiques et historiques sur l'Origine et la Propagation de la fièvre jaune dans les Antilles.* First part. Fort-de-France: Imprimerie du Gouvernement.

Cornu, Jules. 1888. Die Portugiesische Sprache. In Gröber, Gustav, *Grundriss der romanischen philologie.* Strassburg: K. J. Trübner.

Corominas, Juan. 1954–57. *Diccionario crítico etimológico de la lengua castellana.* Bern: Francke.

Cortesâo, Armando. 1954. *The nautical chart of 1424 and the early discovery and cartographical representation of America: A Study of the History of Early Navigation and Cartography.* Coimbra: Univ. of Coimbra.

Cotarelo y Mori, Emilio. 1911. *Colección de entremeses, Loas, Bailes, Jácaras y Mojigangas desde fines del siglo XVI á mediados del XVIII.* Tomo I, Volumen I. Madrid: Casa Editorial Bailly–Bailliére. Nueva Biblioteca de Autores Españoles bajo la dirección del Excmo. Sr. D. Marcelino Menéndez y Pelayo no. 17.

Coulmas, Florian. 1992. *Language and Economy.* Oxford, UK and Cambridge, USA: Blackwell.

*Council Books of Barbados,* Extracts from the. Barbados Public Library, Reel 33.

Cressy, David. 1983. The environment for literacy: Accomplishment and context in seventeenth-century England and New England. In Daniel P. Resnick (ed.) *Literacy in historical perspective.* Washington: Library of Congress.

Croese, Gerard. 1696. *The General History of the Quakers.* London.

Crouch, Nathaniel. 1739. *The English Empire in America.* London.

Crowley, Daniel J. 1955. Festivals of the Calendar in St. Lucia. *Caribbean Quarterly,* vol. 4, no. 2, pp. 99–121.

Cruz, Sor Juana Inés de la. 1952. *Obras completas de sor Juana Inés de la Cruz.* Edition, prologue y notes by Alfonso Méndez Plancarte. Mexico: Fondo de Cultura Económica.

Cruz, Mary. 1974. *Creto Gangá.* La Habana: Instituto del Libro.

Cuervo, Rufino José. 1939. *Disquisiciones filológicas.* 2 vols. Compilación, Introducción, Notas y Dirección de Imprenta por Nicolas Bayona Posada. Bogotá: Editorial Centro, S.A.

————. 1950. *Disquisiciones filológicas.* 2 vols. Edición, prólogo y notas de Rafael Torres Quintero. Bogotá: Instituto Caro y Cuervo.

*Cuffy The Negro's Doggrel. Description of the Progress of Sugar.* 1823. London: E. Wallis.

Cumberland, Richard. 1771. *The West Indian: A comedy.* London: W. Griffin. Reproduced in *The Plays of Richard Cumberland,* vol. 1. Edited with an introduction by Roberta F.S. Borkat. 1982. New York & London: Garland Publishing.

D'Abbeville, Le R.P. Claude. 1614. *Historie de la Mission des peres capucins en l'isle de Maragnan et terres circonvoysines ou est traicte des singularitez admirables & des Meurs merveilleuses des Indiens habitans de ce pais Avec les myssives et advis qui ont este envoyez de nouveau.*

Dallas, R.C. 1803. *The history of the Maroons, from their origin to the establishment of their chief tribe at Sierra Leone: including the Expedition to Cuba, for the purpose of procuring chasseurs; and the state of the island of Jamaica for the last ten years; with a succinct history of the island previous to that period.* London: Longman and Rees.

Dana, Richard Henry. 1840. *Two years before the Mast: A personal narrative.* New York: Harper & Row.

Darwin, Charles. 1859. *On the origin of species by means of natural selection; or, The preservation of favoured races in the struggle for life.* London: John Murray.

————. 1871. *The descent of man, and selection in relation to sex.* London: John Murray.

Dauxion-Lavaysse, Jean François. 1820. *A statistical, commercial and political description of Venezuela, Trinidad, Margarita, and Tobago: containing various* anecdotes and observations, illustrative of the past and present state of these interesting countries; from the French of M. Lavaysse [1813]: with an introduction and explanatory notes by the Editor. Westport, CT: Negro Universities Press. 1969 reprint of 1820 English translation of Voyage aux isles de Trinidad, de Tabago, de la Marguerite, et dans diverses parties de Venezuela.

Davidson, George. 1788? The copy of a letter from a gentleman in the island of St. Vincent to the Reverend Mr. Clarke, one of the Reverend Mr. Wesley's missionaries in the West Indies, containing a short history of the Caribbs. In Coke (1788?)

Davis, N. D. 1887. The Cavaliers and Roundheads of Barbados 1650–2. Guyana: Argosy Press.

Davy, John. [1854] 1971. The West Indies, before and since Slave Emancipation, comprising the Windward and Leeward Islands' Military command; founded on notes and observations collected during a three years' residence. London: W. & F.G. Cash; Dublin: J. M'Glashan and J.B. Gilpin; Barbados: J. Bowen. 1971 edition Frank Cass, London.

Day, Charles William. 1852. Five Years' Residence in the West Indies. In two volumes. London: Colbourn and Co.

D'Estry, Stephen. 1845. Histoire d'Alger, de son territoire et de ses habitants, de ses pirateries, de son commerce et de ses guerres, de ses moeurs et usages: depuis les temps les plus reculés jusqu'à nos jours. Third edition.Tours: Ad. Mame et Cie.

De Laet, Joannes (Jean). 1640. L'histoire du Nouveau Monde ou Description des Indes Occidentales, contenant dix-huict liures, par le sieur Iean de Laet, d'Anuers; Enrichi de nouuelles tables geographiques & figures des animaux, plantes & fruicts. [Translated from the Latin original published in 1633.] A Leyde, Chez Bonaventure & Abraham Elseuiers.

De la Torre, J.M. 1854. Compendio de geografia física, politica, estadistica y comparada de la Isla de Cuba. Habana: Imprenta de Soler.

1868. Nuevos elementos de geografia e historia de la isla de Cuba. Habana: A. Pego.

De Lery, Jean. 1578. Histoire d'un voyage fait en la terre du Bresil, Autrement dite Amerique. Le tout recueilli sur les lieux par Jean de Lery natif de la Margelle, terre de Sainct Sene au Duché de Bourgougne.

De Molinari, M.G. 1887. A Panama: l'isthme de Panama – La Martinique – Haiti. Paris: Librairie Guillaumin et Cie.

De Pareto, Bartholomeus. 1455. Antillia. Illustrazione di una carta geografica del 1455 E delle notizie, che inquel tempo aveansi dell'Antillia.

De Rochefort, Charles. 1658. Histoire Naturelle & Morale des Iles Antilles de l'Amerique. Rotterdam.

1665. Histoire naturelle et morale des iles antilles de l'Amerique. Enrichie d'un grand nombre de belles figures en taille douce, des places & des raretez les plus considerables, qui y sont décrites. Avec un vocabulaire Caraibe. Seconde Edition. Reveü & augmentee de plusieurs descriptions, & de quelques éclaircissemens, qu'on desiroit en la precedente. Roterdam: Arnout Leers.

De Verteuil, Louis. 1858. Trinidad, its geography, natural resources, administration, present condition, and prospects. London: Ward & Lock.

1884. Trinidad, its geography, natural resources, administration, present condition, and prospects. Second edition. London, Paris & New York: Cassell & Co.

De Vries, David Peterson. [1655] 1853. *Voyages from Holland to America A.D. 1632 to 1644.* Translated from the Dutch by Henry C. Murphy. New York: Billin and brothers.

Debret, Jean Baptiste. 1834–9. *Voyage pittoresque et historique au Brésil.* Paris: Firmin Didot Frères.

Defoe, Daniel. 1718. *The Family Instructor,* vol. 2. London: Eman Matthews.

Deive, Carlos Esteban. 1977. *Diccionario de dominicanismos.* Santo Domingo: Politecnia Ediciones.

1978. *El Indio, el Negro y la Vida Tradicional Dominicana.* Santo Domingo: Museo del Hombre Dominicano.

1980. *La esclavitud del negro en Santo Domingo (1492–1844).* Santo Domingo: Museo del Hombre Dominicano.

1985. *Los Cimarrones de Maniel de Nieba: Historia y Etnografía.* Santo Domingo: Banco Central de la Republica Dominicana.

1992. *Vodu y Magia en Santo Domingo.* República Dominicana: Fundación Cultural Dominicana.

Dejean, Yves. 1983. Diglossia Revisited: French and Creole in Haiti. *Word* vol. 34, no. 3, pp. 189–213.

Delafosse, Maurice. 1929. *La langue Mandingue et ses dialectes (Mandinké, Bambara, Dioula)* vol. 1. *Introduction, Grammaire, lexique français–mandingue.* Paris: Librairie Orientaliste Paul Geuthner.

1955. *La langue Mandingue et ses dialectes (Mandinké, Bambara, Dioula)* vol. 2. *Dictionnaire mandingue–français.*

Delgado-Gomez, Angel. 1992. *Spanish Historical Writing about the New World 1493–1700.* Providence, RI: The John Carter Brown Library.

Descourtilz, M.E. [1809] 1935. *Voyage d'un naturaliste en Haiti 1799–1803.* Paris: Librairie Plon. Nouvelle bibliothèque des voyages Publiée sous la direction de Jacques Boulenger.

Despradel, Lil. 1973. República Dominicana: Las Etapas del Antihaitianismo. In (i) *Ahora,* no. 497, 21 de mayo, 1973, pp. 14–17 and (ii) *Ahora,* no. 498, 28 de mayo de 1973, pp. 10–16.

Dessalles, Pierre. 1996. See Forster, Elborg and Robert Foster.

Devas, Raymund. 1964. *The history of the island of Grenada.*

Diamond, Jared. 1997. *Guns, Germs, and Steel: The Fates of Human Societies.* New York and London: W.W. Norton & Co.

Diaz del Castillo, Bernal. 1955. *Historia verdadera de la conquista de la Nueva España.* Tomo 1. Mexico: Editorial Porrua.

*Diccionario de la lengua Castellana (1726–1739) Compuesto por la Real academia española.* Madrid: F. Del Hierro.

*Diccionario enciclopédico Dominicano.* 1988. Santo Domingo: Sociedad Editorial Dominicana.

Dickson, W. 1789. *Letters on Slavery. To which are added, Addresses to the whites and to the Free Negroes of Barbadoes, and accounts of some Negroes eminent for their virtues and abilities.* London: J. Phillips.

Dillard, J.L. 1972. *Black English: Its history and usage in the United States.* New York: Random House.

Doesborowe, John of (Jan van Doesborch). 1511. *Of the newe landes and of ye people founde by the messengers of the kynge of portyngale named Emanuel; Of the .x. dyvers nacyons crystened; Of pope Iohn and his landes and of the costely keyes and wonders melodyes that in that lande is.* Reprinted in Edward Arber (ed.) 1885, pp. xxiii–xxxvi.

Diop, Cheikh Anta. 1981. *Civilisation ou barbarie.* 1991. Translation from French by Yaa-Lengi Meema Ngemi: edited by Harold J. Salemson and Marjolijn de Jager, *Civilization or barbarism: an authentic anthropology.* Brooklyn, NY: Lawrence Hill Books.

Dobal, Carlos. 1984. Hispanidad y dominicanidad. *eme,* vol. xii, no. 71, pp. 89–97.

Domínguez, José de J. 1903. Los Gíbaros. In José González Font (ed.) *Escritos sobre Puerto-Rico.* Barcelona.

Dominguez Compañy, Francisco. 1978. *La vida en las pequeñas ciudades hispanoamericanas de la conquista.* Madrid: Ediciones de cultura Hispanica de Centro Iberoamericano de cooperación.

Dorsainvil, J.-C. 1958. *Manuel d'Histoire d'Haiti.* Avec la collaboration des frères de l'instruction chretienne. Port-au-Prince, Haiti: Éditions Henri Deschamps.

Drake, Sir Francis. 1653. *Sir Francis Drake revived. Who is or may be a pattern to stirre up all heroicke and active spirits of these times, to benefit their countrey and eternize their names by like noble attempts.* London: Nicholas Bourne.

Duany, Jorge. 1996. Imagining the Puerto Rican Nation. In *Latin American Research Review,* vol. 31, no. 3.

———. 2001. Making Indians out of Blacks: The revitalization of Taíno identity in contemporary Puerto Rico. In Gabriel Haslip-Viera (ed.) *Taíno Revival: Critical Perspectives on Puerto Rican Identity and Cultural Politics.* (pp. 83–100). Princeton: Markus Wiener.

Dubuisson, Pierre. 1780. *Nouvelles considerations sur Saint Domingue, en réponse a celles de M.H.D.* Paris.

Du Puis, Le F. Mathias. [1652]. *Relation de l'establissement d'une colonie francoise dans la Gardeloupe isle de l'Amerique, et des moeurs des Sauvages.* 1972 reprooduction of the 1652 edition by the Société d'Histoire de Guadeloupe. Basse-Terre.

Duque de Estrada, Padre Antonio Nicolas. 1823. *Explicacion de la Doctrina Cristiana, Acomodada a la Capacidid de los Negros Bozales.* La Habana.

Du Tertre, le R.P. Jean Baptiste. 1654. *Histoire generale des isles de S. Christophe, de la Guadeloupe, de la Martinique, et autres dans l'Amerique.* Paris.

———. 1667. *Histoire generale des Antilles habitées par les François.* Tome II. *Contenant l'histoire naturelle, enrichy de cartes & de figures.* Paris: Thomas Jolly.

———. 1667–1671. *Histoire generale des Antilles habitées par les François . . . Par le r.p. dv. Tertre . . .* Paris: T. Jolly.

Duke, William. 1741. *Some memoirs of the first settlement of Barbados.* Barbados.

Easel, Theodore. 1840. *Desultory Sketches and tales of Barbados.* London.

Eden, Richard. 1555. (See Anghiera 1555).

Edwards, Bryan. 1793, 1794. *The history, civil and commercial, of the British Colonies in the West Indies. In two volumes.* London: J. Stockdale.

———. 1796. [Introduction in] *The proceedings of the Governor and Assembly of Jamaica, in regard to the Maroon Negroes: published by the order of the Assembly. To which is prefixed An introductory account, containing observations on the disposition, character, manners, and habits of life, of the Maroons, and a detail of the origin, progress, and termination of the late war between those people and the white inhabitants.* London: John Stockdale.

———. [1819] 1966. *The history, civil and commercial, of the British Colonies in the West Indies. With a continuation to the present time.* Fifth edition. New York: AMS Press.

Elcock, W.D. 1960. *The Romance languages*. London: Faber & Faber.

Elder, J.D. 1973. *Song Games from Trinidad and Tobago*. Trinidad: National Cultural Council.

Elkin, Judith. 1992. Imagining idolatry: Missionaries, Indians, and Jews. Occasional paper no. 3, Program in Judaic Studies, Brown Univ. Providence, RI: The Program in Judaic Studies; The Touro National Heritage Trust of Newport; John Carter Brown Library.

Elliott, J.H. 1995. Final Reflections: The Old World and the New revisited. In Kupperman (1995), pp. 391–408.

1998. *Do the Americas have a common history?* An address presented on the occasion of the celebration of the 150th anniversary of the founding of the John Carter Brown Library. 13th November 1996. With a translation by Antonio Feros. Published for the Associates of the John Carter Brown Library, Providence, Rhode Island.

*Enciclopedia Universal Ilustrada Europeo–Americana*. 1927. Tomo IX. Espasa–Calpe, S.A., Madrid.

Enciso, Martin Fernández de. (See Fernández de Enciso)

Entrambasaguas, Joaquín de. 1946. *Estudios sobre Lope de Vega*. Tomo primero. Madrid: Consejo superior de Investigaciones científicas.

Equiano, Olaudah. 1789. *The interesting narrative of Olaudah Equiano, or Gustavus Vassa the African, written by himself*. London.

Exquemelin, Alexandre. 1686. *Histoire des Avanturiers qui se sont signalez dans les Indes, avec la vie, les moeurs, les coutumes des habitants de Saint-Domingue*. Paris.

Fagg, John Edwin. 1965. *Cuba, Haiti & the Dominican Republic*. Englewood Cliffs, NJ: Prentice-Hall.

Feijóo, Samuel (ed.). 1980. *El Negro en la literatura folklorica cubana*. Havana: Editorial Letras Cubanas.

Fermin, Phillippe. 1764. *Traité de Maladies les plus frequentes à Surinam et des remedes les plus propres à les guerir. Suivi d'une dissertation sur le Fameux Crapaud de Surinam, nommé Pipa, & sur sa Generation en particulier, Avec Figures en Taille-douce*. Maestricht: Jacques Lekens.

1769. *Description générale, historique, géographique et physique de la colonie de Surinam*. Amsterdam.

Fernández, Felix. 1984. La estructuración gramatical del español dominicano y la identidad de los dominicanos: una interpretación. *eme*, vol. xii, no. 71, pp. 45–58.

Fernandez de Enciso, Martin. 1519. *Suma de geographia que trata de todas las partidas y provincias del mundo: en especial de las Indias*. Seville.

Fernández de Oviedo y Valdés, Gonzalo. 1526. *Sumario dela natural y general istoria delas Indias que escrivio Gonçalo Fernandez de Oviedo alias de Valdes natural dela villa de Madrid*.

1535. *La historia general de las Indias*. Seville.

[1555] 1959. *Historia general y natural de las Indias*. Biblioteca de Autores Españoles 117–121. Madrid: Ediciones Atlas.

Fernández Méndez, Eugenio. 1976. *Las encomiendas y esclavitud de los indios en Puerto Rico, 1508–1550*. Río Piedras: Editorial de la Univ. de Puerto Rico,

1983. *Los franceses en el Caribe y otros ensayos de historia y antropología*. San Juan, Puerto Rico: Ediciones 'El Cemí'.

1998. *The Sources on Puerto Rico Cultural History: A critical appraisal*. San Juan, Puerto Rico.

Fernández Retamar, Roberto. [1971] 1989. *Caliban and other essays*. Translated by Edward Baker. Univ. of Minnesota Press.

1992. En el centenario de 'Nuestra America', obra del Caribeño José Martí. *Anales del Caribe*, vol. 12.

Fernández Rocha, Carlos. 1974. Aspectos literarios en la obra de Juan Antonio Alix. *eme* no. 11, pp. 99–110.

Fielding, Sir John. 1768. *Extracts from such of the Penal Laws, as particularly relate to the Peace and Good Order of this Metropolis*. New edition. London: T. Cadell.

Fiet, Lowell. 2007. *Caballeros, vejigantes, locas y viejos: Santiago Apóstol y los performeros afropuertorriqueños*. Puerto Rico: Terranova Editores.

Figueiredo, Cândido de. 1949. *Dicionário da língua Portuguesa*. Fourteenth edition. Lisbon: Livraria Bertrand.

Fisher, Richard S. (ed.) 1855. *The Spanish West Indies: Cuba and Porto Rico: geographical, political and industrial. Cuba: from the Spanish of Don J.M. de la Torre. Porto Rico: by J.T. O'Neil, Esq*. New York: J.H. Colton.

Fleury, Capitaine Charles. 1987. *Un Flibustier français dans la mer des Antilles en 1618–1620*. Manuscrit inédit du début du xvii siècle publié par Jean-Pierre Moreau. Préface de Jean Meyer. Clamart: Éditions Jean-Pierre Moreau.

Font, José González. 1903. *Escritos sobre Puerto Rico*. Barcelona.

Foote, Samuel. [1764] *The Patron*.

Forbes, Jack D. 1993. *Africans and Native Americans: The Language of Race and the Evolution of Red-Black Peoples*. Second Edition. Urbana and Chicago, IL: Univ. of Illinois Press.

Ford, John. 1799. Two narratives by female slaves at Barbados, written down there by John Ford, 1799. Manuscript held in Oxford University's Bodleian Library.

Forster, Elborg and Robert Forster (eds). 1996. *Sugar and Slavery, Family and Race. The letters and Diary of Pierre Dessalles, Planter in Martinique, 1808 – 1856*. Baltimore and London: The Johns Hopkins Univ. Press. Editing and Translation of Henri de Frémont and Léo Elisabeth, ed. *Pierre Dessalles (1785–1857). La Vie d'un colon à la Martinique au XIXe siècle*. Courbevoie: private printing, 1984–88.

Forte, Maximilian C. 2005. *Ruins of Absence, Presence of Caribs: (Post)Colonial representations of aboriginality in Trinidad and Tobago*. Gainesville, FL: Univ. Press of Florida.

Foucault, Michel. 2000. *Power. Essential works of Foucault 1954–1984. Selections*. Edited by James D. Faubion; translated by Robert Hurley and others. Volume Three. New York: New Press.

Fowler, John. 1774. *Summary Account of the present flourishing state of the respectable colony of Tobago in the British West Indies*. London.

Fox, George. 1657. *A warning to all teachers of children*. London.

[1671] 1694. *A journal or historical account of the life, travels, sufferings, Christian experiences and labour of love in the work of the ministry of that ancient and faithful servant of Jesus Christ, George Fox, who departed this life in great peace with the Lord, the 13th of the 11th month, 1690*. London.

Franco, Franklyn J. 1969. *Los negros, los mulatos y la nación dominicana*. Santo Domingo, RD: Editora Nacional.

Franco, José L. [1961] 1979. Maroons and slave rebellions in the Spanish territories. In Richard Price (1979), pp. 35–48.

Fraser, Lionel Mordaunt. 1891. *History of Trinidad, vol. 1. (First period) From 1781 to 1813*. Port of Spain, Trinidad: Government Printing Office, vol. 2, 1896. Cass reprint 1971.

Frere, George. 1768. *A short history of Barbados from its first discovery and settlement to the present time. A New Edition corrected and enlarged*. London: J. Dodsley.

Friederici, Georg. 1947. *Amerikanistisches Wörterbuch*. Hamburg: Cram, de Gruyter.

Froger, Sieur. 1698. *Relation d'un voyage fait en 1695, 1696, & 1697 aux Côtes d'Afrique, détroit de Magellan, Brezil, Cayenne & Isles Antilles, par un Escadre des Vaisseaux du Roy, commandeé par M. DeGennes. Faite par le Sieur Froger Ingenieur volontaire sur le Faucon Anglois. Enrichie de grand'nombre de figures dessinées sur les lieux*. Paris.

1699. *Relation d'un voyage fait en 1695, 1696 & 1697*. Amsterdam.

Froude, James Anthony. 1888. *The English in the West Indies, or, the bow of Ulysses*. 1906 reprint by Charles Scribner's Sons, New York. 1969 reprint by Negro Universities Press, New York.

Gage, Thomas. 1648. *The English–American: a new survey of the West Indies*.

1929. *A new survey of the West Indies, 1648. The English–American*. Edited with an Introduction by A.P. Newton. The Argonaut Series, edited by Sir E. Denison Ross and Eileen Power. New York: Robert M. McBride & Co.

Galván, Manuel de Jesús. (1879, 1882) *Enriquillo: Leyenda histórica dominicana (1503–1533), con un estudio de Concha Meléndez*. Third edition 1986. Mexico: Editorial Porrúa.

Gamble, W.H. 1866. *Trinidad: Historical and Descriptive: Being a narrative of nine years' residence in the island. With special reference to Christian missions*. London: Yates and Alexander.

Garaud, Louis. 1895. *Trois ans à la Martinique*. Paris: Librairie d'Education nationale.

Garcilaso de la Vega, El Inca. [1609] 1963. *Obras completas del Inca Garcilaso de la Vega. II. Edición y estudio preliminar del P. Carmelo Saenz de Santa Maria*. Biblioteca de Autores Españoles desde la formación del lenguaje hasta nuestros días. Madrid: Ediciones Atlas.

1966. *Royal Commentaries of the Incas and General History of Peru*. Part One. Translated with an Introduction by Harold V. Livermore. Austin & London: Univ. of Texas Press.

Gardyner, George. 1651. *A Description of the New World. Or, America Islands and Continent*. London.

German de Granda. 1974–5. El repertorio lingüistico de los sefarditas de Curaçao durante los siglos xvii y xviii, y el problema del origen del papiamento. *Romance Philology* 28, (1974–5), pp. 1–16.

1978. *Estudios lingüísticos hispánicos, afrohispánicos y criollos*. El 'habla de negro' en la literatura peninsular del siglo de oro. Estudios y ensayos, 282. Madrid: Biblioteca Románica Hispánica. Editorial Gredos.

1989. Algunos rasgos más de origen africano en el criollo palenquero. In *Estudios sobre Español de América y lingüistica afroamericana. Ponencias presentadas en el 45 congreso internacional de Americanistas (Bogota, Julio de 1985)*. Bogotá: Publicaciones del Instituto Caro y Cuervo.

Gikandi, Simon. 1996. *Maps of Englishness: Writing identity in the culture of colonialism.* New York: Columbia Univ. Press.

Gil, Juan. 1989. *Mitos y utopías del descubrimiento: II. El Pacífico.* Madrid: Alianza

Gilmore, Myron P. 1976. The New World in French and English historians of the Sixteenth Century. In Chiapelli, Fredi. (ed.). *First Images of America,* vol. ii, pp. 519–527. Univ. of California Press.

Girod-Chantrans, Justin. 1785. *Voyage d'un Suisse dans differentes colonies d'Amerique.* Neuchatel: Impr. de la Société typographique.

Godwyn, Morgan. 1680. *The Negro's and Indians Advocate.* London.

Góngora, Luis de. 1609. En la fiesta del Santísimo Sacramento. In Henríquez Ureña, Pedro, 1939, *Luis de Góngora: Romances y letrillas.* Buenos Aires: Editorial Losada.

González, José Luis. [1980] 1993. *Puerto Rico: the four storeyed country and other essays.* Translation by Gerald Guinness of *Pais de cuatro pisos y otros ensayos.* Princeton, N.J. & New York: Markus Wiener Publishing.

Gonzalez, Nancie L. 1988. *Sojourners of the Caribbean: Ethnogenesis and Ethnohistory of the Garifuna.* Urbana and Chicago: Univ. of Illinois Press.

1969. *Black Carib household structure: A study of migration and modernization.* Monograph 48, The American Ethnological Society. Seattle and Washington: Univ. of Washington Press.

Goody, J. 1977. *The Domestication of the Savage Mind.* Cambridge: Cambridge Univ. Press.

Gordon, Shirley. 1963. *A century of West Indian education: A source book.* London: Longmans Green & Co.

Goux, Abbé. 1842. *Catéchisme en langue créole, précédé d'un essai de grammaire sur l'idiome usité dans les colonies françaises.* Paris: H. Vrayet de Surcy.

Grafenstein, Johanna von. 1988. *Haiti 1.* Mexico: Instituto de Investigaciones, Univ. de Guadalajara.

Granier de Cassagnac, Adolphe. 1842. *Voyage aux Antilles, françaises. Première partie.* Paris: Dauvin et Fontaines Libraires.

1844. *Voyage aux Antilles. Deuxieme partie. Les Antilles anglaises, danoises et espagnoles, Saint Domingue et les Etats-Unis.* Paris: Au Comptoir des Imprimeurs-Unis.

Grant, William (ed.). 1931–. *The Scottish National Dictionary.* Edinburgh.

*Great Newes from the Barbados. Or a true and faithful account of the Grand Conspiracy of the Negroes against the English and the happy discovery of the same with the number of those who were burned alive, beheaded, and otherwise executed for their horrid crimes with a short description of that plantation.* 1676. London: L. Curtis.

Greenfield, William. 1830. *A defence of the Surinam Negro-English version of the New testament, in reply to the animadversions of an anonymous writer in the Edinburgh Christian Instructor.* London: Samuel Bagster.

Gregoire, H. 1808. *De la litterature des nègres.* Paris.

Guamán Poma de Ayala, Felipe. 1980. *Nueva corónica y buen gobierno/ Felipe Guamán Poma de Ayala; transcripción, prólogo, notas y cronología, Franklin Pease.* Caracas, Venezuela: Biblioteca Ayacucho.

Guirao, Ramón. 1938. *Orbita de la poesia afrocubana 1928–37. (Antología). Seleccion, notas biograficas y vocabulario.* 1970 Kraus reprint, Nendeln.

Gurney, Joseph John. 1840. *A winter in the West Indies*. London: John Murray.

Gútemberg Bohórquez C., Jesús. 1984. *Concepto de 'Americanismo' en la historia del español*. Bogotá: Instituto Caro y Cuervo.

Haedo, Fray Diego de. 1612. *Topographia, e Historia general de Argel, repartida en cinco tratados, lo se veran casos estraños, muertes espantosas, y tormentos exquisitos, que conviene se entiendan en la Christiandad: con mucha doctrina, y elegancia curiosa*. Valladolid: Diego Fernandez de Cordova y Oviedo.

Hair, P. E.H., Adam Jones and Robin Law (eds). 1992. *Barbot on Guinea: The Writings of Jean Barbot on West Africa 1678–1712*. Volumes I and II. London: The Hakluyt Society.

Hakluyt, Richard. 1589. *Principall navigations, voiages, and discoveries of the English nation*. London: George Bishop and Ralph Newberie.

———. 1598–1600. A voyage of the honourable gentleman M. Robert Duddely to the isle of Trinidad 1594–5. Volume 3, pp. 574–578.

Hall, David and David Grayson Allen (eds). 1984. *Seventeenth-Century New England*. Boston: The Colonial Society of Massachusetts.

Hall, Douglas. 1989. *In Miserable Slavery: Thomas Thistlewood in Jamaica 1750–86*. London: Macmillan.

Halliday, Sir Andrew. 1837. *The West Indies: The natural and physical history of the Windward and Leeward colonies; with some account of the moral, social, and political condition of their inhabitants, immediately before and after the abolition of Negro slavery*. London: J.W. Parker.

Handler, Jerome S. 1969. The Amerindian slave population of Barbados in the seventeenth and early eighteenth centuries. *Caribbean Studies*, vol. 8, no. 4, pp. 38–64.

Handler, J. and F. Lange. 1978. *Plantation slavery in Barbados*. Cambridge, Mass: Harvard Univ. Press.

Handler, J. and R. Corruccini. 1986. Weaning among West Indian Slaves: Historical and bioanthropological evidence from Barbados. *The William and Mary Quarterly*, Third Series, vol. 43.

Handler, J. and JoAnn Jacoby. 1996. Slave names and naming in Barbados 1650–1830. *The William and Mary Quarterly*, Third Series, vol. 52, no. 4 (pp. 685–728).

Harcourt, Robert. [1613] 1928. *A relation of a voyage to Guiana by Robert Harcourt*. Issued by the Hakluyt Society for the year 1926. Second Series, no. lx. With Purchas's transcript of a report made at Harcourt's instance on the marrawini District. Edited with Introduction & Notes by Sir C. Alexander Harris. London.

Harlow, Vincent T. 1926. *A History of Barbados 1625–85*. Oxford. 1969 Reprint by New York Universities Press, New York.

Harris, Leonard. (1993) 1995. Postmodernism and Utopia, an Unholy Alliance. In Hord, Fred Lee (Mzee Lasana Okpara) and Jonathan Scott Lee (eds), *I am because we are: Readings in Black Philosophy*. Amherst: Univ. of Massachusetts Press, pp. 367–82. Reprinted from *Racism, The City and the State*, edited by Malcolm Cross and Michael Keith, 1993. London: Routledge.

Harrison, James A. 1884. Negro English. *Anglia 7*. 232–79.

Haslip-Viera, Gabriel (ed.) 2001. *Taíno Revival: Critical Perspectives on Puerto Rican Identity and Cultural Politics*. Princeton: Markus Wiener.

Hawthorne, Nathaniel. 1850. *The Scarlet Letter*. Boston: Ticknor, Reed and Fields.

Hay, John. 1823. *A narrative of the insurrection in the island of Grenada*.

Hazard, Samuel. 1873. *Santo Domingo, Past and present; with a glance at Hayti.* New York: Harper & Brothers.

Hearn, Lafcadio. 1890. *Two years in the French West Indies.* New York: Harper & Brothers.

——— 1924. *An American Miscellany: Articles and stories now first collected by Albert Mordell, vol.* 1. New York: Dodd, Mead and Co.

Heinl, Robert Debs Jr. and Nancy Gordon Heinl. 1978. *Written in blood: The story of the Haitian people 1492–1971.* Boston: Houghton Mifflin.

Helm, Alex. 1981. *The English mummers' play.* Suffolk, England: Folklore Society.

Hemming, John. 1978. *Red Gold: The conquest of the Brazilian Indians.* London: Macmillan.

Henríquez Ureña, Max. 1963. *Panorama historico de la literatura cubana.* New York: Las Americas Publishing Co.

Henríquez Ureña, Pedro. [1919]. La lengua en Santo Domingo. In *Obra Dominicana* (1988). pp. 429–31. Sociedad Dominicana de Bibliofilos.

——— 1938. *Para la historia de los indigenismos.* Buenos Aires: Univ. de Buenos Aires.

——— 1939 *Luis de Góngora: Romances y letrillas*

——— 1940 *Luis de Góngora: Poemas y sonetos.* Editorial Losada, Buenos Aires. Las cien obras maestras de la literatura y del pensamiento universal publicadas bajo la dirección de Pedro Henríquez Ureña. (No. 15 = 1939, no. 16 = 1940).

——— 1975. *El Español en Santo Domingo.* Second Edition. Santo Domingo, RD: Editora Taller.

Heredia, Jose María. 1970. *Poesías completas.* Selección, estudio y notas por Angel Aparicio Laurencio. Miami: Ediciones Universal.

Herlein, J.D. 1718. *Beschryvinge van de Volk-Plantinge Zuriname.* Leevwarden: Meindert Injema.

Hernandez Aquino, Luis. 1977. *Diccionario de voces indigenas de Puerto Rico.* Rio Piedras, PR: Editorial Cultural.

Herrera, A.L. and R.E. Cicero. 1895. *Catálogo de la colección de antropología del Museo nacional.* Mexico: Museo nacional.

Herrera y Tordesillas, Antonio. 1601–1615. *Historia general de los hechos de los castellanos en las islas, tierra firme del mar Oceano.* Madrid: Emplenta Real.

Hesiod: Works And Days. Translated by Hugh G. Evelyn-White. [1914]. www.sacred-texts.com/cla/hesiod/works.htm.

Hickeringill, Captain E. 1661. *Jamaica Viewed; with all the ports, harbours, and their several soundings, towns, and settlements thereunto belonging. Together, with the nature of its climate, fruitfulness of the soil, and its suitableness to English complexions. With several other collateral observations and reflections upon the island.* Third edition 1705. London.

——— 1705. *Jamaica reviewed.* London.

Higman, B. 1988. Ecological determinism in Caribbean History. Sixth Elsa Goveia Memorial Lecture, Univ. of the West Indies, Cave Hill, Barbados.

Hillary, William. 1759. *Observations on the changes of the air.* London.

Hirsch, Rudolf. 1976. Printed reports on the early discoveries and their reception. In Chiapelli, Fredi (1976), vol. ii, pp. 519–27.

*Histoire naturelle des Indes:* The Drake manuscript in the Pierpont Morgan Library. Preface by Charles E. Pierce, Jr.; Foreword by Patrick O'Brian; Introduction by Verlyn Klinkenborg; Translations by Ruth S. Kraemer. 1996. New York: Norton.

Hobsbawm, E.J. 1990. *Nations and nationalism since 1780: Programme, myth, reality*. Cambridge: Cambridge Univ. Press.

Hobsbawm, E.J. and Terence Ranger (eds). 1983. *The invention of tradition*. Cambridge: Cambridge Univ. Press.

Hoetink, Harmannus. 1967. *The two variants in Caribbean relations*. Oxford: Oxford Univ. Press.

   1982. *The Dominican people 1850–1900: Notes for a historical sociology*. Translated from the 1972 original by Stephen K. Ault. Baltimore and London: The Johns Hopkins Univ. Press.

Hofland, Mrs. [1818] 1871. *The Barbadoes Girl. A Tale*. London: T. Nelson & Sons.

Homer. 1919/1966. *The Odyssey with an English translation by A.T. Murray*. London: William Heinemann; Cambridge, MA: Harvard Univ. Press.

Honychurch, Lennox. 1998. *Dominica, Isle of Adventure: An Introduction and Guide*. Third edition. London: Macmillan Educational.

Howard, Richard A. and Elizabeth S. Howard (eds) 1983. *Alexander Anderson's Geography and History of St. Vincent, West Indies*. Cambridge, MA: Harvard College and The Linnean Society of London.

Howard, Thomas. 1796–8. Journal of a voyage to the West Indies began on Monday Feb. 8 1796. Unpublished manuscript.

Howell, James, 1737. *Epistolæ Ho-Elianæ: familiar letters domestick and foreign, divided into four books: partly historical, political, philosophical: upon emergent occasions*. London.

Huc, Théophile. 1877. *La Martinique: Études sur certaines questions coloniales*. Paris: A. Cotillon et Cie; Challamel ainé.

Hughes, The Rev. Griffith. 1750. *The Natural History of Barbados*. London. Printed for the author.

Hulme, Peter. 1978. Columbus and the cannibals: a study of the reports of anthropophagy in the journal of Christopher Columbus, *Ibero-Amerikanisches Archiv*, IV: 115–39.

   1986. *Colonial encounters: Europe and the native Caribbean, 1492–1797*. London; New York: Methuen.

   1990. The rhetoric of description: The Amerindians of the Caribbean within modern European discourse. *Caribbean Studies* 23 3/4, 35–49.

Hulme, Peter and Neil L. Whitehead (eds). 1992. *Wild Majesty: Encounters with Caribs from Columbus to the present day. An Anthology*. Oxford: Clarendon Press.

Hume, David. 1963. *Of National Characters. Essays: Moral, Political and Literary*. Oxford.

Hutson, J. Edward (ed.). 2001. *The English Civil War in Barbados 1650–1652: Eyewitness Accounts*. Barbados: The Barbados National Trust.

Hymes, Dell (ed.). 1971. *Pidginization and Creolization of Languages*. Cambridge: Cambridge Univ. Press.

Icazbalceta, Joaquin Garcia. [1899] 1975. *Vocabulario de Mexicanismos*. Edición facsímil. Mexico: Ediciones del centenario de la Academia mexicana/5.

ILLACC (Instituto de literatura y linguistica de la academia de ciencias de Cuba). 1983. *Perfil histórico de las letras cubanas desde los orígenes hasta 1898*. La Habana: Editorial Letras cubanas.

*Inkle & Yarico, a legend of Barbados*. By a Barbadian. circa 1840. Barbados.

Isaza Calderon, Baltasar. 1986. *Panameñismos*. Third edition. Panama: Manfer.

Isert, Paul Erdmann. 1793. *Voyages en Guinée et dans les iles Caraibes en Amerique. Tirés de sa correspondance avec ses amis. Traduit de l'allemand*. Paris: Maradan.

James, C.L.R. 1963. *The Black Jacobins: Toussaint L'Ouverture and the San Domingo Revolution*. Second edition, revised. New York: Vintage Books.

1969. The West Indian Intellectual. Introduction in J.J. Thomas. *Froudacity: West Indian fables explained*.

Johnson, J. 1830. *An Historical and descriptive account of the island of Antigua*. London: Henry Baylis.

Johnson, Julie Greer. 1988. *The Book in the Americas: The role of books and printing in the development of culture and society in colonial Latin America. Catalogue of an exhibition*. Providence, RI: The John Carter Brown Library.

Jones, Steve. 1993. *The Language of the Genes: Biology, History and the Evolutionary Future*. London: Flamingo-HarperColllins.

Josefina Guzmán, Daysi. 1974. Raza y lenguaje en el cibao. *eme* no. 11, marzo–abril 1974, pp. 3–45.

Joseph, E.L. [1838] 1970. *History of Trinidad*. Trinidad: Henry James Mills; London: A.K. Newman; Glasgow: F. Orr & Sons. 1970 edition Frank Cass, London.

Kahn, Morton C. 1931. *Djuka: The Bush Negroes of Dutch Guiana*. New York: The Viking Press.

Kamen, Henry. 1980. *Spain in the later Seventeenth century, 1665–1700*. London and New York: Longman.

Kany, Charles E. 1960. *American–Spanish Semantics*. Berkeley and Los Angeles: Univ. of California Press.

Keimer, Samuel. 1741. *Caribbeana*. London.

Kimball, Richard. 1850. *Cuba, and the Cubans: comprising a history of the island of Cuba, its present social, political, and domestic condition; also its relation to England and the United States*. New York: Samuel Hueston; George P. Putnam.

King, William. 1709. *Useful transactions for the months of May, June, July, August and September, 1709. Containing a Voyage to the island of Cajamai in America*. London.

Kingsley, Charles. 1889. *At last, a Christmas in the West Indies*. London: Macmillan.

Klooster, Wim. 1997. *The Dutch in the Americas 1600 – 1800: A Narrative History with the Catalogue of an Exhibition of Rare Prints, Maps, and Illustrated Books from the John Carter Brown Library*. Providence, RI: The John Carter Brown Library.

Knight, Francis. [1640]1747. *A relation of seven years slavery under the Turks of Algiers, suffered by an English captive merchant. Wherein is also contained all memorable passages, fights, and accidents, which happened in that city, and at sea with their ship and gallies during that time*. In Churchill, A. & J. (compilers). *A collection of voyages and travels, consisting of authentic writers in our own tongue, which have not before been collected in English, or have only been abridged in other collections*. Volume VIII, pp. 465–478. 1747. London: Thomas Osborne.

Knight, F.W. 1988. *Slavery and the transformation of society in Cuba, 1511–1760*. Univ. of the West Indies, Mona, Jamaica: Dept. of History.

1990. *The Caribbean: The genesis of a fragmented nationalism*. Second edition. New York: Oxford Univ. Press.

Knight, F.W. and Palmer, C.A. (eds). 1989. *The Modern Caribbean*. Chapel Hill and London: The Univ. of North Carolina Press.

Konetzke, Richard. 1946. El mestizaje y su importancia en el desarrollo de la población hispanoamericana durante la época colonial. *Revista de Indias*, no. 24, pp. 215–37.

Kuethe, Allan. c 1986. *Cuba, 1753–1815: Crown, Military, and Society*. Knoxville: The Univ. of Tennessee Press.

Kupperman, Karen Ordahl (ed.). 1995. *America in European consciousness, 1493–1750*. Chapel Hill and London: Univ. of North Carolina Press.

1995. The changing definition of America. In Kupperman (ed.) 1995, pp. 1–29.

Kutzinski, Vera M. 1993. *Sugar's secrets: Race and the erotics of Cuban nationalism*. Charlottesville and London: Univ. of Virginia Press.

La Courbe, Michel Jajolet de. 1913. *Premier voyage du sieur de la Courbe fait a la coste d'Afrique en 1685. Publié avec une carte de Delisle et une introduction par P. Cultru*. Paris: E. Champion & Larose.

La Faye, Jacques. 1984. Literature and intellectual life in colonial Spanish America. In Leslie Bethell (ed.) 1984.

La Rosa Corzo, Gabino. 2003. La carimba. *Revista Cubana de Ciencias Sociales*, nos 33–34, pp. 35–46.

Labat, le R. Père Jean. 1722. *Nouveau voyage aux isles de l'Amerique*. Paris.

1724. *Nouveau voyage du Père Labat aux isles de l'Amerique*. La Haye.

1742. *Nouveau voyage aux isles de l'Amerique. Contenant l'histoire naturelle de ces pays, l'origine, les mœurs, la religion & le gouvernement des habitans anciens & modernes. Les guerres & les evenemens singuliers qui y sont arrivez pendant le séjour que l'auteur y a fait*. Paris: G. Cavelier.

Lachatañeré, Rómulo. [1939] 1992. *El sistema religioso de los Afrocubanos*. La Habana: Editorial de Ciencias Sociales.

Lafuente, Modesto. 1850. *Historia general de España desde los tiempos mas remotos hasta nuestros días*, vol. 3. Madrid: Establecimiento tipografico de Mellado.

Lalla, Barbara. 2005. Virtual realism: Constraints on validity in textual evidence of Caribbean language history. *Occasional Paper* no. 32, Society for Caribbean Linguistics.

Lalla, Barbara and Jean D'Costa (eds). 1989. *Voices in Exile: Jamaican texts of the 18th and 19th centuries*. Tuscaloosa and London: Univ. of Alabama Press.

*Language in Exile: Three hundred years of Jamaican Creole*. Tuscaloosa and London: Univ. of Alabama Press.

[Langford, Jonas]. 1706. *A brief account of the sufferings of the servants of the Lord called Quakers*.

Lanigan, Mrs. 1844. *Antigua and the Antiguans: A full account of the colony and its inhabitants*. London: Saunders and Otley.

Laqueur, Thomas. 1983. Toward a cultural ecology of literacy in England, 1600–1850. In Daniel P. Resnick (ed.), *Literacy in Historical Perspective*, pp. 43–57. Washington: Library of Congress.

Larousse, Pierre. 1865. *Grand Larousse Universel*.

Larrazabal Blanco, Carlos. 1975. *Los Negros y la esclavitud en Santo Domingo*. Santo Domingo, RD: Julio D. Postigo e hijos.

Las Casas, Fray Bartolome de. [1552–1566] 1957. *Obras escogidas de Fray Bartolome de Las Casas. I. Historia de Las Indias. Texto fijado por Juan Perez de*

*Tudela y Emilio Lopez Oto. Estudio critico preliminar y edicion por Juan Perez de Tudela Bueso.* Biblioteca de Autores Españoles desde la formación del lenguaje hasta nuestros días. Madrid: Ediciones Atlas.

1986. *Historia de las Indias.* Edition, prologue, notes and chronology by André Saint-Lu. Spain: Biblioteca Ayacucho.

**Lasserre, Guy.** 1961. *La Guadeloupe: étude géographique.* 2 vols. Bordeaux: Union Française d'Impression.

**Laurent-Ropa, Denis.** 1993. *Haiti: Une colonie française 1625–1802.* Paris: Editions L'Harmattan.

**Laviana Cuetos, Maria Luisa** (ed.). 1988. *Antología del pensamiento político, social y económico latinoamericano. José Martí.* Edición, selección y notas por María Luisa Laviana Cuetos. Madrid: Ediciones de Cultura Hispánica.

**[Laws of Barbados]** *An Abridgement of the Laws in Force and Use in Her Majesty's Plantations of Virginia, New England, Jamaica, New York, Barbadoes, Carolina, Maryland.* 1704. London.

**Leal, Rine.** 1982. *La Selva Oscura: De los Bufos a la Neocolonia (Historia del teatro cubano de 1868 a 1902).* Havana: Editorial Arte y Literatura.

**Lebrón Saviñón, Mariano.** [1981] 1994. *Historia de la cultura dominicana.* Santo Domingo: Editora Taller.

**Ledru, André-Pierre.** 1810. *Voyage aux îles de Ténériffe, la Trinité, Saint-Thomas, Sainte-Croix et Porto Ricco, exécuté par ordre du gouvernement français, depuis le 30 septembre 1796 jusquau 7 juin 1798, sous la direction du capitaine Baudin, pour faire des recherches et des collections relatives à l'histoire naturelle . . . Ouvrage accompagné de notes et d'additions, par M. Sonnini.* Paris: A. Bertrand.

1863. *Viaje a la isla de Puerto Rico en el año 1797, ejecutado por una comisión de sabios franceses, del orden de su gobierno.* Traducido al castellano por D. Julio L. De Vizcarrondo. Puerto Rico: J.Gonzalez.

**Leger, J.N.** [1907] 1970. *Haiti. Her History and her detractors.* Westport, CT: Negro Universities Press.

**Lemonnier-Delafosse, M.** 1846. *Seconde Campagne de Saint-Domingue du 1er Décembre 1803 au 15 Juillet 1809; précédée de souvenirs historiques & succints de la première campagne – Expédition du Général en chef Leclerc du 14 Décembre 1801 au 1er Décembre 1803.* Havre: H. Brindeau & Cie.

*Segunda Campaña de Santo Domingo; Guerra Dominico–Francesa de 1808.* 1946 Translation by C. Armando Rodriguez. Santiago, RD: El Diario.

**León, Nicolás.** 1924. *Las castas del México colonial o Nueva España.* Mexico: Talleres Gráficos del Museo Nacional de Arqueología, Historia y Etnografía.

**Le Page, R.B.** (ed.). 1960. *Creole language studies I: Jamaican Creole.* London: Macmillan & Co; New York: St. Martin's Press.

**Le Page, R.B. and Andrée Tabouret-Keller.** 1985. *Acts of identity: Creole-based approaches to language and ethnicity.* Cambridge: Cambridge Univ. Press.

**Leslie, Charles.** 1739. *A new and exact account of Jamaica.* Edinburgh.

1740. *A new history of Jamaica from the earliest accounts, to the taking of Porto Bello by Vice-Admiral Vernon. In thirteen letters from a gentleman to his friends.* Second edition. London: J. Hodges.

**Lespinasse, Beauvais.** 1882. *Histoire des Affranchis de Saint-Domingue.* Paris: Joseph Kugelmann.

**Lewis, G.** 1983. *Main currents in Caribbean thought.* Baltimore: The Johns Hopkins Univ. Press.

Lewis, Matthew G. 1834. *Journal of a West Indian planter in Jamaica*. London: John Murray. 1969 Reprint by Negro Universities Press, N.Y.

Lewis Galanes, Adriana. 1988. El Album de Domingo del Monte (Cuba, 1838/39). *Cuadernos Hispanoamericanos* 451–2, pp. 255–65.

Ligon, R. 1657. *A true and exact history of the island of Barbados*. London. 1970 Reprint by Frank Cass.

Linschoten, Jan. [1596] 1619. *Description de l'Amerique & des parties d'icelle, comme de la Nouvelle France, Floride, des Antilles, Lucaya, Cuba, Jamaica, &c. Avec une carte geographique de l'Amerique australe, qui doit estre inseree en la page suivante.* French translation of the American part of *Beschryvinghe van gantsche custe* (1596). Amsterdam: Jean Evertsz Cloppenburch.

Lipski, John. 1987. African influence on Hispanic dialects. In Lenard Studerus (ed.) *Current trends and issues in Hispanic linguistics*, pp. 33–68. Arlington, Texas: The Summer Institute of Linguistics and the Univ. of Texas at Arlington.

    1994. *Latin American Spanish*. London and New York: Longman.

Lirus, Julie. 1979. *Identité Antillaise: Contribution à la connaissance psychologique et anthropologique des Guadeloupéens, et des Martiniquais*. Paris: Editions Caribéennes. Second edition 1982.

Littleton, Edward. 1689. *The groans of the Plantations or a true Account of their grievances and extreme sufferings by the heavy impositions upon sugar, and other Hardships. Relating more particularly to the Island of Barbados*. London: M. Clark.

Lloyd, William. 1839. *Letters from the West Indies during a visit in the autumn of MDCCCXXXVI and the spring of MDCCCXXXVII*. London: Darton and Harvey.

Long, Edward. 1774. *The history of Jamaica, or General Survey of the Antient and Modern state of that island: with reflections on its Situations, Settlements, Inhabitants, Climate, Products, Commerce, Laws and Government*. 1970, New Edition with a New Introduction by George Metcalf of King's College, London. Frank Cass.

Look Lai, Walton. 1993. *Indentured Labor, Caribbean Sugar: Chinese and Indian Migrants to the British West Indies, 1838–1918*. Baltimore and London: The Johns Hopkins Univ. Press.

López Cantos, Ángel. 2001. *Los puertorriqueños: Mentalidad y Actitudes (Siglo XVII)*. San Juan, PR: Ediciones Puerto.

Lopes de Gómara, Francisco. 1552. *History of the Indies and Mexico*. Included in Anghiera (1555).

López de Cogolludo, Fr. Diego. [1633]. *Los tres siglos de la dominación española en Yucatan o sea Historia de esta provincia. II.* 1971 Reprint by Akademische Druck, Graz.

    [1688] 1954–5. *Historia de Yucatán*. Campeche: Comisión de Historia.

López de Velasco, Juan. [1571–74] 1894. *Geografía y descripción universal de las indias*. Madrid: Establecimiento tipográfico de Fontanet.

    [1571–74] 1971. *Biblioteca de Autores Españoles desde la formación del lenguaje hasta nuestros días*. Edición de Don Marcos Jiménez de la Espada. Estudio preliminar de Doña María del Carmen González Muñoz. Madrid: Ediciones Atlas.

López Morales, Humberto. 1978. (ed.) *Corrientes actuales en la dialectología del Caribe hispanico: Actas de un simposio*. San Juan: Editorial Universitaria, Univ. de Puerto Rico.

    1980. Sobre la pretendida existencia del 'Criollo' cubano. In *Anuario de Letras*, vol. 18, pp. 85–116.

1991. *Investigaciones lexicas sobre el español antillano.* Santiago, Republica Dominicana: Editorial Pontificia Univ. Catolica Madre y Maestra.

López Morales, Humberto, and Vaquero, María. (eds) 1987. *Actas del 1 congreso internacional sobre el español de América.* Puerto Rico: Academia puertorriqueña de la lengua española.

López Prieto, D. Antonio (Compiler). 1881. *Parnaso Cubano – Coleccion de poesias selectas de autores cubanos con una introduccion històrica critica sobre el desarrollo de la poesia en la Isla de Cuba; Biografías y Notas criticas literarias de reputados literatos.* Habana: Miguel de Villa – Libreria.

Lorenzo Sanz, Ricardo. 1978. *Simon Bolivar.* Madrid: Editorial Hernando.

Lovell, Langford. 1818. *A letter to a friend relative to the present state of the island of Dominica.* Winchester: James Robbins.

Lucas Papers. 1803. Notes of Meeting of the Board of Council, Barbados, November 1st 1803. Microfilm 14, Barbados Public Library.

Luffman, John. 1789. *A brief account of the Island of Antigua . . . written in the years 1786, 7, 8.* London: Printed for T. Cadell.

Maceo, Patin. 1989. *Obras lexicograficas.* Santo Domingo, RD: Sociedad Dominicana de Bibliofilos.

Mackenzie, Charles. 1830. *Notes on Haiti made during a residence in that republic.* London: Henry Colburn and Richard Bentley.

Madden, R.R. 1835. *A Twelvemonth's Residence in the West Indies, during the transition from Slavery to Apprenticeship; with incidental notices of the state of the society, prospects, and natural resources of Jamaica and other islands. In two volumes.* Philadelphia, PA: Carey, Lea and Blanchard. 1970 Reprint by Negro Universities Press, Westport, CT.

1849. *The island of Cuba its resources, progress and prospects, considered in relation especially to the influence of its prosperity on the interests of the British West India Colonies.* London: C. Gilpin.

[1840] 1981. *Poems by a slave in the island of Cuba. The life and poems of a Cuban Slave: Juan Francisco Manzano.* Edited by Edward Mullen. Hamden, CT: Archon Books.

Madiou, Thomas. [1847] 1989. *Histoire d'Haiti.* Port-au-Prince: J. Courtois.

1848. *Histoire d'Haiti.* Port-au-Prince, Haiti: Henri Deschamps.

Major, R.H. 1847. (ed.) *Select Letters of Christopher Columbus, with other original documents, relating to his four voyages to the New World.* Translated and edited by R.H. Major, Esq. of the British Museum. London: Hakluyt Society.

Maldonado-Denis, Manuel. 1992. *Eugenio María de Hostos y el pensamiento social iberoamericano.* Mexico: Fondo de Cultura económica.

Malkiel, Yakov. 1976. Changes in the European languages under a new set of sociolinguistic circumstances. In Chiapelli (ed.) 1976, pp. 581–93.

Marees, Pieter de. [1602] 1987. *Description and Historical Account of the Gold Kingdom of Guinea (1602).* Translated from the Dutch and edited by Albert van Dantzig and Adam Jones. Oxford: Oxford Univ. Press.

*Marly; Or a Planter's Life in Jamaica.* 1828. Glasgow: Richard Griffin & Co; London: Hunt and Clarke.

Mármol Carvajal, Luis del. 1599. *Segunda Parte y libro septimo de la descripcion general de Africa, donde se contiene las Provincias de Numidia, Libia, la tierra de los negros, la baxa y alta Etiopia, y Egipto cõ todas las cosas memorables della.* Malaga: Juan Rene.

Marques, António Henrique R. de Oliveira. 1976. *History of Portugal*. Second edition. New York: Columbia Univ. Press.

Marrero, Levi. 1973. *Cuba: Economía y Sociedad. El siglo XVII (I)*. Madrid: Editorial Playor.

Martí, José. 1973. *Cuba, Nuestra América, Los Estados Unidos*. Selección y prólogo de Roberto Fernández Retamar. Mexico: Siglo xxi editores.

Martin, Samuel. 1750. *Essay on plantership*. London.

1765. *An Essay upon Plantership, humbly inscribed to his Excellency George Thomas, Esq. Chief Governor of All the Leeward Islands*. Fourth edition. London.

1775. *A short treatise on the Slavery of Negroes in the British colonies. Shewing that they are much happier than in their Native Country, happier than the Subjects of Arbitrary Governments, and at least as happy as the Poor Labourers of Great Britain and Ireland*. Antigua: Robert Mearns.

Martinez Vergne, Teresita. 1989. Politics and Society in the Spanish Caribbean during the Nineteenth century. In Knight and Palmer (1989).

Martinus, Frank. 1989. West African connection (The influence of the Afro-Portuguese on the Papiamentu of Curaçao). In *Estudios sobre Español de América y lingüistica afroamericana. Ponencias presentadas en el 45 congreso internacional de Americanistas (Bogota, Julio de 1985)*. Bogotá: Publicaciones del Instituto Caro y Cuervo.

Martyr, Peter. *See* Anghiera, Pietro Martire d'.

Mathews, Samuel. 1793. *The Lying Hero or an Answer to J.B. Moreton's Manners and Customs in the West Indies*. St Eustatius: Edward Low & Co.

1822. *The Willshire Squeeze, to which are added specimens of the Negro familiar dialect and proverbial sayings with songs*. Demerara.

Mathison, Gilbert. 1811. *Notices respecting Jamaica in 1808–1809–1810*. London: John Stockdale.

Mauleon Benitez, Carmen Cecilia. 1974. *El español de Loíza Aldea*. Madrid: Ediciones Partenón.

Maurile de S. Michel, Fray. 1652. *Voyage des isles camercanes en l'Amerique qui font partie des Indes occidentales*. Le Mans.

M'Callum, Pierre. 1805. *Travels in Trinidad (1803)*. Liverpool.

McKenzie, Charles. 1830. *Notes on Haiti made during a residence in that republic*. London: Henry Colburn and Richard Bentley.

McKinnen, Daniel. 1804. *A tour through the British West Indies in the years 1802 and 1803*. London: Printed for J. White.

McLean, James J. and T. Pina Chevalier. 1921. *Datos historicos sobre la frontera dominico-haitiana*. Santo Domingo, RD.

Megenney, William W. 1978. *A Bahian Heritage: An ethnolinguistic study of African influences on Bahian Portuguese*. Chapel Hill, NC: North Carolina Studies in Romance Languages and Literatures.

Meignan, Victor. 1878. *Aux Antilles*. Paris: Plon &Cie.

Méndez Plancarte, A. (ed.). 1952. *Obras completas de Sor Inés de la Cruz*. Tomo 2. Villancicos.

Merrell, David J. 1994. *The Adaptive Seascape*. Minneapolis: Univ. of Minnesota Press.

Meyer-Lübke, W. 1926. *Introducción a la linguistica románica. Versión de la tercera edición alemania, con notas y adiciones por Américo Castro*. Madrid.

Midas, André. 1949. *La Martinique vue par les voyageurs et les romanciers* (Conférence faite à la Mutualité le 25 février 1949). Fort-de-France: Groupe d'Études Martiniquais et de la Fédération Mutualiste du Département.

Mintz, Sidney W. 1971. The socio-historical background to pidginization and creolization. In Hymes, Dell. 1971. (ed.) *Pidginization and Creolization of Languages.* Cambridge: Cambridge Univ. Press.

Mintz, Sidney W. and Richard Price. 1976/1992. *The Birth of African–American Culture: an Anthropological perspective.* Boston: Beacon Press.

Mir, Pedro. 1984. *Tres leyendas de colores: Ensayo de interpretacion de las tres primeras revoluciones del nuevo mundo.* Third edition. Santo Domingo, RD: Editora Taller.

1977. Carta anti-prólogo. In Julio César Mota Acosta, *Los Cocolos en Santo Domingo.* Santo Domingo: La Gaviota.

M'Neill, Hector. 1788/89. *Observations on the treatment of the Negroes, in the island of Jamaica, including some account of their temper and character, with remarks on the importation of slaves from the coast of Africa. In a letter to a physician in England.* London: Robinson, Robinson and Gore.

Mohan, Peggy. 1978. Non-lexical *say* and language universals. Paper presented at the conference of the Society for Caribbean Linguistics, Univ. of the West Indies, Barbados.

Moister, Rev. William. 1883. *The West Indies, Enslaved & Free.* London: T. Woolmer.

Monardes, Dr. Nicolás. 1565. *Dos libros.*

1577. *Joyfull Newes out of the Newe Founde Worlde. Englished by Thom Frampton.* London.

Montaigne, Michel de. circa 1578. Des Cannibales. In Tilley, Arthur and Boase, A.M. (eds). 1954. *Montaigne Selected Essays.* Manchester: Manchester Univ. Press.

Montanus, Arnoldus. 1671. *America: Being the latest and most accurate description of the New World.* London.

Montlezun, Baron de. 1818. *Souvenirs des Antilles: Voyage en 1815 et 1816, aux États-Unis, et dans l'archipel Caraibe.* Paris: Gide Fils.

*Montserrat code of laws: from 1668, to 1788.* 1790. London.

Moore, Francis. 1738. *Travels into the inland parts of Africa: Containing a description of the several nations for the space of six hundred miles up the River Gambia; their trade, habits, customs, languages, manners, religion and government; the power, disposition and characters of some Negro princes; with a particular account of Job Ben Solomon, a Pholey, who was in England in the year 1733, and known by the name of the African. To which is added, Capt. Stibbs's voyage up the Gambia in the year 1723, to make discoveries; with an accurate map of that river taken on the spot; and many other copper plates.* London: E. Cave.

Moore, John. 1941. *The geography of Gulliver's Travels.* Urbana, IL

Moran Arce, Lucas. 1982. *Ser o no ser, o La angustia existencial puertorriqueña.* San Juan, Puerto Rico: Imprenta Universitaria.

More, Hannah. 1795. *The Sorrows of Yamba; Or, the Negro Woman's Lamentation.*

Moreau de Saint-Méry, M.L.E. 1788. *Fragment sur les moeurs de Saint-Domingue.* Port-au-Prince.

1796. *Description topographique, physique, civile, politique et historique de la partie française de l'isle Saint-Domingue.* Philadelphia, PA: The Author.

1796a. *Description topographique et politique de la partie espagnole de l'isle Saint-Domingue. Tome premier.* Philadelphia, PA: Moreau de Saint-Méry.

1797. *Description topographique, physique, civile, politique et historique de la partie française de l'isle Saint-Domingue.* 1958. *Nouvelle édition entièrement revue et completée sur le manuscrit accompagnée de plans d'une carte hors-texte suivie d'un index des noms de personnes, par Blanche Maurel et Etienne Taillemite.* Paris: Société de l'histoire des colonies françaises et Librairie Larose.

Moreton, J.B. 1793. *West India Customs and Manners: Containing strictures on the soil, cultivation, produce, trade, officers, and inhabitants; with the method of establishing and conducting a sugar plantation. To which is added, the practice of training new slaves.* London: Printed for J. Parsons, W. Richardson, H. Gardner, and J. Walter.

Mörner, Magnus. 1967. *Race Mixture in the History of Latin America.* Boston: Little, Brown and Company.

M'Queen, James. 1825. *The West India colonies; the calumnies and misrepresentations circulated against them by the Edinburgh Review, Mr. Clarkson, Mr. Cropper, & &.* London: Longman, Hurst, and Co.

Munster, Sebastian. 1541. *Cosmographia.*

Murillo Velarde, P. 1752. *Geographia historica. Libro IX. De la America, y de las islas adyacentes, y de las tierras arcticas, y antarcticas, y islas de las mares del norte, y sur.* Madrid: Don Agustin de Gorduela y Sierra.

Musgrave, T. 1891. *Historical and descriptive sketch of the colony of St. Vincent.* Compiled under the direction of the Commissioners for the Winward Islands by T.B.C. Musgrave. Printed at Gardner's.

Nicolson, Jean. 1776. *Essai sur l'histoire naturelle de l'isle de Saint Domingue, avec des figures en taille-douce.* Paris.

Ober, Frederick A. 1880. *Camps in the Caribbees: The adventures of a naturalist in the Lesser Antilles.* Boston: Lee and Shepard.

1894. *Aborigines of the West Indies.* AAS April 1894.

Obregón, Mauricio. 1991. *The Columbus papers: The Barcelona letter of 1493, the Landfall Controversy, and the Indian Guides.* New York: Macmillan.

Oexmelin. See Exquemelin.

Ogilby, John. 1671. *America: being the latest, and most accurate description of the New World. Collected from most authentick authors, augmented with later observations, and adorned with maps and sculptures.* London.

O'Kelly, James J. 1874. *The Mambi-Land or Adventures of a Herald Correspondent in Cuba.* London: Sampson Low, Marston & Co.

1968. *La Tierra del Mambi.* Havana: Instituto del Libro.

Oldendorp, C.G.A. 1770, 1777. *History of the evangelical Brethren on the Caribbean islands of St. Thomas, St. Croix, and St. John.* Edited by Johann Jakob Bossard. 1987 English edition and translation by Arnold Highfield and Vladimir Barac. Ann Arbor, MI: Karoma.

Oldmixon, John. 1708. *The British Empire in America.* London.

O'Neil, J.T. 1855. In Fisher, Richard S. (ed.) 1855. *The Spanish West Indies: Cuba and Porto Rico: geographical, political and industrial. Cuba: from the Spanish of Don J.M. de la Torre.* Porto Rico: by J.T. O'Neil, Esq.

Orderson, John. 1827. *Leisure hours at the Pier; or, a treatise on the education of the poor of Barbados.* Liverpool: Thomas Kaye.

1835. *The fair Barbadian and faithful Black; Or, A Cure for the Gout. A Comedy in Three Acts.* Liverpool: Ross and Nightengale.

1842. *Creoleana: or, Social and domestic scenes and incidents in Barbados in days of yore.* London: Saunders and Otley.

Ormond, Richard and Carol Blackett-Ord. 1987. *Franz Xaver Winterhalter and the Courts of Europe 1830–70.* London: National Portrait Gallery.

Ortiz, Fernando. 1986. *Los Negros Curros.* Texto establecido con prólogo y notas aclaratorias – Diana Iznaga. La Habana: Editorial de Ciencias Sociales.

1947. *Cuban counterpoint: Tobacco and sugar.* Translated from Spanish by Harriet de Onís. Introduction by Bronislaw Malinowski. Prologue by Herminio Portell Vilá. New York: Alfred A. Knopf.

1974. *Nuevo catauro de cubanismos.* La Habana: Editorial de Ciencias Sociales.

Oviedo, Gonzalo Fernández de. *See* Fernández de Oviedo.

Pares, Richard. 1950. *A West India Fortune.* London, New York, Toronto: Longmans, Green and Co.

Pastor Bodmer, Beatriz. 1983. *Discurso narrativo de la conquista de América.* Havana: Casa de las Américas.

Paton, William Agnew. 1896. *Down the islands: a voyage to the Caribbees.* New York: Charles Scribner's Sons.

Patterson, Orlando. 1967. *The Sociology of Slavery.* Jamaica: Sangsters.

1970. Slavery and slave revolts: a socio-historical analysis of the first Maroon War, 1665–1740. *Social and Economic Studies* vol. 19, no. 3, pp. 289–325.

Peguero, Luis Joseph. [1762, 1763] 1975. *Historia de la conquista de la isla española de Santo Domingo. Trasumptada el año de 1762.* Edición, Estudio preliminar y Notas de Pedro J. Santiago. 2 vols. Santo Domingo: Museo de las Casas Reales.

Pelleprat, Le Pere Pierre. 1655. *Relation des missions des pp. de la Compagnie de Jesus Dans les Isles, & dans la terre ferme de l'Amerique Meridionale. Divisee en deux parties: Avec une introduction à la langue des Galibis Sauvages de la terre ferme de l'Amerique.* Paris.

Perissinotto, Giorgio. 1987. Las primeras documentaciones de 'Caribe' ~ 'Caníbal' en las lenguas occidentales. In López Morales and Vaquero (eds) 1987. pp. 289–301.

Philippe, Jean-Baptiste. 1824. *Free Mulatto: An address to the Right Hon. Earl Bathurst.* London. 1987 Reprint by Paria Publishing Co. Port-of-Spain, Trinidad.

Phillippo, James. 1843. *Jamaica: Its past and present state.* 1969 Reprint with a new introduction by Philip Wright. London: Dawsons & Sons.

Phillips, T. 1732. Journal of a voyage made in the Hannibal of London, Ann. 1693, 1694, from England, to Cape Monseradoe, in Africa; and thence along the coast of Guiney to Whidaw, the island of St. Thomas, and so forward to Barbadoes with a cursory account of the country, the people, their manners, forts, trade, &c. By Thomas Phillips of the said ship. In Churchill, Awnsham. Comp. 1732. *A collection of voyages and travels,* vol. vi. London: T. Osborne.

Pichardo y Tapia, Esteban. [1836, 1849, 1862, 1875] 1976. *Diccionario provincial casi-razonado de vozes y frases cubanas.* La Habana: Editorial de Ciencias Sociales [based on the 1875 edition].

Pike, Ruth. 1967. Sevillian Society in the Sixteenth Century: Slaves and Freedmen. *Hispanic American Historical Review,* vol. 47, pp. 344–59.

Pinckard, George. 1806. *Notes on the West Indies*. London: Longman, Hurst, Rees, and Orme.

Pitman, Frank Wesley. 1917. *The Development of the British West Indies 1700–1763*. New Haven, CT: Yale Univ. Press; Oxford: Oxford Univ. Press.

Polo, Marco. 1959. *The travels of Marco Polo*. London: André Deutsch.

Pompilus, Pradel. 1961. *La langue française en Haiti*. Translated in Johanna von Grafenstein. 1988–9. *Haití 2*. Mexico, DF: Instituto de Investigaciones Dr. José María Luis Mora.

Poole, Robert. 1753. *The beneficient bee: or, traveller's companion*. London: E. Duncomb.

Poyer, John. 1808. *The History of Barbados, from the first discovery of the island, in the year 1605, till the accession of Lord Seaforth, 1801*. London: J. Mawman.

Price, Richard (ed.) 1979. *Maroon Societies: Rebel slave communities in the Americas*. Second edition. Baltimore, MD; London: The Johns Hopkins Univ. Press.

Prost, A. 1953. *Les langues mandé-sud du groupe mana-busa*. Dakar: Mémoires de l'institut français d'Afrique noire.

1956. *La langue sonay et ses dialectes*. Dakar: Mémoires de l'institut français d'Afrique noire.

Prudent, Lambert-Félix. 1980. *Des baragouins à la langue antillaise: analyse historique et sociolinguistique du discours sur le créole*. Paris: Editions Caribéennes.

Puckrein, Gary A. 1984. *Little England: plantation society and Anglo-Barbadian politics 1627–1700*. New York: New York Universities Press.

Purchas, Samuel. [1625] 1905. *Hakluytus Posthumus or Purchas His Pilgrimes. Contayning a history of the worlde in sea voyages and lande travells by Englishmen and others*. Glasgow: James MacLehose and Sons.

Quint, David. 1995. A reconsideration of Montaigne's *Des Cannibales*. In Kupperman 1995, pp. 166–91.

Quintero, I.A. [nineteenth century]. Lyric poetry in Cuba. Boston Public Library manuscript.

Ramsay, Rev. James. 1784. *Essay on the treatment and conversion of African slaves in the British sugar colonies*. London: James Phillips.

1786. *A letter from Capt. J.S. Smith to the Rev'd Mr. Hill on the state of the Negroe Slaves. To which are added an Introduction, and remarks on Free Negroes*. London: J. Phillips.

1788 *Objections to the abolition of the slave trade*. London: J. Phillips.

Raynal, Abbé. 1770. *Histoire philosophique et politique des établissemens & du commerce*. Amsterdam.

1773. *Histoire philosophique et politique des établissemens & du commerce*. Amsterdam.

1784. *Histoire philosophique et politique des établissemens et du commerce*. Lausanne.

1798. *A philosophical and political history of the settlements and trade of the Europeans in the East and West Indies. Revised, augmented, and published in ten volumes. Newly translated from the French, by J.O. Justamond, with a new set of maps adapted to the Work, and a copious index. In six volumes*. 1969 Reprint by New York Universities Press, New York.

[Real Academia Española]. 1729, 1734. *Diccionario de la lengua Castellana, en que se explica el verdadero sentido de las voces, su naturaleza y calidad, con las phrases o modos de hablar, los proverbios o refranes, y otras cosas convenientes al uso de la lengua.* Madrid: Francisco del Hierro.

**Redhead, W.A.** 1970. Truth, Fact and Tradition in Carriacou. *Caribbean Quarterly,* vol. 16, no. 3, pp. 61–3.

**Remos, Juan J.** 1958. *Proceso historico de las letras cubanas.* Madrid: Ediciones Guadarrama.

**Renard, Raymond.** 1966. *Sepharad: le monde et la langue judéo-espagnole des Séphardim.* Mons, Belgium: Annales Universitaires de Mons.

**Renny, Robert.** 1807. *An History of Jamaica with observations on the climate, scenery, etc. . . . to which is added an illustration of the advantages which are likely to result, from the abolition of the Slave Trade.* London: J. Cawthorn.

**Resident.** 1828. *Sketches and Recollections of the West Indies.* London.

**Resnick, Daniel P.** (ed.) 1983. *Literacy in Historical Perspective.* Washington: Library of Congress.

**Rey-Delmas, Jn.** 1797. *Exposé rapide de la Révolution de Saint-Domingue, Depuis 1789 jusqu'en 1797.*

**Richard, Renaud.** (coordinador). 1997. *Diccionario de Hispanoamericanismos no recogidos por la Real Academia.* Madrid: Juan Ignacio Luca de Tena; Ediciones Cátedra, S.A.

**Ridpath, George.** 1703. *The case of Scots-Men residing in England and in the English plantations.* Edinburgh.

**Rivas, Mercedes.** 1990. *Literatura y esclavitud en la novela cubana del siglo XIX.* Seville: Escuela de estudios hispano-americanos.

**Rivera de Álvarez & Manuel Álvarez Nazario.** 1982. *Antología general de la literatura puertorriqueña.* Madrid: Ediciones Partenón.

**Rivera Rivera, Modesto.** 1980. *Concepto y expresion del costumbrismo en Manuel A. Alonso Pacheco (El Gibaro).* San Juan de Puerto Rico: Instituto de Cultura puertorriqueña.

**Roberts, Peter A.** 1971. The verb in Grenadian French Creole. MA thesis, Univ. of the West Indies.

1996. Samuel Augustus Mathews. In Christie, P. (ed.). *Caribbean language issues old and new,* pp. 63–85. Mona, Jamaica: Univ. of the West Indies Press.

1997. *From Oral to Literate Culture: Colonial experience in the English West Indies.* Mona, Jamaica: Univ. of the West Indies Press.

1997. The (re)construction of the concept of *indio* in the national identities of Cuba, Puerto Rico and the Dominican Republic. In Fiet, L. and Becerra, J. (eds). *Caribbean 2000: Regional and/or National definitions, identities and cultures,* pp. 99–120. Caribe 2000, Univ. of Puerto Rico.

1998. The concept of **ladino** and the melting pot process. In Fiet, L. and Becerra, J. (eds). *Segundo simposio de Caribe 2000: Hablar, nombrar, pertenecer,* pp. 154–169. Caribe 2000, Univ. of Puerto Rico.

1999. The changing face of 'jíbaros'. In Fiet, L. and Becerra, J. (eds). *A Gathering of Players and Poets. Caribbean 2000. Symposium III (1998),* pp. 47–58. Caribe 2000, Univ. of Puerto Rico.

2000. Changing values and national identities in the Caribbean and their effect on language education policy. In Leicester, Mal, Modgil, Celia and Modgil, Sohan (eds). *Education, Culture and Values Series,* vol. I. *Systems of Education: Theories, Policies and Implicit Values.* Chapter 14, pp. 152–63. London & New York: Falmer Press.

2001. What's in a name, an Indian name? In Haslip-Viera, Gabriel (ed.). *Taíno Revival: Critical Perspectives on Puerto Rican Identity and Cultural Politics*, pp. 83–100. Princeton, NJ: Markus Wiener.

2003. Continuidad y adaptación de una figura de mascarada. *Conjunto. Revista de teatro latinoamericano. Casa de las Américas, no.* 130 octubre–diciembre de 2003, pp. 8–19.

2005. Labat's *calenda*: origin and associations. In *La Torre*, Año IX, Núm. 32, pp. 239–49.

2006. The odyssey of *criollo*. In Thornburg, Linda and Fuller, Janet (eds). *Studies in Contact Linguistics: Essays in Honor of Glenn G. Gilbert*, pp. 3–25. New York: Peter Lang.

2006. Calenda: The rise and decline of a cultural image. *Sargasso 2005–2006*, 1, pp. 51–68.

2007. *West Indians and their language*. Second edition. Cambridge: Cambridge Univ. Press.

**Robertson, Rev.** 1730. *A letter to the Right Reverend the Lord Bishop of London, from an inhabitant of His Majesty's Leeward-Caribbee-Islands. Containing some considerations on His Lordship's two letters of May 19, 1727, the first to the masters and mistresses of families in the English plantations abroad; the second to the missionaries there. In which is inserted, A short essay concerning the conversion of the negro-slaves in our sugar colonies: written in the month of June, 1727, by the same inhabitant.* London: J. Wilford.

**Rodriguez, C. Armando.** [1915] 1929. *La frontera dominico-haitiana: estudio geographico, juridico, historico.* Santo Domingo, RD: Imprenta de J.R. Vda. García.

**Rodriguez Demorizi, Emilio.** [1938] *Poesía popular dominicana.* Third edition 1979. Santiago, RD: Univ. Católica Madre y Maestra.

1944. *Vicisitudes de la lengua española en Santo Domingo.* Ciudad Trujillo, RD: Editora Montalvo.

1955. *Invasiones haitianas de 1801, 1805 y 1822.* Ciudad Trujillo, RD: Editora del Caribe, C. por A.

1975. *Lengua y Folklore de Santo Domingo.* Santiago, RD: Univ. Católica Madre y Maestra.

**Rodríguez O, Jaime E.** 1975. *The emergence of Spanish America: Vincente Rocafuerte and Spanish Americanism 1808–1832.* Berkeley, Los Angeles, London: Univ. of California Press.

**Rojas, Maria Teresa de.** 1947, *Indice y extractos del archivo de protocolos de la Habana*, vol. I, 1578–1585, pp. xvi, 462. Habana: Ucar, Garcia y Cia.

**Romanet, J.** 1804. *Voyage à la Martinique; vues et observations politiques sur cette isle, avec un aperçu de ses productions végétales et animales.* Paris: L. Pelletier.

**Rosario, Rubén del.** 1965. *Vocabulario puertorriqueño.* Sharon, CT: The Troutman Press.

1974. El español antillano. In *Actas del Simposio de San Juan. Programa Inter-Americano de Linguistica y Enseñanza de Idiomas.* San Juan, PR: Departamento de Instrucción Publica.

**Rosario Candelier, Bruno.** 1974. Los valores negros en la poesía dominicana. *eme*, no. 15, pp. 30–66.

**Rosenblat, Angel.** [1962] 1991. Origen e historia del 'che' argentino. In *Biblioteca Angel Rosenblat IV: Estudios dedicados a la Argentina.* Caracas: Monte Avila.

Rouse, Irving. 1992. *The Tainos: Rise & decline of the people who greeted Columbus*. New Haven, CT and London: Yale Univ. Press.

Rufz, E. 1850. *Etudes historiques et statistiques sur la population de la Martinique*, vol. 1. Saint-Pierre, Martinique: Imprimerie de Carles.

Russell, Thomas. 1868. *The etymology of Jamaica grammar, by a young gentleman*. Kingston, Jamaica: M. DeCordova, MacDougall & Co.

Russell, William. 1778. *The History of America*. 2 vols. London: Fielding and Walker.

Rymer, James. 1775. *A description of the island of Nevis: with an account of its principal diseases*. London: printed for T. Evans.

Saco, José Antonio. [1879; second edition 1938] 1982. *Historia de la esclavitud africana en el nuevo mundo y en especial en los países américo-hispanos* Selección e introducción de Eduardo Torres-Cuevas y Arturo Sorhegui. La Habana: Editorial de Ciencias Sociales.

Saintard, P. 1754. *Essai sur les colonies francoises, ou Discours politiques sur la nature du gouvernement, de la population & du commerce de la colonie de S.D.* Paris.

1756. *Roman politique sur l'etat present des affaires de l'Amerique*. Amsterdam; Paris: Duchesne.

Saint-Quentin, Alfred de. 1872. *Introduction à l'histoire de Cayenne suivie d'un recueil de contes, fables & chansons en créole avec traduction en regard, notes & commentaires, par Alfred de St-Quentin. Étude sur la grammaire créole par Auguste de St-Quentin*. Antibes: J. Marchand.

Sánchez Valverde, Antonio. 1785. *Idea del valor de la isla Española, y utilidades, que de ella puede sacar su monarquia*. Madrid: Pedro Marin.

1988. *Ensayos. Biblioteca de clásicos dominicanos, vol. v.* Ediciones de la fundación Corripio.

Sang, Mu-Kien A. 1987. *Ulises Heureaux: biografía de un dictador*. Santo Domingo: Instituto Tecnológico de Santo Domingo.

Santamaría, Francisco Javier. [1959] 1978. *Diccionario de mejicanismos: razonado, comprobado con citas de autoridades, comparado con el de americanismos y con los vocabularios provinciales de los más distinguidos diccionaristas hispanoamericanos*, third edition. Méjico, DF: Editorial Porrua.

1942. *Diccionario general de Americanismos*. Tomo 1. Méjico, DF: Editorial Pedro Robredo.

Santos, Danilo de los. 1980. Reflexiones sobre la identidad nacional y cultural de los Dominicanos. *eme*, vol. viii, no. 47, pp. 3–16.

Sartre, Jean-Paul. 1966. *Essays in Aesthetics*. Selected and translated by Wade Baskin. New York: Washington Square Press.

Satineau, Maurice. 1928. *Histoire de la Guadeloupe sous l'ancien régime 1635–1789*. Paris: Bibliothèque historique.

Sauer, Carl Ortwin. 1966. *The Early Spanish Main*. Berkeley and Los Angeles: Univ. of California Press.

Savage, Jay M. 1969. *Evolution*. Second edition. New York: Holt, Rinehart and Winston.

Saviñon, M. 1994. See Lebrón Saviñón, Mariano. [1981] 1994.

Scarano, Francisco A. 1993. *Puerto Rico: Cinco siglos de historia*. Santafé de Bogotá, Colombia: McGraw-Hill Interamericana.

Schäffer, Dagmar. 1988. *Portuguese Exploration to the West and the Formation of Brazil 1450–1800: Catalogue of an Exhibition.* Providence, RI: The John Carter Brown Library.

Schaw, Janet. 1934. *Journal of a lady of quality; being the Narrative of a Journey from Scotland to the West Indies, North Carolina, and Portugal, in the years 1774 to 1776.* Edited by Evangeline Walker Andrews, in collaboration with Charles Mclean Andrews. New Haven, CT: Yale Univ. Press.

Schoelcher, Victor. 1842. *Des colonies françaises: Abolition immédiate de l'esclavage.* Paris: Pagnerre.

1847. *Histoire de l'esclavage pendant les deux dernières années. Deuxième partie.* Paris: Pagnerre.

1948. *Esclavage et colonisation. Avant-propos de Ch. A. Julien. Introduction par Aimé Cesaire. Textes choisis et annotés par Émile Tersen.* Paris: Presses Universitaires de France.

Schomburgk, Robert. 1848. *The history of Barbados.* London. Longmans.

Schuchardt, Hugo. 1883. Sur le créole de la Réunion. *Romania* xi (1883), pp. 589–93.

1979. *The Ethnography of Variation: Selected writings on Pidgins and Creoles.* Edited and translated by T. L. Markey. Introduction by Derek Bickerton. Ann Arbor, MI: Karoma.

Sengova, Joko. 1994. Recollections of African language patterns in an American speech variety: An assessment of Mende influences in Lorenzo Dow Turner's Gullah data. In Montgomery, Michael (ed.). *The Crucible of Carolina: Essays in the development of Gullah language and Culture,* pp. 175–200. Athens, GA & London: The Univ. of Georgia Press.

Senior, Olive. 2003. *Encyclopedia of Jamaican Heritage.* Jamaica: Twin Guinep Publishers.

Sharpe, Kenneth. 1977. *Peasant politics: Struggle in a Dominican village.* Baltimore, MD: Johns Hopkins Univ. Press.

Shephard, Charles. 1831. *An historical account of the island of St. Vincent.* London: W. Nicol.

Sheppard, Jill. 1977. *The "Redlegs" of Barbados: Their Origins and History.* Millwood, NY: KTO Press.

Sheridan, Richard. 1961. The rise of a colonial gentry: a case study of Antigua, 1730–1775. *The Economic History Review,* 2nd series, vol. 13, no. 3, pp. 342–57.

Sheridan, Thomas. 1783. *A rhetorical grammar of the English language.* Philadelphia.

Sherwell, Guillermo. 1951. *Simon Bolivar. The Liberator, Patriot, Warrior, Statesman, Father of Five Nations: A sketch of his life and his work.* The Bolivarian Society of Venezuela.

Shimose, Pedro. 1993. *Historia de la literatura latinoamericana.* Madrid: Editorial Playor.

Shoemaker, Nancy. 1997. How Indians got to be red. *American Historical Review,* vol. 102, no. 3, pp. 625–44.

Simón, Fray Pedro. 1627. *Noticias historiales de las conquistas de Tierra Firme en las Indias Occidentales.*

[1627] 1987. *Noticias historiales de Venezuela,* vol. 2, second edition. Caracas: Academia Nacional de la Historia.

Singleton, John. 1767. *A general description of the West-Indian Islands, as far as relates to the British, Dutch, and Danish Governments, from Barbados to Saint Croix. Attempted in Blank verse in four books.* Barbados: Esmand and Walker.

1777. *Description of the West Indies. A poem in four books. The second edition.* London: James Marks.

Sloane, Sir Hans. 1707. *A voyage to the islands Madera, Barbadoes, Nieves, S. Christophers and Jamaica.* London.

Smith, Anthony D. 1986. *The ethnic origins of nations.* Oxford: Blackwell.

1991. *National identity.* Reno, NV: Univ. of Nevada Press.

1998. *Nationalism and modernism: a critical survey of recent theories of nations and nationalism.* London: Routledge.

Smith, John. 1630. *The true travels, adventures and observations of Captaine John Smith, In Europe, Asia, Affrica, and America, from Anno Domini 1593 to 1629. All written by actuall authours, whose names you shall finde along the History.* London: Thomas Slater.

Smith, M.G. 1971. A note on truth, fact and tradition in Carriacou. *Caribbean Quarterly*, vol. 17, nos 3 & 4, pp. 128–38.

Smith, William. 1744. *A new voyage to Guinea: Describing the customs, manners, soil, climate, habits, buildings, education, manual arts, agriculture, trade, employments, languages, ranks of distinction, habitations, diversions, marriages, and whatever else is memorable among the inhabitants.* London: J. Nourse.

1745. *Natural History of Nevis.* London.

Snelgrave, Captain William. 1754. *A New Account of Guinea and the Slave Trade.* London.

Some Gentlemen of St. Christopher. 1784. *An answer to the Reverend James Ramsay's Essay on the Treatment and Conversion of Slaves, in the British Sugar Colonies.* Basterre, St Christopher: Edward L. Low.

St. Clair, Lieut. Col. Thomas Staunton. 1834. *A soldier's recollections of the West Indies and America. With a narrative of the expedition to the island of Walcheren.* London: Richard Bentley.

St. John, Sir Spenser. 1884. *Hayti, or the Black Republic.* London: Smith, Elder, & Co.

Stalin, Joseph. [1935] 2003. *Marxism and the National and Colonial Question.* Honolulu: Univ. Press of the Pacific.

Stedman, John Gabriel. [1794] 1806. *Narrative of a five years' expedition against the revolted negroes of Surinam, in Guiana, on the wild coast of South America.* London.

Stewart, John. 1808. *An account of Jamaica and its inhabitants. By a Gentleman long resident in the West Indies.* London: Longman, Hurst, Rees and Orme.

Stinchcomb, Dawn F. 2004. *The development of literary blackness in the Dominican Republic.* Gainesville, FL: Univ. Press of Florida.

Street, Brian V. 1984. *Literacy in theory and practice.* Cambridge studies in oral and literate culture. Cambridge: Cambridge Univ. Press.

Stuart, Villiers. 1891. *Adventures amidst the equatorial forests and rivers of South America; also in the West Indies and the wilds of Florida. To which is added "Jamaica Revisited".* London: John Murray.

Sued Badillo, Jalil. 1978. *Los caribes, realidad o fábula: ensayo de rectificación histórica.* Río Piedras, PR: Editorial Antillana.

1995. The Island Caribs: New approaches to the question of ethnicity in the early colonial Caribbean. In Whitehead (1995).

Sued Badillo, Jalil and López Cantos, Angel. 1986. *Puerto Rico Negro*. Rio Piedras, PR: Editorial Cultural.

Svalesen, Leif. 2000. *The Slave ship Fredensborg*. Translated from Danish/Norwegian by Pat Shaw and Selena Winsnes. Kingston: Ian Randle Publishers.

Swift, Jonathan. 1726. *Travels into several remote nations of the world. In four parts. By Lemuel Gulliver, first a Surgeon, and then a Captain of Several Ships*. London.

Sypher, Wylie. 1939. The West-Indian as a 'Character' in the Eighteenth Century. *Studies in Philology*, vol. 36, pp. 503–20.

Taylor, Douglas. 1951. *The Black Carib of British Honduras*. New York: Wenner–Gren Foundation for Anthropological Research. Reprinted by Johnson Reprint Corporation, New York. Viking Fund Publications in Anthropology, no. 17.

1958. Carib, Caliban, Cannibal. *International Journal of American Linguistics* 24, no. 2, pp. 156–7.

1958. Lines by a Black Carib. *International Journal of American Linguistics* 24, no. 4, pp. 324–5.

1977. *Languages of the West Indies*. Baltimore, MD and London: The Johns Hopkins Univ. Press.

Taylor, Patrick. 2001. Dancing the Nation: An introduction. In Taylor, Patrick (ed.). *Nation Dance: Religion, identity, and cultural difference in the Caribbean*. Kingston: Indiana Univ. Press & Ian Randle Publishers.

Thevet, André. 1558. *Singularitez de la France antarctique*. Paris.

1575. *La cosmographie universelle d'André Thevet, cosmographe du rey*. Paris.

Thomas, Hugh. 1998. *The Slave Trade: The history of the Atlantic slave trade 1440–1870*. London: Papermac.

Thomas, J.J. [1869] 1969. *The Theory and Practice of Creole Grammar*. With an introduction by Gertrud Buscher. London and Port of Spain: New Beacon Books.

1889. *Froudacity: West Indian fables explained*. Second edition 1969. London: New Beacon Books.

Thome, James A. and Kimball, J. Horace. 1838. *Emancipation in the West Indies. A six months' tour in Antigua, Barbadoes, and Jamaica, in the year 1837. The Anti-Slavery Examiner no. 7*. New York: The American Anti-Slavery Society.

Thompson, Edward. 1770. *Sailor's letter's*. Dublin: J. Hoey.

Thompson, Robert Wallace. 1956. Préstamos lingüísticos en tres idiomas trinitarios. In *Estudios Americanos*, vol. XII, no. 61, pp. 249–54.

Thurloe, John, 1742. *A collection of the state papers of John Thurloe . . . Containing authentic memorials of the English affairs from the year 1638, to the restoration of King Charles II. Published from the originals, formerly in the library of John lord Somers . . . and since in that of Sir Joseph Jekyll . . . Including also a considerable number of original letters and papers, communicated by . . . the Archbishop of Canterbury from the library at Lambeth . . . the Earl of Shelburn, and other hands. The whole digested into an exact order of time. To which is prefixed, The life of Mr. Thurloe: with a complete index to each volume. By Thomas Birch*. London, Printed for the executor of F. Gyles.

Tilley, Arthur and A.M. Boase. (eds) 1954. *Montaigne: Selected Essays*. Manchester: Manchester Univ. Press.

TLEC. 1992. *Tesoro lexicográfico del español de Canarias*. Madrid: Real Academia Española and Gobierno de Canarias, Consejería de Educación, Cultura y Deportes.

Tobin, James. 1785. *Cursory remarks upon the Reverend Mr. Ramsay's Essay on the Treatment and Conversion of African Slaves in the Sugar Colonies*. London.

Todorov, Tzvetan. 1982. *La conquête de l'Amérique: la question de l'autre*. Paris: Seuil.

——— 1982. *The conquest of America: The question of the other*. New York: Harper and Row.

Tolentino Dipp, Hugo. 1992. *Raza e historia en Santo Domingo: los orígenes del prejuicio racial en América*. Santo Domingo, Republica Dominicana: Fundación Cultural Dominicana.

Tonkin, Elizabeth. 1992. *Narrating our pasts: The social construction of oral history*. Cambridge Studies in Oral and Literate Culture 22. Cambridge: Cambridge Univ. Press.

Towne, Richard. 1726. *A treatise of the diseases most frequent in the West Indies, and herein more particularly of those which occur in Barbadoes*. London

Trapham, Thomas. 1679. *A Discourse of the state of health in the island of Jamaica With a provision therefore calculated from the air, the place, and the water; The customs and manners of living &c*. London.

Trollope, Anthony. 1860. *The West Indies and the Spanish Main*. New York: Harper & Brothers.

Turiault, J. 1874. *Étude sur le langage Créole de la Martinique*. Brest: J.B. Lefournier ainé.

Turnbull, D. 1840. *Travels in the West: Cuba; with Notices of Porto Rico and the Slave Trade*. 1969 reprint by Negro Universities Press, New York.

[Turnbull, G.] Eyewitness. 1795. *A Narrative of the Revolt and Insurrection of the French Inhabitants in the Island of Grenada*. London.

Turner, Lorenzo D. 1949. *Africanisms in the Gullah dialect*. Ann Arbor, MI: Univ. of Michigan Press.

Tuttle, Edward F. 1976. Borrowing versus semantic shift: New World nomenclature in European languages. In Chiappelli, F. (ed.). 1976, vol. 2. pp. 598–611.

Ulloa, Don Antonio de. 1772, 1792. *Noticias americanas: Entretenimientos físico-históricos sobre la América meridional, y septentrional Oriental*. Madrid.

Underhill, Edward Bean. [1862] 1970. *The West Indies: Their social and religious condition*. Westport, CT: Negro Universities Press.

Uring, Nathaniel. 1725. *A relation of the late intended settlement of the islands of St. Lucia and St. Vincent . . . 1722*. London.

Uslar-Pietri, Arturo. 1950. Lo criollo en la literatura. In *Cuadernos Americanos*, no. 1, pp. 266–78.

Vallejo de Paredes, Margarita (ed.). 1981. *Antología Literaria Dominicana 1: Poesía*. Instituto Tecnológico de Santo Domingo.

Van Name, Addison, 1870. *Contributions to Creole grammar*. New Haven.

Varela, Consuelo. 1992. *Cristóbal Colón: Textos y documentos completos. Nuevas cartas: Edición de Juan Gil*. Madrid: Alianza Editorial.

Vasconcellos, J. Leite de. 1928. *Antroponimia portuguesa; tratado comparativo da origem, significação, classificação, e vida do conjunto dos nomes proprios, sobrenomes, e apelidos, usados por nos desde a idademedia ate hoje*. Lisbon: Imprensa nacional.

Vaugelas, Claude Favre de, baron de Pérouges. 1647. *Remarques sur la langue françoise.* 1934 facsimile of the original edition, published under the patronage of the Société des textes français modernes. Paris: E. Droz.

Vespucci, Amerigo. 1505–6. *The first four voyages.* Reprinted in facsimile and translated from the rare original edition. In Young, George (ed.) .1893, pp. 65–167.

Viau, Alfred. 1955. *Negros, Mulatos, Blancos o Sangre, nada más que sangre.* Ciudad Trujillo, RD: Editora Montalvo.

Vicioso, Abelardo. 1979. *Santo Domingo en las letras coloniales (1492–1800).* Santo Domingo: Editora de la Univ. Autónoma de Santo Domingo.

Viola, Herman J. and Carolyn Margolis. (eds) 1991. *Seeds of Change.* Washington and London: Smithsonian Institution Press.

Virey, J.J. 1824. *Histoire naturelle de genre humain, nouvelle édition, augmentée et entièrement refondue, avec figures.* Tome deuxième. Paris: Crochard.

Viterbo, Joaquim de Santa Rosa de. 1798. *Elucidario das palavras, termos e frases que em Portugal antigamente se usaram e que hoje regularniente se ignoram.* Second edition revised and corrected 1865. Lisbon: A. J. Fernandes Lopes.

Voltaire. 1767. *L'Ingénu.*

Voorhoeve, Jan and Lichtveld, Ursy M. (eds). 1975. *Creole Drum: An Anthology of Creole Literature in Surinam. With English translations by Vernie February.* New Haven, CT and London: Yale Univ. Press.

Waddell, Rev. Hope Masterton. 1863. *Twenty-nine years in the West Indies and Central Africa. A Review of Missionary work and Adventure 1829–1858.* 1970 Reprint by Frank Cass, London.

Waldseemüller, Martin. [1507] 2003. *Waldseemuller world map (1507).* Seattle, WA: Educare Press.

Walker, William. 1704. *Marry or do worse. A Comedy. As it is now Acted at the New Theatre in Little-Lincolns-Inn-Fields, by Her Majesty's Servants.* London: Richard Basset.

Waller, John A. 1820. *A voyage in the West Indies: containing various observations made during a residence in Barbadoes, and several of the Leeward Islands; with some notices and illustrations relative to the city of Paramarabo, in Surinam . . .* London: Printed for Sir R. Phillips and co.

Walton, William. 1810. *Present state of the Spanish colonies: including a particular report of Hispañola, or the Spanish part of Santo Domingo; with a general survey of the settlements on the south continent of America, as relates to history, trade, population, customs, manners, &c., with a concise statement of the sentiments of the people on their relative situation to the mother country, &c.* London: Longman, Hurst, Rees, Orme, and Brown.

Ward, Edward. 1700. *A trip to Jamaica: with a true character of the people and island. By the Author of Sot's Paradise. The Seventh Edition.* London.

Warren, George. 1667. *An Impartial Description of Surinam upon the Continent of Guiana in America. With a History of several strange Beasts, Birds, Fishes, Serpents, Insects, and Customs of that Colony, &c.* London: Nathaniel Brooke.

Warren, Henry. 1740. *A Treatise concerning the Malignant Fever in Barbados and Neighbouring islands: with An Account of the Seasons there, from the Year 1734 to 1738. In a letter to Dr. Mead.* London.

Waterton, Charles. 1825. *Wanderings in the Antilles.* London.

Welles, Sumner. 1926. *Naboth's vineyard – The Dominican Republic 1844–1924.* 1966 edition published by Paul P. Appel, Mamaroneck, NY.

Wells, H.G. [1920] 1951. *The Outline of History: Being a plain history of life and mankind.* London: Cassell and Co.

Wells, Spencer. 2002. *The Journey of Man: A genetic odyssey.* Princeton, NJ and Oxford: Princeton Univ. Press.

Wentworth, Trelawney. 1834. *The West India Sketch Book.* London: Whittacker.

*West India Merchant, Being a Series of Papers Originally printed under that signature in the London Evening Post.* 1778. London: J. Almon.

White, Hayden. 1978. *Tropics of discourse: Essays in cultural criticism.* Baltimore, MD and London: The Johns Hopkins Univ. Press.

Whitehead, Neil. 1995. (ed.) *Wolves from the sea: Readings in the Anthropology of the Native Caribbean.* Leiden: KITLV Press.

Wilk, Richard and Mac Chapin. 1990. *Ethnic minorities in Belize: Mopan, Kekchi and Garifuna.* Belize City, Belize: SPEAR (Society for the Promotion of Education and Research).

Willyams, Cooper. 1796. *An account of the campaign in the West Indies in the year 1794.* 1990 Reprint of the 1796 edition. *Bibliotheque d'histoire Antillaise. Collection publiée par la Société d'histoire de la Guadeloupe* 12. Basse-terre: Société d'histoire de la Guadeloupe.

Williams, Cynric. 1827. *A tour through the island of Jamaica, from the western to the eastern end in the year 1823.* Second edition. London: Thomas Hurst, Edward Chance & Co.

Williams, Eric. 1970. *From Columbus to Castro: The history of the Caribbean 1492–1969.* London: André Deutsch.

Williams, Joseph. 1932. *Whence the 'Black Irish' of Jamaica?* New York: L. MacVeagh, Dial Press.

Wilson, Samuel. 1999. *The indigenous people of the Caribbean.* Gainesville, FL: Univ. of Florida Press.

Winer, Lise. 1984. Early Trinidadian Creole: The Spectator Texts. *English World-Wide,* vol. 5, no. 2, pp. 181–210.

Wood, Donald. 1968. *Trinidad in transition.* Oxford: Oxford Univ. Press.

Wright, Joseph. 1898–1905. *The English Dialect Dictionary.* London.

Wright, Louis B. 1939. The classical tradition in colonial Virginia. In *The Papers of the Bibliographical Society of America,* vol. 33, pp. 85–97.

Wyld, Henry Cecil. 1920. *A History of Modern Colloquial English.* London: T. Fisher Unwin.

Yearling, Elizabeth. 1978. Cumberland, Foote, and the Stage Creole. *Notes and Queries* 25, Feb. 1978, pp. 59–60.

Young, George (ed.) 1893. *The Columbus Memorial: containing the first letter of Columbus descriptive of his voyage to the New world; the Latin letter to his royal patrons, and a narrative of the four voyages of Amerigo Vespucci. Reproduced in fac-simile from the unique and excessively rare originals. With illustrations, introductions, and notes.* Philadelphia: Jordan Bros.

Young, Robert J.C. 1995. *Colonial Desire: Hybridity in Theory, Culture and Race.* London and New York: Routledge.

Young, Sir William. 1764. *Considerations which may tend to promote the settlement of our West Indian colonies.* London.

Young, Sir William. 1795. *An account of the black charaibs in the island of St. Vincent's.* 1971 Reprint by F. Cass, London.

**Zayas y Alfonso, Alfredo.** 1931. *Lexicografía antillana; diccionario de voces usadas por los aborígenes de las Antillas mayores y de algunas de las menores y consideraciones acerca de su significado y de su formación.* Havana: Tipos.-Molina y cía.

**Zenon Cruz, Isabelo.** 1975. *Narciso descubre su trasero (El negro en la cultura puertorriqueña).* Second edition. Humacao, PR: Editorial Furidi.

# Index

Printed in the United States
By Bookmasters